INTERNATIONAL SERIES
OF
MONOGRAPHS ON PHYSICS

International Series of Monographs on Physics

138. I. M. Vardavas, F. W. Taylor: *Radiation and climate*
137. A. F. Borghesani: *Ions and electrons in liquid helium*
136. C. Kiefer: *Quantum gravity, Second edition*
135. V. Fortov, I. Iakubov, A. Khrapak: *Physics of strongly coupled plasma*
134. G. Fredrickson: *The equilibrium theory of inhomogeneous polymers*
133. H. Suhl: *Relaxation processes in micromagnetics*
132. J. Terning: *Modern supersymmetry*
131. M. Mariño: *Chern-Simons theory, matrix models, and topological strings*
130. V. Gantmakher: *Electrons and disorder in solids*
129. W. Barford: *Electronic and optical properties of conjugated polymers*
128. R. E. Raab, O. L. de Lange: *Multipole theory in electromagnetism*
127. A. Larkin, A. Varlamov: *Theory of fluctuations in superconductors*
126. P. Goldbart, N. Goldenfeld, D. Sherrington: *Stealing the gold*
125. S. Atzeni, J. Meyer-ter-Vehn: *The physics of inertial fusion*
123. T. Fujimoto: *Plasma spectroscopy*
122. K. Fujikawa, H. Suzuki: *Path integrals and quantum anomalies*
121. T. Giamarchi: *Quantum physics in one dimension*
120. M. Warner, E. Terentjev: *Liquid crystal elastomers*
119. L. Jacak, P. Sitko, K. Wieczorek, A. Wojs: *Quantum Hall systems*
118. J. Wesson: *Tokamaks, Third edition*
117. G. Volovik: *The Universe in a helium droplet*
116. L. Pitaevskii, S. Stringari: *Bose-Einstein condensation*
115. G. Dissertori, I. G. Knowles, M. Schmelling: *Quantum chromodynamics*
114. B. DeWitt: *The global approach to quantum field theory*
113. J. Zinn-Justin: *Quantum field theory and critical phenomena, Fourth edition*
112. R. M. Mazo: *Brownian motion—fluctuations, dynamics, and applications*
111. H. Nishimori: *Statistical physics of spin glasses and information processing—an introduction*
110. N. B. Kopnin: *Theory of nonequilibrium superconductivity*
109. A. Aharoni: *Introduction to the theory of ferromagnetism, Second edition*
108. R. Dobbs: *Helium three*
107. R. Wigmans: *Calorimetry*
106. J. Kübler: *Theory of itinerant electron magnetism*
105. Y. Kuramoto, Y. Kitaoka: *Dynamics of heavy electrons*
104. D. Bardin, G. Passarino: *The Standard Model in the making*
103. G. C. Branco, L. Lavoura, J. P. Silva: *CP Violation*
102. T. C. Choy: *Effective medium theory*
101. H. Araki: *Mathematical theory of quantum fields*
100. L. M. Pismen: *Vortices in nonlinear fields*
99. L. Mestel: *Stellar magnetism*
98. K. H. Bennemann: *Nonlinear optics in metals*
97. D. Salzmann: *Atomic physics in hot plasmas*
96. M. Brambilla: *Kinetic theory of plasma waves*
95. M. Wakatani: *Stellarator and heliotron devices*
94. S. Chikazumi: *Physics of ferromagnetism*
91. R. A. Bertlmann: *Anomalies in quantum field theory*
90. P. K. Gosh: *Ion traps*
88. S. L. Adler: *Quaternionic quantum mechanics and quantum fields*
87. P. S. Joshi: *Global aspects in gravitation and cosmology*
86. E. R. Pike, S. Sarkar: *The quantum theory of radiation*
83. P. G. de Gennes, J. Prost: *The physics of liquid crystals*
82. B. H. Bransden, M. R. C. McDowell: *Charge exchange and the theory of ion-atom collision*
73. M. Doi, S. F. Edwards: *The theory of polymer dynamics*
71. E. L. Wolf: *Principles of electron tunneling spectroscopy*
70. H. K. Henisch: *Semiconductor contacts*
69. S. Chandrasekhar: *The mathematical theory of black holes*
51. C. Møller: *The theory of relativity*
46. H. E. Stanley: *Introduction to phase transitions and critical phenomena*
32. A. Abragam: *Principles of nuclear magnetism*
27. P. A. M. Dirac: *Principles of quantum mechanics*
23. R. E. Peierls: *Quantum theory of solids*

Radiation and Climate

Atmospheric Energy Budget from Satellite Remote Sensing

I.M. VARDAVAS

Department of Physics, University of Crete

F.W. TAYLOR

Atmospheric, Oceanic, and Planetary Physics, Clarendon Laboratory, University of Oxford

OXFORD
UNIVERSITY PRESS

OXFORD
UNIVERSITY PRESS

Great Clarendon Street, Oxford, OX2 6DP,
United Kingdom

Oxford University Press is a department of the University of Oxford.
It furthers the University's objective of excellence in research, scholarship,
and education by publishing worldwide. Oxford is a registered trade mark of
Oxford University Press in the UK and in certain other countries

First published in 2007

Published in the United States of America by Oxford University Press
198 Madison Avenue, New York, NY 10016, United States of America

British Library Cataloguing in Publication Data
Data available

ISBN 978–0–19–922747–1

PREFACE

Interest in the Earth's climate is at an all–time high, due in large part to concerns that it may be changing to a degree that will soon adversely affect the habitability of the planet. This realization coincides with rapid improvements in the tools needed to study and understand the physical processes that produce and maintain the climate, including remote–sensing instruments and observing platforms in space, and fast computers and sophisticated algorithms to analyse large amounts of data and to run advanced theoretical models.

The radiation that reaches the Earth from the Sun and its interaction with the Earth's atmosphere plays a critical role in determining the climatic conditions that prevail over the globe. The Earth's surface–radiation budget drives the water content of the atmosphere and with it the dominant component of the natural greenhouse effect that serves to maintain the benign climate on our planet. Variations in the incoming solar radiation and in the composition and cloud content of the atmosphere and ice cover of the surface result in climatic variability on timescales of billions of years to less than a century. Recently, human activity has begun to modify the Earth's climate, through emissions of greenhouse gases, like carbon dioxide and methane, and of aerosols, that alter the radiative balance and generally warm the planet, but can also cool it depending on the exact circumstances.

The net radiation at the top of the atmosphere is now extensively monitored by satellites but the radiation budget at the Earth's surface (which determines the surface temperature, the single most important climate variable) can only be indirectly derived from radiation–transfer models that require detailed climatological data, such as cloud and aerosol radiative properties, water and temperature profiles in the atmosphere, and surface properties. These quantities can be measured in different ways, but the only approach that results in global and seasonal coverage with sufficient detail is the use of remote–sensing instruments on satellites, i.e. by the detailed spectral analysis of the outgoing radiation field as well as its energy flux.

Climate models aim to quantify the radiation changes or forcings that arise from natural variations, for example in solar luminosity, and anthropogenic effects such as pollution–induced changes in atmospheric composition, cloudiness and aerosol distribution. Long–term prediction and monitoring of the Earth's radiation budget and climatic change therefore require reliable radiation-transfer computations. The methods by which this can be achieved require not only a detailed understanding of the quantum-mechanical interactions of molecules and photons of various kinds, but also the development of computationally efficient algorithms that quantify the transfer of energy as radiation throughout the atmosphere, especially in the vertical direction and in the presence of clouds and aerosols.

Atmospheric composition itself controls both the solar and terrestrial radiation fields and the thermal structure, and in turn the radiation field determines the photochemistry and composition of the atmosphere. There is thus a strong coupling between atmospheric photochemistry, atmospheric composition and climate.

The purpose of this book is to provide advanced students and researchers with an understanding of the radiation processes that control the Earth's climate and climatic change, the methods used by remote–sensing systems for climate measurements and monitoring, and the theoretical principles involved in incorporating radiative processes into climate models. It aims to provide a picture of the recent and current state of the Earth's radiation budget and the physico–chemical processes that determine it.

Radiation and climate is a field of research that encompasses many subdisciplines in Physics, Chemistry, and Astronomy. Lessons learned for the Earth are also valuable to the understanding of climate on other planets, and, as we can now study the latter in detail, vice versa. The book concludes by setting the study of the contemporary terrestrial climate in context with the conditions that are thought to have developed over the history of the planet, and with those recently revealed on our neighbours in the Solar System.

I.M. Vardavas and F.W. Taylor
Heraklion
October 2006

ACKNOWLEDGEMENTS

The writing of a textbook with the wide scope embraced here involves drawing on the work of a very large number of researchers, past and present. We have acknowledged our main sources in the text and in the notes and references at the end of each chapter, to these and others too numerous to mention go our gratitude and recognition. Any errors or omissions are of course the responsibility of the present authors.

FWT would like to thank Jesus College, Oxford, for a travel grant and the Foundation for Research and Technology–Hellas for hospitality at various times during the writing of the book. IMV would like to thank the Foundation for Research and Technology–Hellas (FORTH) for its support during the writing of the book, and the FORTH Environment Research Laboratory team for their assistance. We are grateful to Dr D.J. Taylor for drawing many of the original figures.

CONTENTS

1

THE CLIMATE SYSTEM

1.1 Introduction

We define climate as the mean state of the atmosphere, averaged over several years and all seasons, with particular emphasis on those factors that affect the temperature of the Earth's surface, in both a local and a global mean sense. The *climate system* includes the Sun, the Earth's surface and some aspects of its interior, for example that which gives rise to outgassing and volcanism, as well as the atmosphere, hydrosphere, biosphere and cryosphere. In this chapter, we provide an overview of the climate system and the main ways in which it is observed and predicted to behave. The overall goal of climate research, so far as the physical sciences are concerned, is to refine observational techniques and the theoretical underpinning of climate models to the point where we fully understand the present climate and are able to make meaningful predictions of its future evolution.

Much of the current level of interest in the Earth's climate has to do with the fact that it is perceptibly changing at present, and that the accumulated change over the next century is thought likely to have serious consequences for the habitability of the planet as a whole. In considering this scenario, it is usual to distinguish between *natural* and *anthropogenic* effects on the climate. In this book, where we are mainly concerned with the physics of the problem and in particular the role of radiation in the climate system, the distinction is more one of degree than of kind. Popular concern about 'global warming', for instance, is based mainly on relatively small (although still serious in social terms) changes in the energy balance at the surface due to atmospheric compositional changes. This anthropogenic contribution to the 'greenhouse effect' is smaller than the natural 'greenhouse', which renders the Earth habitable in the first place, and largely indistinguishable in terms of the physical processes at work.

In this chapter, we give an overview of these processes, and the principal engines of change, in order that the theory of atmospheric radiation set out in the following pages may be set in the context of current interest in, and concern about, the climate on the Earth. Figure 1.1 presents a well–known diagram from the Intergovernmental Panel on Climate Change, or IPCC, an authoritative, global body charged with obtaining a balanced consensus on the problem of global change for use ultimately by politicians and other policymakers seeking to define and tackle the problem. It summarizes the main factors involved, and estimates the

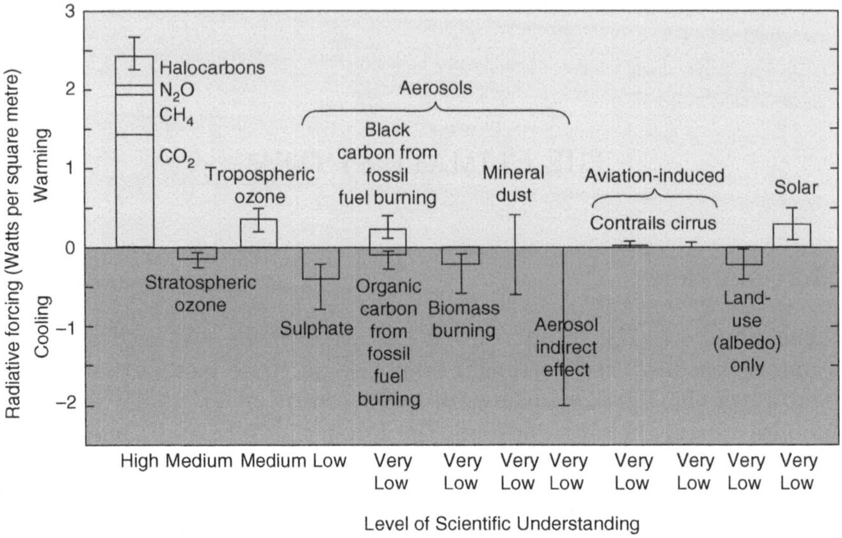

FIG. 1.1. The global mean radiative forcing of the climate system for the year 2000, relative to 1750. (Source: IPCC 2001)

magnitude (and direction) of their contribution to climate change in terms of a factor called radiative forcing (see §1.5). In the following sections we discuss each of these in enough detail to provide the necessary background to the later chapters in which the relevant theory and experimental application of atmospheric radiation are developed.

1.2 Solar radiation

The Earth's climate is primarily determined by the amount of solar radiation flux (W m^{-2}) incident at the top of the atmosphere (TOA) and its spectral distribution, which for the present Sun is about 5% in the ultraviolet, 55% in the visible, and 40% in the infra–red part of the electromagnetic spectrum. The amount of energy from the Sun ultimately determines the surface temperature of the Earth, while the solar spectral flux, especially the ultraviolet flux, affects the atmospheric molecular composition.

Over a span of billions of years, the Earth's climate is affected by the evolution of the Sun, which like other stars changes its luminosity and the spectral distribution of its radiation output with age. The solar energy that warms the Earth has increased by about 30% from the time the Sun became a main–sequence star some 4.6 billion years ago, to a phase where the hydrogen is converted to helium in the stellar core resulting in long–term stable solar radiative emission. However, small changes associated with the solar cycle and the number of sunspots are seen to occur, and a change of just 1% in the solar luminosity can alter the Earth's mean surface temperature by 1 degree, a significant amount in human terms.

In addition to any changes in the output from the Sun itself, changes in the solar energy that is absorbed by the Earth's atmosphere–surface system can arise from many factors related to the reflectivity (albedo) of the planet, and from slow variations in the Earth's orbit around the Sun. Croll (1867a,b) and Milankovitch (1941), proposed theories linking the Earth's climate record, particularly the ice ages, with periodic changes in its elliptical orbit around the Sun. Milankovitch proposed that variations related to the obliquity of the Earth's axis of rotation and the eccentricity of its orbit could change the climate on timescales with periods ranging from 41000 years for obliquity variations, as high as 413000 years for eccentricity variations, and about 21000 years for precession of the equinoxes or variations in the Earth's longitude of perihelion (§5.6). The short–term variations affect the distribution of solar energy over the globe, while the longer eccentricity variations affect the amount of solar energy reaching the Earth. Milankovitch considered the combination of variations in the amount and distribution patterns of incoming radiation to be critical factors for the formation of continental ice sheets that could trigger ice ages.

1.3 The atmosphere

The pressure and composition of the present–day atmosphere were determined by outgassing from the Earth's interior, primarily through volcanic activity, with an unknown admixture of mainly volatile material due to cometary impacts. The result, the atmosphere we see today, consists almost entirely of three *major constituents*, nitrogen (78.1%), oxygen (20.9%), and argon (0.93%), which have not varied significantly in abundance over the last several millennia at least, and are not expected to vary much in the foreseeable future. Furthermore, these three gases each contain atoms of a single element and so possess molecular symmetries that prevent them from absorbing or emitting infra–red radiation under most conditions of temperature and pressure found in the Earth's atmosphere. These gases do affect the propagation of solar radiation through the process of Rayleigh scattering, and hence contribute to global cooling of the planet despite the fact that they do not contribute directly to the greenhouse effect. They do moderate the greenhouse effect indirectly, as the atmospheric pressure they produce affects the efficiency by which the infra–red–active molecules absorb the Earth's thermal radiation emitted to space. For instance, although Mars has more of the greenhouse gas CO_2 in its atmosphere, its greenhouse effect is weaker than on the Earth, due to Mars' very low atmospheric pressure of only about 7 mbar, less than 1% of that on the Earth.

In the present terrestrial atmosphere, the *trace gases*, water vapour (H_2O), carbon dioxide (CO_2), methane (CH_4), nitrous oxide (N_2O) and ozone (O_3), are

Table 1.1 *Pre–industrial (1750) and present (1998) abundances of well–mixed greenhouse gases, their annual increase rate and atmospheric lifetime. (After IPCC 2001)*

	CO_2	CH_4	N_2O	CFC-11
Pre-industrial concentration	≈ 280 ppm	≈ 700 ppb	≈ 270 ppb	zero
Concentration 1998	365 ppm	1745 ppb	314 ppb	268 ppt
Concentration change per year[b]	1.5 ppm [a]	7.0 ppb [a]	0.8 ppb	−1.4 ppt
Atmospheric lifetime (year)	5 to 200 [c]	12	114	45

[a] Rate has fluctuated between 0.9 ppm/year and 2.8 ppm/year for CO_2 and between 0 and 13 ppb/year for CH_4 over the period 1990 to 1999.
[b] Rate is calculated over the period 1990 to 1999.
[c] No single lifetime can be defined for CO_2 because of the different rates of uptake by different removal processes.

the main *greenhouse gases.* The man–made chloroflurocarbons (CFCs), associated with cosmetic sprays and refrigeration, also contribute to the present greenhouse effect. These are minor constituents of the atmosphere in terms of abundance (e.g. in units ppm, i.e. parts per million by volume), and, unlike the major constituents, these abundances are prone to change. Carbon dioxide, methane, nitrous oxide and CFCs have increased systematically within the past century as shown in Table 1.1. Water–vapour abundance is largely controlled by the ocean–surface temperature and transport processes, and the global distribution of its atmospheric concentration is highly variable.

The presence of liquid water on the Earth is a primary determinant of the establishment of life. Life itself has changed the climate, for instance, by modifying the Earth's albedo by covering the land surface with plants that have a variety of reflecting properties. Biological decay has changed the absorption of ultraviolet radiation within the oceans, pure water being mainly a scatterer of solar radiation. Liquid oceans are weak reflectors of solar radiation whereas the presence of the condensed phase in the form of snow and ice rapidly changes the surface to an efficient solar reflector, with dramatic consequences for the climate.

The oceans are the main source of atmospheric water vapour, through evaporation, which depends on the net heat (both solar and terrestrial) absorbed at the surface and stored at all depths. The relatively high heat–storage capacity of the oceans buffers the Earth against rapid climatic changes, while the atmospheric water vapour they produce plays a crucial role in the control of climatic variability and change. It absorbs significant amounts of solar ultraviolet and near–infra–red radiation, and of the infra–red radiation emitted by the Earth towards space. As the atmospheric temperature falls with increased height in the troposphere, water vapour condenses to form a variety of cloud types that increase the Earth's albedo through multiple scattering of photons. Clouds also act to warm the Earth by preventing direct surface and lower–atmospheric thermal emission to space. Overall, the role of clouds is complex and the radiative role of clouds is one of the most difficult processes to simulate in climate models.

Carbon dioxide is an important atmospheric minor constituent because its strong infra–red absorption bands make it a major contributor to the energy budget of the atmosphere and surface of the Earth. Because it has a long lifetime, the abundance of CO_2 tends to be nearly constant to a considerable height, although a small discontinuity in the mixing ratio has been measured at the tropopause, and its vertical distribution remains rather poorly known in the mesosphere and above where dissociation takes place. Measurements have been made in the mesosphere and lower thermosphere using rocket–borne mass spectrometers that show a sharp decline in CO_2 abundance between about 90 and 110 km, from a value not greatly different from that at the tropopause to more than an order of magnitude less. This high–altitude decline in CO_2 is expected on theoretical grounds due to photolysis by solar ultraviolet energy of wavelength 230 nm and shorter, leading to the production of carbon monoxide and atomic oxygen. Carbon monoxide, CO, is also produced in the stratosphere, along with water vapour, by the photochemical oxidation of methane. The role of CO in the heating and cooling of the atmosphere is much less than for CO_2, but its abundance is obviously of interest because of its relevance to an understanding of the CO_2 budget, and as a tracer of the atmospheric circulation at stratospheric and mesospheric levels, for which little data exists.

Changes in the mean annual concentration of CO_2 with time are one of the most cited causes of recent climate change. Measurements at Mauna Loa in Hawaii, made far from the main anthropogenic sources and therefore considered globally representative, show an increase from 315.98 parts per million by volume (ppmv) of dry air in 1959 to 377.38 ppmv in 2004, an average annual increase of 1.4 ppmv per year. Over the whole of the industrial era, i.e. from about 1750 to the present day, the atmospheric concentration of carbon dioxide has increased by 31%; the rate of increase is accelerating and has been about 0.4% per year over the past two decades. The rise is largely attributed to anthropogenic emissions due to fossil–fuel burning and deforestation. About half of the anthropogenic CO_2 emissions are sequestered by processes in the oceans and the land surface, which reduces or at least delays the potential greenhouse effect of these emissions. The net result is that the current concentration of atmospheric CO_2 has not been exceeded during the past 420000 years, while the present rate of increase is unprecedented during at least the past 20000 years.

Ozone is formed in the atmosphere by photochemical processes involving a number of natural and athropogenic species. Its residence time in the atmosphere is relatively short, varying from weeks to months, and it exhibits significant spatial variability. In the ultraviolet spectral region, where it has strong absorption, there are no other molecules with significant atmospheric concentration to control the ultraviolet radiation that reaches the Earth's surface. Since it is formed from molecular and atomic oxygen, ozone was most likely not present in significant levels in the anoxic early terrestrial atmosphere (some 4 billions years ago) to provide a shield for early biological evolution.

The total amount of O_3 in the lower atmosphere is estimated to have increased by 36% since 1750, due primarily to anthropogenic gas emissions that result in the production of ozone. The atmospheric concentrations of many of the halocarbon gases that are both ozone–depleting and greenhouse gases (e.g. $CFCl_3$ and CF_2Cl_2), are either increasing more slowly or decreasing since 1995, in response to reduced emissions under the regulations of the Montreal Protocol. Their replacements (e.g. CHF_2Cl and CF_3CH_2F) and some other synthetic compounds (e.g. perfluorocarbons, PFCs, and sulphur hexafluoride, SF_6) are also greenhouse gases, and their concentrations are currently increasing.

Methane has increased by 151% since 1750 and its abundance continues to grow; the present CH_4 concentration has not been exceeded during the past 420000 years. Methane is an important greenhouse gas, which may have had a much higher abundance in the atmosphere of the early Earth. Its role is made complicated by photochemical processes, as can be appreciated from its present role in the atmosphere of Titan, a moon of Saturn that has a thick atmosphere containing a mixture of aerosols or haze. The haze is generated by the photodissociation of methane that results in complex hydrocarbons that eventually polymerize to form large aerosol particles. These particles have the ability to absorb and scatter solar radiation from the ultraviolet all the way to the infra–red. On the Earth, methane also controls the amount of water vapour in the upper atmosphere and so plays an important indirect role in the temperature of the upper atmosphere.

Globally representative measurements of the concentration of CH_4 in the atmosphere have been made since 1983, and the record of atmospheric concentrations has been extended to earlier times from air extracted from ice cores and fern layers. About half of the current CH_4 emissions are anthropogenic, from fossil fuels, cattle, rice agriculture, and landfills. Carbon monoxide (CO) emissions can result in an increase in methane, while increases in NO and NO_2, together referred to as NO_x, result in a decrease in methane, and in an increase in tropospheric ozone. Emission rates from biospheric sources are inherently highly variable. Methane is removed from the atmosphere by photochemical reactions.

Nitrous oxide has increased by 17% since 1750 and continues to increase; the present N_2O concentration has not been exceeded during at least the past thousand years. About a third of current N_2O emissions are anthropogenic (agriculture, biomass burning, and industrial activities). Natural sources include soils (about 65%) and the oceans (about 30%).

1.4 Clouds and aerosols

Clouds cover on average about 60% of the globe (Fig. 1.2) and play a crucial role in determining the climate by acting both as reflectors to solar radiation and absorbers of terrestrial radiation. They constitute the biggest uncertainty in climate–change predictions because of difficulties in simulating changes in the

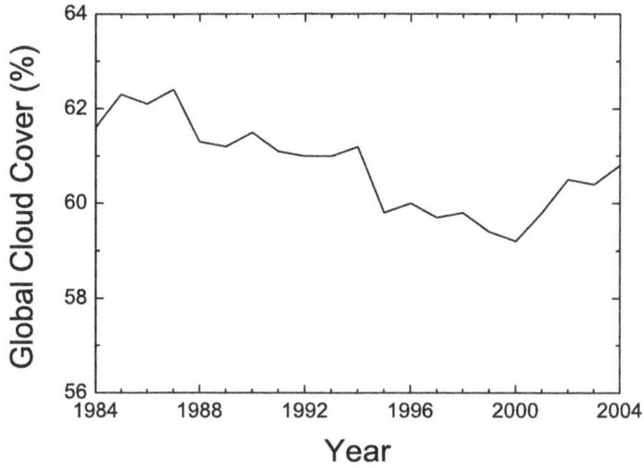

FIG. 1.2. The variation of the global cloud cover based on ISCCP data for the period 1984–2004.

cloud cover and the radiative properties of the many types of cloud. Cloud cover is highly variable both temporally and spatially. Cloud overlap can modify rapidly the resultant scattering and absorbing properties of the cloudy sky. The radiative properties of droplets are different from ice crystals, and different condensation processes can occur rapidly.

In addition to the airborne droplets or ice crystals that make up clouds, lower concentrations of much smaller solid and liquid particles (aerosols) exist everywhere and can affect climate both directly and indirectly. Some have a complex chemical composition that results in a variety of radiation absorbing and scattering properties. Those generated naturally by volcanism and sea–spray evaporation, consisting mainly of sulphate and chlorides, are scatterers that tend to produce a cooling of the planet, while those that arise from biomass and fossil–fuel burning also contain carbonaceous materials that are strongly absorbing and thus warming agents. Other types of aerosols include irregularly shaped particles of various types of minerals, produced by aeolian processes, especially dust storms. The water content of aerosols, which depends on the ambient relative humidity, plays an important role in their radiative properties. Their atmospheric residence time is of the order of days to weeks, and they have a highly variable global distribution.

Tropospheric aerosol loads are thought to have increased over recent years due to increased anthropogenic emissions of particles and their precursor gases. The inclusion of aerosols is now necessary in climatic change studies that examine changes in surface and atmospheric temperatures, snow and ice–cover extent, sea–level rise, precipitation changes, frequency and intensity of extreme weather events, and desertification, since aerosols crucially affect the radiation budget at the top of the atmosphere, in the atmosphere, and at the surface, and hence atmospheric dynamics as well as evaporation and surface energy balance. Aerosol particles affect the Earth's climate by influencing its radiation energy budget in two ways; i) directly by scattering and absorbing solar radiation, and ii) indirectly, by modifying the cloud microphysical properties. Aerosols can act as cloud condensation nuclei, affecting cloud optical and radiative properties (cloud albedo and optical thickness) or cloud amount, lifetime and precipitation efficiency. In climate research, these are known as the first and second indirect aerosol effects, respectively.

Modern climate models must attempt to incorporate both the direct and indirect radiative forcing effects of aerosols. However, despite significant progress in understanding the role of aerosols in climate, it still remains a large uncertainty. Aerosols exhibit large spatial–temporal variability and heterogeneity associated with their short atmospheric lifetime and complex interactions with clouds in terms of both their physical and chemical properties. In order to improve the aerosol parameterization schemes in climate models, and the accuracy of estimated aerosol radiative forcings, monitoring of the spatial and temporal distributions of aerosol physical, chemical, optical and radiative properties is required both on global and local scales. To this aim, a worldwide effort has been undertaken since the early 1990s, in order to produce a global aerosol climatology by combining satellite–based observations (such as Total Ozone Mapping Spectrometer, TOMS, Moderate Resolution Imaging Spectro–Radiometer, MODIS, Multiangle Imaging Spectroradiometer, MISR, Polarization and Directionality of the Earth's Reflectance, POLDER, with measurements from ground–based monitoring networks such as the Aerosol Robotic Network (AERONET). To retrieve aerosol properties accurately, these data sets are complemented by field experimental campaigns, both ground based and airborne, conducted in climatically important regions. These serve at the same time as calibration and validation tools for satellite data.

1.5 Radiative equilibrium and radiative forcing

The dominant factor determining the surface temperature of a planet is the balance between net incoming solar radiation and outgoing thermal infra–red radiation. That is, *radiative equilibrium* can be assumed for most practical purposes, although strictly speaking it is never achieved, as time-dependent and non-radiative processes are always present. The processes that control the net solar radiation absorbed by the planet are complex and, as we have seen, involve the multiple scattering and absorption of solar radiation by atmospheric molecules, clouds, aerosols and the surface. The outgoing thermal infra–red radiation is controlled by the temperature of the surface and by the gases, clouds, and other particles in the atmosphere. The temperature is determined in turn by the local energy balance.

In the absence of an atmosphere and without significant internal heat sources, it is straightforward to calculate the solar radiation absorbed by the surface of a planet and its re-emission as infra–red radiation to space, and hence the surface temperature. If an atmosphere is present that contains infra–red active molecules (which means, in general, those composed of different types of atoms, such as H_2O and CO_2) these control the amount of infra–red radiation absorbed by the atmosphere, preventing it from escaping directly to space. Radiation from the atmosphere back towards the surface raises its temperature above that of an airless body with the same albedo, giving rise to what we (somewhat inaccurately) call the greenhouse effect. The atmosphere thus plays a crucial role in determining the climate of a planet, particularly since the greenhouse gases are present on the Earth in quite small proportions that are prone to change due to natural and man-made phenomena (e.g. large volcanic eruptions, and extensive deforestation, respectively).

The response to any change affecting the radiative properties of the atmosphere or the surface is ultimately a change in the surface temperature of the planet and hence its climate. The term *radiative forcing* is used to denote an externally imposed perturbation in any component of the radiative energy budget of the climate system. Radiative forcing is a measure of the change in radiative fluxes within the atmosphere–surface system, especially those of the net solar and terrestrial fluxes at the top of the atmosphere and at the Earth's surface. In making assessments of climate change, bodies such as the IPCC have focused on the changes in radiative forcing between pre-industrial times and the present, the recent past, and the future. The ability to monitor the TOA fluxes and climatological parameters over the globe by satellites, e.g. ERBE (Earth Radiation Budget Experiment) for radiative fluxes and ISCCP (International Satellite Cloud Climatology Project) for cloud cover and albedo, has provided climate modellers the opportunity to validate their predictions and improve their understanding of radiative process against measurements on a planetary scale. The radiative forcing due to increases of the well-mixed greenhouse gases from 1750 to 2000 is estimated by IPCC to be 2.43 W m^{-2}, broken down by species in Table 1.2. The net forcing can be compared to a value of 1366 W m^{-2} for the total solar irradiance (TSI) (also known as the solar constant), defined as the total radiative energy flux from the Sun at the mean distance of the Earth's orbit.

The tropospheric increase in ozone corresponds to a positive radiative forcing of 0.35 Wm^{-2}. Ozone forcing varies considerably by region and responds much more quickly to changes in emissions than the long-lived greenhouse gases, such as CO_2. The observed depletion of the stratospheric ozone layer from 1979 to 2000 is estimated to have caused a negative radiative forcing (–0.15 Wm^{-2}). Assuming

Table 1.2 *Pre–industrial (1750) and present (1998) abundances of well–mixed greenhouse gases and the radiative forcing due to the change in abundance. Volume mixing ratios for CO_2 are in ppm, for CH_4 and N_2O in ppb, and for the rest in ppt. (Source: IPCC 2001)*

Gas	Abundance 1750	Abundance 1998	Radiative forcing (W m^{-2})
CO_2	278	365	1.46
CH_4	700	1745	0.48
N_2O	270	314	0.15
CFC-11	0	268	0.07
CFC-12	0	533	0.17
Other			0.10

full compliance with current halocarbon regulations, the positive forcing of the halocarbons will be reduced, as will the magnitude of the negative forcing from stratospheric ozone depletion as the ozone layer recovers over the twenty–first century.

1.6 Atmospheric circulation

Much of the radiative behaviour of the atmosphere is a response to the transfer of energy within the climate system by dynamics. Of particular importance for the global radiative energy balance is the poleward transfer of energy, to balance the decrease with latitude of the input from the Sun. About half of this transfer, amounting to several petawatts (1 PW = 10^{15} W) occurs in the atmosphere, and the rest in the oceans. The ocean circulation is coupled to that of the atmosphere by momentum and heat transfer between near–surface winds and the water surface.

The circulation of the atmosphere transports momentum, cloud, water, ozone and pollution, as well as heat, around the globe. The mean or 'general' circulation consists primarily of rising air in the strongly heated low–latitude regions, with descent over the poles where radiative cooling exceeds heating. The warm air cools during its movement polewards, descends and moves at low levels back towards the equator, completing the large circulation pattern known as a Hadley cell. The rapid rotation of the Earth means the Coriolis effect adds a component parallel to the equator to the poleward flow in the Hadley cell, which closes the cell, by forcing the north–south flow into an east–west direction, before it can reach high latitudes. Further cells develop polewards of this and the mean circulation consists of three circulation cells, the Hadley, Ferrel, and Polar cells, in each hemisphere (Fig. 1.3).

The rising air at the equator creates a band called the intertropical convergence zone, which draws in low-level air from the subtropics. In this region, also known as the doldrums, horizontal pressure gradients are weak and winds are light. From there air rises towards the tropopause and flows polewards, gradually deflecting

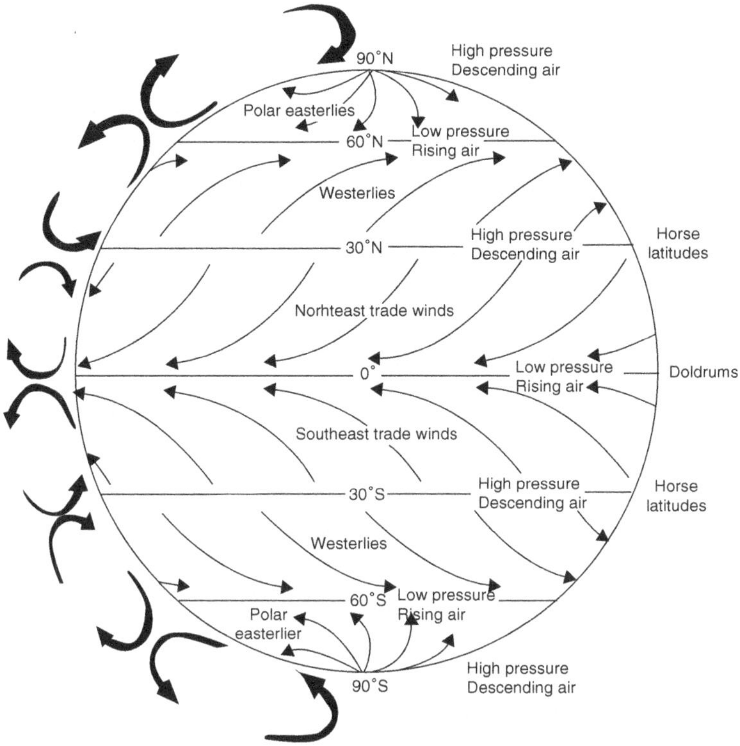

FIG. 1.3. A schematic diagram of the general circulation of the atmosphere of the Earth, as described in the text. On a fast-rotating planet like the Earth, the thermally driven equator–to–pole flow is deflected by the Coriolis acceleration, towards the west for equatorwards motion, and vice versa, in both hemispheres. This renders a single equator–to–pole cell unstable, and instead three cells, known as the Hadley, Ferrel and Polar cells, are found to dominate the general circulation when wave and chaotic motions are averaged out.

under the influence of the Coriolis force, towards the east, until by about 30° latitude the wind is predominantly zonal, forming the subtropical jet stream. The upper air then sinks back to the surface creating a high–pressure zone with low surface winds known as the horse latitudes. From there the air moving back towards the equator to complete the Hadley cell is deflected west, creating the trade winds.

The surface air moving polewards from the horse latitudes is deflected westwards by the Coriolis force, producing the westerlies. In these latitudes, roughly 30 to 60°, upper air winds tend to blow towards the poles and are deflected eastwards to the polar jet stream at the high–latitude limit of the Ferrel cell. Poleward of that, the rapidly decreasing distance of the surface from the axis of rotation means the atmosphere in the polar cell tends to spin up to conserve its angular momentum, forming the polar vortex. Inside this, air descends rapidly before commencing its journey equatorwards. Where it meets the mild air moving polewards in the Ferrel cell, it forms the subpolar low, a region noted for storm formation.

The long–term mean atmospheric circulation accounts for only about 25% of the polewards transport of heat required to balance the solar input and the global radiative cooling of the Earth. Another ~25% is by eddies, including midlatitude storm systems and waves and turbulence on a wide range of temporal and spatial scales, and the rest in the ocean. The rate at which this transfer takes place seems to be such as to maximize the rate of entropy production, dS/dt. This would be zero if the transport stopped, and also if the motions were very rapid. The actual situation is somewhere in between and is postulated, with some support from studies using general circulation models, to be that for which dS/dt finds a maximum value.

Thus it appears that dynamical energy transfer within the climate system not only tends towards a state of maximum entropy, as required by the Second Law of Thermodynamics, but also that it moves in this direction at the fastest possible rate. The explanation, broadly speaking, is that the processes involved are complex and chaotic, and so offer an essentially infinite number of multiple pathways towards the state of maximum disorder. The absorption, transfer and emission of radiation by the climate system, which involves a larger rate of entropy production than the dynamical transport of heat, does not have this property and does not behave in this way.

1.7 The hydrosphere and the cryosphere

The hydrosphere and cryosphere are the liquid and solid water components of the climate system. The hydrosphere comprises the surface–water bodies such as rivers, lakes, and oceans that cover about 70% of the Earth's surface, and groundwater in aquifers. Most of the precipitation on land returns to the atmosphere via evaporation, a smaller part runs into the sea, and some soaks into underground aquifers. Runoff to the sea and oceans delivers nutrients and salts that affect their heat capacity and hence modify the oceanic circulation. The cryosphere comprises the ice sheets of Greenland and Antarctica, continental glaciers, snow cover, sea ice in the Arctic and Southern oceans and permafrost. Its high reflectivity to solar radiation provides a cooling mechanism that can change relatively rapidly as condensation and melting processes have the potential to alter the planetary albedo from 5% to 80%. Ice sheets store a large amount of water and so are a potential source of sea–level variations.

1.7.1 *Oceanic circulation*

Any understanding of the role of the atmospheric circulation in global energy balance is not complete without a corresponding appreciation of the energy budget of the ocean. However, a crucial difference between the two, apart from the obvious differences in mass, density and heat capacity, is the almost completely negligible role played in ocean energetics by radiative transfer. Liquid water is effectively opaque to electromagnetic radiation at virtually all wavelengths, a further consequence of which is the difficulty of gathering data by remote sensing.

The ocean circulation has two fairly distinct components. The top few hundred metres, containing the mixed and thermocline layers, is driven primarily by drag between the water surface and the prevailing winds. The deep–ocean circulation, however, is dominated by differences in density resulting from temperature and salinity gradients. In both cases, the long–term average motions (that is, with waves and turbulence averaged out) are determined mainly by the balance between the main driving force (wind drag and density gradients, respectively) and the Coriolis force that enters as a result of the Earth's rotation.

The deep–ocean layer constitutes more than 90% of the mass of the ocean, i.e. all but the top few hundred metres of depth on average. It extends all the way to the surface near the poles, where the water is particularly cold and dense, resulting in strong downwelling towards the ocean floor. This rapid sinking motion is focused in relatively small areas near Greenland in the North Atlantic, and in the Weddell Sea in the Antarctic. Here, particularly cold and dense surface waters are found due to low solar heating, strong evaporative cooling aided by the prevailing high winds, and brine enrichment by freezing. The last process occurs in reverse when polar icecaps melt, and become a source of relatively fresh, low-density water, like melting snowfields on land or heavy rainfall. One of the concerns in a world undergoing global warming is that the fresh water released by melting ice caps will reduce the salinity, the density, and hence the downwelling in the polar seas, and so slow down or even reverse the deep–ocean circulation.

Over the rest of the globe, the vertical motions in the ocean are generally much weaker, and generally upwards to balance the strong downward flow in the Arctic and Antarctic. The resulting slow, global, overturning motion known as the thermohaline circulation transfers heat, salt and tracers around the globe. The motions near the surface are mainly the result of wind drag, transferring momentum from the atmosphere to the surface and near-surface layers of the water. The wind also generates surface waves that further affect the coupling between wind and water, and produce vertical as well as horizontal motions. These cause turbulent mixing of the top ∼10–100 m, producing the nearly isothermal mixed layer.

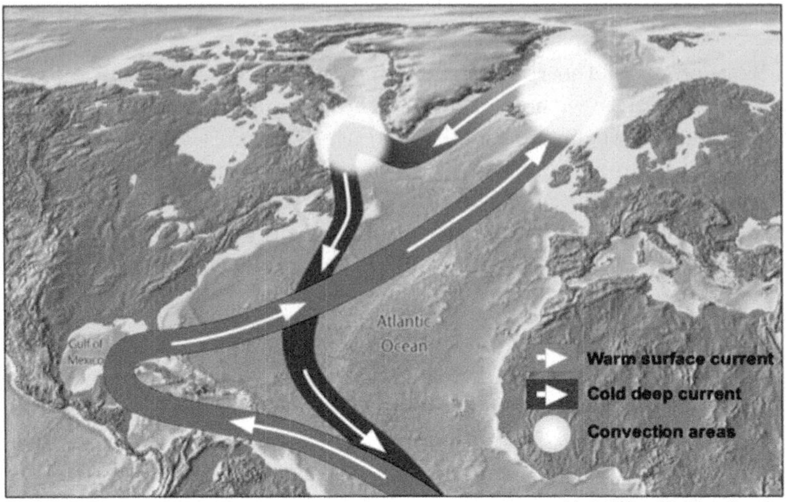

Met Office Hadley Centre

FIG. 1.4. The North Atlantic Drift, or Gulf Stream, showing the flow of warm tropical surface currents northwards to the North Sea, and the flow of deep cold currents southwards. Two convection areas convert warm surface water to cold deep water that drives the Gulf Stream circulation. This circulation is responsible for the milder climate in the Eastern North Atlantic region compared to the Western. ©Crown copyright (UK Met Office/Hadley Centre 2005)

The general pattern of surface currents in the ocean is well known from centuries of navigation and measurement. In the Northern Hemisphere, each of the ocean basins contains two counter–rotating gyres (oceanic current systems of planetary scale driven by global wind systems) in equilibrium with the wind forcing. The Southern Hemisphere circulation has a single, large gyre, because of the existence of the strong Antarctic circumpolar current, where the water can travel right around the globe without meeting a continental barrier. The gyres are characterized by currents that are much stronger on the western boundary than on the eastern. The western boundary current in the North Atlantic is the North Atlantic Drift or Gulf Stream, which transports 100 million cubic metres of warm water per second in a stream about 100 km wide, off the east coast of North America (Fig. 1.4).

Solar energy may be stored locally by the sea or ocean during summer and released back to the atmospheric environment during the winter months (§8.9). Absorbed solar energy can also be transferred regionally, affecting climate on a subglobal scale, as done by the Gulf Stream. As shown in Fig. 1.4, heat is transported by surface waters flowing north and eastward, with cold saline waters from the North returning at depth, and so conditions at the same latitude are very different either side of the Atlantic basin.

The oceanic large heat capacity provides a buffer against rapid climatic changes. However, oceans control the climate system by redistributing heat from the warm equatorial regions to the cold polar regions. Ocean circulation can be modified by long–term changes in precipitation, continental runoff, evaporation, sea–ice formation, cloud cover and prevailing winds. Variations in oceanic heat transport can alter global distributions of heat and sea–surface temperatures. There is thus a feedback since sea–surface temperatures control evaporation, atmospheric water vapour, clouds, precipitation and surface albedo (snow cover), and hence global climate.

1.7.2 *El Niño southern oscillation (ENSO)*

The important effect that ocean circulation changes can have on the global climate is illustrated by the role of certain natural circulation patterns in the interannual and longer–term variability of the climate. The strongest natural fluctuation of climate on interannual timescales is the ENSO phenomenon, a large, quasiperiodic climatic variation that arises from an interaction between the tropical Pacific Ocean and the overlying atmosphere.

ENSO is a natural cycle that couples the ocean–atmosphere system over the tropical Pacific and operates on a timescale of 2–7 years. Once developed, it causes a shift in the seasonal temperature and precipitation patterns in many different regions of the world, since heating of the tropical atmosphere creates changes in the global atmospheric circulation. Thus, ENSO is a dominant source of inter–annual climate variability around the world. Normally, the equatorial Pacific Ocean is characterized by warm waters in the west and cold waters in the east. The ENSO warm phase (El Niño) is associated with an unusual warming of the eastern and central equatorial Pacific. La Niña, the ENSO cold phase, is the counterpart to El Niño, often following it. It is characterized by cooler than normal sea surface temperatures across the equatorial eastern Pacific. Thus, ENSO is an oscillation between warm and cold events with a peak that typically occurs late in the calendar year (late December to early January). Both El Niño and La Niña events last for about a year, but they can last for as long as 18 months. During ENSO, a feedback between atmospheric and ocean properties is observed. Sea–surface temperature (SST) anomalies induce wind–stress anomaly patterns that in turn produce a positive feedback on the SST. Variations of the above properties cause significant changes in other oceanic and atmospheric variables, e.g. the mean depth of the thermocline, the water–vapour content of the atmosphere and the relative distributions of low, middle and high clouds.

Water vapour and clouds are the main regulators of the radiative heating of the planet since changes in these parameters modulate the variability in the radiation fluxes that control the heating or cooling of the Earth's surface and atmosphere. The radiation field, in turn, influences SST and atmospheric water vapour. Thus ENSO involves complex climatic processes and feedbacks that make its onset time, duration, strength and spatial structure difficult to predict. International monitoring programmes of the coupled atmosphere–ocean system started in the Pacific around 1985 and led to the TAO-TRITON (Tropical Atmosphere Ocean project/Triangle Trans Ocean Buoy Network) array of moored buoys. The aim of this programme is to provide real–time measurements of winds, sea–surface temperature, subsurface temperature, sea level and ocean flow that help in the understanding of the physical processes responsible for ENSO. The variability and the spatial distribution of the ocean and atmospheric variables are not the same for all ENSO events. Thus, a definition of ENSO is necessary for the study of this phenomenon. The phase and strength of ENSO events are defined by an index. Several different indices have been used in the literature, mostly based on SST, although there is one index, the Southern Oscillation index, which is related to air–pressure differences at sea level, between Darwin (Australia) in the west and Tahiti in the east. The SST–based indices are obtained from the SST anomalies with respect to average values over some specified region of the ocean. There has also been an effort to combine several atmospheric–oceanic variables into a single index like the multivariate ENSO index. Averages of 850 mb wind, outgoing longwave radiation (OLR) at the top of the atmosphere as well as precipitation over specific regions are also used, although not often, to monitor ENSO.

The Earth's climate system is driven by the radiative energy balance between the solar shortwave radiation (SW) absorbed by the atmosphere and the surface of the Earth and the thermal longwave radiation (LW) emitted by the Earth to space. In this respect, ENSO events are expected to be associated with the spatial and temporal variability of the radiative energy balance over the tropical and subtropical Pacific. The net heat flux into the ocean plays a key role in ENSO evolution and it is a significant variable in the models that have been developed to make ENSO predictions. The variation of the net heat flux during ENSO events is of paramount importance to the dynamics of the system. The net heat flux into the ocean is a small residual of four terms, the downward shortwave radiation at the surface (DSR), the latent heat loss, the sensible heat transfer and the net longwave radiation to the Earth's surface (NSL). The NSL is the difference between the downward longwave radiation (DLR) at the Earth's surface and the Earth's surface thermal emission. The DLR at the Earth's surface is a very important component of the surface radiation budget with variations arising from increases in greenhouse gases or from changes in other atmospheric properties that occur during ENSO events (Intergovernmental Panel on Climate Change, IPCC 2001).

The resulting extreme phases of the cycle, the warm El Niño and cold La Niña events, manifest themselves primarily in variations (anomaly) around the average values of sea–surface temperature, as shown in Fig. 1.5 for the central Pacific region (5N-5S, 170-120 W). Also shown is the time series of the computed downwelling longwave radiation (DLR) anomaly using a radiation–transfer model and

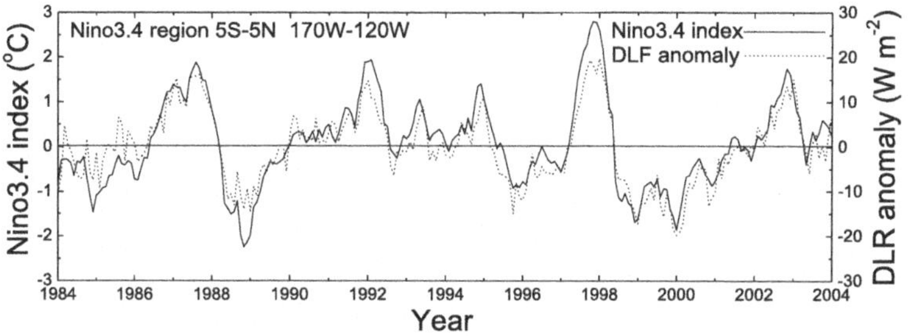

FIG. 1.5. El Niño index based on the sea–surface temperature anomaly in the central Pacific region 5N-5S, 170-120 W (data from NOAA). Negative indices correspond to La Niña events. Also shown is the computed downwelling longwave radiation (DLR) anomaly for the region. (Pavlakis *et al.* 2007)

ISCCP cloud data and NCEP/NCAR water vapour and atmospheric temperature data. The DLR anomaly follows very closely the SST anomaly. During the El Niño there are large enhancements in precipitation in the central and eastern Pacific region, and reductions in the western Pacific. The SST is warmer over most of the Pacific region and pressure gradients are smaller, so the trade winds weaken. During the La Niña the signs of the anomalies are reversed, the water in the eastern Pacific region is colder than normal, the surface pressure and easterly trade winds increase, and rainfall in the central and eastern Pacific decreases as a result of the colder SST conditions. Indices of El Niño evolution have been used to monitor the strength of the events. El Niño has been quantified in terms of simple indices corresponding to times when the sea–surface temperature anomalies in central Pacific regions exceed some value. For example, El Niño events are correlated with SST anomalies in the (5N-5S, 170-120W) region when they exceed 0.4 °C, and are enough to produce perceptible impacts in Pacific Rim countries. However, such a definition does not discriminate between major, moderate, and minor El Niño events.

1.7.3 *North Atlantic oscillation (NAO)*

The North Atlantic Oscillation is a well–defined pressure pattern that consists of opposing variations of barometric pressure near Iceland and near the Azores. It has a strong influence via so–called 'teleconnections' on the climate of Europe and parts of Asia. Its pronounced influence on the climate of the Atlantic basin has been known for more than two centuries. Its positive phase is associated with a westerly current, between the Icelandic low-pressure area and the Azores high-pressure area, which carries cyclones and their associated frontal

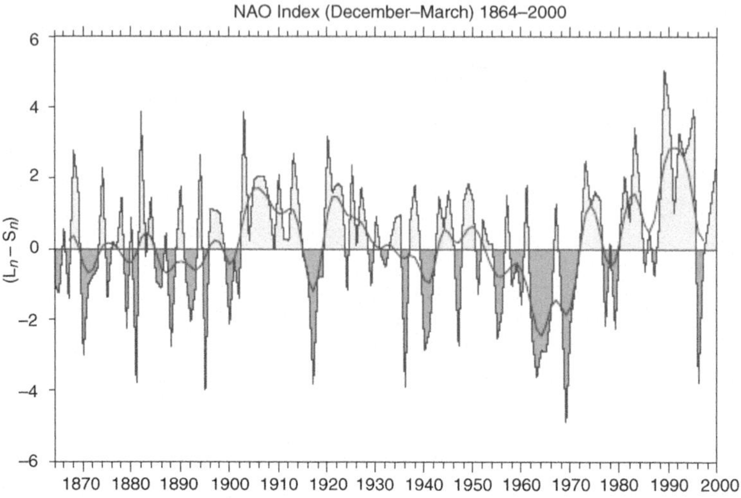

FIG. 1.6. Winter (December–March) index of the NAO based on the difference of normalized sea-level pressure (SLP) between Lisbon, Portugal, and Stykkisholmur/Reykjavik, Iceland from 1864 through 2000. The indicated year corresponds to January (e.g. December 1949–March 1950). The average winter SLP data at each station were normalized by division of each seasonal pressure by the long-term mean (1864–1983) standard deviation. The heavy solid line represents the index smoothed to remove fluctuations with periods less than 4 years. (Source: Hurrell and Dickson 2005)

systems towards Europe. However, the pressure difference between Iceland and the Azores fluctuates on timescales of days to decades, and can be reversed at times. The variability of NAO has considerable influence on regional climate variability in Europe, in particular in wintertime (Hurrell and Dickson 2005). Surface air temperature and sea-surface temperature across wide regions of the North Atlantic Ocean, North America, the Arctic, Eurasia, and the Mediterranean are significantly correlated with NAO variability (Fig. 1.6).

When the NAO index is positive, there are enhanced winter westerlies across the North Atlantic carrying relatively warm moist maritime air over Europe and across Asia. Also, stronger northerlies over Greenland and northeastern Canada carry cold air southward and decrease land temperatures and SST over the northwest Atlantic. Over North Africa and the Middle East there is cooling and warming of North America, associated with the stronger clockwise flow around the subtropical Atlantic high–pressure centre. Anomalously low precipitation rates occur over much of Greenland and the Canadian Arctic during high NAO index winters, as well as over much of central and southern Europe, the Mediterranean, and parts of the Middle East. In contrast, more precipitation than normal falls on Iceland through to Scandinavia.

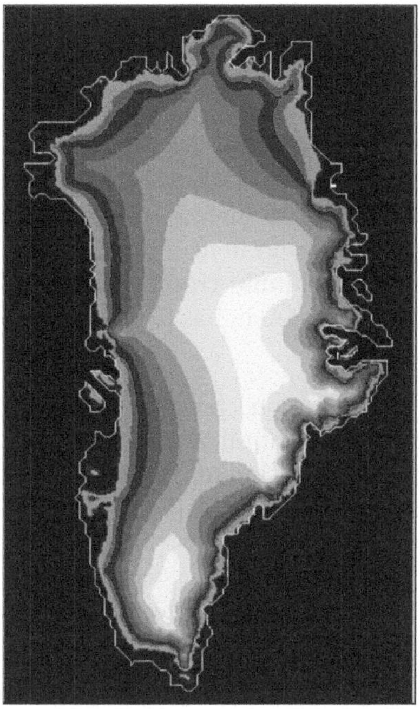

FIG. 1.7. The Greenland Ice Sheet contains just over 7 m of sea–level equivalent. The contours represent 250 m differences in ice–sheet depth, with the highest contour at 3000 m. The ice sheet covers about 1.7 million km^2 and represents about 10% of the world's total fresh water reserves. Greenland extends between 60-85 N and 70-20 W. ©Crown copyright (UK Met Office/Hadley Centre 2005)

1.7.4 *Ice sheets*

Global warming due to a strengthened greenhouse effect has the potential of melting important large ice sheets, such as the West Antarctic and Greenland Ice Sheets. The former contains ice that would, if melted, be equivalent to 6 m of global sea–level rise, while the latter contains 7 m sea–level equivalent. Much larger amounts of water are locked up as ice in Antarctica, but for this to become free would require a warming of more than 20 °C. The Greenland Ice Sheet, on the other hand, is expected to start to decrease in size as summer temperatures rise (Fig. 1.7). This was until recently thought to be a slow process, taking about 1000 years to decrease the ice by 50%, but evidence is accumulating that suggests it may be considerably faster, and already underway. A reduction in ice sheets, in snow cover, and in sea ice would alter the Earth's surface albedo, leading to enhanced global warming.

1.8 The land surface and biosphere

1.8.1 *Land–surface albedo*

The plant life covering the land surface affects the moisture content of the soil and aquifers and the surface albedo. Deforestation and desertification lead to a lower water–holding capacity of the surface, to spectral reflectance increases, to increases in surface temperature and hence increased thermal emission to space. Large–scale deforestation in the humid tropics, e.g. South America (particularly Amazonia), Africa, and Southeast Asia has been identified as the most important ongoing land-surface process.

Desertification in subtropical regions has led to land degradation and to the reduction of groundwater levels in aquifers that in turn have led to a cessation of river discharge to the sea, thus affecting marine ecosystems. If the depletion of soil moisture is severe, plant root zones can separate from the deeper water sources, such as aquifers, thus initiating processes of desertification. This is especially acute in regions with distinct wet and dry seasons (e.g. Mediterranean, tropical Australia). The driving force of land-surface processes is the radiative energy input. Temperature gradients that are created in the soil and between the surface and the atmosphere control water vapour fluxes. Photosynthesis is controlled by the visible radiation and through it biological activity and the carbon cycle. The albedo of bare soils is relatively low for moist fertile soils rich in organic matter, while degraded soils become brighter as in deserts. The albedo of desert soils then depends on the mineral components, the brightest being quartz, gypsum and salt deposits. Contamination with iron oxides makes the surface appear yellowish or red (e.g. Australian deserts). Eventually, winds can expose the underlying rock further modifying the land-surface albedo (e.g. Sahara). The magnitude of the land–surface albedo from the ultraviolet to the infra–red depends also on the vegetative canopy structure. Broad–leaf plants and low plants have high reflectance in the near infra–red, while tall and dense canopies are able to scatter light more efficiently, increasing solar–radiation absorption by the land surface. Reflectance by plants is low (below 10%) in the visible and higher in the near infra–red (as high as 60%).

1.8.2 *Carbon dioxide sequestering*

Carbon dioxide is sequestered from the atmosphere both by the oceans, the land surface and the biosphere. Carbon dioxide is very soluble in the oceans due to carbonate chemistry. The total dissolved CO_2 in the oceans is about fifty times that in the atmosphere and the rate of uptake is limited by vertical mixing. However, in the upper oceanic layers where mixing occurs down to about 100 m due to winds, the total mass of inorganic carbon dissolved as CO_2 is about equal to that in the atmosphere. This mixed layer plays a crucial role in the exchange of CO_2 between the atmosphere and oceans on timescales of about 10 years. In the terrestrial biosphere, photosynthesis and the uptake of CO_2 from the atmosphere provide the bulk of the organic carbon that is stored. On land, carbon dioxide is taken up by ecosystems, increasing with biological activity. This uptake is limited by the relatively small fraction of plant carbon that can enter long–term storage as wood and humus. On the other hand, according to the IPCC, the CO_2 released by deforestation in the tropics between 1980 and 1990 was more than compensated by other terrestrial sinks. Such storage can range between 10 and 200 years. Geochemical cycles that lead to sedimentation of carbonates from weathering of volcanic rocks by reactions of CO_2 dissolved in surface waters have characteristic timescales of thousands of years.

1.9 The climate record

Table 1.3 is a summary of the changes to the climate system that have occurred in the twentieth century, as recorded by the IPCC. Most striking are the dramatic increases in the greenhouse gases CO_2, CH_4 and N_2O that have occurred since pre-industrial times. While tropospheric ozone has also increased, the abundance of the same gas in the stratosphere has decreased, most significantly between 1970 and 2000. There has been a rise in global mean surface temperature of 0.6 ± 0.2 °C over the twentieth century, accompanied by a mean sea level rise of about 1–2 mm/yr. The Arctic sea-ice extent and thickness have declined, while El Niño events have become more frequent, persistent and intense, particularly during the last 20–30 years.

1.9.1 *Temperature trends*

Figure 1.8 shows the current best estimate of the change in global average surface temperature from 1861 to 2004, both in terms of interannual variability and the overall trend. The data show an upward trend from 1920 to 1940, followed by a cooling until 1975, and a marked rise since then, with the warmest years in the period since 1981. The measurements are taken from thousands of weather stations over the globe, on land, ships, buoys, and satellites. Clearly, the data tends to be less reliable the earlier it gets; particular care is needed in using the data before 1920.

1.9.2 *Sea–ice extent*

Figure 1.9 shows the decrease in Arctic sea–ice extent since 1970, amounting to about 1 million km^2 or about 2.5% per decade. Trends in sea-ice thickness are more difficult to measure but are estimated at as much as 40% in the Arctic. In Antarctica, there appears to have been no significant long–term trend in sea-ice extent.

Table 1.3 *Changes in the twentieth century to the climate system. (Source: IPCC 2001)*

Indicator	Observed Changes
Atmospheric CO_2	280 ppm between 1000–1750 AD, 368 ppm in 2000 AD, 31 ±4% increase
Terrestrial biospheric CO_2 exchange	A source of 30 GtC between 1800 and 2000; in 1990s a net sink of about 14 ±7 GtC
Atmospheric CH_4	700 ppbv from 1000 to 1750; 1750 ppbv in 2000 AD, 151 ±25% increase
Atmospheric N_2O	270 ppbv from 1000 to 1750; 316 ppbv in 2000 AD, 17 ±5% increase
Tropospheric O_3	Increase of 35 ±15% between 1750 and 2000; regional variations
Stratospheric O_3	decreased between 1970 and 2000; varies with altitude and latitude
Global mean temperature	Increased by 0.6 ±0.2 °C over the twentieth century
Northern Hemisphere surface temperature	Increase over the twentieth century greater than any other century in last 1000 years
Continental precipitation	Increased by 5–10% over the twentieth century in Northern Hemisphere; decreased in North Africa and Mediterranean
Heavy precipitation events	Increased in mid and high northern latitudes
Frequency and severity of drought	Increase in parts of Asia and Africa in recent decades
Global mean sea level	Increased at a rate of 1–2 mm/year during twentieth century
Duration of river and lake ice cover	Decrease by about 2 weeks in the twentieth century in mid to high northern latitudes
Arctic sea-ice extent and thickness	Thinned by 40% in recent decades late summer–early autumn; decrease in extent by 10–15% since 1950s spring–summer
Non-polar glaciers	Widespread retreat during the twentieth century
Snow cover	Decreased in extent by 10% as measured since 1960 by satellites
Permafrost	Thawed and degraded in parts of polar, subpolar regions
El Niño events	More frequent, persistent and intense the last 20–30 years

1.9.3 *Extreme events*

'Extreme' weather events are those that deviate significantly from the mean, like heat waves, droughts and floods, particularly those with impacts that can be catastrophic to ecosystems, agriculture and society. According to the IPCC, extreme events (defined to be within the upper or lower ten percentiles) have

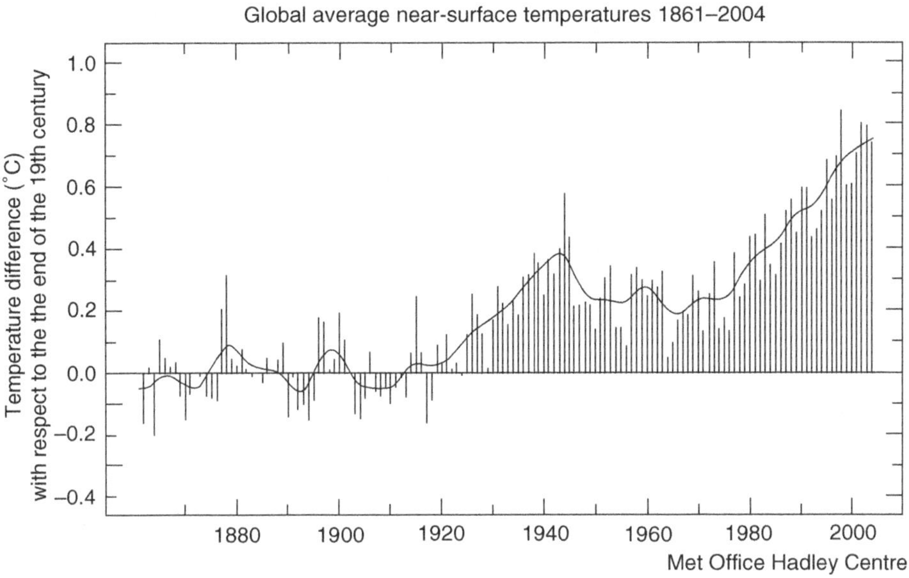

FIG. 1.8. Global average near–surface temperatures from 1861 to 2004, shown relative to the last decade of the nineteenth century. The observations are a combination of near–surface air and sea–surface temperatures, corrected to minimize errors in measurement practices and artifacts. Annual averages are shown as bars and the solid line shows the smoothed trend. ©Crown copyright (UK Met Office/Hadley Centre 2005)

increased despite the fact that total precipitation has decreased or remained constant. Since most extreme weather involves precipitation, the implication is that precipitation events are less frequent but are heavier on the average. As with temperature rise, there is emerging evidence that the incidence of both severe drought and severe wetness was roughly constant during the first part of the twentieth century but has increased in recent decades, correlated with a shift towards more warm ENSO events. In many mid and high latitude regions, the day–to–day temperature variability has decreased, the daily minimum temperature has increased, and the freeze-free period is longer. Since 1950 there has been a significant reduction in unseasonably low temperatures across much of the globe, with a smaller increase in the frequency of abnormally high temperatures.

Figure 1.10 shows the change in the rate of occurrence of 3–day rainfall intensity events, which have been shown to be important precursors to flooding. Hadley Centre predictions suggest that man-made climate change will tend to intensify the water cycle, increasing the rainfall at higher latitudes, a trend that is consistent with recent observations of increases in the rate of discharge from Eurasian rivers into the Arctic.

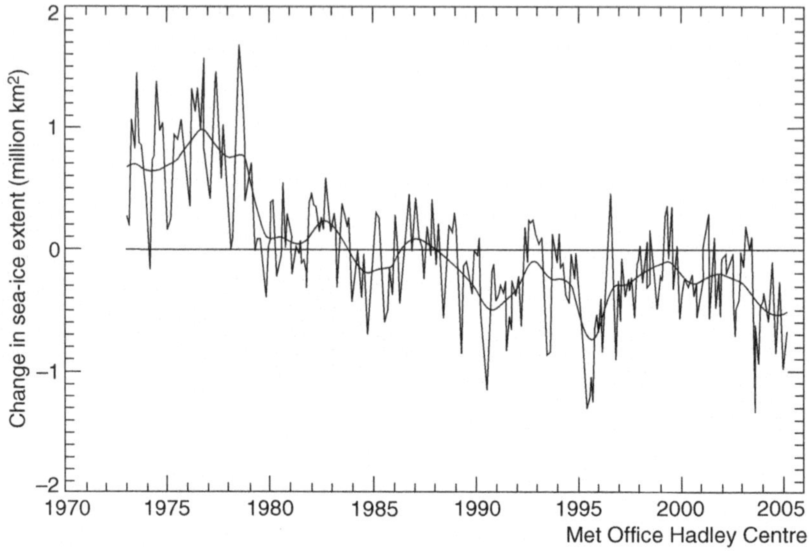

FIG. 1.9. Change in Arctic sea–ice extent since 1970. In deriving the trend, the extent of sea ice was defined as the area within which the concentration of sea ice is greater than 15%. ©Crown copyright (UK Met Office/Hadley Centre 2005)

1.9.4 *Sea–level trends*

The longest record of direct measurements of sea level comes from tide gauges. The IPCC has estimated that the rate of global mean sea–level rise during the twentieth century was in the range 10–20 cm, and that this rise was greater than that during the nineteenth century. However, no significant acceleration in the rate of rise was detected during the course of the twentieth century. Satellite measurements over the past decade indicate a rise of 2.5 mm/year in the global mean, but with large regional variations.

Over the last ice age cycle, sea level ranged from 5 m higher than today's, at the time of the last interglacial about 120 thousand years ago, to 120 m below today's, at the depth of the last ice age 21000 years ago when glaciers were at their maximum extent. Figure 1.11 shows relative sea level from measurements dating back to 1700.

1.10 Projections of future climate

In its 2001 report, the IPCC assessed the predictions of global climate models for the period 1990–2100 to illustrate the possible changes to the climate expected under future greenhouse–gas concentrations and sulphate aerosol loadings. The calculations were made by a number of different models and research groups, and

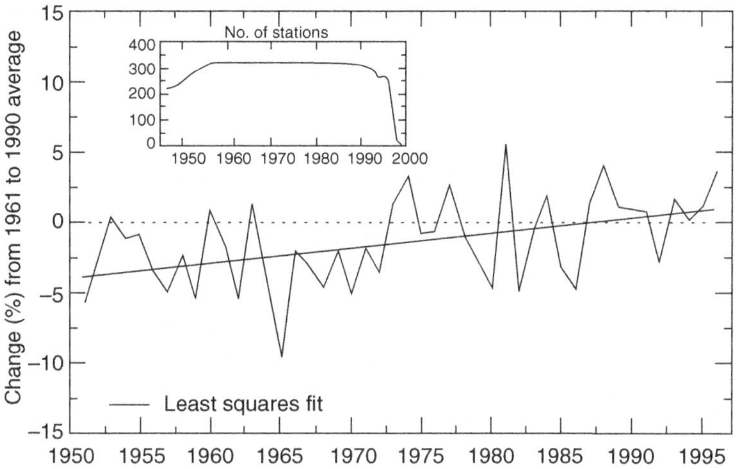

FIG. 1.10. The change in maximum annual 3-day rainfall events in northern midlatitudes from the average, over the period 1961 to 1990. (Source: IPCC 2001)

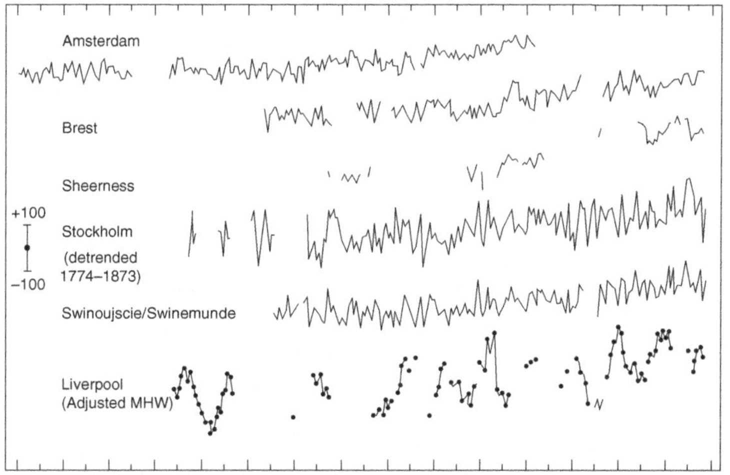

FIG. 1.11. Relative sea level since 1700, from stations in Northern Europe. The scale bar on the left indicates a change of ±100 mm. (Source: IPCC 2001)

based on a range of scenarios for projected changes in atmospheric composition. All are based on runs or integrations of atmosphere–ocean general circulation models (AOGCMs), the results of which apply to spatial scales of hundreds of kilometres and larger (see Chapter 11). Some of the simulations also included the effects of ozone and/or the indirect effects of aerosols, but most did not attempt to include the less well understood effects arising from land-use changes, mineral dust, black carbon, etc. None of the simulations included estimates of changes in the radiation from the Sun, or of the effect of future volcanic eruptions on atmospheric aerosol concentrations.

1.10.1 *Emission scenarios and global warming*

As part of its work the IPCC produced a Special Report on Emissions Scenarios, known as SRES, which made a range of projections of the greenhouse–gas emissions expected from human activity in the next 100 years. These ranged from optimistic (controlled population levels, use of new, cleaner technology and so forth) to pessimistic ('business as usual'). The following is an account of their main conclusions and recommendations.

1.10.1.1 *Greenhouse gases* Fossil–fuel burning is expected to remain the dominant influence on trends in atmospheric CO_2 concentration during the twenty–first century. As the amount in the atmosphere increases, the ocean and the land will take up a decreasing fraction of anthropogenic CO_2 emissions and the net effect will be to further increase the atmospheric CO_2 concentration. By 2100, carbon–cycle models project a CO_2 mixing ratio of 540 to 970 ppm, 90 to 250% above the value of 280 ppm in the year 1750. Since uncertainties, especially about the feedback from the biosphere, introduce a variation of about −10 to +30% around each scenario, the total predicted range is 490 to 1260 ppm (75 to 350% above the 1750 concentration).

Changing land use could influence the atmospheric CO_2 concentration. If all of the carbon released by historical land-use changes could be restored to the terrestrial biosphere over the course of the century (e.g. by reforestation), the CO_2 concentration would be reduced by 40 to 70 ppm. Without this, stabilization of atmospheric CO_2 concentrations at 450, 650 or 1000 ppm would require global anthropogenic CO_2 emissions to drop below 1990 levels, within a few decades, about a century, or about two centuries, respectively, and continue to decrease steadily thereafter. Eventually CO_2 emissions would need to decline to a very small fraction of current emissions.

Estimates of the future concentrations of the non-CO_2 greenhouse gases vary considerably across the SRES illustrative scenarios, with CH_4 changing by +190 to +1970 ppb from a present concentration of 1760 ppb, N_2O changing by +38 to +144 ppb (present concentration 316 ppb), total tropospheric O_3 changing by −12 to +62%, and a wide range of changes in concentrations of HFCs, PFCs and SF6. In some scenarios, total tropospheric O_3 would become as important a radiative forcing agent as CH_4.

1.10.1.2 *Aerosols* The SRES scenarios include the possibility of either increases or decreases in anthropogenic aerosols (e.g. sulphate aerosols, biomass aerosols, black and organic carbon aerosols) depending on the extent of fossil–fuel use and the success or failure of policies to abate polluting emissions. In addition, natural aerosols (e.g. sea salt, dust and emissions leading to the production of sulphate and carbon aerosols) are projected to increase as a result of changes in climate.

1.10.1.3 *Marker scenarios* The IPCC selected four main 'marker' emission scenarios, illustrated in Fig. 1.12. Scenario A1 (high emissions) describes a future world of very rapid economic growth, a global population that peaks in midcentury and declines thereafter, and the rapid introduction of new and more efficient technologies. A1FI is a variant of this which remains fossil–fuel intensive. A2 (medium–high emissions) describes a very heterogeneous world, with an underlying theme of self-reliance and preservation of local economic and technological identities. In B2 (medium-low emission) the emphasis is on local solutions to economic, social and environmental sustainability, a continuously increasing global population, intermediate levels of economic development, and less rapid and more diverse technological change than in the B1 and A1 storylines. B1 (low emissions) describes a convergent world with the same global population, that peaks in midcentury and declines thereafter, as in the A1 storyline, but with rapid change in economic structures toward a service and information economy, with reductions in material intensity and the introduction of clean and resource–efficient technologies. IPCC stresses that it is not possible to attach probabilities to each scenario.

1.10.2 *Climate projections for the twenty-first century*

The IPCC climate projections are meant to give the breadth of the impacts of increasing greenhouse emissions over the next century, over a realistic range of emission scenarios and models. Although models continue to improve, at present the choice of model is found to make a difference comparable to the choice of scenario. Of course, many uncertainties remain; our understanding of the physical processes that govern climate and our ability to simulate these highly non-linear processes are imperfect and some of the most important determinants of climate, such as clouds and aerosols and their radiative properties, are certainly inadequately simulated. The fact that the models do not show a consensus over future changes in interannual variability as a result of the El Niño-Southern Oscillation (ENSO) points to other limitations in the current generation of models. Nevertheless, a number of key conclusions emerged from the IPCC study that, if not totally robust, certainly require attention and precautionary or preventative action.

1.10.2.1 *Projections of temperature rise* In general, the troposphere and surface warms, while the stratosphere cools. The land warms faster than the ocean, and there is greater relative warming at high latitudes. The increase in mean

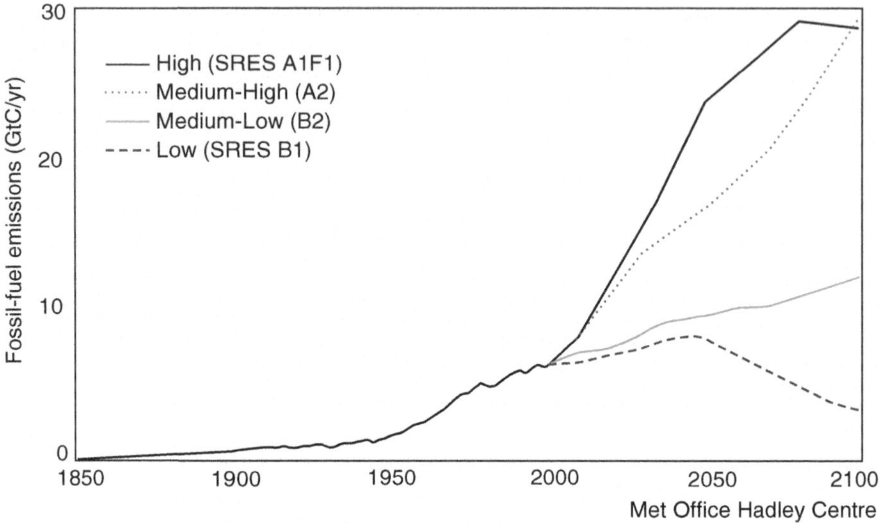

FIG. 1.12. Projected emissions of carbon dioxide (in units of 10^{12} tonnes of carbon per year) from fossil–fuel burning for four future emission scenarios, together with estimated emissions from 1850–2000. ©Crown copyright (UK Met Office/Hadley Centre 2005).

surface air temperature is accompanied by more frequent extreme high and less frequent extreme low temperatures. There is a decrease in diurnal temperature range in many areas, with nighttime lows increasing more than daytime highs.

Results for the four SRES emission scenarios give a mean global surface temperature warming between 2000 and 2100 that ranges between 1.5 and 6.0 C, depending on the scenario (Fig. 1.13). The B1 scenario, representing perhaps the most likely world view, at least as seen from 2006, gives about a 2 C warming according to the Hadley Centre HadCM3 climate model. Warming over the next three or four decades is similar for all scenarios, due to the thermal inertia of the climate system and the effective long lifetime of carbon dioxide. Temperatures over land are expected to be higher than over the oceans, with a warming between 3 and 8 °C. Overall, the projected rate of warming is much larger than the observed changes during the twentieth century, and, according to palaeoclimate data, is very likely to be without precedent during at least the last 10000 years.

1.10.2.2 *Sea–level rise* Sea levels can rise under global warming through the thermal expansion of the oceans (the main contribution, see Chapter 11), the influx of glacier, snow and ice–melt water, and melting of the Greenland and Antarctic ice sheets. The IPCC examined sea–level rise from seven climate models driven by the above emission scenarios and found a range of predicted sea–level rise from 0.1 to about 0.9 m (Fig. 1.14). All models give regional variations

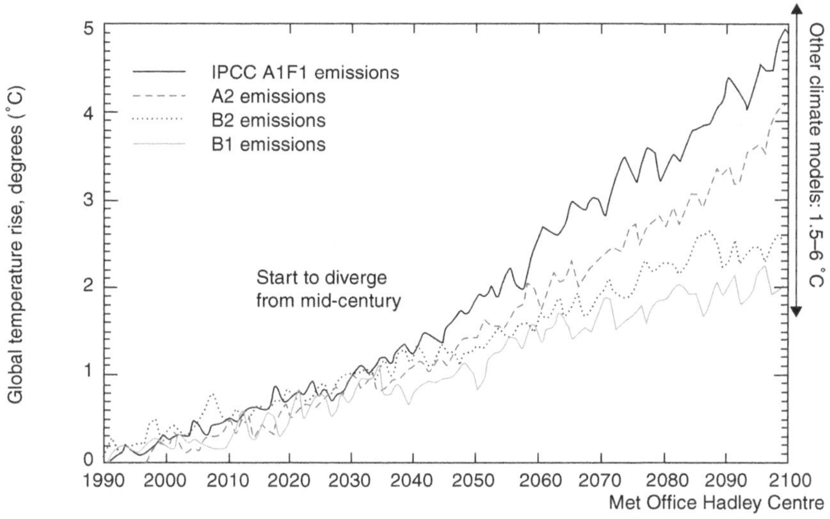

FIG. 1.13. Global temperature rise according to the four SRES emission scenarios, predicted by the Hadley Centre climate model (HadCM3). The range of results from other climate models, according to IPCC, are also given. ©Crown copyright (UK Met Office/Hadley Centre 2005)

of sea-level rise that can be twice the global.

1.10.2.3 *Precipitation* As the climate warms, the Northern Hemisphere sea–ice extent decreases. The globally averaged mean water vapour, evaporation and precipitation increase. Snowfall is predicted to increase over Antarctica according to the Hadley Centre model; the actual extent of this effect, which represents negative feedback since it tends to cool the global climate by increasing the planetary albedo, varies between models. Most agree that tropical areas and high latitudes have increased mean precipitation, while subtropical areas experience a decrease. The intensity of rainfall events increases everywhere, with more interannual variability of northern summer monsoon precipitation. A general decrease in soil moisture in the midcontinental areas during summer is expected.

1.10.2.4 *Thermohaline circulation* Most models predict a weakening of the Northern Hemisphere thermohaline circulation, which contributes to a reduction in the surface warming in the northern North Atlantic. However, there is still a warming over Europe due to increased greenhouse gases. In experiments where the atmospheric greenhouse–gas concentration is stabilized at twice its present–day value, the North Atlantic circulation recovers from initial weakening within one to several centuries. However, beyond 2100, a complete and possibly irreversible shutdown of the thermohaline circulation becomes a possibility, with

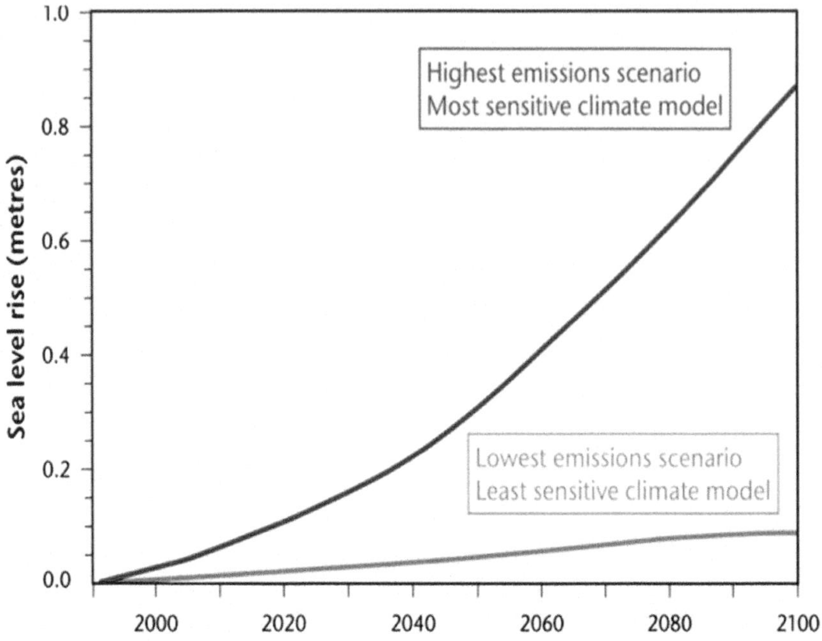

FIG. 1.14. Projected rise in global sea level for the highest– and lowest–emission scenarios based on IPCC estimates. ©Crown copyright (UK Met Office/Hadley Centre 2005)

consequences that have been little explored but are likely to be severe.

1.11 Bibliography

1.11.1 *Notes*

Most aspects of the climate system summarized in this chapter, including basic definitions of its components, are covered more fully in the references IPCC 2001 and Hadley Centre 2005 below. See also the Intergovernmental Panel on Climate Change Website, at http://www.ipcc.ch/, and the review by Houghton.

For more detailed coverage on the North Atlantic Oscillation (NAO) see Hurrel and Dickson, and Wang. The El Niño Southern Oscillation ENSO is discussed by Hanley *et al.*; Philander; and Trenberth and Stepaniak; there are good reviews by Wang and Fiedler; Fedorov *et al.*; and Dijkstra. Recent results on radiation measurements linked to ENSO are presented by Pavlakis *et al.*

For a review of direct and indirect radiative forcing due to tropospheric aerosols see Haywood and Boucher.

1.11.2 *References and further reading*

Bolle, H-J. (2003). *Mediterranean climate. Variability and trends.* Springer, Berlin.

Croll, J. (1867a). On the change in the obliquity of the ecliptic, its influence on the climate of the polar regions and on the level of the sea. *Philos. Mag.*, **33**, 426–445.

Croll, J. (1867b). On the eccentricity of the Earth's orbit, and its physical relations to the glacial epoch. *Philos. Mag.*, **33**, 119–131.

Dijkstra, H. A. (2006). The ENSO phenomenon: theory and mechanisms. *Adv. Geosc.*, **6**, 3–15.

Fedorov, A. V., Harper, S. L., Philander, S. G., Winter, B. and Wittenberg, A. (2003). How predictable is El Niño?, *BAMS*, **84**, 911–919.

Hadley Centre (2005). *Climate change and the greenhouse effect. A briefing from the Hadley Centre.* UK Met Office, Exeter.

Hanley, D. E., Bourassa, M. A., O'Brien, J. J., Smith, S. R. and Spade, E. R. (2003). A quantitative evaluation of ENSO indices. *J. Climate*, **6**, 1249–1258.

Haywood, J. M. and Boucher, O. (2000). Estimates of the direct and indirect radiative forcing due to tropospheric aerosols: a review. *Rev. Geophys.*, **38**, 513–543.

Hurrell, J. W. and Dickson, R. R. (2005). Climate variability over the North Atlantic, in *Marine ecosystems and climate variation. The North Atlantic - A comparative perspective*, ed. N.C. Stenseth, G. Ottersen, J.W. Hurrell and A. Belgrano. Oxford University Press, Oxford.

Houghton, J. T. (2005). Global warming. *Rep. Prog. Phys.*, **68**, 1343–1403.

IPCC, 2001: *Climate change 2001: The scientific basis. Contribution of working group I to the third assessment report of the Intergovernmental Panel on Climate Change.* Houghton, J. T., Ding, Y., Griggs, D. J., Noguer, M., van der Linden, P. J., Dai, X., Maskell, K. and Johnson, C. A. (ed.), Cambridge University Press, Cambridge, United Kingdom and New York, NY, USA.

Milankovitch, M. M. (1941). Canon of insolation and the ice age problem. Königlich Serbische Academie, Belgrade. English translation by the Israel Program for Scientifc Translations, United States Department of Commerce and the National Science Foundation, Washington D.C.

Pavlakis, K. G, Hatzidimitriou, D., Drakakis, E., Matsoukas, C., Fotiadi, A., Hatzianastassiou, N., and Vardavas, I. M. (2006). ENSO surface longwave radiation forcing over the tropical Pacific. *Atm. Chem. Phys. Diss.*, **6**, 1–34.

Philander, S. G. (1990). *El Niño, La Niña, and the Southern Oscillation.* Academic Press, San Diego.

Rossow, W. B. and Schiffer, R. A. (1999). Advances in understanding clouds from ISCCP, *Bull. Am. Meteorol. Soc.*, **80**, 2261–2287.

Trenberth, K. E. and Stepaniak, D. P. (2001). Indices of El Niño evolution. *J. Climate*, **14**, 1697–1701.

Wang, C. and Fiedler P. C. (2006). ENSO Variability and the Eastern Tropical Pacific: A review. *Prog. Oceanogr.*, **69**, 239–266.

Wang, C. (2005). ENSO, Atlantic climate variability, and the Walker and Hadley circulations. In *The Hadley Circulation: Present, Past and Future.* H.F Diaz and R.S. Bradley (ed) Kluwer Academic Publishers, Amsterdam.

2

ATMOSPHERIC PHYSICS AND THERMODYNAMICS

2.1 Introduction

The atmosphere is a thin shell of gas held gravitationally to the planet and having a thickness of only about 1% of the radius of the solid body. Its original composition, drawn from the solar nebula from which the Sun and the planets formed, will have included many more light elements, especially hydrogen and helium, than are found today, and a complicated evolutionary process, still not fully understood, was involved in the progression from one to the other.

Key stages in the evolution of the atmosphere were the fairly rapid loss of the original atmosphere, probably in its entirety, followed by its gradual replacement by the exhalation of volcanic gases from the interior. These were supplemented to an unknown degree by an influx of volatile material similar to that which we see in present–day comets. The original loss processes were driven by radiation and particle fluxes from the young Sun, perhaps in a 'T–Tauri' phase, which were much more efficient than current levels at dissociating and driving off gases from the inner planets, although the process is still going on at a much reduced level. So are volcanic emissions and cometary infall, although again both will have been much more vigorous in the early Solar System. The emissions from the interior have probably always been of similar composition to that from modern volcanoes, so mostly water vapour, carbon dioxide, and sulphur dioxide. Comets vary a great deal in composition, but are mainly water ice with significant amounts of ammonia and methane. The nitrogen in the modern atmosphere may have originally accrued as ammonia, which was dissociated and the hydrogen content lost to space, leaving the stable N_2 molecule behind. The other major constituent, O_2, is of course a product of life on the planet and evolved in a very complex manner that we have yet to understand.

All of the sources and sinks became sufficiently consistent at some time in the past 100 million years for the atmosphere to have stabilized at something like its present composition and surface pressure. For climate studies, spanning the last few millennia and the next few centuries, the abundances of molecular nitrogen and oxygen are taken to be invariant. This is not the case for the minor constituents, water vapour, ozone, and carbon dioxide, all of which show important variations on quite short timescales and are very important for understanding climate and climate change. Variations in the trace constituents, such as oxides of nitrogen and chlorine compounds, are also large and can be important for their

Table 2.1 *Mean composition of dry, ozone-free air (% by volume) in the troposphere. Nitrogen and oxygen are commonly referred to as major constituents; water vapour, ozone, argon and carbon dioxide as minor constituents, and the others as trace constituents.*

Species	Mixing ratio
N_2	78.08
O_2	20.95
CO_2	0.037
Ar	0.934
Ne	1.82×10^{-3}
He	5.24×10^{-4}
Kr	1.14×10^{-4}
Xe	8.7×10^{-6}
H_2	5.3×10^{-5}
CH_4	1.7×10^{-4}
N_2O	3.1×10^{-5}

effects on the ozone distribution and as greenhouse gases in their own right.

The accurate calculation of radiative transfer in the atmosphere under given conditions requires a knowledge of the concentration of the radiatively active constituents and their temperature, both as a function of altitude (pressure). In the following sections, we recap briefly the normal composition of the atmosphere, and simple expressions that describe the vertical variation of pressure (hence density) and of temperature under various conditions.

2.2 Atmospheric composition

2.2.1 *Well-mixed species*

A mean atmospheric composition can be defined for the main constituents that do not condense or undergo a significant level of chemical activity, and which therefore have essentially the same proportion everywhere, in the lower atmosphere (Table 2.1). As already noted in Chapter 1, the principal 'greenhouse' gases are carbon dioxide, water vapour, ozone, methane, nitrous oxide and the chlorofluorocarbon family, which play a major role in infra–red radiative transfer in the atmosphere and affect the energy balance at all levels, including the surface.

Carbon dioxide is released by the combustion of fossil fuels (oil, coal, and natural gas), deforestation, and manufacturing, and removed by plant photosynthesis and by virtue of its solubility in the ocean. It accounts for about half of the global–warming potential caused by human activity.

Principal sources of methane are landfills, marshes and wetlands, agriculture and livestock, natural gas, biomass burning, and life forms such as termites. It is removed by oxidation and photodissociation in the stratosphere. Individual molecules of methane have 20–30 times the global–warming potential of CO_2 and they are accumulating in the atmosphere much more rapidly, meaning that methane could overtake carbon dioxide as the principal anthropogenic contribution to global warming.

Nitrous oxide and chlorofluorocarbons are noted for their role in stratospheric ozone destruction as well as global warming. N_2O originates in the burning of coal and wood, and microbial activity in the soil, and biological decay in oceans. CFCs are manufactured for use in refrigerators and air conditioners, solvents, aerosol propellants and manufacturing. Both are removed only slowly by chemical processes and rainout. Some CFCs achieve global–warming potentials in excess of 10000 times that of CO_2.

2.2.2 *Water vapour and ozone*

To the list in Table 2.1, we must add two important species that have relatively large abundances and key roles in atmospheric processes (for instance, see Fig. 2.1), but no single well-defined value for their abundance. Water vapour and ozone are both highly variable, the former because of phase changes – evaporation, condensation, and freezing – and the latter because of chemical exchange, even in the troposphere. The water–vapour (H_2O) content of the lower atmosphere varies from almost zero to 4% or occasionally more, with an average value of about 0.8%, while the ozone abundance is a few parts per million in the troposphere, rising to 10 ppm in the stratospheric ozone layer at an altitude of around 20 km.

Figure 2.2 shows some vertical profile measurements of water vapour, synthesized into a global cross–section for one particular longitude and time. The extreme variability that can be seen in this data results from the sensitivity of the mixing ratio to temperature and to the availability of liquid water in cloud droplets and on the surface, plus the effects of dynamical transport.

Nevertheless, the general trend is that the largest values are near the surface, falling with height to very low mixing ratios above the tropopause. For radiative–transfer modelling, it is necessary to have a mean profile for water vapour, just as for other, less variable species. Figure 2.3 shows such a profile, obtained by averaging a very large number of individual profiles obtained under a range of conditions. It shows a steady fall from just below 1% at the surface to a few parts per million in the stratosphere, where the mixing ratio stays nearly constant with height. The low value in the stratosphere is a consequence of the cold-trapping effect of the tropopause, since most of the stratospheric water vapour originates at the surface, although the small increase with height that is seen in the mean

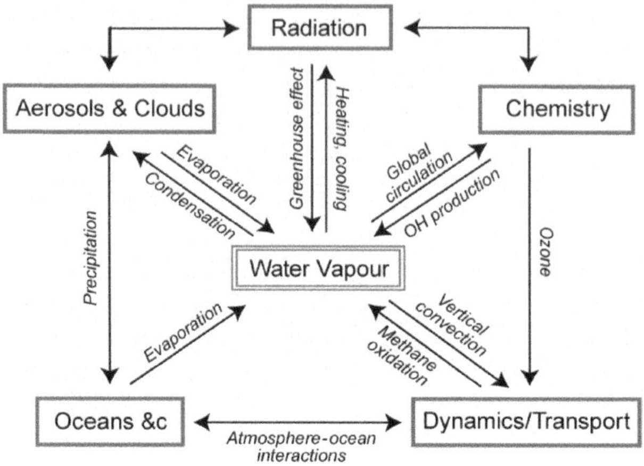

FIG. 2.1. A schematic diagram showing the various interactions within the climate system that involve water vapour, and their interactive nature. Included with the oceans are other water bodies. (Revised after ESA)

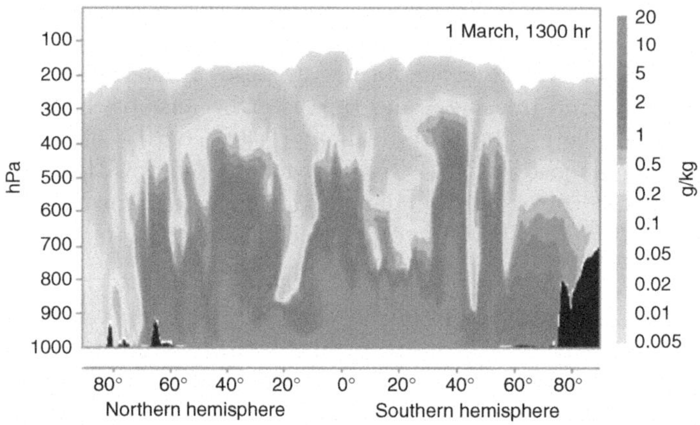

FIG. 2.2. Global water–vapour mass–mixing ratio measurements, made using lidar, showing the large variability in space and time, as well as vertically, of this key species. (Data from ESA)

profile indicates a secondary source in the stratosphere itself, believed to be the photochemical oxidation of methane.

A similar problem exists when trying to define a mean ozone profile, which, like water vapour, is very variable, although for different reasons. Ozone has a fairly short lifetime in the atmosphere, of the order of a day, and therefore the amount present at any time and place is sensitive to the balance between production and

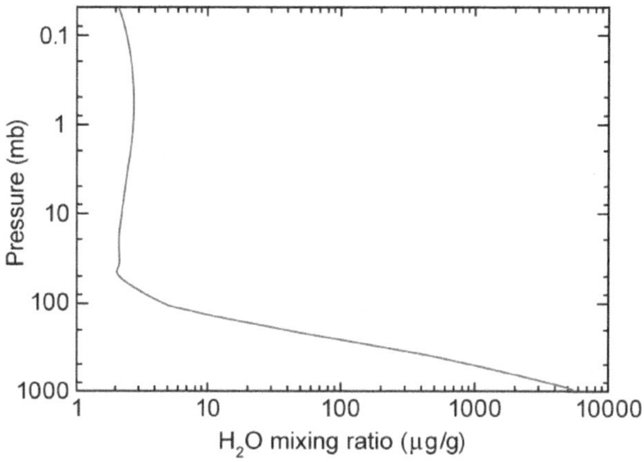

FIG. 2.3. Mean water–vapour profile, synthesized from 80000 radiosonde ascents. Since water is freely available in the troposphere, the mixing ratio tends to follow the temperature profile, which provides a cold trap at the tropopause that tends to limit the stratospheric mixing ratio to the minimum at the top of the troposphere. However, the stratosphere has a small internal source of water due to the photochemical oxidation of methane, and this produces the shallow maximum that can be seen at about 1 mb. (Data from Laboratoire de Météorolgie Dynamique du CNRS)

loss processes, and to dynamical transport. Not only is the profile very variable with season, but there are also secular changes due to the build–up of man–made pollution that affects ozone photochemistry and hence the efficiency of the various sources and sinks (see Chapter 7). These changes are illustrated in Fig. 2.4, which shows two profiles at the same location obtained by averaging measurements over two twelve–month periods thirty years apart. The averaging removes most of the daily and seasonal fluctuations, and reveals the overall decline in stratospheric ozone, and corresponding increase in tropospheric ozone, driven by complex human–induced changes in ozone chemistry, even at this remote location (McQuarie Island, in the Southern Ocean south of New Zealand).

Near and within the Antarctic, the seasonal variation is particularly large, with mixing ratios in the region around 20 km altitude, where the peak is normally found, falling almost to zero. Figure 2.5 illustrates the more moderate variations that are found at midlatitudes, more typical of most of the globe.

2.2.3 *Trace constituents*

The atmosphere also contains a large number of species that are present only in extremely small amounts, typically measured in parts per trillion. While most are so rare their presence can be ignored when considering the radiative com-

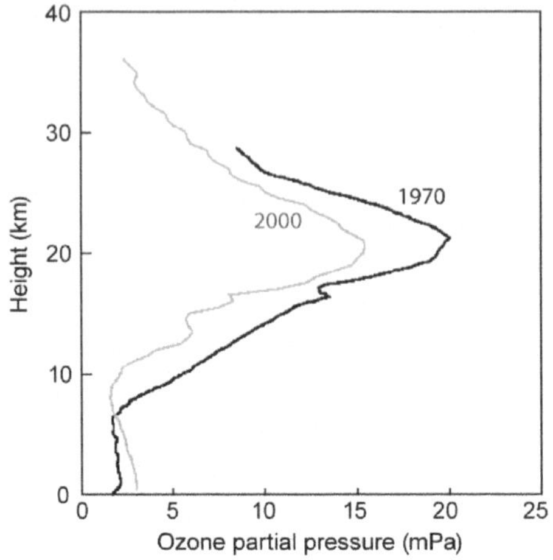

FIG. 2.4. Ozone profiles at McQuarie Island, between Australia and Antarctica. These are averages over the seasons for the years shown, and show how most of the atmospheric ozone is concentrated in a layer in the stratosphere. They also reveal the effect of global pollution, which has depleted stratospheric ozone by around 20% over the 30–year period, while enhancing the smaller amounts found near the surface by about the same factor. (Data from the Australian State of Environment Committee 2001)

ponent of climate, this is not always the case. A few species are increasing so rapidly, and have absorption bands in atmospheric–window regions where more abundant species are transparent, that their contribution to greenhouse warming is not negligible, especially when looking to the future. Most of these are chlorofluorocarbons (CFCs), which are manufactured for use in refrigerators and air conditioners, solvents, aerosol propellants and manufacturing and which are removed from the atmosphere only slowly by chemical processes and rainout.

A relatively common example is CFC–12 (CCl_2F_2), dichlorodifluoromethane, which has a global–warming potential (defined as how much a given mass contributes to global warming over a given time period compared to the same mass of carbon dioxide) of around 10000 times that of CO_2 (Table 2.2). Thus, a mixing ratio of $\sim 3 \times 10^{-8}$ or about 30 ppb would be sufficient to produce a similar threat to that of carbon dioxide itself.

Other trace species affect radiation in the atmosphere indirectly, by their effect on more common greenhouse gases. Here the most common examples are the ozone–depleting species, of which there are many. Their importance can be estimated by combining their effectiveness as catalysts in the reactions that remove ozone with their mean mixing ratios to obtain an index of ozone–destroying potential (Table 2.2).

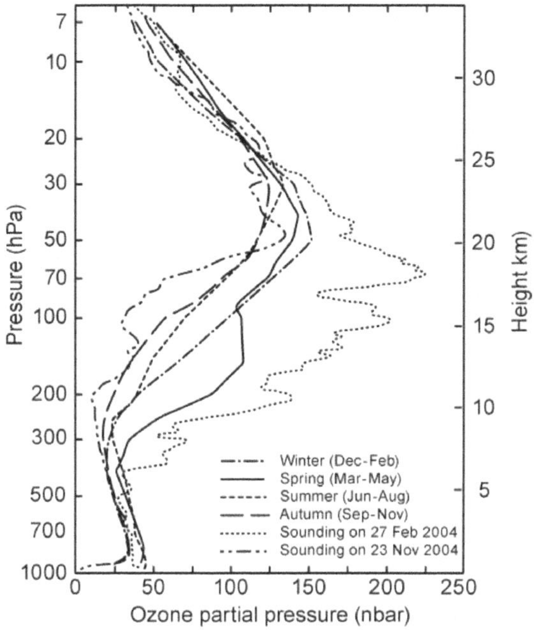

FIG. 2.5. The altitude distribution of ozone over Switzerland in the year 2004, showing both seasonal averages and two individual profiles. The ozone peak is more pronounced, but at a lower height, early in the year. Note the layering that occurs in the single profiles. (Source: MeteoSweitz)

2.3 Pressure and hydrostatic equilibrium

Although it strictly does not apply to atmospheric gases in motion, as they commonly are, the laws of hydrostatics are a good starting point for understanding the vertical structure of the atmosphere. The balance between the pressure gradient and gravity gives the hydrostatic equation

$$\mathrm{d}p = -\rho g \mathrm{d}z \tag{2.1}$$

where p is pressure, ρ is density, and z is height. The ideal gas law, another useful approximation, for one mole occupying a volume V at temperature T is

$$pV = RT, \tag{2.2}$$

whence

$$\rho = \frac{Mp}{RT}, \tag{2.3}$$

where M is the mass of one mole and R is the universal gas constant. Combining the two gives

Table 2.2 *Lifetime (years) in the atmosphere, ozone destroying potential (ODP) relative to CFC–11 and global warming potential (GWP) relative to CO_2, for some important trace constituents of the atmosphere. (After IPCC 2001)*

Species	Years	ODP	GWP
Trichlorofluoromethane (CFC-11) CCl_3F	45	1	4000
Dichlorofluoromethane (CFC-12) CCl_2F_2	100	1	10000
Trichlorofluoroethane (CFC-113) $C_2Cl_3F_3$	85	1	6000
Dichlorotetrafluoroethane (CFC-114) $C_2Cl_2F_4$	300	1	1000
Monochloropentafluoroethane (CFC-115) $C_2Cl_2F_4$	1700	0.5	7200
Bromochlorodifluoromethane (Halon-1211) CF_2ClBr	16	3	1300
Bromotrifluoromethane (Halon-1301) CF_3Br	65	10	6900
Dibromotetrafluoromethane (Halon-2402) $C_2F_4Br_2$	20	6	-
Chlorofluoromethane (CFC-13) CCl_3F	640	1	1400
Carbon tetrachloride CCl_4	26	1	1400
Trichloroethane (methyl chloroform) $C_2H_3Cl_3$	5	0.1	140

$$\frac{\mathrm{d}p}{p} = -\frac{\mathrm{d}z}{H} \qquad (2.4)$$

where $H = RT/Mg$ is the scale height, the height over which the pressure drops by $1/\mathrm{e}$. Integrating from the surface ($z = 0$) to height z

$$p = p_0 \exp\left(-\int_0^z \frac{\mathrm{d}z}{H}\right). \qquad (2.5)$$

Since $R = 8.314$ J K^{-1} mole^{-1} and $M = 28.96\times10^{-3}$ kg mole^{-1} for dry air, $H = 8.45$ km at 290 K and 5.84 km at $T = 200$ K.

2.4 Vapours and ideal gases

The pressure–temperature (P–T) diagram of molecules gives three regions where the solid, liquid or vapour phase exists for a real gas. The curves on the P–T diagram give values of pressure and temperature where liquid–solid, solid–vapour and gas–liquid phases coexist, called the melting, sublimation and vaporization curves, respectively. The triple point, on the P–T diagram corresponds to the pressure, P_o, and temperature, T_o, where all three phases coexist, as shown in Fig. 2.6. We note that the melting curve of H_2O has a negative slope, while for other molecules it has a positive gradient. For a given mass of water at some specific temperature, if we decrease the pressure the volume increases, and so the density decreases. Thus, if the melting curve has a negative slope, a decrease in pressure (at constant temperature) results in a change of phase from liquid to solid, hence ice has a lower density than water and so floats on water. In contrast, the density of solid CO_2 is higher than that of the liquid phase. This irregular thermodynamic characteristic of H_2O is of fundamental importance to life on Earth, as freezing of water means that the ice floats and so aquatic life can exist below the surface of frozen rivers, lakes and oceans. It also means that

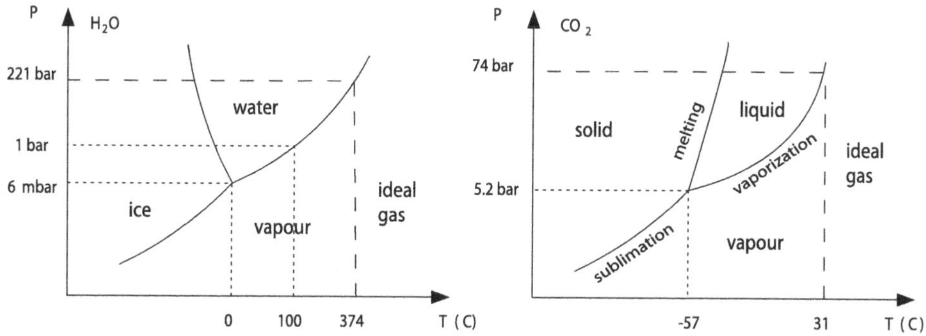

FIG. 2.6. Schematic P–T curves of H_2O and CO_2. Shown are the triple points and the critical points of each molecule. We note that the melting curve of H_2O has a negative slope (so ice has lower density than water), while for other molecules it has a positive gradient and the solid phase is denser than the liquid phase.

ice can melt easier with an increase in solar heating, preventing water bodies from remaining solid at depth for a long time.

Also shown is the boiling point of water (100 °C), and we note that it corresponds to a water–vapour pressure equal to 1 atmosphere, i.e. the ambient pressure. There exist also a critical temperature, T_c and pressure, P_c, beyond which the molecule cannot be liquified. In this region, the molecule behaves like an ideal gas and at sufficiently high temperature above T_c the pressure and temperature of the gas are related by the ideal gas law $PV = RT$. The critical pressure and temperature for molecules of interest for Earth-like planetary atmospheres are given in Table 2.3. The critical pressure is defined to be the pressure that is just sufficient to liquify the gas at temperature T_c. From the table we see that N_2 and O_2 cannot be liquified for the range of pressures and temperatures in the Earth's atmosphere. Water vapour, however, is a vapour in the Earth's atmosphere and so is a condensible gas, easily forming oceans, clouds, rain, snow or ice. For Venus, with surface pressure of about 92 bar, of which 96.5% is CO_2, and surface temperature of 730 K, we see that $T > T_c = 304$ K, and the atmospheric temperature is sufficiently high at any given atmospheric pressure so CO_2 behaves as an ideal gas. On Titan, where the surface pressure is 1.5 bar and the temperature 94 K, methane behaves like H_2O does on Earth, and so it is a condensible gas that constitutes about 5% of the atmospheric pressure. On Mars, with a surface pressure of 7 mbar and temperature of 220 K, CO_2 (95% of the pressure) is condensible in the phase region of vapour–solid, with triple point at $T_o = 216$ K and $P_o = 5.2$ bar. Thus, at the Martian poles, CO_2 can condense out and return back to the atmosphere by sublimation (solid to vapour phase change).

Table 2.3 *Triple–point temperature, T_o, and pressure, P_o, and critical temperature, T_c, and pressure, P_c, of molecules relevant to Earth-like planetary atmospheres.*

Molecule	T_o (K)	P_o (mbar)	T_c (K)	P_c (bar)
N_2	63.2	125.3	126.3	34.0
O_2	54.4	1.5	154.6	50.4
CO_2	216.6	5180.0	304.2	73.8
CH_4	90.7	116.9	191.1	46.4
NH_3	195.4	60.8	405.6	112.8
H_2O	273.2	6.1	647.3	221.2

We note that the triple point of H_2O is 273.16 K and the triple–point pressure is 6.105 mbar without the Earth's atmospheric pressure. Under the Earth's atmospheric pressure the triple–point temperature of water with dissolved air becomes 273.15 K (0 °C) (e.g. Iribarne and Godson 1981) and the triple–point pressure becomes 6.132 mbar. Thus, at the atmospheric pressure of 1.01325 bar, the triple–point temperature and pressure are only slightly modified. Further, we also note that if we define a critical molar volume, V_c, then the ratio RT_c/P_cV_c does not equal unity but ranges between 3.4 and 4.5, for gases of atmospheric interest.

2.5 Vertical temperature structure

Figure 2.7 shows the mean vertical temperature structure of the atmosphere and the names commonly given to the regions, demarked by the vertical temperature gradient. The lowest ten km or so are characterized by a temperature profile that drops off with height at a roughly constant rate, equal for dry air to the temperature trajectory of a parcel of air that is raised adiabatically from the surface to the tropopause. The implication of this is that dynamical processes such as convection, waves and turbulence dominate, acting to drive the temperature gradient towards the stable value at which vertical motions would be suppressed. The value of the adiabatic temperature gradient is obtained from the First Law of Thermodynamics when the exchange of heat between a vertically–moving parcel and its surroundings is zero

$$C_v \mathrm{d}T = -p\mathrm{d}V \tag{2.6}$$

and C_v is the heat capacity per mole of air at constant volume, the ideal gas law

$$p\mathrm{d}V + V\mathrm{d}p = R\mathrm{d}T = (C_p - C_v)\mathrm{d}T \tag{2.7}$$

and the hydrostatic equation

$$V\mathrm{d}p = -V\rho g\mathrm{d}z = -Mg\mathrm{d}z, \tag{2.8}$$

to obtain for dry air

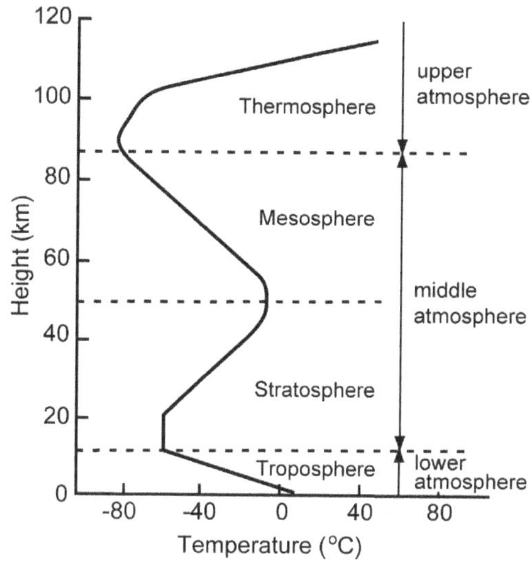

FIG. 2.7. The mean vertical temperature profile of the atmosphere, and the nomenclature used for regions of different vertical temperature gradient. Two conventions are in common use: the classical 'spheres' separated by inflexions in the profile where $dT/dz = 0$, and the more modern convention which refers simply to the lower, middle and upper atmosphere.

$$\frac{dT}{dz} = -\frac{Mg}{C_p} = -\frac{g}{c_p} \approx -9.7 \text{ K km}^{-1}. \tag{2.9}$$

where c_p is the specific heat capacity (or heat capacity per unit mass) at constant pressure.

At some level, the density falls to the point where radiative transfer of heat between regions of substantially different temperature, including between atmospheric layers and cold space, becomes significant and eventually dominates. This change manifests itself in a transition from an adiabatic gradient of temperature versus height of about 7 K km^{-1}, for typical humidity, to one that is invariant with height. The temperature T_S in the stratosphere can be estimated by assuming radiative balance between the planet (including the largely opaque troposphere) below at an effective temperature T_E, and cold space above, whence

$$e\sigma(T_E)^4 = 2e\sigma(T_S)^4, \tag{2.10}$$

where T_E is known from theory and observation to be close to 255 K, from which it follows that

$$T_S = \frac{T_E}{2^{1/4}} = \frac{255}{2^{1/4}} = 215 \text{ K}. \tag{2.11}$$

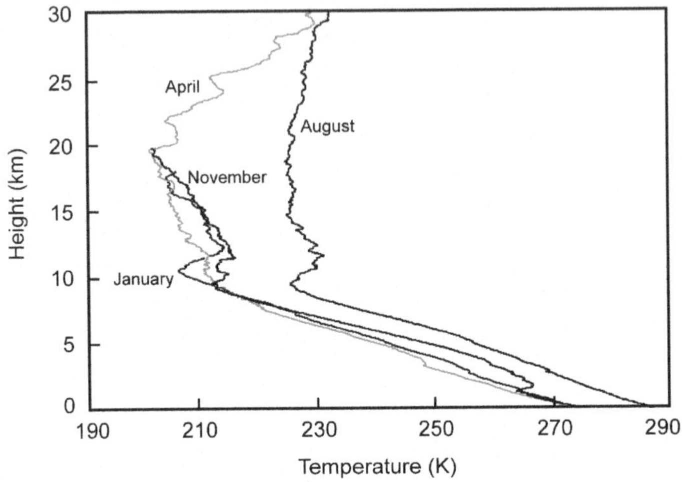

FIG. 2.8. Selected vertical temperature profiles measured by individual radiosonde balloon ascents over Scandinavia, in the months indicated, representing the four seasons.

In the real atmosphere, the transition region between troposphere and stratosphere, the tropopause, is often remarkably sharp (Fig. 2.8).

The nature of the observed vertical temperature profile validates the concept of the lower atmosphere as a region where the energetics are dominated by dynamics, while in the middle atmosphere radiation holds sway. Of course, the detailed picture is more complex than this. Although all but the shortest atmospheric paths are opaque to radiation on the average at tropospheric air densities, the atmosphere is quite transparent to radiation at visible wavelengths. Even in the infrared, where there is molecular vibrational-rotational band absorption (see Chapter 4), 'window' regions in the thermal infra–red spectrum exist where photons can travel from the surface to space without being absorbed. Conversely, motions occur in the stratosphere, including high winds, waves and turbulence, although vertical propagation of heat by dynamics is not as vigorous as it is below the tropopause.

Temperature increases with height in the upper stratosphere due to the absorption of solar ultraviolet radiation by ozone. Above the ozone layer, in the mesosphere, the temperature falls again, reaching below 200 K near an altitude of about 80 km. The upper atmosphere exhibits very high temperatures because of the energy released during dissociation of atmospheric molecules and atoms by energetic solar photons and solar and cosmic ray particles. Charged–particle layers (the ionosphere) are also present. Although some interesting physics occurs in the thermosphere, this region is too low in density to be of much significance

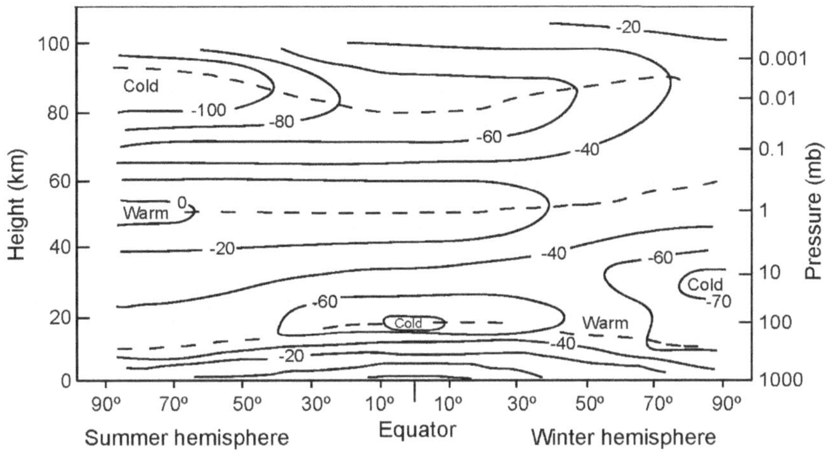

FIG. 2.9. The zonal mean atmospheric temperature (C) versus height and latitude, at the time of the solstice (midsummer in one hemisphere, midwinter in the other). The approximate heights of the tropopause, stratopause and mesopause are indicated by the dashed lines, showing their variation with latitude.

to the energy budget of the Earth or to affect the climate at the surface.

Figure 2.9 shows the latitude–altitude cross-section of mean atmospheric temperature structure at the solstice. The tropopause height varies with latitude from about 8 to as much as 18 km above the surface, being higher in the equatorial regions and lowest at high latitudes in the winter. The stratopause is warmest in summer, as would be expected since both ozone abundance and heating rate are proportional to the photon flux, but the mesopause behaves in the opposite sense, being coldest in summer. Both levels, in fact, experience large heating rates due to dynamical processes in the winter, when high circumpolar winds dissipate and their kinetic energy degrades to heat. This factor has a greater tendency to dominate in the mesosphere where the radiative heating is feeble compared to the ozone–rich, and relatively dense, stratosphere, although 'sudden warmings' of as much as 100 K or more due to the collapse of the polar vortex are a well–known stratospheric phenomenon, occurring spasmodically in midwinter.

2.6 Thermodynamics of moist air

Figure 2.10 shows some typical measurements of temperature and water–vapour mixing ratio, both of which fall with increasing height in the upper part of the troposphere, and then become nearly constant in the lower stratosphere. The slight rise in stratospheric temperature, following a line of constant frost point,

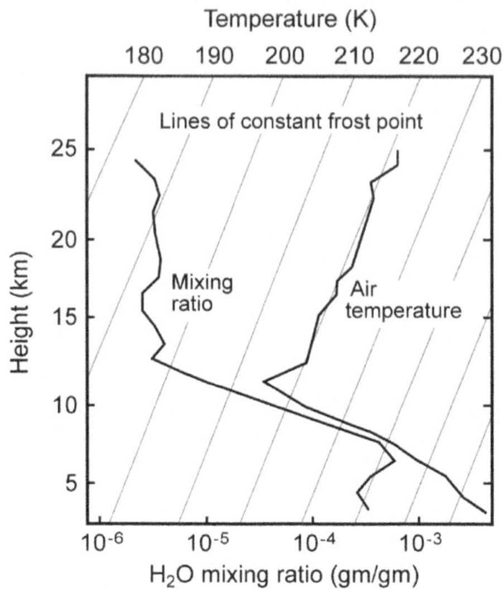

FIG. 2.10. Profiles of air temperature and water–vapour mass–mixing ratio, measured by balloon–borne instruments at 52°N. (Courtesy of Dr E.J. Williamson)

suggests that the air is saturated at the very low temperatures found just above the tropopause, even though the mixing ratio is only about 4 parts per million by volume (ppmv).

Tropospheric air is often close to saturation and latent–heat effects have an important effect on the vertical temperature profile. The First Law of Thermodynamics expression for a parcel of air rising adiabatically becomes

$$dQ = C_v dT + p dV + \frac{L}{M_{H_2O}} dm = 0, \tag{2.12}$$

where m is the mass of water in 1 mole of air and L is the molar latent heat of water. Using the ideal gas law

$$pdV + Vdp = C_p dT - C_v dT \tag{2.13}$$

and the Clausius–Clapeyron equation

$$\frac{dp_{H_2O}}{dT} = \frac{L}{T(V_v - V_l)} = \frac{Lp_{H_2O}}{RT^2}, \tag{2.14}$$

where p_{H_2O} is the partial pressure of water and V_v and V_l are the molar volumes of water vapour and liquid, respectively, and the hydrostatic equation $dp/dz = -\rho g$, we can obtain an expression for the adiabatic lapse rate in saturated air as

$$\left(\frac{dT}{dz}\right)_{sat} = f\left(\frac{dT}{dz}\right)_{dry} \tag{2.15}$$

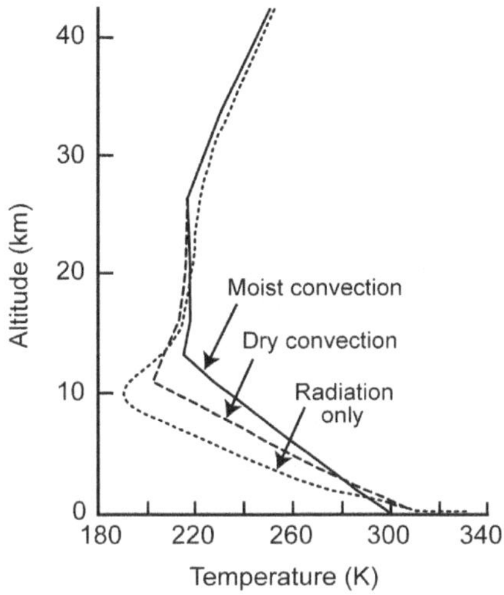

FIG. 2.11. Representative vertical temperature profiles calculated assuming (top curve) convective equilibrium in a moist atmosphere, (middle) convective equilibrium in a dry atmosphere, and (bottom) radiative equilibrium. (See Manabe and Wetherald 1967)

$$f = \frac{1+a}{1+b}, \tag{2.16}$$

$$a = \frac{Lp_{\mathrm{H_2O}}}{pRT}, \tag{2.17}$$

$$b = \frac{L^2 p_{\mathrm{H_2O}}}{C_{\mathrm{p}} pRT^2}. \tag{2.18}$$

The saturated lapse rate is always less than the dry value, and the ratio f can fall as low as 0.35 in a warm, humid atmosphere.

A third model for the vertical temperature profile is one that assumes dynamical processes are negligible compared to radiation, so the profile is determined by the equilibrium between the heating caused by the absorption of incoming solar radiation, the cooling to space by thermal infra–red emission, and the radiative exchange between atmospheric layers at different heights. For an approximate calculation, the atmosphere can be separated into layers, each of which is assumed to be a blackbody absorber and therefore to emit according to the Stefan–Boltzmann law. The nth layer with temperature T_n is in equilibrium with its neighbours when

$$\sigma T_{n-1}^4 + \sigma T_{n+1}^4 = 2\sigma T_n^4, \tag{2.19}$$

while the energy–balance equation for the Earth's surface is

$$S_0 + \sigma T_1^4 = \sigma T_0^4, \tag{2.20}$$

where S_0 is the solar irradiance at the Earth's surface and σ is the Stefan–Boltzmann constant, 5.670×10^{-8} J m^{-2} K^{-4} s^{-1}. Figure 2.11 shows examples of profiles calculated according to assumptions of: (i) radiative equilibrium, (ii) dry convection, and (iii) moist convection; the last of these being the most realistic, so most measured profiles in the real atmosphere fall closest to this. Note the convergence of all three profiles in the stratosphere, where radiative balance applies in all cases.

2.7 Condensation processes, clouds and aerosols

The vertical temperature profile of the atmosphere enters into radiative transfer models not only through the source function for the emission of radiation, and through the temperature dependence of the absorption coefficients, but also indirectly, through its relationship to density variations that force vertical and horizontal motions. These redistribute spatially variable absorbers such as ozone and water vapour, but are even more important for the formation and distribution of clouds, a key element in any radiative model of the climate system.

2.7.1 *Criteria for vertical stability*

The radiative equilibrium temperature profile (Fig. 2.11) represents the hypothetical case of an atmosphere with no vertical motions, in which heat is transferred by radiation and (negligible amounts of) conduction only. The temperature gradient is so steep that successive vertical layers increase in density with increasing height, a situation that is obviously unstable. Convection and other forms of dynamical transport such as wave propagation and turbulence will therefore occur and transfer heat vertically until stability is restored, along with a smaller vertical temperature gradient, which we would expect to be close to the adiabatic value calculated above. The real atmosphere is seldom low enough in humidity for this to be the simple g/c_p gradient we obtained for the dry case, but in general we expect that

$$\left(\frac{\mathrm{d}T}{\mathrm{d}z}\right)_\mathrm{sat} > \frac{\mathrm{d}T}{\mathrm{d}z} > \left(\frac{\mathrm{d}T}{\mathrm{d}z}\right)_\mathrm{dry} > \left(\frac{\mathrm{d}T}{\mathrm{d}z}\right)_\mathrm{rad}. \tag{2.21}$$

The atmosphere is then *conditionally stable* against convection. However, if air at some level is forced upwards by an outside force such as a propagating wave, it may reach the level where condensation sets in. Then it will become unstable and continue to rise, with further condensation of water vapour, leading to the formation of clouds.

2.7.2 *Growth of cloud particles*

The conditions under which clouds form, and the rate at which they grow, can be studied by considering the microphysics of a single droplet. These do not form spontaneously, even in supersaturated air, since the random congregation of water molecules can, at best, produce only microscopically small droplets that are not stable against re–evaporation. Instead, condensation nuclei are required to initiate droplet growth. These must consist of hygroscopic, water–soluble material, the commonest being sea–salt crystals produced by the action of the wind on waves and spray, and sulphates of volcanic or anthropogenic origin.

The rate of growth of a cloud therefore depends in the first instance on the number of suitable nuclei that are available to initiate growth. The rate at which each nucleus grows into a droplet large enough to interact with radiation (i.e. roughly of comparable size to the wavelength of the photons, or greater) depends on the relative humidity of the surrounding air, which is of course a function of temperature and therefore altitude. As unstable air rises the degree of saturation, and then supersaturation, tends to increase and to increase the rate of droplet growth. Once growth is initiated on the available nuclei, new drops will not form in a rising air parcel and the water available as vapour (effectively that in excess of the mixing ratio corresponding to saturation) is consumed by the growth of the existing particles.

The classical expression for the vapour pressure $p_\mathrm{v}(r)$ over a convex surface of radius r is Kelvin's formula

$$p_\mathrm{v}(r) = p_\mathrm{v}(\infty)\exp\left(\frac{2\sigma}{r}\frac{M}{\rho_\mathrm{l}RT}\right) \approx p_\mathrm{v}(\infty)\exp\left(\frac{a}{r}\right), \tag{2.22}$$

where $p_\mathrm{v}(\infty)$ is the equivalent value over a plane surface, σ and M are the surface tension and molecular weight of the liquid and ρ_l the density of the liquid. The constant a has the same units as r, and temperature T is taken to be constant in the vicinity of the drop. According to this, the air is always supersaturated above a small drop so it will tend to evaporate unless the surrounding air is even more supersaturated, the value required being unreasonably high. For example, to make a droplet with a radius of only 0.5 nm requires the simultaneous collision of about 17 molecules, and around 750% supersaturation of the ambient air.

However, a dissolved solvent reduces the vapour pressure adjacent to the surface of the droplet according to Raoult's law, which states that the saturated vapour pressure (SVP) of a drop of radius r depends directly on the concentration of solute, $n/(n_0 + n)$, where n and n_0 are the numbers of molecules of solute and of water, respectively. Incorporating this with Kelvin's law gives

$$\frac{p_\mathrm{v}(r)}{p_\mathrm{v}(\infty)} = \exp\left(\frac{a}{r}\right)\left(1+\frac{b}{r^3}\right)^{-1} \approx 1 + \frac{a}{r} - \frac{b}{r^3}, \tag{2.23}$$

known as the Kohler curve (Fig. 2.12). This has a maximum at the critical

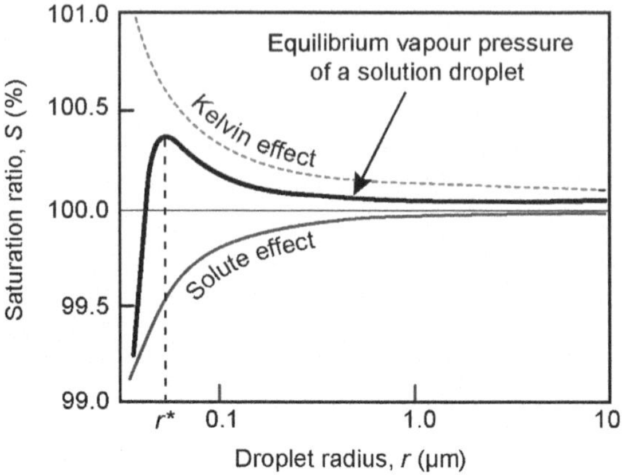

FIG. 2.12. The Kohler curve (thick line) for the vapour pressure at the surface of a droplet containing solute, which reduces the vapour pressure, as a function of droplet radius, on which the vapour pressure depends inversely through the Kelvin effect. The net effect is a saturation ratio (SVP relative to a flat surface) with a strong peak at the critical radius r^* for droplet growth. Once the droplet is larger than this, adding water reduces the SVP at its surface and encourages further growth, and viceversa.

values of r and S above which the particle will always grow, and below which it can shrink, given in this approximation by

$$r^* = \sqrt{\frac{3b}{a}} \quad \text{and} \quad S^* = 1 + \sqrt{\frac{4a^3}{27b}}. \tag{2.24}$$

For a droplet containing 10^{-17} kg of NaCl, the constants a and b have values of 6×10^{-3} μm and 2×10^{-5} μm^3, respectively, and r^* is about 0.1 μm. The rate at which a drop of mass m and radius r will grow is given by the diffusion law

$$\frac{\mathrm{d}m}{\mathrm{d}t} = 4\pi R^2 D \frac{\mathrm{d}\rho}{\mathrm{d}R}, \tag{2.25}$$

where R is the radial dimension, D is the diffusion coefficient of water vapour in air, and ρ is the vapour density. This integrates to give

$$\frac{\mathrm{d}r}{\mathrm{d}t} = \frac{D}{r}\left(\frac{\rho_\infty - \rho_r}{\rho_\mathrm{l}}\right), \tag{2.26}$$

where ρ_r and ρ_∞ are the vapour densities at, and far from, the surface of the drop and ρ_l is the density of liquid water, for the rate of growth of the droplet at constant temperature.

The actual temperature depends upon the balance between the heat gain due to release of latent heat L and the heat loss due to the thermal conductivity of air, λ, which occurs when the droplet is warmer than its surroundings by an amount

$$(T_{\mathrm{r}} - T_{\infty}) = \frac{LD(\rho_{\mathrm{r}} - \rho_{\infty})}{\lambda}. \tag{2.27}$$

The numerical solution of these equations, along with the Clausius–Clapeyron and Kohler expressions, gives the growth rate of a drop under a given set of conditions. In the real atmosphere, the growth rate of a droplet by diffusion of vapour is very slow. A submicrometre droplet can take many hours to grow to 100 μm, the size of a representative cloud particle. Droplets larger than this are prone to sedimentation, although this depends on the support offered by the vertical velocity of the surrounding air. Should the air begin to subside to warmer, lower levels, the evaporation of droplets of all sizes will occur.

If, on the other hand, the air continues to rise to colder levels, continued growth, freezing and coagulation are all possible. The vapour pressure over ice is less than that over supercooled water, and so freezing can lead to the very rapid growth of small particles. Once these are large enough to begin to fall, collisions lead to coalescence and even more rapid growth, finally producing precipitation. Because freezing accelerates particle growth, a lot of the rain that reaches the surface in midlatitudes starts out as frozen droplets in clouds, that melt during their descent to the surface. Depending on temperature and humidity conditions in the air below the cloud, the droplets often may not reach the surface if the evaporation rate is high.

2.7.3 *Cloud particle size distributions*

The radiative–transfer properties of clouds depend strongly on the size distribution of the particles they contain. As the previous discussion indicates, this can be a very dynamic factor, and the best that can be done in climate models or retrievals from remotely sensed data is to assume some sort of mean or representative distribution based on statistical studies of measurements made in real clouds under various conditions.

In modelling clouds the focus is generally on cirrus of moderate optical thickness at relatively high altitudes. Cumulus and similar clouds, because they are near the surface and generally quite optically thick, can generally be represented as black or grey bodies without the need to consider their detailed microphysics. The particles in cirrus are made of ice and are non–spherical, usually with at least one dimension that is large compared to midinfra–red wavelengths, typically measured in hundreds of micrometres. If these are assumed to be randomly oriented then they can be treated, obviously with the introduction of additional uncertainties, as spheres of some representative mean radius $\bar{r}$. This allows the use of Mie scattering theory to compute the optical depth and directional phase function of model clouds relatively easily (see Chapter 6).

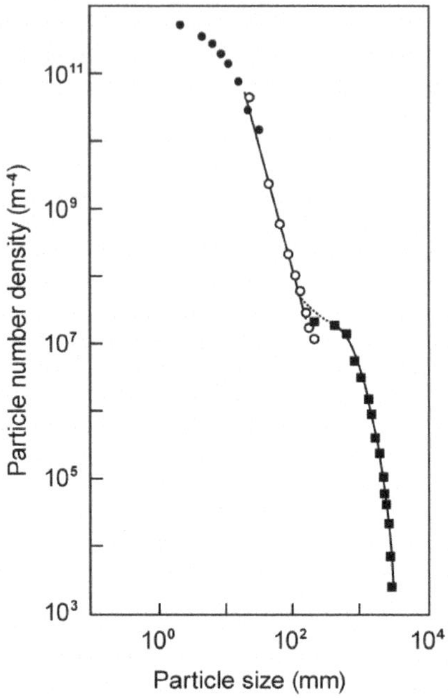

FIG. 2.13. Cloud ice particle number density $n(D)$ versus particle size, showing a bimodal distribution with a secondary peak at ~500 μm. (From Platt 1977)

Ice cloud parameterisations in state–of–the–art models generally use three variables, the ice–water content (IWC), the effective particle radius $\bar{r}$, and a factor describing the shape of the particle–size distribution (PSD). The last may just be a Gaussian width or other measure of the spread of sizes in the cloud, or set to zero by adopting a monodisperse distribution in which all particles are treated as being of the same effective radius. However, measurements show that ice clouds often have a bimodal (double-peaked) distribution (Fig. 2.13) and for these, in particular, the monodisperse assumption is a poor approximation, which can lead to substantially different optical thicknesses for model clouds with the same IWC and $\bar{r}$. PSD can also vary greatly with cloud type, height, latitude and IWC. A large number of formulae that address this problem in various ways may be found in the literature (see also Chapter 6).

2.7.4 *Aerosols*

In addition to the soluble particles involved in cloud formation, long-lived aerosols of solid or liquid of many types are constantly present in large numbers everywhere in the atmosphere, including the stratosphere. These are often invisible to the eye, or at most appear as a fine haze or 'turbidity', but have radiative effects that are important within the climate system. For instance, they scatter

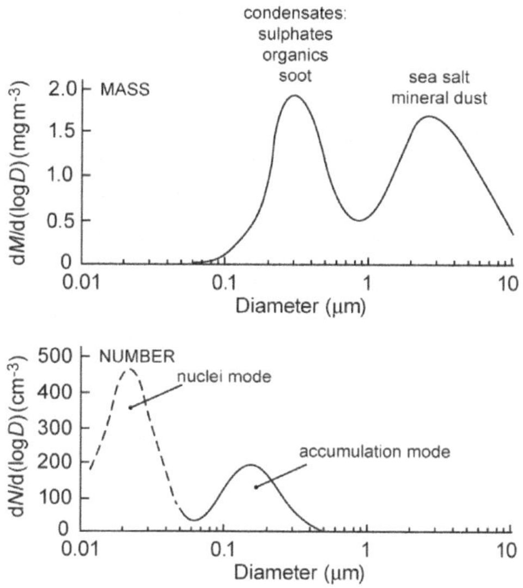

FIG. 2.14. Typical aerosol concentrations by mass (M) and by number density (N), as a function of the logarithm of diameter (D). Most of the mass is in the largest particles, produced by the effect of wind on dusty surfaces or on sea–spray, but very small particles (smoke, bacteria, etc.) are much more numerous. The central peak, consisting mainly of accumulated particles (for instance, sulphuric acid drops produced by chemical reactions involving gases produced from volcanoes and fossil–fuel combustion) tends to be the most important for radiative effects in the climate system, although individual situations vary greatly.

incoming sunlight, increasing the albedo of the Earth, and absorb both solar and longwave radiation. Some are of natural origin, from volcanoes, dust storms, forest and grassland fires, pollen, and sea–spray. Human activities such as the burning of fossil fuels, and agricultural or industrial processes are significantly adding to the natural aerosol burden.

The primary aerosol types are

1. Sulphate particles, both solid (mostly ammonium sulphate) and liquid (sulphuric acid). These originate in the combustion of the sulphur content of fuels used in cars, power stations and industry, and to a lesser extent in natural processes like volcanic eruptions and biomass burning.
2. Sea–salt particles, lifted by the wind as sea–spray evaporates.
3. Fine dust, raised by winds from deserts and other dry plains.
4. Carbon particles (soot) and other organics that form the bulk of the visible part of smoke plumes from combustion processes of many kinds.

Figure 2.14 shows a representative plot of the size distribution of aerosols (see also Chapter 6) that might be found under typical conditions in the mid-troposphere. Like ice clouds, a bimodal distribution is common, although in this case the reason is less elusive since aerosol particles have a number of different sources and compositions. The difference between the distribution by mass and by number is striking; most of the mass tends to be in the largest particles, but smaller particles are by far the most numerous, and hence tend to have the more important role in affecting radiative energy balance.

The composition is also an important factor when considering the radiative properties of particles. Sulphate aerosols are particularly good at scattering visible light, and the buildup of these in the atmosphere is a major factor in global change, tending to work in the opposite direction to greenhouse–gas accumulation and produce global cooling. Dust and soot, on the other hand, absorb both solar and thermal infrared radiation and tend to heat the atmosphere, resulting in a net warming of the climate system. Some commonly found aerosols, such as mineral dust or weakly absorbing organic species, can either cool or warm the atmosphere depending on their particle–size distribution and their chemical composition, which is so variable it affects the ratio of absorption to scattering. Location on the globe is also important, since prevailing weather conditions affect aerosol formation and behaviour. Stratospheric aerosols have long lifetimes, measured in years and even decades, but tropospheric aerosols tend to survive only for days or possibly weeks, resulting in large spatial and temporal variability over the surface of the Earth.

2.8 Bibliography

2.8.1 *Notes*

The topics in this chapter are introduced at an elementary level by Taylor and developed at a more advanced level by Andrews, Houghton, and Salby. The books by Wells and by Curry and Webster also include overviews of basic physical oceanography, while Rogers and Yau is a standard reference on cloud and aerosol physics.

2.8.2 *References and further reading*

Andrews, D. G. (2000). *An introduction to atmospheric physics.* Cambridge University Press, Cambridge.

Curry, J. A. and Webster, P. J. (1999). *Thermodynamics of atmospheres and oceans.* Academic Press, New York.

Houghton, J. T. (2002). *The physics of atmospheres*, 3rd edn. Cambridge University Press, Cambridge.

Iribarne, J. V. and Godson, W. L. (1981). *Atmospheric thermodynamics.* D. Reidel, Dordrecht.

Manabe, S. and Wetherald, R. T. (1967). Thermal equilibrium of the atmosphere with a given distribution of relative humidity. *J. Atmos. Sci.*, **24**, 241–259.

Platt, C. M. R. (1977). Lidar observations of a mixed-phase altostratus cloud. *J. Appl. Meteorol.*, **16**, 339–345.

Rogers, R. R. and Yau, M. K. (1996). *A short course on cloud physics.* Butterworth-Heinemann, Boston.

Salby, M. L. (1996). *Fundamentals of atmospheric physics.* Academic Press, New York.

Taylor, F. W. (2005). *Elementary climate physics.* Oxford University Press, Oxford.

Wells, N. (1997). *The atmosphere and ocean: A physical introduction.* Wiley, New Jersey.

3

RADIATION–TRANSFER THEORY

3.1 Introduction

The production of radiation by bodies around us is usually from what are essentially blackbody surfaces, or surfaces that absorb most of the thermal radiation incident on them. Planetary atmospheres are usually non–ideal blackbodies, unless they contain high amounts of absorbing molecules. Radiation that is emitted by a surface or an atmospheric layer undergoes absorption and/or scattering, by the molecules that constitute the atmosphere, as it is transferred through the atmosphere. Over very large distances radiation is also diluted by the enlargement of space surrounding the radiation source, according to the inverse square law. Absorption and emission by a surface take place over the projected area of the surface in the direction of the radiation. Hence, a planet absorbs solar radiation basically as a disc and emits thermal infra–red radiation as a sphere.

The emission from a blackbody is described by the laws of Planck, Wien and Stefan–Boltzmann. The first gives the spectral distribution of the photons, the second the location of the wavelength of maximum emission and the third gives the dependence of the total amount of radiation emitted on temperature. As the temperature increases, photons at progressively higher energies are emitted. In planetary atmospheres, natural emission is thus restricted to the infra–red whilst absorption can occur at all wavelengths from gamma rays to radiowaves. For thermal infra–red radiation we need to include absorption and emission in the equation of radiation transfer. For the incoming solar radiation we include only absorption and scattering by atmospheric molecules and particles (clouds, aerosols).

We now present some simple classical methods for solving the radiation–transfer equation, as these methods form the basis for obtaining and understanding the radiation field within a planetary atmosphere.

3.2 Basic definitions

3.2.1 *The inverse square law*

Let us assume that at the centre of a virtual sphere of surface area A there is an infinitesimally small body that is radiating isotropically in all directions. Then, the fraction of the total emitted radiation that is transferred through an element

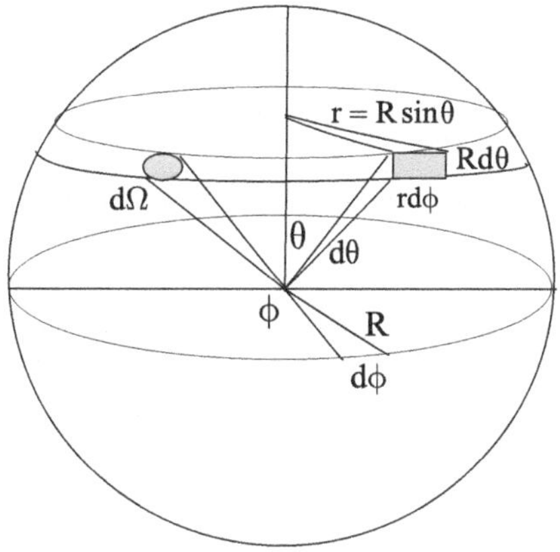

FIG. 3.1. Radiation sphere with radius R, zenith angle θ, azimuthal angle ϕ, minor–circle radius r, and element of solid angle $\mathrm{d}\Omega$.

of area $\mathrm{d}A$ on the sphere is $\mathrm{d}A/A$. The emitted radiation can be regarded as consisting of small conical bundles of rays with the base of the cone having an element of area $\mathrm{d}A$. The cone of radiation comprises a solid angle, $\mathrm{d}\Omega$, measured in steradians (sr), and given by

$$d\Omega = \sin\theta d\theta d\phi, \tag{3.1}$$

where θ is the zenith angle and ϕ the azimuth angle, as defined in Fig. 3.1. In polar co–ordinates, the polar axis defines the zenith angle, whilst the equatorial axis defines the azimuth angle. The radius of the sphere is R, whilst that of a minor circle on the sphere is denoted by r. The element of area $\mathrm{d}A$ is then given by

$$\mathrm{d}A = (r\mathrm{d}\phi)(R\mathrm{d}\theta) \tag{3.2}$$

and hence

$$\mathrm{d}A = R^2 \sin\theta \mathrm{d}\theta \mathrm{d}\phi \tag{3.3}$$

so that the total area of the sphere is

$$A = \int_0^{2\pi} \mathrm{d}\phi \int_0^{\pi} R^2 \sin\theta \mathrm{d}\theta = 4\pi R^2. \tag{3.4}$$

Thus the fraction of the total radiation transferred through $\mathrm{d}A$ is given by

$$\frac{\mathrm{d}A}{A} = \frac{1}{4\pi} \sin\theta \mathrm{d}\theta \mathrm{d}\phi = \frac{\mathrm{d}\Omega}{4\pi}. \tag{3.5}$$

Thus, the surface of the radiation sphere can be divided into units of steradians with a total 4π (sr) for the whole sphere.

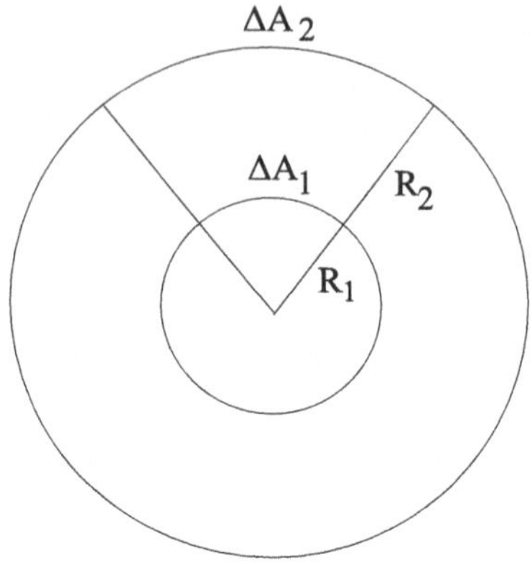

FIG. 3.2. The inverse square law, based on an emitting central sphere of radius R_1 and a receiving sphere of radius R_2. Surface–area elements for each sphere are also shown.

At the centre of a virtual sphere of radius R_2 let us place a spherical body of radius R_1 that is emitting isotropically in all directions, as shown in Fig. 3.2. Then the emitted radiation from an element of area ΔA_1 of the central sphere is incident on an element of surface area ΔA_2 on the external virtual sphere, where $\Delta A_2 > \Delta A_1$. If the number of emitted photons per unit area is conserved within the solid angle subtended by ΔA_1 and ΔA_2, then the radiation per unit area at a distance R_2 from the centre of the virtual sphere is diluted according to the ratio

$$\frac{\Delta A_1}{\Delta A_2} = \frac{R_1^2}{R_2^2}. \tag{3.6}$$

3.2.2 *Radiance*

The monochromatic radiance, I_λ is defined as the energy (J) emitted by or transferred through an element of area δA (m^2) in a direction defined by the unit vector $\hat{n}$ within an element of solid angle $\mathrm{d}\Omega$ (sr) per second (s) within a wavelength interval $\delta\lambda$ about λ. If the wavelength is measured in μm then the units of the radiance are

$$\frac{\mathrm{J}}{\mathrm{m}^2 \cdot \mathrm{s} \cdot \mathrm{sr} \cdot \mu\mathrm{m}_\lambda},$$

where we include the subscript λ on the wavelength to distinguish it from the dimensions of the emitting element of area. If we denote the direction that is perpendicular to the infinitesimal element of area, taken to be planar, by the unit vector $\hat{k}$, then the projection of the element of area in the direction of the radiation is $\delta A(\hat{k} \cdot \hat{n})$. Hence, the projected area is reduced as the radiation direction moves towards the horizon according to $\cos\theta$, where θ is the angle

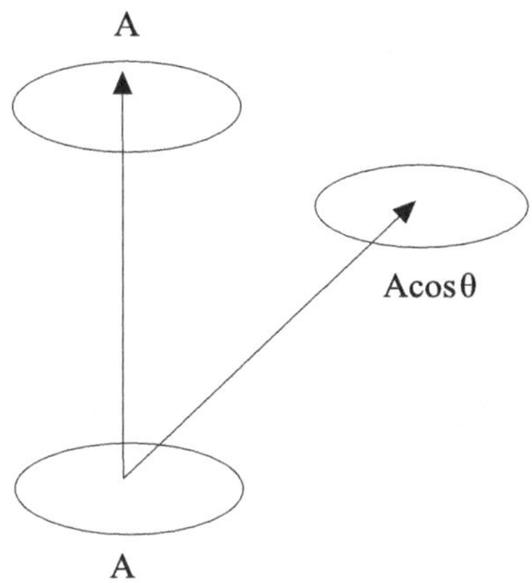

FIG. 3.3. Projection of the emitting area in a direction that makes an angle θ to the normal.

between the perpendicular direction and the direction of the radiation, as shown in Fig. 3.3. The total energy (J) that is emitted or transferred through the element of area in the direction $\hat{n}$ is then

$$\delta E_\lambda = I_\lambda \delta A(\hat{k} \cdot \hat{n})\delta\Omega\delta\lambda\delta t \tag{3.7}$$

in a time interval δt, within the spectral interval $\delta\lambda$, within solid angle $\delta\Omega$.

3.2.3 *Mean radiance and flux*

The monochromatic mean radiance at a point in space $\vec{r}$ is defined as the radiance averaged over all directions according to

$$J_\lambda(\vec{r}, t) = \frac{1}{4\pi}\oint I_\lambda(\vec{r}, \hat{n}, t)\mathrm{d}\Omega \qquad \frac{\mathrm{J}}{\mathrm{m}^2 \cdot \mathrm{s} \cdot \mathrm{sr} \cdot \mu\mathrm{m}_\lambda}. \tag{3.8}$$

The monochromatic *radiation flux* (or *irradiance*) is the radiance emitted by or transferred through an element of area, δA, located at a point $\vec{r}$ averaged over all directions, accounting for the projection of the surface area according to

$$f_\lambda(\vec{r}, t) = \oint I_\lambda(\vec{r}, \hat{n}, t)\cos\theta\mathrm{d}\Omega \tag{3.9}$$

and has units W m^{-2} μm$_\lambda^{-1}$. We can perform the above integral by using eqn (3.1) to express $d\Omega$ in polar co-ordinates and on setting $\mu = \cos\theta$ we have

$$f_\lambda(\vec{r}, t) = \int_0^{2\pi}\int_{-1}^{1} I_\lambda(\vec{r}, \hat{n}, t)\mu\mathrm{d}\mu\mathrm{d}\phi. \tag{3.10}$$

We note that if the radiance is isotropic then the integral vanishes for an element of area that is simply transferring radiation, as in the case within an atmosphere, and we have no net flux crossing the area. If we have a surface emitting isotropically to the upper hemisphere only, then the integral reduces to

$$f_\lambda(\vec{r}, t) = \pi I_\lambda(\vec{r}, t), \tag{3.11}$$

where the π has units of steradians. We may integrate the monochromatic flux over all wavelengths to obtain the total flux (or total irradiance, W m^{-2}) emitted by a surface

$$f(\vec{r}, t) = \pi \int_0^\infty I_\lambda(\vec{r}, t)\mathrm{d}\lambda. \tag{3.12}$$

Within an atmosphere the total flux incident on an element of volume is obtained by integrating over the surrounding sphere

$$f(\vec{r}, t) = 4\pi \int_0^\infty I_\lambda(\vec{r}, t)\mathrm{d}\lambda. \tag{3.13}$$

3.2.4 *Luminosity*

The monochromatic luminosity L_λ, of an element of surface area emitting or transferring radiation, is the monochromatic flux times the surface area according to

$$L_\lambda(\vec{r}, t) = \int I_\lambda(\vec{r}, \hat{n}, t)\delta A \cos\theta \mathrm{d}\Omega \tag{3.14}$$

For an element of area that emits or transfers radiation homogeneously

$$L_\lambda(\vec{r}, t) = \delta A \int I_\lambda(\vec{r}, \hat{n}, t) \cos\theta \mathrm{d}\Omega \tag{3.15}$$

and the total luminosity, L, for isotropic radiation is

$$L(\vec{r}, t) = f(\vec{r}, t)\delta A. \tag{3.16}$$

In the case of an emitting spherical surface of radius R, the total surface luminosity is

$$L(R, t) = 4\pi R^2 f(R, t). \tag{3.17}$$

3.3 Blackbody radiation

A blackbody is defined to be a perfect emitter and absorber. For a perfect blackbody solid surface or a totally opaque atmospheric layer, the radiance is isotropic and emission is homogeneous over the entire surface area, characterized by a homogeneous surface temperature T(K). For a planet or star with an absorbing–emitting–scattering atmosphere the radiation emitted to space may not be ideal blackbody radiation.

3.3.1 *Planck's law*

The radiance of an ideal blackbody is given by the Planck function, in terms of wavelength λ as

$$B_\lambda(T) = \frac{2hc^2/\lambda^5}{\mathrm{e}^{hc/\lambda kT} - 1}, \tag{3.18}$$

where c is the speed of light in a vacuum, h is Planck's constant and h is Boltzmann's constant. In terms of frequency ($\nu = c/\lambda$) it takes the form

$$B_\nu(T) = \frac{2h\nu^3/c^2}{\mathrm{e}^{h\nu/kT} - 1}. \tag{3.19}$$

We note that the two forms of the Planck function are related by

$$B_\nu \mathrm{d}\nu = -B_\lambda \mathrm{d}\lambda, \tag{3.20}$$

which expresses the fact that the emitted energy per sr per unit area per s is the same over the same energy interval.

3.3.2 *Stefan–Boltzmann law*

The total flux (W m^{-2}) emitted by a blackbody is given by eqn (3.12) by replacing the radiance by the Planck function

$$f(T) = \pi \int_0^\infty B_\lambda(T)\mathrm{d}\lambda. \tag{3.21}$$

On setting $hc/\lambda kT = y$ and integrating we have

$$f(T) = \frac{2\pi(kT)^4}{h^3c^2} \int_0^\infty \frac{y^3\mathrm{d}y}{\mathrm{e}^y - 1}, \tag{3.22}$$

where the integral is equal to $\pi^4/15$, which gives the simple result

$$f(T) = \sigma T^4 \tag{3.23}$$

known as the Stefan–Boltzmann Law, where σ is the Stefan–Boltzmann constant

$$\sigma = \frac{2\pi^5 k^4}{15h^3c^2}, \tag{3.24}$$

which is equal to 5.67×10^{-8} J m^{-2} K^{-4} s^{-1}. For a star or planet we can assign an effective blackbody temperature, T_{eff}, so that the total luminosity of the body is given by

$$L = 4\pi R^2 \sigma T_{\mathrm{eff}}^4, \tag{3.25}$$

where R is its radius.

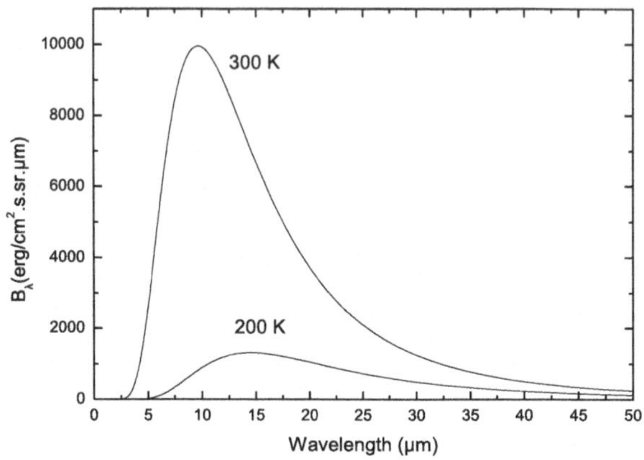

FIG. 3.4. The variation of the Planck function with wavelength and temperature.

3.3.3 *Radiation energy density*

From the conservation of continuity, the radiation flux can be expressed as

$$\text{flux (W m}^{-2}) = \text{density (J/m}^3) \times \text{velocity (m/s)}, \tag{3.26}$$

within a blackbody the flux of photons into an element of volume from the surrounding sphere is 4 σT^4 hence the radiation density, ε, is

$$\varepsilon = \frac{4\sigma T^4}{c}. \tag{3.27}$$

3.3.4 *Wien's displacement law*

The variation of the Planck function with wavelength and temperature is shown in Fig. 3.4. The wavelength, λ_*, in μm, corresponding to the maximum emission of radiation, is inversely proportional to the blackbody temperature according to Wien's Law

$$\lambda_* = \frac{2898}{T(\text{K})}. \tag{3.28}$$

The wavelength λ_* is given by the condition that $\partial B_\lambda(T)/\partial\lambda = 0$. As the blackbody temperature increases, photons at higher energies become more numerous, since the photon energy $h\nu = hc/\lambda$. The peak in the Planck function moves to higher wavelengths as the temperature decreases, and we note from Fig. 3.4 that the Planck function at the lower temperature is always below that at the higher temperature. We also note the rapid rise in the peak value of the Planck function

Table 3.1 *Radiation type, wavelength range and corresponding order of magnitude blackbody temperature according to Wien's Law.*

Radiation	$\lambda_a - \lambda_b$ (μm)	T(K)	Energy–level transitions
γ-rays	$10^{-6} - 10^{-4}$	10^8	nuclear
X-rays	$10^{-4} - 10^{-2}$	10^6	internal electron
ultraviolet	$10^{-2} - 0.39$	10^4	external electron
visible	$0.39 - 0.75$	10^3	external electron
near–infra–red	$0.75 - 5$	10^3	molecular vibration
infra–red	$5 - 40$	10^2	molecular vibration
far–infra–red	$40 - 400$	10	molecular rotation
microwave	$400 - 10^4$	10^{-1}	molecular rotation
radiowave	$10^4 - 10^7$	10^{-4}	spin reversal

with temperature. Table 3.1 gives the type of radiation and the corresponding typical blackbody temperature for its production.

3.4 Absorption, emission and scattering

Atmospheric molecules can absorb, emit and scatter radiation. Particles such as cloud droplets or ice crystals and aerosols (dust of various chemical composition including water) also play an important role in radiation transfer. During scattering, a photon can change both its direction and wavelength. For planetary atmospheres, absorption and re–emission is coherent in wavelength, as is scattering by particles. Scattering plays a major role in the transfer of solar radiation through a planetary atmosphere by changing the photon direction. An atmosphere may have a physical depth but for photons it is the physical depth divided by the mean free path before absorption and scattering that determines their transfer through an atmosphere.

3.4.1 *Absorption and emission*

We define *absorption* as the process by which a particle (electron, atom or molecule) interacts with a photon resulting in the photon energy being transferred to the thermal energy of the atmosphere containing the particle. The inverse process is *emission*. This process allows energy from one part of the atmosphere to be deposited in another part, possibly far in optical terms. This transfer of energy between layers of an atmosphere has the implication that the system is not closed and so at best only local thermodynamic equilibrium (LTE) can prevail. Strict thermodynamic equilibrium (TE) requires the system to be closed to the outside environment. We shall discuss concepts of TE, LTE and non–LTE in a later section.

Examples of absorption and emission processes include:

Photoionization, or bound–free electronic transitions, whereby a photon is absorbed by an atom resulting in a free electron. The photon energy is taken up by the energy required to free it from its bound state within the atom with the remaining energy going to the kinetic energy of the electron. Photoionization can also occur in stages, whereby an electron is first excited to a higher bound state followed by *collisional excitation* beyond the ionization energy limit. A free electron can undergo the inverse process to photoionization, *radiative recombination*, by the emission of a photon and its capture in an atomic bound state. This emission process results in atmospheric thermal energy being converted to radiative energy.

Bound–bound electronic transitions followed by collisional de-excitation. The electron is excited to a higher energy level, the highest being the ionization limit, followed by collision of the atom with another particle resulting in the transfer of the initial photon energy to the thermal energy of the atmosphere. The reverse process is radiative de-excitation, whereby an electron falls to a lower–energy bound state via the emission of a photon.

Free–free electronic transitions whereby a free electron absorbs a photon and changes its trajectory relative to an ion so as to move further from the ion. This is in fact another example of a change in an electric dipole required for photon–particle interaction. The result is that the electron is accelerated because it has gained energy. This is a result of quantum mechanics, as classically an accelerating charge emits photons. The reverse process is *free–free emission* or *Bremsstrahlung*, whereby an electron is decelerated along its trajectory relative to an ion with the loss of energy appearing as a photon.

3.4.2 *Scattering*

We define *scattering* as the process by which a photon, through its interaction with another particle, can alter either its energy or direction, or both. If the photon energy is not altered significantly then we refer to the photon–particle interaction as being *elastic* with insignificant transfer of energy to the thermal energy of the atmosphere.

Examples of *elastic scattering* processes include:

Redistribution scattering, whereby a bound electron is photo–excited to a higher bound state followed by radiative de–excitation back to its original bound state. The emitted or outgoing photon may have undergone a change in energy and direction relative to the absorbed or incoming photon, but the energy change is insignificant. This is essentially elastic scattering, as the energy difference arises from the spread or *broadening* of the electronic bound states that are not discrete energy states. If they were, then the redistribution scattering would be *coherent*, i.e. the incoming photon energy and the outgoing photon energy would be identical. However, there is normally a small spread of allowed energies of an electronic bound state due to such broadening mechanisms as natural, Doppler or collisional, which will be discussed in a later section. An extreme example of redistribution scattering is *fluorescence*, whereby a photon is absorbed by an atom resulting in an electron excited to a higher bound state followed by radiative de-excitation in stages down to some lower–energy level with the emission of several photons of lower energy to that of the incoming photon.

Rayleigh scattering, whereby a photon interacts with a relatively small particle, such as an atom or molecule, as though the particle is a solid body that scatters the photon without the particle undergoing recoil. The scattering process is coherent in energy but the photon direction is modified. The scattering becomes more effective towards smaller wavelengths.

Mie scattering, whereby a relatively large particle, such as a water droplet or dust particle, scatters the photon in a different direction to its incoming direction without a change in photon energy.

Thomson scattering, whereby a photon interacts with an electron without the electron undergoing recoil. This holds for photons with low energies, usually below that of X–rays.

Examples of *inelastic scattering* processes include:

Compton scattering, whereby a high–energy (X–rays and above) photon interacts with an electron that undergoes recoil. The photon–scattering process is thus incoherent and the outgoing photon energy is significantly different (lower) from that of the incoming photon.

Inverse Compton scattering, whereby an electron of extremely high energy (relativistic) interacts with a photon and transfers part of its energy to the photon. The outgoing photon has thus increased its energy significantly, for example conversion of an ultraviolet photon to an X–ray photon.

3.4.3 *The extinction coefficient and optical depth*

Let us consider an element of volume V (Fig. 3.5) containing particles each having an effective cross-sectional area, s_λ, for absorption and scattering of photons of wavelength λ. We then define the *monochromatic extinction coefficient*, χ_λ as the total effective cross-sectional area, ns_λ, for absorption and scattering of photons of wavelength λ by all particles, n, per volume V at some location $\vec{r}$ at time t along the direction of the incoming photons. If the photons traverse a distance $\mathrm{d}l$, then the *effective volume for extinction*, or the reduction in the volume within which photons of wavelength λ can be transmitted, is $\mathrm{d}V_\lambda$ given by

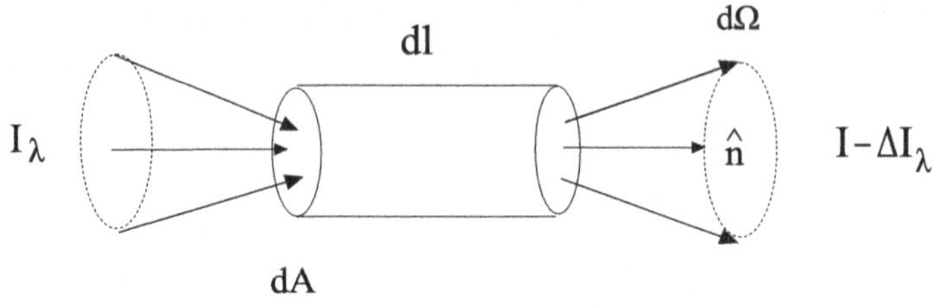

FIG. 3.5. Volume extinction by an element of volume. The fraction ΔI_λ of the radiance represents those photons that have been either scattered out of the incoming solid angle, $\mathrm{d}\Omega$, about the original direction, $\hat{n}$, or absorbed by the particles.

$$\frac{\mathrm{d}V_\lambda}{V} = n s_\lambda \mathrm{d}l = \chi_\lambda \mathrm{d}l. \tag{3.29}$$

Thus, the radiance, I_λ, which is a measure of photons per unit area, is reduced by $\mathrm{d}I_\lambda$ after a distance $\mathrm{d}l$ within the volume V according to

$$\frac{\mathrm{d}I_\lambda}{I_\lambda} = -\frac{\mathrm{d}V_\lambda}{V} = -\chi_\lambda \mathrm{d}l. \tag{3.30}$$

If the photons travel a distance L then the above equation can be integrated to give Lambert's exponential decay law for photons

$$I_\lambda = I_{o\lambda} t_\lambda, \tag{3.31}$$

where $I_{o\lambda}$ is the incoming radiance, t_λ is the *transmission coefficient* given by

$$t_\lambda = \mathrm{e}^{-\tau_\lambda} \tag{3.32}$$

and τ_λ is the *optical depth* defined by

$$\tau_\lambda = \int_0^L \chi_\lambda(l)\mathrm{d}l, \tag{3.33}$$

or

$$d\tau_\lambda = \chi_\lambda(l)\mathrm{d}l \tag{3.34}$$

in differential form.

Thus, if photons of wavelength λ traverse a distance L then the probability of not being removed from the initial direction is given by t_λ. Thus, $1/\chi_\lambda$ represents a *characteristic length*, or *length constant* for the removal of photons of wavelength λ along their path, and is thus a measure of the mean free path of photons traversing this element of volume. Thus, the optical depth is the physical distance traversed by the photons in units of mean free path. We thus define the location of the effective *optical surface of an atmosphere* at an optical depth near unity (e.g. the solar photosphere).

For atmospheres without significant macroscopic motions to modify the random thermal motions of the absorbing/scattering particles, χ is *isotropic*. It has two components, an absorption component, κ and a scattering component, σ

$$\chi_\lambda(\vec{r}, t) = \kappa_\lambda(\vec{r}, t) + \sigma_\lambda(\vec{r}, t). \tag{3.35}$$

Hence, the probability of a photon undergoing a scattering event as opposed to an absorption event is called the single–scattering albedo, ω, given by

$$\omega_\lambda(\vec{r}, t) = \sigma_\lambda(\vec{r}, t)/\chi_\lambda(\vec{r}, t). \tag{3.36}$$

3.4.4 *Volume emission and Kirchoff's law*

The volume emission, η_λ is defined to be the energy emitted at wavelength λ, in direction $\hat{n}$, per unit volume per steradian per second per unit of wavelength, so that the energy emitted by an element of volume is

$$\mathrm{d}E = \eta_\lambda(\vec{r}, \hat{n}, t)\mathrm{d}A\mathrm{d}l\mathrm{d}\Omega\mathrm{d}\lambda\mathrm{d}t. \tag{3.37}$$

For a blackbody, this can be written also in terms of the *emission coefficient*, ϵ, representing the effective emission surface per unit volume

$$\mathrm{d}E = \epsilon_\lambda(\vec{r}, t)B_\lambda(\vec{r}, \hat{n}, t)\mathrm{d}A\mathrm{d}l\mathrm{d}\Omega\mathrm{d}\lambda\mathrm{d}t. \tag{3.38}$$

In the absence of scattering, the amount of energy absorbed by an element of volume can be written equivalently

$$\mathrm{d}E = \kappa_\lambda(\vec{r}, t)I_\lambda(\vec{r}, \hat{n}, t)\mathrm{d}A\mathrm{d}l\mathrm{d}\Omega\mathrm{d}\lambda\mathrm{d}t. \tag{3.39}$$

If the element of volume is within a closed system in thermodynamic equilibrium (TE), then $I_\lambda = B_\lambda$, and the radiance is given by the Planck function, the radiance of a blackbody. In a state of TE all processes are in equilibrium including *radiative equilibrium* (RE), and so the emitted and absorbed energy by the element of volume are equal

$$\eta_\lambda(\vec{r}, \hat{n}, t) = \kappa_\lambda(\vec{r}, t)I_\lambda(\vec{r}, \hat{n}, t). \tag{3.40}$$

Thus, for a blackbody, the emission coefficient is identical to the absorption coefficient

$$\epsilon_\lambda(\vec{r}, t) = \kappa_\lambda(\vec{r}, t). \tag{3.41}$$

This is known as *Kirchoff's Law*. In the presence of blackbody volume emission, the total volume emission can be written as the sum of a thermal and a scattering component

$$\eta_\lambda(\vec{r}, \hat{n}, t) = \kappa_\lambda(\vec{r}, t)B_\lambda(\vec{r}, t) + \eta_{\lambda\mathrm{scat}}(\vec{r}, \hat{n}, t). \tag{3.42}$$

The thermal component is isotropic, whilst the scattering component in general is non–isotropic.

3.4.5 *The source function and redistribution*

The ratio of the total volume emission to the total volume extinction is called the *source function*

$$S_\lambda(\vec{r}, \hat{n}, t) = \eta_\lambda(\vec{r}, \hat{n}, t)/\chi_\lambda(\vec{r}, t). \tag{3.43}$$

In the absence of scattering, we only have thermal emission and the source function is equal to the Planck function. Thus, for thermal radiation $S_\lambda = B_\lambda$. More generally, we can write the source function as the sum of two components in terms of the single-scattering albedo

$$S_\lambda(\vec{r}, \hat{n}, t) = (1 - \omega_\lambda(\vec{r}, t))B_\lambda(\vec{r}, t) + \omega_\lambda(\vec{r}, t)S_{\lambda\text{scat}}(\vec{r}, \hat{n}, t). \tag{3.44}$$

The scattering source function is generally non-isotropic and depends on the incoming radiance and the scattering mechanism or *photon–redistribution* process. Photon redistribution in energy and direction can occur as a result of a bound–bound absorption–emission process or scattering of a photon by a particle. This process is described by the redistribution function, R, which gives the joint probability density for redistribution of photon energy and direction. It is thus meaningful to discuss photon redistribution in terms of photon frequency, ν, rather than photon wavelength. Thus, the joint probability of an incoming photon in direction $\hat{n}'$ and frequency ν' being absorbed and re-emitted or scattered in direction $\hat{n}$ with frequency ν is

$$R(\nu', \hat{n}'; \nu, \hat{n})\mathrm{d}\nu'\mathrm{d}\nu\frac{\mathrm{d}\Omega'}{4\pi}\frac{\mathrm{d}\Omega}{4\pi},$$

with the normalization condition

$$\frac{1}{(4\pi)^2}\oint \mathrm{d}\Omega' \oint \mathrm{d}\Omega \int_0^\infty \mathrm{d}\nu' \int_0^\infty \mathrm{d}\nu R(\nu', \hat{n}'; \nu, \hat{n}) = 1. \tag{3.45}$$

This strictly holds if the scattering particle does not undergo inelastic collisions with the other particles of the atmosphere, between the time of absorption and emission.

The scattering–volume emission in terms of the redistribution function can be written as

$$\eta_{\nu\text{scat}}(\vec{r}, \hat{n}, t) = \sigma_\nu(r)\oint \frac{\mathrm{d}\Omega'}{4\pi}\int_0^\infty \mathrm{d}\nu' R(\nu', \hat{n}'; \nu, \hat{n})I_{\nu'}(\vec{r}, \hat{n}', t). \tag{3.46}$$

3.4.6 *Limiting forms of the redistribution function*

Generally, the redistribution function is very complex but there are two limiting simple forms used, based on its physical application. For atmospheric molecules (Rayleigh scattering), aerosols and cloud droplets (Mie scattering), frequency redistribution is not important and only photon redirection is considered. For spectral line absorption–emission, arising from bound–bound transitions, the reverse is true. Particle scattering is thus considered to be coherent in frequency and the scattering coefficient, σ_ν can be a complex function of frequency, depending on the size and physico–chemical properties of the particle. For bound–bound absorption–emission electronic transitions the scattering process is usually taken to be isotropic with only frequency redistribution considered.

3.4.6.1 *Isotropic scattering* In a static atmosphere, we may average the redistribution function over all incoming or outgoing photon directions to obtain the *direction–averaged* frequency redistribution function

$$R(\nu', \nu) = \frac{1}{4\pi} \oint R(\nu', \hat{n}'; \nu, \hat{n}) \mathrm{d}\Omega' \tag{3.47}$$

$$= \frac{1}{4\pi} \oint R(\nu', \hat{n}'; \nu, \hat{n}) \mathrm{d}\Omega. \tag{3.48}$$

The direction averaging needs only to be performed either over the incoming photon directions or the outgoing photon directions, as the redistribution function depends only on the angle ϑ, between incoming and outgoing photon directions given by

$$\hat{n} \cdot \hat{n}' = \cos\vartheta, \tag{3.49}$$

and we have the normalization condition

$$\int_0^\infty \mathrm{d}\nu' \int_0^\infty \mathrm{d}\nu R(\nu', \nu) = 1. \tag{3.50}$$

For an atmosphere with significant macroscopic motions care must be taken to obtain a direction–averaged redistribution function, as Doppler effects become important and one needs to average in the reference frame of the particles and not that of the observer. This is because in the observer's frame, the Doppler effects are dependent strongly on direction, resulting in highly anisotropic scattering.

For isotropic scattering, the redistribution function is independent of ϑ and so it can be written also as $R(\nu', \nu)$ and the scattering component of the volume emission can be expressed in terms of the mean radiance $J_{\nu'}$, defined by eqn (3.8), as

$$\eta_{\nu\mathrm{scat}}(\vec{r}, t) = \sigma_\nu(\vec{r}, t) \int_0^\infty R(\nu', \nu) J_{\nu'}(\vec{r}, t) \mathrm{d}\nu'. \tag{3.51}$$

We may average the redistribution function, $R(\nu', \nu)$, over all outgoing photon frequencies, ν, to obtain the absorption profile or probability density for absorption of a photon with frequency ν', $\varphi(\nu')$ from

$$\varphi(\nu') = \int_0^\infty R(\nu', \nu) \mathrm{d}\nu, \tag{3.52}$$

with the normalization condition

$$\int_0^\infty \varphi(\nu')\mathrm{d}\nu' = 1. \qquad (3.53)$$

The emission profile, $\psi(\nu)$, or probability density for emission of a photon with frequency ν, is defined by

$$\psi_\nu(\vec{r},t) = \frac{\eta_\nu(\vec{r},t)}{\int_0^\infty \eta_\nu(\vec{r},t)\mathrm{d}\nu} = \frac{\int_0^\infty R(\nu',\nu)J_{\nu'}(\vec{r},t)\mathrm{d}\nu'}{\int_0^\infty \varphi(\nu')J_{\nu'}(\vec{r},t)\mathrm{d}\nu'}. \qquad (3.54)$$

This expression holds if the scattering is elastic (no change in the velocity of the scattering particle) and there is no coupling to other bound states, that is we consider transitions only between two electronic energy levels.

Thus, generally the absorption and emission profiles are not the same. There are special cases where they are the same. From the above equation we see that if the mean radiance is a very weakly varying function of frequency and the redistribution function is symmetric $R(\nu',\nu) = R(\nu,\nu')$, then $\psi(\nu) = \varphi(\nu)$. Another case arises when collisions between the absorbing particles and all other particles are very frequent (high pressures) so that we have *complete redistribution* in frequency. This situation will be further discussed later when we consider spectral–line broadening mechanisms.

3.4.6.2 *Coherent scattering* When frequency redistribution is not significant, we can separate the redistribution function into a probability density for photon redirection, described by a phase function, p, and a probability density for coherent scattering in frequency. For bound–bound electronic transitions the redistribution function has the form

$$R(\nu',\hat{n}';\nu,\hat{n}) = p(\hat{n}',\hat{n})\varphi(\nu')\delta(\nu-\nu'), \qquad (3.55)$$

where $\delta(\nu-\nu')$ is the Dirac delta function, while for particle scattering, such as Rayleigh and Mie, it has the form

$$R(\hat{n}',\hat{n}) = p(\hat{n}',\hat{n}), \qquad (3.56)$$

and the phase function is a function of $\cos\vartheta$ and is normalized according to

$$\frac{1}{4\pi}\oint p(\cos\vartheta)\mathrm{d}\Omega' = 1. \qquad (3.57)$$

For isotropic scattering $p(\cos\vartheta) = 1$ and we can define an *asymmetry factor*, g, which quantifies the extend of the non-isotropy of the scattering process by

$$g = \frac{1}{2}\int_{-1}^{1} p(\cos\vartheta)\cos\vartheta\ \mathrm{d}\cos\vartheta. \qquad (3.58)$$

The asymmetry factor has the following values according to the scattering process

$$
\begin{aligned}
\mathrm{g} &= 1 && \text{forward scattering} \\
&= 0 && \text{isotropic or symmetric scattering} \\
&= -1 && \text{backward scattering.}
\end{aligned}
$$

If, for convenience, we set $\mu = \cos\vartheta$, then for isotropic scattering $p(\mu) = 1$, or for symmetric scattering (Rayleigh) $p(\mu) = \frac{3}{4}(1+\mu^2)$, the asymmetry factor is zero. For Mie scattering or scattering by large particles, defined by the parameter $x = 2\pi r/\lambda \geq 1$, where r is the radius of the particle, the asymmetry factor has typical values near 0.8. For water droplets such scattering represents essentially forward scattering due to refraction. For such scattering the phase function is complex, however, there are approximate expressions such as the Henyey–Greenstein phase function

$$p_{\mathrm{HG}}(\mu) = \frac{1-g^2}{(1+g^2-2g\mu)^{\frac{3}{2}}}. \tag{3.59}$$

The above phase function gives a good approximation for the forward–scattering peak, so that an approximation to the Mie phase function, which for water droplets has a strong forward peak due to refraction and a weak backward peak due to internal reflection, is given by

$$p(\mu) = b p_{\mathrm{HG}}(\mu; g_1) + (1-b) p_{\mathrm{HG}}(\mu; g_2). \tag{3.60}$$

For example, the above equation gives realistic values for scattering by marine aerosols at $\lambda = 0.7\mu\mathrm{m}$, for the values $g_1 = 0.824$, $g_2 = -0.55$ and $b = 0.9724$.

3.4.7 *Forms of the source function*

3.4.7.1 *Isotropic and coherent scattering* For bound–bound electronic transitions with isotropic directional and coherent frequency scattering, the volume–emission scattering component is given by

$$\eta_{\nu\mathrm{scat}} = \sigma_\nu \int_0^\infty \phi(\nu')\delta(\nu-\nu')J_{\nu'}\,\mathrm{d}\nu', \tag{3.61}$$

and the total volume emission has the form

$$\eta_\nu = \kappa_\nu B_\nu + \sigma_\nu \varphi(\nu) J_\nu, \tag{3.62}$$

while the source function is written as

$$S_\nu = (1-\omega_\nu)B_\nu + \omega_\nu \varphi(\nu) J_\nu, \tag{3.63}$$

where ω_ν is the single-scattering albedo. For particle scattering, the total volume emission has the form

$$\eta_\nu = \kappa_\nu B_\nu + \sigma_\nu J_\nu, \tag{3.64}$$

and the source function is

$$S_\nu = (1-\omega_\nu)B_\nu + \omega_\nu J_\nu. \tag{3.65}$$

3.4.7.2 *Complete frequency redistribution* In the case of complete redistribution in frequency, $R(\nu^{'},\nu) = \phi(\nu^{'})\phi(\nu)$, and we have

$$\eta_\nu = \kappa_\nu B_\nu + \sigma_\nu \varphi(\nu) \int_0^\infty \varphi(\nu^{'}) J_{\nu^{'}} \mathrm{d}\nu^{'} \tag{3.66}$$

$$S_\nu = (1-\omega_\nu) B_\nu + \omega_\nu \varphi(\nu) \int_0^\infty \varphi(\nu^{'}) J_{\nu^{'}} \mathrm{d}\nu^{'} . \tag{3.67}$$

3.4.7.3 *Non–isotropic coherent redistribution* The volume emission for the case of coherent scattering in frequency but non–isotropic redistribution in photon direction, as defined by the angle ϑ between incoming and outgoing photon direction, is

$$\eta_\nu = \kappa_\nu B_\nu + \frac{\sigma_\nu}{4\pi} \int_0^{2\pi} \int_{-1}^{1} p(\cos\vartheta) I_\nu(\mu^{'},\phi^{'}) \mathrm{d}\mu^{'} \mathrm{d}\phi^{'} , \tag{3.68}$$

where the phase function $p(\cos\vartheta) = p(\mu,\phi;\mu^{'},\phi^{'})$, $\mu = \cos\theta$, θ is the zenith angle and ϕ is the azimuth angle in polar co-ordinates, and

$$\cos\vartheta = \mu\mu^{'} + (1-\mu^2)^{1/2}(1-\mu^{'2})^{1/2}\cos(\phi-\phi^{'}). \tag{3.69}$$

The source function can then be written as

$$S_\nu = (1-\omega_\nu) B_\nu + \frac{\omega_\nu}{4\pi} \int_0^{2\pi} \int_{-1}^{1} p(\mu,\phi,\mu^{'},\phi^{'}) I_\nu(\mu^{'},\phi^{'}) \mathrm{d}\mu^{'} \mathrm{d}\phi^{'} . \tag{3.70}$$

3.5 The equation of radiation transfer

The equation of radiation transfer describes the modification of the radiation field as it traverses an atmosphere. It has terms related to absorption, emission and scattering of photons by the atmospheric molecules, but can include surface effects. Planetary atmospheres can, as a first approximation, be considered as plane–parallel or simple layers. It is better to simplify and solve the transfer equation to deal with specific problems rather than attempt to formulate a very general solution in three dimensions. We thus focus on the classical solutions for specific cases that were developed by the early workers in the field. These solutions serve as a basis for the development of more complicated methods (see Chapter 6) and for an initial understanding of the key radiation processes in planetary atmospheres.

3.5.1 *General form*

The general form of the equation of radiation transfer for the radiance, $I_\lambda(\vec{r},\hat{n},t)$, at a point $\vec{r}$ and time t, within an atmosphere without significant macroscopic motions, is

$$\frac{1}{c}\frac{\partial I_\lambda(\vec{r},\hat{n},t)}{\partial t} + \frac{\partial I_\lambda(\vec{r},\hat{n},t)}{\partial l} = \eta_\lambda(\vec{r},\hat{n},t) - \chi_\lambda(\vec{r},t) I_\lambda(\vec{r},\hat{n},t), \tag{3.71}$$

where c is the speed of light, l is the distance travelled by a photon of wavelength λ and direction $\hat{n}$. Each term has the units of the volume emission, η, watts per cubic meter per steradian per wavelength interval, and χ is the extinction coefficient. The first term on the left-hand side represents the time variation of the radiance, while the second term its variation along the photon direction. The first term on the right-hand side represents the volume emission in the direction of the incoming photon, while the second represents the photons that are removed from the incoming direction due to absorption and scattering.

The displacement, dl, is related to a change in position, d$\vec{r}$, by the equality $\hat{n}\mathrm{dl} = \mathrm{d}\vec{\mathrm{r}}$. In Cartesian co-ordinates the radius vector can be written as

$$\vec{r} = x\hat{i} + y\hat{j} + z\hat{k}, \tag{3.72}$$

and so the direction of the photon can be written as

$$\hat{n} = \frac{\mathrm{d}x}{\mathrm{d}l}\hat{i} + \frac{\mathrm{d}y}{\mathrm{d}l}\hat{j} + \frac{\mathrm{d}z}{\mathrm{d}l}\hat{k} \tag{3.73}$$

$$= \cos\theta_{xl}\hat{i} + \cos\theta_{yl}\hat{j} + \cos\theta_{zl}\hat{k}, \tag{3.74}$$

in terms of the direction cosines between each Cartesian axis and the photon direction. Thus, the equation of transfer can be rewritten as

$$\frac{1}{c}\frac{\partial I_\lambda(\vec{r},\hat{n},t)}{\partial t} + \hat{n}\cdot\vec{\nabla} I_\lambda(\vec{r},\hat{n},t) = \eta_\lambda(\vec{r},\hat{n},t) - \chi_\lambda(\vec{r},t) I_\lambda(\vec{r},\hat{n},t), \tag{3.75}$$

where the nabla operator, $\vec{\nabla}$, is given by

$$\vec{\nabla} = \frac{\partial}{\partial x}\hat{i} + \frac{\partial}{\partial y}\hat{j} + \frac{\partial}{\partial z}\hat{k}. \tag{3.76}$$

3.5.2 *Plane–parallel atmosphere*

For a one–dimensional plane–parallel atmosphere, and neglecting temporal and horizontal variations of the radiance, the equation of transfer simplifies to

$$\mu\frac{\partial I_\lambda(z,\mu)}{\partial z} = \eta_\lambda(z,\mu) - \chi_\lambda(z) I_\lambda(z,\mu), \tag{3.77}$$

where $\mu = \cos\theta_{zl}$ and z defines the outward vertical direction of the atmosphere. If the volume emission, η, is a known function of z, then the equation is a simple differential equation that can be solved for each direction μ and height z in the atmosphere. If the volume emission involves scattering, as in planetary atmospheres, then the above equation is an integrodifferential equation, since

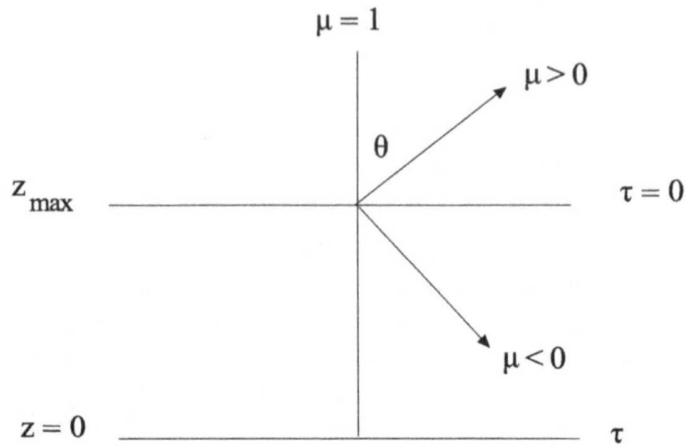

FIG. 3.6. Atmospheric layer of optical thickness τ and physical thickness z_{max}. The upwelling direction corresponds to $\mu > 0$, where $\mu = \cos\theta$.

the volume emission, η, depends on the radiance, I. If we divide the transfer equation by the volume extinction coefficient, χ_λ, then we obtain it in terms of the source function and optical depth

$$\mu \frac{\partial I_\lambda(\tau_\lambda, \mu)}{\partial \tau_\lambda} = I_\lambda(\tau_\lambda, \mu) - S_\lambda(\tau_\lambda, \mu). \tag{3.78}$$

To solve the above equation we need to specify two boundary conditions, one at the top of the atmosphere (TOA) and one at the base.

3.5.3 *Solution for pure absorption*

A planetary atmosphere does not emit visible or ultraviolet radiation given its range of temperatures (see Table 3.1). If we neglect scattering, we can solve the transfer equation to obtain its simplest solution, the absorption exponential decay law of Lambert. In this case, the volume emission, η_λ is zero and so the source function is zero, as is the scattering coefficient, σ_λ. If we consider a slab or single–layer atmosphere without surface reflection, then we solve

$$\mu \frac{\partial I_\lambda(\tau_\lambda, \mu)}{\partial \tau_\lambda} = I_\lambda(\tau_\lambda, \mu). \tag{3.79}$$

Note that we have defined $\mu = \cos\theta$ such that $\mu > 0$ in the outward direction while the optical depth is zero at TOA increasing towards the surface. The solution for a layer defined by the boundaries $z_{\max}$ to z, as shown in Fig. 3.6, is then

$$I_\lambda(\tau_\lambda, \mu) = I_\lambda(0, \mu) e^{\tau_\lambda(z, z_{\max})/\mu}, \tag{3.80}$$

where $\mu < 0$ for incoming radiation with the upper boundary value $I_\lambda(0, \mu)$ at $z_{\max}$.

3.5.4 *Solution for thermal emission*

A planetary atmosphere can absorb and emit thermal infrared radiation, given that it has heteronuclear molecules (e.g. H_2O, CH_4, O_3, CO_2) that can absorb thermal infra–red radiation. These are molecules that have an electric dipole moment due to residual positive and negative charges on the different types of atoms that form the molecule. For homonuclear molecules (e.g. N_2 and O_2) there is no dipole moment and hence these molecules do not interact with the thermal radiation field. The equation of transfer when we have both absorption and emission, can be written as

$$\mu \frac{\partial I_\lambda(\tau_\lambda, \mu)}{\partial \tau_\lambda} = I_\lambda(\tau_\lambda, \mu) - B_\lambda(\tau_\lambda), \tag{3.81}$$

since the source function, S_λ is equal to the Planck function (i.e. is isotropic) for thermal radiation at sufficiently high pressures, and hence not valid near TOA. If we consider a single layer with a homogeneous temperature, T, and optical depth τ_λ, then the emission from either the top (or base) of the layer is given by the solution

$$I_\lambda(\tau_\lambda, \mu) = B_\lambda(T)(1 - \mathrm{e}^{-\tau_\lambda/\mu}), \tag{3.82}$$

with $\mu > 0$. We immediately see that if the optical depth is large, the radiance from the top of the layer is

$$I_\lambda(\tau_\lambda) = B_\lambda(T), \tag{3.83}$$

and hence an optically thick ($\tau_\lambda \gg 1$) atmospheric layer emits like a blackbody. If the layer is optically thin ($\tau_\lambda \ll 1$) then we can use a Taylor–series expansion of the exponential function, and keep the first two terms to obtain

$$I_\lambda(\tau_\lambda, \mu) \approx B_\lambda(T)\tau_\lambda/\mu. \tag{3.84}$$

3.5.5 *The diffusivity approximation*

Let us consider the case of atmospheric absorption of thermal radiation above a planetary surface taken to emit isotropically as a blackbody at temperature T_o. The total upward flux arriving at the top of atmosphere, TOA, that was emitted by the surface is

$$f_\lambda = 2\pi B_\lambda(T_o) \int_0^1 \mathrm{e}^{-\tau_\lambda/\mu} \mu \mathrm{d}\mu \tag{3.85}$$

where $\tau_{o\lambda}$ is the total optical depth of the atmosphere and $\mu > 0$ defines the upward non-vertical direction of the emitted radiation. If we perform the transformation

$$w = \frac{1}{\mu} \qquad \mathrm{d}w = -\frac{1}{\mu^2}\mathrm{d}\mu \qquad \mathrm{d}\mu = -\frac{\mathrm{d}w}{w^2}$$

then

$$\int_0^1 e^{-\tau_\lambda/\mu}\mu d\mu = \int_\infty^1 \omega^{-3}e^{-\tau_\lambda\omega}d\omega,$$

and the flux can be computed from

$$f_\lambda = \pi B_\lambda 2E_3(\tau_{o\lambda}), \tag{3.86}$$

where πB_λ is the *surface spectral emission* (W m$^{-2}\mu$m), integrated over all directions, and the *special function* E_n is the exponential integral defined by

$$E_n(x) = \int_1^\infty t^{-n}e^{-xt}dt = x^{n-1}\int_x^\infty t^{-n}e^{-t}dt, \tag{3.87}$$

which obeys the recurrence relations

$$\frac{dE_n}{dz} = -E_{n-1}(z) \qquad E_n(0) = \frac{1}{n-1} \qquad (n > 1)$$

$$\text{for} \qquad z > 0 \qquad \text{and} \qquad E_n(\infty) = 0.$$

The *diffusivity approximation* is to define a characteristic direction by μ_c such that the total flux arriving at TOA in all directions μ is equivalent to radiation traversing the atmosphere in one effective direction. This characteristic direction is defined by the approximation

$$2\pi E_3(\tau_{o\lambda}) \approx \pi e^{-\tau_{o\lambda}/\mu_c}$$

$$E_3(\tau_{o\lambda}) \approx \frac{1}{2}e^{-\tau_{o\lambda}/\mu_c}$$

where $\mu_c = 0.6$, and hence it is customary to multiply the atmospheric optical depth by 1.66 to account for the transverse transfer of radiation through the atmosphere. As can be seen, when μ approaches zero the effective optical depth approaches infinity, i.e. we have totally tangentially directed radiation that of course is never transmitted to TOA. On the other hand, as μ approaches zero, the projection of the emitting surface area approaches zero in any case.

3.5.6 *The Eddington and diffusion approximations*

Deep (optically) within a static planetary atmosphere the radiation field tends towards isotropy, $I_\lambda(\mu) = I_\lambda$, if the scattering is coherent and isotropic. Near the surface directional effects become important. Eddington (1926) allowed for the anisotropy of the radiation field by expanding the radiance $I_\lambda(\mu)$ in terms of Legendre polynomials $P_n(\mu)$ so that

$$I_\lambda(\mu) = a_\lambda P_0(\mu) + b_\lambda P_1(\mu) + c_\lambda P_2(\mu) + ..., \tag{3.88}$$

where $\mu = \cos\theta$, with the first three Legendre polynomials given by

$$P_0(\mu) = 1 \qquad P_1(\mu) = \mu \qquad P_2(\mu) = \frac{1}{2}(3\mu^2 - 1)$$

, and showed that deep within an atmosphere a good approximation is to neglect higher-order terms beyond the second. Thus, the Eddington approximation is to set

$$I_\lambda(\mu) \approx a_\lambda + b_\lambda \mu. \tag{3.89}$$

For coherent and isotropic scattering, by particles within a static atmosphere, the source function is isotropic and given by

$$S_\lambda = (1 - \omega_\lambda)B_\lambda + \omega_\lambda J_\lambda, \tag{3.90}$$

where ω_ν is the single-scattering albedo. The equation of transfer for a plane–parallel atmosphere can then be written as

$$\mu \frac{\partial I_\lambda(\tau_\lambda, \mu)}{\partial \tau_\lambda} = I_\lambda(\tau_\lambda, \mu) - S_\lambda(\tau_\lambda). \tag{3.91}$$

We now introduce the *three moments of the radiation field* used by Eddington

$$J_\lambda = \frac{1}{2}\int_{-1}^{1} I_\lambda \mathrm{d}\mu, \tag{3.92}$$

$$H_\lambda = \frac{1}{2}\int_{-1}^{1} \mu I_\lambda \mathrm{d}\mu = \frac{f_\lambda}{4\pi}, \tag{3.93}$$

$$K_\lambda = \frac{1}{2}\int_{-1}^{1} \mu^2 I_\lambda \mathrm{d}\mu, \tag{3.94}$$

with the corresponding equations of transfer

$$\mu \frac{\partial I_\lambda(\mu)}{\partial \tau_\lambda} = I_\lambda(\mu) - S_\lambda, \tag{3.95}$$

$$\frac{\partial H_\lambda}{\partial \tau_\lambda} = J_\lambda - S_\lambda, \tag{3.96}$$

$$\frac{\partial K_\lambda}{\partial \tau_\lambda} = H_\lambda. \tag{3.97}$$

On using the Eddington approximation we obtain

$$J_\lambda = a_\lambda(\tau_\lambda), \tag{3.98}$$

$$H_\lambda = \frac{b_\lambda(\tau_\lambda)}{3}, \tag{3.99}$$

$$K_\lambda = \frac{a_\lambda(\tau_\lambda)}{3} = \frac{J_\lambda(\tau_\lambda)}{3}, \tag{3.100}$$

with

$$\frac{\partial K_\lambda}{\partial \tau_\lambda} = H_\lambda = \frac{1}{3}\frac{\partial J_\lambda}{\partial(\tau_\lambda)}, \tag{3.101}$$

thus

$$\frac{4\pi}{3}\frac{\partial J_\lambda(\tau_\lambda)}{\partial \tau_\lambda} = f_\lambda, \tag{3.102}$$

and

$$\frac{1}{3}\frac{\partial^2 J_\lambda(\tau_\lambda)}{\partial \tau_\lambda^2} = J_\lambda(\tau_\lambda) - S_\lambda(\tau_\lambda). \tag{3.103}$$

Substituting for the source function for coherent and isotropic scattering we have

$$\frac{1}{3}\frac{\partial^2 J_\lambda(\tau_\lambda)}{\partial \tau_\lambda^2} = (1-\omega_\lambda)(J_\lambda(\tau_\lambda) - B_\lambda(\tau_\lambda)), \tag{3.104}$$

which has the form of a steady–state diffusion equation. The above equation can then be solved for the mean radiance J by any standard technique given the boundary conditions at the surface and deep in the atmosphere (see Chapter 6).

3.5.7 *The Schuster–Schwarzschild approximation*

Another approach for removing the need to solve the transfer equation without explicitly taking into account the directional dependence of the radiation field is the Schuster–Schwarzschild or *two–stream approximation*. Here, the radiation field can be anisotropic but only two directions are considered, outgoing and incoming directions, corresponding to the radiances I^+ and I^-, respectively. For clarity we do not include the subscript for wavelength. If the scattering is symmetric, as in the case of Rayleigh scattering by atmospheric molecules, then the mean radiance $J = \frac{1}{2}(I^+ + I^-)$ and the net outgoing flux is given by $f = \pi(I^+ - I^-)$. For solar radiation absorbed and scattered by cloud particles, the Mie scattering is highly asymmetric. In both cases the source function is not isotropic (Chapter 6). However, in the two–stream approximation we consider two isotropic components of the source function for the outgoing and incoming directions, S^+ and S^-, respectively.

If we consider the scattering of visible solar radiation in the atmosphere, then the source function has no thermal emission component and the two components can be written as

$$S^+ = \omega J^+ = \omega(aI^+ + (1-a)I^-) \tag{3.105}$$

$$S^- = \omega J^- = \omega(aI^- + (1-a)I^+), \tag{3.106}$$

where $a = \frac{1}{2}(1+g)$ is the fraction of the radiation that is scattered in the initial direction of the photons, i.e. scattered forwards (Sagan and Pollack 1967,

Irvine 1968). For Rayleigh scattering, $g = 0$ and the parameter a takes the value $\frac{1}{2}$, while for Mie scattering with $g = 0.85$, for example, $a = 0.925$. In the original Schuster–Schwarzschild approximation $g = 0$ and $\omega = 1$, pure scattering, resulting in the two transfer equations

$$\mu_e \frac{\partial I^+}{\partial \tau} = I^+ - S^+, \tag{3.107}$$

and

$$-\mu_e \frac{\partial I^-}{\partial \tau} = I^- - S^-, \tag{3.108}$$

with $\mu_e = \frac{1}{2}$ to allow for the effect of non-vertical radiation transfer on the optical depth, and we note that μ_e is negative for incoming (downwelling) solar radiation. For asymmetric scattering Sagan and Pollack (1967) *modified* the two–stream approximation by introducing the above parameter a to preserve anisotropies and set $\mu_e = \frac{1}{\sqrt{3}}$, as suggested by Chandrasekhar (1960). Another approach is to set $\mu_e = \mu_o$ where μ_o is the cosine of the angle at which the photons are incident on the scattering layer. The above equations for I^+ and I^- are coupled linear first–order differential equations with constant coefficients and can be solved in closed form by well–known techniques. Sagan and Pollack give the solution of these equations for $\omega \neq 1$ in terms of the fractions of the incident radiation reflected, R, transmitted, T, and absorbed, A, by such a scattering atmospheric layer. These fractions are

$$R = \frac{(u+1)(u-1)(e^{\tau_e} - e^{-\tau_e})}{(u+1)^2 e^{\tau_e} - (u-1)^2 e^{-\tau_e}} \tag{3.109}$$

$$T = \frac{4u}{(u+1)^2 e^{\tau_e} - (u-1)^2 e^{-\tau_e}}, \tag{3.110}$$

with the fraction absorbed, $A = 1 - R - T$ and where

$$u^2 = \frac{1 - \omega + 2(1-a)\omega}{1 - \omega}, \tag{3.111}$$

and the effective optical depth is given in terms of the actual vertical optical depth τ of the scattering layer by

$$\tau_e = u(1 - \omega)\tau/\mu_e. \tag{3.112}$$

For a purely scattering ($\omega = 1$) layer $A = 0$ and $T = 1 - R$, where R is given by

$$R = \frac{(1-g)\tau/\mu_e}{2 + (1-g)\tau/\mu_e}. \tag{3.113}$$

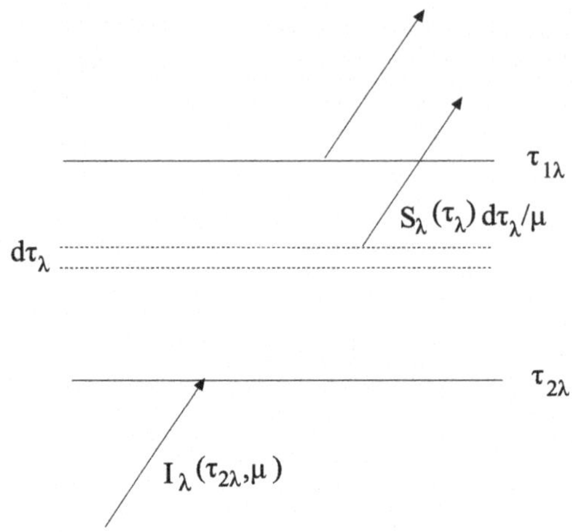

FIG. 3.7. Atmospheric layer with upper boundary, τ_1, and lower boundary τ_2. Incident at the base of the layer is the radiance $I_\lambda(\tau_{2\lambda}, \mu)$. The upwelling radiation at the top of the layer comprises the transmitted radiance plus radiation emitted by the layer itself.

3.5.8 *General solutions for upwelling radiation*

We now consider the general solution of the plane–parallel equation of transfer

$$\mu \frac{\partial I_\lambda(\tau_\lambda, \mu)}{\partial \tau_\lambda} = I_\lambda(\tau_\lambda, \mu) - S_\lambda(\tau_\lambda) \tag{3.114}$$

for upwelling radiation when the source function is a known function, $S(\tau_\lambda)$, of the optical depth. If we multiply both sides of the equation by the integration factor, $e^{-\tau_\lambda/\mu}$, then the transfer equation can be integrated directly given a boundary condition. The equation of transfer then has the form

$$\mu \frac{\partial I_\lambda(\tau_\lambda, \mu) e^{-\tau_\lambda/\mu}}{\partial \tau_\lambda} = -\frac{1}{\mu} S_\lambda(\tau_\lambda) e^{-\tau_\lambda/\mu}. \tag{3.115}$$

Let us consider an atmospheric layer whose base is defined by the optical depth $\tau_{2\lambda}$ and top is defined by $\tau_{1\lambda}$, as shown in Fig. 3.7. Integrating both sides of the above equation from $\tau_{1\lambda}$ to $\tau_{2\lambda}$ with the boundary condition that the radiance at the base of the layer is $I_\lambda(\tau_{2\lambda}, \mu)$, then the upwelling ($\mu > 0$) radiance at the top of the layer is given by

$$I_\lambda(\tau_{1\lambda}, \mu) = I_\lambda(\tau_{2\lambda}, \mu) e^{-(\tau_{2\lambda} - \tau_{1\lambda})/\mu} + \frac{1}{\mu} \int_{\tau_{1\lambda}}^{\tau_{2\lambda}} S(t_\lambda) e^{-(t_\lambda - \tau_{1\lambda})/\mu} dt_\lambda, \tag{3.116}$$

where $\tau_{2\lambda} > \tau_{1\lambda}$. The first term on the right-hand side of the above equation represents the transmitted radiance incident at the base of the layer, while the second term represents the emission from the layer itself. If we now set $\tau_{1\lambda} = 0$ and $\tau_{2\lambda} \to \infty$, then the emergent or upwelling radiance at the top of the atmosphere (TOA) that arises from the emission of an optically thick atmosphere is given by

$$I_\lambda(0,\mu) = \frac{1}{\mu}\int_0^\infty S(t_\lambda)\mathrm{e}^{-t_\lambda/\mu}\mathrm{d}t_\lambda. \tag{3.117}$$

The upwelling radiance at TOA is thus the Laplace transform of the atmospheric source function. If, for simplicity, we take the source function to be a linear function of the optical depth, $S(\tau_\lambda) = a + b\tau_\lambda$, then the upwelling radiance $I_\lambda(0,\mu) = a + b\mu = S(\mu)$. This is known as the *Eddington–Barbier relation*. Thus, the upwelling radiance at TOA is characteristic of the source function within the atmosphere where $\tau_\lambda/\mu = 1$. If the atmosphere is in local thermodynamic equilibrium (LTE) (valid at sufficiently high pressures) then the source function is given by the Planck function, $B_\lambda(T)$, and the above results apply to the thermal radiation emitted by the atmosphere. For a planetary atmosphere this corresponds to thermal infra–red radiation.

3.5.9 *Schwarzschild–Milne equations*

At some level, τ_λ, within a planetary atmosphere, we can obtain solutions for the mean radiance, J_λ, and net upward radiation flux, f_λ, if the source function is a known function of the optical depth. The mean radiance for a plane–parallel atmosphere is given by

$$J_\lambda(\tau_\lambda) = \frac{1}{2}\int_{-1}^{1} I_\lambda(\tau_\lambda,\mu)\mathrm{d}\mu. \tag{3.118}$$

The upwelling radiance at τ_λ, arising solely from atmospheric emission, can be written as

$$I_\lambda^+(\tau_\lambda,\mu) = \frac{1}{\mu}\int_{\tau_\lambda}^\infty S_\lambda(t_\lambda)\mathrm{e}^{-(t_\lambda-\tau_\lambda)/\mu}\mathrm{d}t_\lambda \tag{3.119}$$

where $\mu > 0$ and the downwelling radiance can be written as

$$I_\lambda^-(\tau_\lambda,\mu) = \frac{1}{\mu}\int_0^{\tau_\lambda} S_\lambda(t_\lambda)\mathrm{e}^{-(t_\lambda-\tau_\lambda)/\mu}\mathrm{d}t_\lambda \tag{3.120}$$

where $\mu < 0$. The mean radiance is then given by

$$J_\lambda(\tau_\lambda) = \frac{1}{2}\int_0^1 I_\lambda^+(\tau_\lambda,\mu)\mathrm{d}\mu + \frac{1}{2}\int_{-1}^0 I_\lambda^-(\tau_\lambda,\mu)\mathrm{d}\mu. \tag{3.121}$$

Performing the integration over all angles gives

$$J_\lambda(\tau_\lambda) = \frac{1}{2}\int_0^\infty S_\lambda(t_\lambda)E_1(|t_\lambda - \tau_\lambda|)\mathrm{d}t_\lambda. \tag{3.122}$$

The net upwelling flux is given by

$$f_\lambda(\tau_\lambda) = 2\pi \int_{-1}^{1} I_\lambda(\tau_\lambda, \mu)\mu \mathrm{d}\mu, \tag{3.123}$$

and can be calculated from

$$f_\lambda(\tau_\lambda) = 2\pi \int_{\tau_\lambda}^{\infty} S_\lambda(t_\lambda)E_2(|t_\lambda - \tau_\lambda|)\mathrm{d}t_\lambda - 2\pi \int_0^{\tau_\lambda} S_\lambda(t_\lambda)E_2(|t_\lambda - \tau_\lambda|)\mathrm{d}t_\lambda. \tag{3.124}$$

For a planetary atmosphere with an emitting surface, the net upwelling flux at optical depth, τ_λ, includes a component from the surface (here taken as a blackbody at temperature T_g) and so

$$\begin{aligned} f_\lambda(\tau_\lambda) = {} & 2\pi B_\lambda(T_\mathrm{g})E_3(|\tau_{\mathrm{g}\lambda} - \tau_\lambda|) + 2\pi \int_{\tau_\lambda}^{\tau_{\mathrm{g}\lambda}} S_\lambda(t_\lambda)E_2(|t_\lambda - \tau_\lambda|)\mathrm{d}t_\lambda \\ & -2\pi \int_0^{\tau_\lambda} S_\lambda(t_\lambda)E_2(|t_\lambda - \tau_\lambda|)\mathrm{d}t_\lambda. \end{aligned} \tag{3.125}$$

3.5.10 *Grey atmospheres in radiative equilibrium*

In radiative equilibrium the emission and extinction processes are balanced

$$\oint \mathrm{d}\Omega \int_0^{\infty} \mathrm{d}\lambda[\eta_\lambda - \chi I_\lambda] = 0. \tag{3.126}$$

Neglecting scattering, the source function will be isotropic and we can write

$$\int_0^{\infty} \kappa_\lambda S_\lambda \mathrm{d}\lambda = \int_0^{\infty} \kappa_\lambda J_\lambda \mathrm{d}\lambda \tag{3.127}$$

and if we can replace the wavelength–dependent absorption coefficient, κ_λ, by a grey or wavelength–independent value, κ, then $S = J$, where

$$I = \int_0^{\infty} I_\lambda \mathrm{d}\lambda \quad \text{and} \quad S = \int_0^{\infty} S_\lambda \mathrm{d}\lambda, \tag{3.128}$$

and the transfer equation in terms of the grey optical depth, τ, has the form

$$\mu\frac{\mathrm{d}I}{\mathrm{d}\tau} = I - J \tag{3.129}$$

which has the general solution

$$J(\tau) = \frac{1}{2}\int_0^{\infty} J(t)E_1(|t - \tau|)\mathrm{d}t \tag{3.130}$$

known as the *Milne equation* and can be solved through iteration. In planetary atmospheres, where we consider thermal radiation, a common approach to obtaining a wavelength–independent absorption coefficient is to use the *Planck mean absorption coefficient*, κ, defined by

$$\kappa \int_0^\infty B_\lambda \mathrm{d}\lambda = \int_0^\infty \kappa_\lambda B_\lambda \mathrm{d}\lambda \tag{3.131}$$

which gives the correct thermal emission.

Under the Eddington approximation, the mean intensity, J, is related to the net upwelling flux, f, for a grey atmosphere by

$$\frac{4\pi}{3}\frac{\partial J(\tau)}{\partial \tau} = f, \tag{3.132}$$

where

$$\frac{\partial f}{\partial \tau} = 4\pi(J - S), \tag{3.133}$$

and

$$f = \int_0^\infty f_\lambda \mathrm{d}\lambda. \tag{3.134}$$

For an atmosphere that is in radiative equilibrium $J = S$ and hence f is constant and so

$$J(\tau) = \frac{3}{4\pi} f\tau + c. \tag{3.135}$$

At the upper boundary, $\tau = 0$, it can be shown that under the Eddington approximation, $f = 2\pi J$ so that

$$J(\tau) = \frac{3}{4\pi} f\tau + \frac{1}{2\pi} f. \tag{3.136}$$

If the atmosphere is also in local thermodynamic equilibrium (LTE), then $S = B$ where

$$B(T(\tau)) = \int_0^\infty B_\lambda \mathrm{d}\lambda = \frac{1}{\pi}\sigma T^4, \tag{3.137}$$

and $J = B$ for both LTE and RE. We can now solve for the temperature structure of the atmosphere under these conditions. At the upper boundary we have $f_\mathrm{o} = 2\pi J_\mathrm{o} = 2\pi B_\mathrm{o}$ so that

$$B(T(\tau)) = B(T_\mathrm{o})(1 + \frac{3}{2}\tau), \tag{3.138}$$

or

$$T^4(\tau) = T_\mathrm{o}^4(1 + \frac{3}{2}\tau). \tag{3.139}$$

The *effective temperature* of the planet–atmosphere system is defined by

$$f = \pi B(T_\mathrm{eff}) = 2\pi B(T_\mathrm{o}), \tag{3.140}$$

where T_o is called the atmospheric *skin temperature*, related to the effective temperature through

$$T_o = (1/2)^{\frac{1}{4}} T_{\text{eff}}. \tag{3.141}$$

For the Earth, $T_{\text{eff}} \approx 255$ K so that $T_o \approx 214$ K, corresponding to the temperature of the tropopause. Thus, as $\tau \to 0$ the atmospheric temperature becomes isothermal and approaches the skin–temperature value. We also note that the effective temperature of the planet–atmosphere system is located near where $\tau = 2/3$. Near the Earth's surface convective equilibrium determines the temperature structure and not radiative equilibrium. These concepts are discussed in more detail in Chapters 2 and 11.

3.5.10.1 *Limb–darkening law* Since $J = S$ we can use eqn (3.136) for the source function in eqn (3.117) to obtain for the emergent radiance at zenith angle, z, where $\mu = \cos z$, so that

$$I(0, \mu) = (1 + \frac{3}{2}\mu) f / 2\pi, \tag{3.142}$$

so that the ratio of the emergent radiation in the direction μ and that emerging normally ($\mu = 1$) to the atmospheric surface is

$$\frac{I(0, \mu)}{I(0, 1)} = \frac{2}{5} + \frac{3}{5}\mu. \tag{3.143}$$

Thus, the ratio of the emergent radiance from near the limb ($\mu = 0$) of the atmosphere (i.e. from layers near the top of the atmosphere) and that emerging normally from the atmosphere (i.e. from deep in the atmosphere) is $\frac{2}{5}$. We remember that the optical depth in the direction of the radiation is τ/μ.

3.6 Bibliography

3.6.1 *Notes*

Atmospheric radiative–transfer theory had its origin in early work on stellar atmospheres, as described by Eddington and more recently by Mihalas. The book by Chandrasekhar is the classical reference for the mathematics of radiation transfer, and gives analytic methods for the solution of the radiative–transfer equation that were particularly important in the days before digital computers. The Goody and Yung book, itself an update of R.M. Goody's classic text dating from 1964, is solidly based in the physics and chemistry of atmospheric radiation, and is a standard reference for approximate methods such as spectral–band models that took advantage of early computers of limited power. The book by Liou is recommended for its treatment of radiation in cloudy atmospheres.

3.6.2 *References and further reading*

Chamberlain, J. W. and Hunten, D. M. (1978). *Theory of planetary atmospheres.* Academic Press, New York.

Eddington, A. S. (1926). *The internal constitution of the stars.* Cambridge University Press, Cambridge.

Chandrasekhar, S. (1960). *Radiative transfer.* Dover Publications Inc., New York.

Goody, R. M. and Yung, Y. L. (1989). *Atmospheric radiation.* Oxford University Press, Oxford.

Irvine, W. M. (1968). Multiple scattering by large particles. *Astrophys. J.*, **152**, 823–834.

Liou, K. (1980). *An introduction to atmospheric radiation.* Academic Press, New York.

Mihalas, D. (1978). *Stellar atmospheres.* W. H. Freeman and Company, San Francisco.

Sagan, C. and Pollack, J. (1967). Anisotropic non-conservative scattering and the clouds of Venus. *J. Geophys. Res.*, **72**, 469–477.

Vardavas, I. M. (1976). Non-Isotropic redistribution effects on spectral line formation. *J. Quant. Spectrosc. Radiat. Transfer*, **16**, 1–13.

Vardavas, I. M. (1976). Redistribution effects on line formation in moving stellar atmospheres II. *J. Quant. Spectrosc. Radiat. Transfer*, **16**, 781–788.

4

THERMAL INFRA–RED TRANSFER IN THE ATMOSPHERE

4.1 Introduction

Absorption of terrestrial thermal infra–red radiation by molecules in the atmosphere depends on their concentration, type and whether they are heteronuclear (e.g. CO_2, H_2O, CH_4, O_3, NH_3) or homonuclear (e.g. N_2 and O_2), that is whether or not they have an electrical dipole that can interact with infrared radiation. The atmosphere consists of mainly homonuclear molecules and hence it is the trace molecules, which strongly absorb infra–red radiation emitted by the Earth's surface to space, that determine the strength of the *greenhouse effect*. The absorption of the heteronuclear molecules arises mainly through bound–bound transitions between quantized rotational levels that are affected by the vibration of the molecules. We thus speak of vibrational–rotational lines that form *molecular bands* of spectral lines that absorb infra–red radiation, particularly between 5 μm and 100 μm, the spectral region where the Earth primarily emits blackbody radiation.

Absorption by rotational spectral lines through bound–bound transitions depends crucially on the broadening of the bound levels. There are primarily three line–broadening mechanisms; natural broadening that arises from the Heisenberg Uncertainty Principle, collisional broadening and Doppler broadening. The most important line broadening in planetary atmospheres arises through collisions that depend on the atmospheric pressure. The broadening mechanisms allow photons that have energies greater or less than the difference in energy between the levels of the bound–bound transition to be absorbed. This enhances the ability of the molecule to absorb photons over a larger spectral interval with the result that the greenhouse effect becomes stronger as the pressure of the planetary atmosphere increases. This effect enhances the greenhouse effects on Venus (surface pressure of about 100 bar) and Earth (1 bar) compared to the weak greenhouse effect on Mars (7 mbar).

4.2 Spectral lines

The rate of photon absorption by an atom (or molecule) resulting in the excitation of a bound electron from a lower to an upper energy level depends on the strength of the upwards electronic transition that is given by the Einstein coefficient for absorption, a quantity intrinsic to the particular transition and that is usually measured in spectroscopy laboratories. The absorption also depends

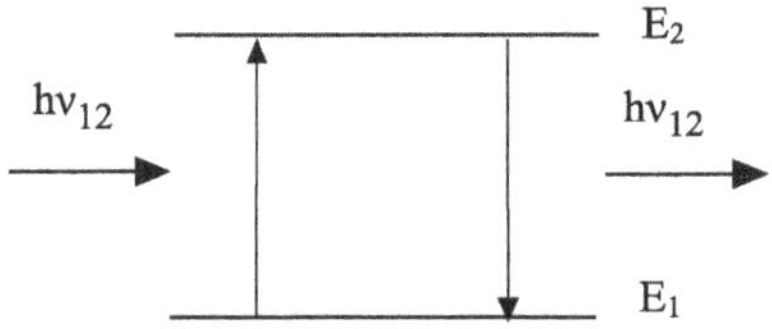

FIG. 4.1. Coherent scattering of a photon with frequency ν_{12} by absorption and re-emission between discrete atomic energy levels.

on the broadening of the bound energy levels, as this determines the available energy states that an electron can be excited to. Emission, on the other hand, not only depends on the available energy states of the upper level of the transition but also on the population of atoms that have electrons in this excited state. In thermodynamic equilibrium, TE, a system is closed and all processes are in equilibrium, the population of the excited state is given by the Boltzmann distribution, which depends on the atmospheric temperature. The radiation field and the thermodynamic state of the atmosphere can be closely coupled through collisions that excite and de–excite atoms and so govern their populations at each level. When collisions control the populations of the energy levels in a particular part of an atmosphere we have only local thermodynamic equilibrium, LTE, as the system is open to radiation loss. When collisions become infrequent then there is a decoupling between the radiation field and the thermodynamic state of the atmosphere and emission is determined by the radiation field itself, and we have no local thermodynamic equilibrium, or Non–LTE.

4.2.1 *Line broadening mechanisms*

4.2.1.1 *Natural line broadening* Atomic transitions between discrete energy levels, E_1 and E_2 result in the absorption and emission of photons of energy exactly equal to the difference in energy between the two levels, as shown in Fig. 4.1, where

$$E_2 - E_1 = h\nu_{12}. \tag{4.1}$$

According to the Heisenberg Uncertainty Principle, an atom (or molecule) has a definite lifetime, Δt, at an energy level with uncertainty, ΔE, in its energy, given by

$$\Delta E \sim \hbar/\Delta t \tag{4.2}$$

where $\hbar = h/2\pi$, and h is the Planck constant. The lifetime of an atom in its ground state is large with the result that this natural energy–level broadening is negligible. The first excited state has a smaller lifetime with the result that the energy–level broadening is significant and measurable (Fig. 4.2). The probability density (absorption profile) that an atom is excited to the upper level of the transition by absorbing a photon of energy corresponding to a wave number ω, where $\omega = \nu/c$, is given by the Lorentz profile

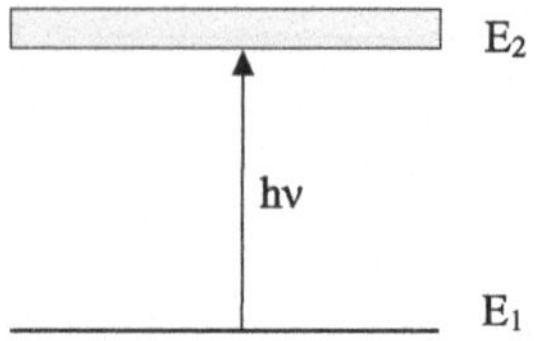

FIG. 4.2. Broadening of the energy of the upper subordinate level of a bound–bound transition arising from the ground state.

$$\phi(\omega) = \frac{b_{\mathrm{N}}/\pi}{(\omega - \omega_{\mathrm{o}})^2 + b_{\mathrm{N}}^2}, \tag{4.3}$$

which is depicted in Fig. 4.3. The natural broadening parameter, b_{N} corresponds to the half–width of the Lorentz profile at half–maximum height, and the profile is centred at ω_{o}, the spectral line centre. The absorption profile is normalized according to

$$\int_0^\infty \phi(\omega)\mathrm{d}\omega = 1. \tag{4.4}$$

4.2.1.2 *Collisional line broadening* Collisions between the absorbing atom and other atoms decrease the lifetime of the absorbing atom in the excited state with the result that the upper energy level is significantly broadened. The absorption profile is Lorentzian and so it is the same as that arising from natural line broadening. However, the absorption profile half–width parameter, b_{c}, is very large compared to that for natural broadening and is both pressure and temperature depended, having a typical dependence

$$b_{\mathrm{c}} = b_{\mathrm{c_o}} \left(\frac{p}{p_{\mathrm{o}}}\right) \left(\frac{T_{\mathrm{o}}}{T}\right)^{\frac{1}{2}}, \tag{4.5}$$

where p and T are the atmospheric pressure and temperature, respectively, while p_{o} and T_{o} are the reference pressure and temperature of the measurement $b_{\mathrm{c_o}}$. We may combine the natural and collisional half–widths into one Lorentzian profile half–width $b_{\mathrm{L}} = b_{\mathrm{N}} + b_{\mathrm{c}}$, on the reasonable assumption that the two broadening processes are independent. For example, at T=300 K, p=1 bar, $b_{\mathrm{L}} \approx 0.1$ cm^{-1} for rotational lines of CO_2 with ω located near 1000 cm^{-1}. If the transition is between subordinate levels that are broadened then it can be shown that the absorption profile is Lorentzian with half–width equal to the sum of the half–widths of the two levels of the transition.

4.2.1.3 *Doppler line broadening* Line broadening also arises owing to the random thermal motions of atoms that result in Doppler shifts in photon frequency as seen by the absorbing atom. A photon of frequency ν travelling in direction

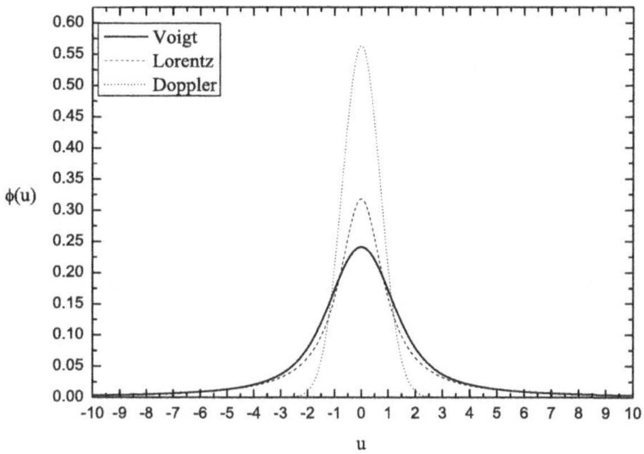

FIG. 4.3. Normalized Voigt, Lorentz and Doppler profiles of spectral lines, for $a = 1$, as functions of the dimensionless wave number parameter u, see eqn (4.12).

$\hat{n}$ relative to an atom with velocity component v, in the direction of the photon, appears to have a frequency that is Doppler shifted by

$$\Delta\nu = \frac{\nu}{c}v. \tag{4.6}$$

If v is positive, the atom is approaching the photon and the frequency of the photon is increased (blueshifted) in the rest frame of the atom. The photon frequency is redshifted if the atom is receding.

For an atmosphere in thermodynamic equilibrium, the velocity distribution of the atoms is Maxwellian and the probability of an atom having a component within dv of v, in any given direction, is

$$\phi(v)\mathrm{d}v = \frac{\exp(-v^2/\alpha^2)}{\alpha\sqrt{\pi}}\mathrm{d}v, \tag{4.7}$$

and arises from a Boltzmann distribution of particle kinetic energies, where α represents the most probable speed (in 3D) of the atoms given by

$$\alpha = \left(\frac{2kT}{m}\right)^{1/2} = \left(\frac{2RT}{M}\right)^{1/2}, \tag{4.8}$$

where m is the mass of the atom (or molecule), M its molecular weight (g mole^{-1}), and $R = kN_{\mathrm{A}} = 8.314 \times 10^7$ erg K^{-1} mole^{-1} is the universal gas constant, where $k = 1.3806 \times 10^{-16}$ erg K^{-1} is Boltzmann's constant and $N_{\mathrm{A}} =$

6.022×10^{23} mole^{-1} is Avogadro's number. The velocity distribution is thus Gaussian with standard deviation equal to $\alpha/\sqrt{2}$.

An atom that absorbs a photon through transitions between two discrete states with energy difference $h\nu_o$ can now absorb photons at other frequencies besides ν_o. If we transform to wave number then we can convert the atom's velocity v to a Doppler shift in wavenumber via

$$v = c(\omega - \omega_o)/\omega_o \tag{4.9}$$

and the velocity distribution to a probability density for absorption, or Gaussian absorption profile

$$\phi(\omega)\mathrm{d}\omega = \frac{\mathrm{e}^{-((\omega-\omega_o)/\gamma_D)^2}}{\gamma_D\sqrt{\pi}}\mathrm{d}\omega, \tag{4.10}$$

with standard deviation equal to $\gamma_D/\sqrt{2}$, where $\gamma_D = \alpha\omega_o/c$ is the *Doppler shift corresponding to the most probable velocity*, while the Doppler half width $b_D = \gamma_D (\ln 2)^{1/2}$ and is thus given by

$$b_D = \frac{\omega_o}{c}\left(\frac{2RT}{M}\right)^{1/2} (\ln 2)^{1/2}, \tag{4.11}$$

and depends solely on the atmospheric temperature. For rotational lines of CO_2 we have seen that, for $T = 300$ K at one atmosphere, the collisional broadening half-width is about 0.1 cm^{-1}, while for Doppler broadening of rotational lines with $\omega_o = 1000$ cm^{-1} the half-width is only 0.001 cm^{-1}. Thus, not only are the absorption profiles of collisionally broadened lines wider than those of Doppler-broadened lines within the troposphere, they also have Lorentzian profiles with significant absorption for frequencies far from the line centre ν_o. In the stratosphere, where the pressure is low, Doppler line broadening dominates over collisional.

4.2.1.4 *Mixed line broadening* We have seen that collisional line broadening is significant at high pressures, while Doppler line broadening is significant at high temperatures. At intermediate conditions, we have mixed collisional and Doppler line broadening and an associated absorption profile in terms of the Voigt function, $H(a, u)$

$$\phi(a, u)\mathrm{d}u = \frac{1}{\sqrt{\pi}} H(a, u)\mathrm{d}u \tag{4.12}$$

with the normalization condition

$$\int_{-\infty}^{\infty} \phi(a, u)\mathrm{d}u = 1, \tag{4.13}$$

where $a = b_{\rm L}/\gamma_{\rm D}$ is the *mixing parameter* and

$$H(a,u) = \frac{a}{\pi}\int_{-\infty}^{\infty}\frac{{\rm e}^{-y^2}{\rm d}y}{(u-y)^2+a^2}, \tag{4.14}$$

and $u = (\omega - \omega_{\rm o})/\gamma_{\rm D}$. In terms of u we also have

$$\phi_{\rm L}(u){\rm d}u = \frac{a/\pi}{u^2+a^2}{\rm d}u \tag{4.15}$$

$$\phi_{\rm D}(u){\rm d}u = (1/\sqrt{\pi}){\rm e}^{-u^2}{\rm d}u. \tag{4.16}$$

When $a \ll 1$ the Voigt profile becomes Doppler, while for $a > 1$ it becomes Lorentzian. The Voigt profile is similar to the Doppler profile for $u < 2$, referred to as the Doppler core, and has Lorentzian–type wings outside this region, as shown in Fig. 4.3, for $a = 1$.

4.2.2 *The Einstein coefficients*

For an atmosphere in thermodynamic equilibrium, the populations of atoms (or molecules) in each energy level is given by the Boltzmann distribution

$$\frac{n_{i,r}}{n_i} = \frac{g_{i,r}{\rm e}^{-h\nu_{i,r}/kT}}{u_i(T)}, \tag{4.17}$$

where $n_{i,r}$ is the number of atoms per unit volume (number density), in level r, n_i corresponds to the total population of atoms type i, $g_{i,r}$ is the statistical weight of level r arising, for example, from the degeneracy (different quantum states with the same energy) of angular momentum. The partition function u_i is given by

$$u_i(T) = \sum_r g_{i,r}{\rm e}^{-h\nu_{i,r}/kT}. \tag{4.18}$$

For a particular type of atom, the ratio of the populations in level r and the ground state is given by

$$\frac{n_r}{n_1} = \frac{g_r}{g_1}{\rm e}^{-h\nu_0/kT}. \tag{4.19}$$

There are three radiative processes, as shown in Fig. 4.4, that control the populations of the upper (u) and lower (l) level of a bound–bound transition of energy difference $h\nu_0$

1. Spontaneous emission of a photon with de-excitation from the upper level to the lower at a rate $A_{\rm ul}$ per second.
2. Absorption of a photon at a rate $B_{\rm lu}J_{\rm a}$ per second with excitation from the lower level to the upper.

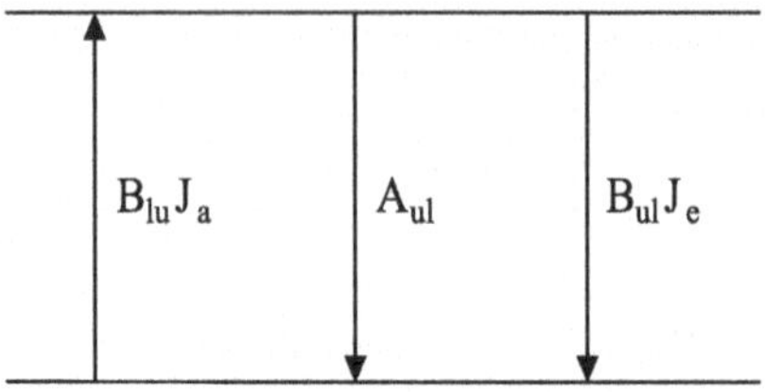

FIG. 4.4. The Einstein coefficients describing the rate of absorption, spontaneous emission and stimulated emission arising from bound–bound radiative transitions.

3. Stimulated emission at a rate of $B_{\mathrm{ul}}J_{\mathrm{e}}$ per second with de-excitation from the upper level to the lower level.

The coefficients A_{ul}, B_{lu} and B_{ul}, are the Einstein coefficients for the above radiative processes, and J_{a} is the total effective mean radiance for absorption given by

$$J_{\mathrm{a}} = \int_0^\infty \phi(\nu) J_\nu \mathrm{d}\nu, \tag{4.20}$$

where J is the mean radiance, defined by eqn (3.8), and we have introduced the absorption profile, $\phi(\nu)$, to allow for the probability of absorption of photons with frequency ν, away from the line centre of the transition, ν_{o}, due to line–broadening mechanisms. Similarly, the total effective mean radiance for stimulated emission is given by

$$J_{\mathrm{e}} = \int_0^\infty \psi(\nu) J_\nu \mathrm{d}\nu \tag{4.21}$$

where ψ is the stimulated emission profile, which for unpolarized radiation is identical to the spontaneous emission profile (Dirac 1958, Oxenius 1965). In thermodynamic equilibrium, TE, all processes are balanced and the radiance I_ν is given by the Planck function B_ν, the radiance of a blackbody. The radiation field is then isotropic and hence $I_\nu = J_\nu$ and hence $J_\nu = B_\nu$. If n_{l} and n_{u} are the total populations of atoms in the lower and upper state, respectively, then the total number of atoms that are radiatively excited to the upper level per second per unit volume of atmosphere is given by

$$\frac{\partial n_{\mathrm{ex}}}{\partial t} = n_{\mathrm{l}} B_{\mathrm{lu}} J_{\mathrm{a}}, \tag{4.22}$$

with

$$J_{\mathrm{a}} = \int_0^\infty \phi(\nu) B_\nu \mathrm{d}\nu. \tag{4.23}$$

However, the width of a spectral line is insignificant compared to the width of the Planck function and hence we can assume that B_ν is constant over the width of the spectral line and so $J_a = B_{\nu_o}$. In a similar way we can obtain the total number of atoms undergoing radiative de-excitation from the upper level

$$\frac{\partial n_{\rm dex}}{\partial t} = n_{\rm u}(A_{\rm ul} + B_{\rm ul}J_{\rm e}), \tag{4.24}$$

where

$$J_{\rm e} = \int_0^\infty \psi(\nu)B_\nu {\rm d}\nu. \tag{4.25}$$

In TE, collisions are effective in maintaining complete redistribution of the populations of atoms in the upper state and so preserve the line–broadening distribution which is that given by the absorption profile, hence $\psi_\nu = \phi_\nu$ and $J_{\rm e} = B_{\nu_o}$. Further, in TE there must be a balance between the radiation processes so that $\partial n_{\rm ex}/\partial t = \partial n_{\rm dex}/\partial t$ and hence

$$n_{\rm l}B_{\rm lu}B_{\nu_o} = n_{\rm u}(A_{\rm ul} + B_{\rm ul}B_{\nu_o}). \tag{4.26}$$

On rearranging we obtain

$$B_{\nu_o} = \frac{A_{\rm ul}/B_{\rm ul}}{n_{\rm l}B_{\rm lu}/n_{\rm u}B_{\rm ul} - 1}, \tag{4.27}$$

which can only be true for the Planck function if

$$A_{\rm ul} = 2h\nu_{\rm o}^3 B_{\rm ul}/c^2, \tag{4.28}$$

$$g_{\rm l}B_{\rm lu} = g_{\rm u}B_{\rm ul}, \tag{4.29}$$

on using the TE Boltzmann distribution for $n_{\rm u}/n_{\rm l}$. These are known as the *Einstein relations* that are independent of temperature and pressure and so they hold whether or not we have TE.

4.2.3 *Line absorptance or equivalent width*

The line absorption coefficient κ_ν, defined in Chapter 3, can now be written in terms of the absorption profile and the Einstein coefficient for absorption

$$\kappa_\nu = \kappa_o\phi(\nu), \tag{4.30}$$

where

$$\kappa_{\rm o} = h\nu_{\rm o}n_{\rm l}B_{\rm lu}/4\pi, \tag{4.31}$$

and $\kappa_{\rm o}$ is the *frequency integrated line absorption coefficient*, also known as *line strength* or *line intensity.* We can rewrite the line absorption coefficient to include a correction that takes into account stimulated emission, regarded as negative absorption. We can do this since stimulated emission results in photons that have the same direction as the stimulating photons. On the other hand, spontaneous emission takes place isotropically. Thus we have

$$\kappa_\nu = h\nu_o \left[n_l B_{lu}\phi(\nu) - n_u B_{ul}\psi(\nu)\right] / 4\pi. \tag{4.32}$$

Under TE we have complete redistribution and so $\psi(\nu) = \phi(\nu)$ and we can use the Einstein relations plus the Boltzmann distribution to obtain

$$\kappa_o = h\nu_o n_l B_{lu} \left(1 - e^{-h\nu_o/kT}\right) / 4\pi. \tag{4.33}$$

This can be expressed also in terms of the *oscillator strength* of the transition, f_{lu} that is related to B_{lu} via

$$h\nu_o B_{lu}/4\pi = \pi e^2 f_{lu}/m_e c, \tag{4.34}$$

where m_e is the rest mass of the electron. Thus, given the absorption profile and either the Einstein coefficient or the oscillator strength we can calculate the line absorption coefficient and hence the line optical depth if we know the absorber amount. The optical depth can now be written as

$$\tau_\nu = \kappa_\nu X, \tag{4.35}$$

if we can assume that the absorption coefficient is constant in the layer that we have absorber amount or path X (cm), κ_ν then has units of cm^{-1}, as ϕ has units of Hz^{-1} and n_l has units of cm^{-3}. The line strength κ_o then has units of cm^{-1}Hz, in frequency units. We note that as an integral in wavenumber, the line strength has units of cm^{-2}.

In the absence of reflection, the *transmissivity*, or transmission coefficient, t_ν, of the atmospheric layer, is related to its *absorptivity*, α_ν, through

$$t_\nu = e^{-\tau_\nu} = 1 - \alpha_\nu. \tag{4.36}$$

We can now generally define the *line absorptance*, or *line equivalent width*, A_o, in terms of line strength κ_o, for a line centred at ν_o, by

$$A_o(a, X) = \int_0^\infty (1 - e^{-\tau_\nu}) d\nu, \tag{4.37}$$

where $\tau_\nu(a, X) = \kappa_o \phi(a, \nu) X$, where a is the mixing parameter for the Voigt profile. We note that frequency can only vary between zero and infinity. As we have seen in §3.5.4, an atmospheric layer of uniform temperature emits thermal radiation according to $\epsilon_\nu B_\nu$ where the *emissivity of the layer* is given by

$$\epsilon_\nu = 1 - e^{-\tau_\nu} \tag{4.38}$$

based on Kirchoff's Law that states that the absorption coefficient, κ_ν is equal to the emission coefficient, ε_ν. Thus, the spectral line emissivity, ϵ_ν, is equal to the spectral line absorptivity, α_ν, for the atmospheric layer. For a blackbody, we have that $\alpha_\nu = \epsilon_\nu = 1$.

For rotational lines, it is convenient to work with wavenumber, $\omega = \nu/c$ in units of cm^{-1}, and so we can now define the *mean absorptivity*, α_o, over the spectral line via

$$A_o(a, X) = \alpha_o \Delta\omega = \int_{\Delta\omega} (1 - e^{-\tau_\omega}) d\omega \tag{4.39}$$

where $\Delta\omega$ is the integration interval of the spectral line. Thus α_o is dimensionless, while the line absorptance has units of cm^{-1}. We can also convert to the dimensionless wave number parameter $u = (\omega - \omega_o)/\gamma_D$ and so obtain

$$A_o(a, X) = 2\gamma_D \int_0^\infty (1 - e^{-\tau(a,u)}) du, \tag{4.40}$$

if the line is symmetric about line centre, ω_o. A useful dimensionless form of the equivalent width is $w(a, X) = A_o(a, X)/2\gamma_D$.

4.2.4 *The curve of growth*

The curve of growth gives the variation of the line absorptance, or equivalent width, as a function of absorber amount, X. If we introduce the dimensionless absorber–amount parameter y, defined by

$$y = \frac{\kappa_o X}{\gamma_D \sqrt{\pi}} \tag{4.41}$$

where the line strength, κ_o, is in cm^{-2} (integrated in wave number units), X is in cm and γ_D is in cm^{-1}, then in terms of dimensionless quantities, the curve of growth is given by $w(a, y)$ versus y.

4.2.4.1 *Limiting forms* Let us denote w_D, w_L and w_V, the dimensionless equivalent width for Doppler, Lorentz and Voigt absorption profiles, respectively. If $a \to 0$, or $y \to 0$, then $w_V \to w_D$. When $a > 1$ or $y > 10/a$ then $w_V \to w_L$. For $y \leq 10$ the Doppler equivalent width can be computed from (Struve and Elvey, 1934)

$$w_D = \frac{y\sqrt{\pi}}{2} \sum_{n=0}^{\infty} \frac{(-y)^n}{(n+1)!(n+1)^{1/2}}, \tag{4.42}$$

with a relative error of $< 1\%$. The above series converges very slowly for $y > 10$ and for this range of values, w_D can be evaluated, with a relative error $< 1\%$, from the truncated series

$$w_D(v(y)) = v + \frac{0.2886}{v} - \frac{0.1335}{v^3} + \frac{0.0070}{v^5}, \tag{4.43}$$

where $v(y) = \sqrt{\ln y}$. The Lorentz equivalent width can be computed from

Table 4.1 *The coefficients b and c used to evaluate the Ladenburg–Reiche function L(x) for x ≤ 7 using eqn (4.45). (Vardavas 1993)*

n	b_n	c_n
1	2.2499997	0.56249985
2	1.2656208	0.21093573
3	0.3163866	0.03954289
4	0.0444479	0.00443319
5	0.0039444	0.00031761
6	0.0002100	0.00001109

$$w_{\rm L}(a,y) = \pi a L[(x(a,y)], \tag{4.44}$$

where $x(a,y) = y/(2a\sqrt{\pi})$ and $L(x)$ is the Ladenburg–Reiche function given by

$$L(x) = x{\rm e}^{-x}[J_0(ix) - iJ_1(ix)], \tag{4.45}$$

which can be computed rapidly, with a relative error $< 1\%$, for $x \le 7$ from

$$L(x) = x{\rm e}^{-x}[1 + \frac{x}{2} + \sum_{n=1}^{6}\left(\frac{x}{3}\right)^{2n}(b_n + c_n x)] \tag{4.46}$$

using series expansions for the Bessel functions J_0 and J_1 (see Abramowitz and Stegun 1965). The coefficients b and c are given in Table 4.1. For $x > 7$, the simple expression

$$w_{\rm L} = \pi^{\frac{1}{4}}\sqrt{ay}\left[1 - \exp(-y\sqrt{\pi}/4a)\right]^{\frac{1}{2}} \tag{4.47}$$

can be used with a relative error of $< 1\%$. Near $x = 7$, the relative error rises to just 2% for both of the above expressions.

The Voigt equivalent width can be evaluated from the expression

$$w_{\rm V}(a,y) = \frac{w_{\rm D}(y)}{g(a,y)} + \frac{w_{\rm L}(a,y)}{f(a,y)}, \tag{4.48}$$

where

$$f(a,y) = 1 + \frac{c_0}{a^{m(a)}y^{c_1}}, \tag{4.49}$$

$$g(a,y) = 1 + b_0 a y^{b_1}, \tag{4.50}$$

and

$$m(a) = 1 - 0.24\exp[-(\log(a) + 1)^2], \tag{4.51}$$

with $c_0 = 1.12$, $c_1 = 1.22$, $b_0 = 0.47$, and $b_1 = 0.98$. The error is generally $< 1\%$ with a maximum value of about 5%. The equivalent width $w_{\rm V}$ can be set equal

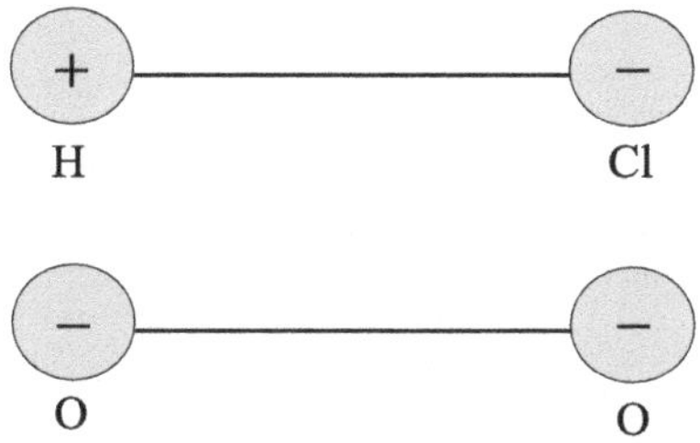

FIG. 4.5. Examples of heteronuclear (HCl) and homonuclear (O_2) diatomic molecules. The heteronuclear molecules have an electric dipole moment.

to w_D within 1% error for $a \leq 10^{-6}$ while for $a \geq 1.5$ the equivalent width w_V can be set equal to w_L within a 2% error. In the limit of $y \ll 1$, both the Doppler and Lorentz equivalent widths vary linearly with the absorber amount

$$w_D(y) = w_L(y) = \frac{y\sqrt{\pi}}{2}, \tag{4.52}$$

whereas for $y \gg 1$, the Lorentz equivalent width varies as the square root of the absorber amount

$$w_L(a, y) = \pi^{\frac{1}{4}}\sqrt{ay}. \tag{4.53}$$

For large y, the Doppler equivalent width varies as the square root of the logarithm of the absorber amount

$$w_D(y) = \sqrt{\ln y}. \tag{4.54}$$

4.3 Rotational lines and bands

Let us consider a simple model of a rotating diatomic molecule composed of solid spheres and joined by a rigid bond. We first note that for the molecule to interact with the radiation field it must have a permanent electric dipole moment (Fig. 4.5), and this requires that the molecule is made up of heteronuclear atoms, such as HCl, and not homonuclear diatomic molecules, such as O_2, which have no dipole moment and thus there are no vibration–rotation transitions and so they do not absorb or emit in the infra-red. In the rigid-rotator model, the molecule can rotate about an axis perpendicular to the bond axis going through the centre of gravity of the system, as shown in Fig. 4.6 The moment of inertia, I, is then given by

$$I = \bar{m}r_o^2, \tag{4.55}$$

where r_o is the distance between the atoms and the *reduced mass*, $\bar{m}$, is given by

$$\bar{m} = \frac{m_1 m_2}{m_1 + m_2}. \tag{4.56}$$

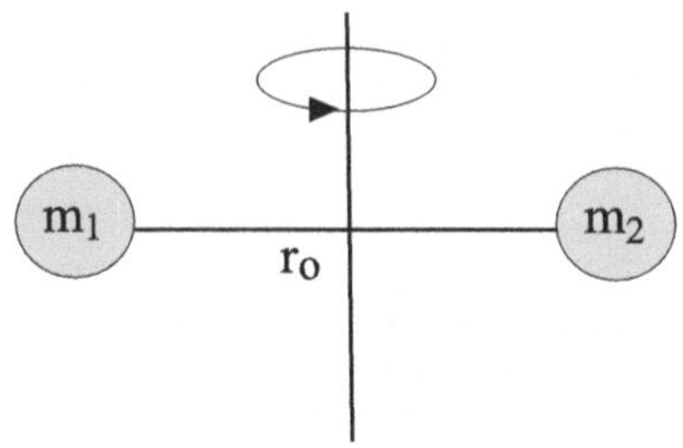

FIG. 4.6. Rigid–rotating diatomic molecule of radius r_o.

Typical values for r_o, indicative of the length of the molecule, is about 1 Å. Solution of the Schrödinger equation results in quantized rotational energy levels

$$E_J = \frac{h^2}{8\pi^2 I} J(J+1), \quad J = 0, 1, 2, 3, \cdots, \tag{4.57}$$

where J is the *rotational quantum number* and h is Planck's constant. If we introduce the *rotational constant*, B, given by

$$B = \frac{h}{8\pi^2 Ic}, \tag{4.58}$$

then the allowed rotational energy levels are given by

$$E_J = hcBJ(J+1), \quad J = 0, 1, 2, 3, \cdots, \tag{4.59}$$

with energy difference between rotational energy levels

$$\Delta E_{J,J+1} = 2hcB(J+1), \tag{4.60}$$

or, in terms of wavenumber

$$\Delta\omega_{J,J+1} = 2B(J+1) \quad J = 0, 1, 2, 3, \cdots, \tag{4.61}$$

with rotational transitions allowed according to the *selection rule* $\Delta J = \pm 1$, where here $+1$ corresponds to excitation and -1 to de-excitation. We thus have rotational energy levels according to the series $2B, 6B, 12B, ...$ and so we have spectral lines with line centre at $2B, 4B, 6B, ...$, as shown in Fig. 4.7. We thus have an infinite number of rotational levels, however, the populations of molecules that have been excited by collisions to such energy levels above the ground state ($J = 0$) are given by the Boltzmann distribution, if we have thermodynamic equilibrium,

$$\frac{n_J}{n_0} = \exp(-E_J/kT), \tag{4.62}$$

where n_0 corresponds to the population of the ground state. A typical value for B is 2 cm^{-1} and so for Earth-type temperatures, say $T = 300$ K, $n_1/n_0 \simeq 0.98$.

FIG. 4.7. Allowed rotational levels for the rigid–rotator model and corresponding spectral line series.

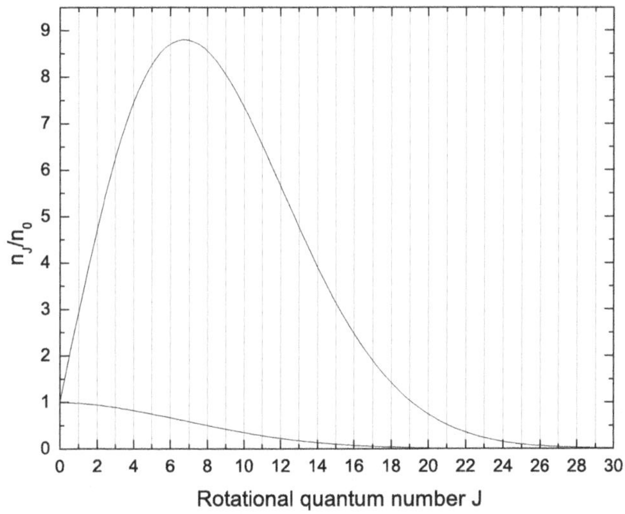

FIG. 4.8. The envelope of the ratio of the populations of rotational level J and ground level (J=0), for $B = 2$ cm^{-1} and temperature $T = 300$ K. Upper curve demonstrates the effect of degeneracy.

We see that the population of the first excited state is not very different from that of the ground state. Since the line intensity of a transition, κ, is linearly proportional to the population of the lower state of the transition, we see that the line intensity for rotational lines decreases slowly with J, as shown in Fig. 4.8, lower curve. The lower curve gives the shape of the envelope of the variation of line intensity of the rotational lines with lower level J, relative to the intensity

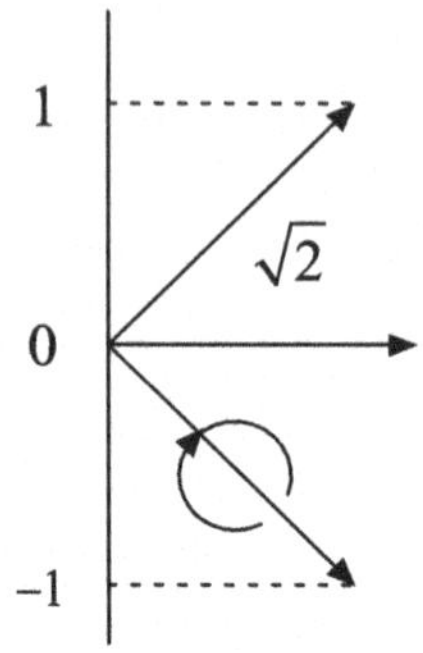

FIG. 4.9. The three projections of the angular momentum vector for $J = 1$.

of the line with the ground level ($J = 0$) as its lower level. We notice that by $J = 18$ the intensity has become insignificant. Thus, we have a *band* of molecular rotational lines with a characteristic envelope that describes the distribution of the strength of each line. Here, the first line, corresponding to $J = 0 \leftrightarrow J = 1$ transitions, is centred at $2B$ cm^{-1}.

4.3.1 *Effect of degeneracy*

Atomic or molecular energy levels that correspond to more than one allowed quantum state are termed *degenerate*. In the case of rotational energy levels, there is a law of quantum mechanics that allows the angular momentum vector, $\vec{P}$, along the axis of rotation, to have components, along some specified reference direction, whose magnitude is an integral (including zero) multiple of $\hbar$. The magnitude of the angular momentum vector, P, is related to the energy of level J by

$$E_J = P_J^2/2I, \tag{4.63}$$

where

$$P_J = \sqrt{J(J+1)}\hbar, \tag{4.64}$$

so that $P_1 = \sqrt{2}\hbar$. We can obtain three projections of the vector $\vec{P}$ of magnitude $\sqrt{2}$ along the vertical axis that have magnitudes –1, 0 and +1, as shown in Fig. 4.9. Thus, we say that the rotational level $J = 1$ is three–fold degenerate, i.e. three allowed quantum states with the same energy E_J. Generally, a level J has $(2J + 1)$–fold degeneracy and this increases the population of molecules in each rotational level J by the factor $2J + 1$, so that

$$\frac{n_J}{n_0} = (2J+1)\exp(-E_J/kT). \tag{4.65}$$

The resultant band envelope is shown in Fig. 4.8 (upper curve), which exhibits a peak in n_J/n_0 at $J = 7$, and hence in line intensity, for the transition $J = 7 \leftrightarrow$

$J = 8$, corresponding to the rotational spectral line centred at $14B$ cm^{-1}. It can be shown that the *band peak* occurs for the transition whose lower level, J, is closest to the value $\sqrt{T/2.86B} - 1/2$, where T is in Kelvin and B is in cm^{-1}.

4.4 Vibrational lines and bands

4.4.1 *The harmonic vibrator*

The diatomic molecular rigid rotator is an idealization that does not take into account the fact that the atoms are not rigidly bound and that the distance between the atoms is not fixed. A diatomic molecule can, to a first approximation, be regarded as a harmonic vibrator–rotator. If we consider only the vibrational energy transitions then the allowable vibrational levels are given by

$$E_v = \left(v + \frac{1}{2}\right) h\omega_o c \qquad v = 0, 1, 2, \cdots, \tag{4.66}$$

where ω_o is the wave number of the fundamental vibration and v is the vibrational quantum number. The selection rule for the harmonic vibrator is $\Delta v = \pm 1$. Thus, all vibrational transitions correspond to the same energy difference and hence to a single spectral line with wave number ω_o (Fig. 4.10). If the molecule is either compressed or stretched, the energy required to do this increases as the radius is decreased or increased from its equilibrium value, just like a spring. As in the case of the simple rotator, a vibrating molecule is radiatively (infra–red) active if it has a residual electric dipole moment. This is not possible for homonuclear diatomic molecules and some vibration modes of linear polyatomic molecules, such as CO_2, as we will discuss later.

We saw that the lowest energy level of a rotator was zero whereas for the harmonic vibrator it is $\frac{1}{2}h\omega_o$. This is a consequence of the fact that the atoms that constitute the molecule can never be completely at rest with respect to each other, a result that is also expected from the Heisenberg Uncertainty Principle.

4.4.2 *The anharmonic vibrator*

In the case of the diatomic harmonic vibrator model, there is a restoring force that increases as the separation of the atoms increases. In reality, if the atoms are at a sufficient distance apart, the bond between them will weaken and the molecule will dissociate into its constituent atoms. We expect that for small amplitudes of vibration the molecule will behave like a harmonic oscillator. At sufficiently large internuclear distance the molecule will proceed towards dissociation as show in Fig. 4.11. To account for molecular dissociation, the anharmonic vibrator model is introduced where vibrational transitions become progressively smaller as the bond is stretched, so that the energy difference between vibrational levels decreases as the vibrational quantum number, v, increases. To a first approximation the wave number of the energy levels is given by

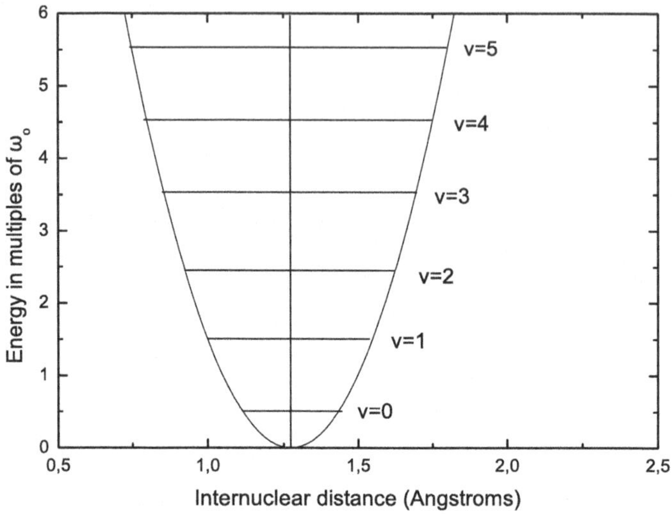

FIG. 4.10. Potential curve and energy levels of the diatomic harmonic vibrator. All transitions involve the same energy difference, and hence correspond to the same spectral line.

$$\omega_v = \left(v + \frac{1}{2}\right)\omega_{\rm e} - \left(v + \frac{1}{2}\right)^2 \omega_{\rm e} x_{\rm e} \qquad v = 0, 1, 2, \cdots, \tag{4.67}$$

where we have introduced a first–order correction of anharmonicity through the constant $x_{\rm e}$, and $\omega_{\rm e}$ is the harmonic oscillator frequency or equilibrium frequency. The constant $x_{\rm e}$ is always positive and small so that the separation of the energy levels of the anharmonic oscillator decreases slowly with increasing v, as shown in Fig. 4.11. When the energy of vibration is equal to or greater than the dissociation energy, $D_{\rm e}$, the molecule breaks up into its constituent atoms. The selection rules for the anharmonic oscillator are $\Delta v = \pm 1, \pm 2, \pm 3, \cdots$. Although larger energy transitions are now allowed, unlike the rigid rotator the populations of the first and second excited states are very small compared with the ground–state population. For vibration, the energy separation of the energy levels is of the order of 1000 cm^{-1}. Thus, assuming thermodynamic equilibrium and using the Boltzmann distribution for the populations of the excited states relative to the ground state, we find that at atmospheric temperatures the population of the first excited level is typically 1% of the ground state. Thus, the three most significant vibrational transitions originate from the ground state with jumps to the first three excited states. Absorption of radiation for these transitions

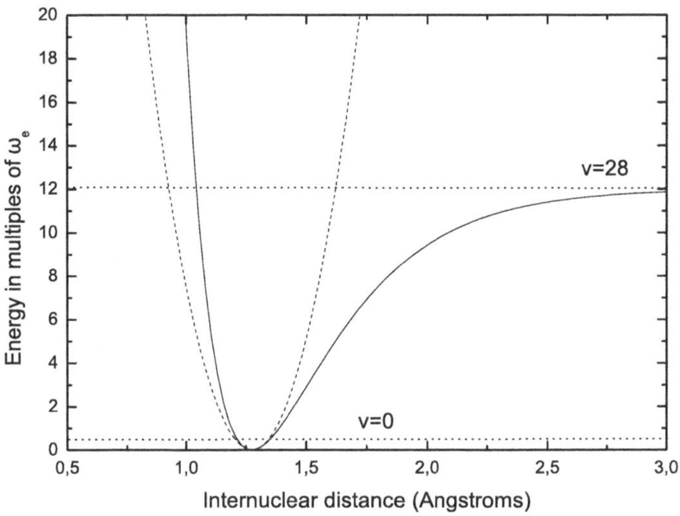

FIG. 4.11. Potential curve and energy levels of the diatomic anharmonic vibrator. Values for the dissociation energy and internuclear distance are representative for the molecule HCl. The equilibrium oscillator frequency is $\omega_e = 2990.6$ cm^{-1}, $r_{eq} = 1.274$ Å, and $D_e = 12\omega_e$. We note that the spacing between energy levels must decrease for dissociation to occur for v=28.

becomes rapidly weaker with increasing vibrational quantum number. We note that weaker transitions may become more significant as the temperature rises to 1000 K.

The corresponding vibrational spectral lines for each transition occur at the following energies:

1. $\omega_1 = \omega_e(1 - 2x_e)$ cm^{-1} for $v = 0 \to v = 1$;
2. $\omega_2 = 2\omega_e(1 - 3x_e)$ cm^{-1} for $v = 0 \to v = 2$;
3. $\omega_3 = 3\omega_e(1 - 4x_e)$ cm^{-1} for $v = 0 \to v = 3$.

For example, the molecule HCl has $\omega_e = 2990.6$ cm^{-1} and $x_e = 0.0174$, resulting in:

1. $\omega_1 = 2886.5$ cm^{-1} strong
2. $\omega_2 = 5669.0$ cm^{-1} weak
3. $\omega_3 = 8347.4$ cm^{-1} negligible.

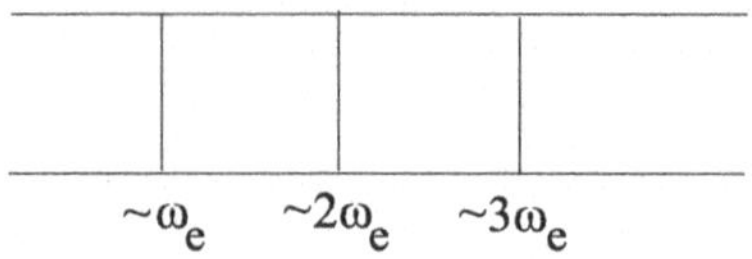

FIG. 4.12. Vibrational spectral lines of the anharmonic vibrator. The lines are centred near multiples of the equilibrium vibrational frequency ω_e. Only the spectral lines corresponding to the fundamental frequency of oscillation and the first two overtones are significant.

Thus, for the anharmonic vibrator we observe three spectral lines, of different strengths, at wave numbers close to ω_e, $2\omega_e$, and $3\omega_e$, as shown in Fig. 4.12

4.4.3 *The anharmonic vibrator–rotator*

The molecule can undergo both vibration and rotation and so can absorb photons of a larger variety in frequency than the simple vibrator or rigid rotator. The rotational and vibrational states of a diatomic molecule are coupled even though to a first approximation the combined rotational-vibrational energy of the molecule can be determined from the sum of its rotational and vibrational components. This is a good approximation because the energies of rotation and vibration are so different, typically of order 10 cm^{-1} and 1000 cm^{-1}, respectively. Thus, the molecule performs at least 100 rotations for each vibration. In this, the Born–Oppenheimer approximation, the energy of each state is simply the sum of the allowed rotational and vibrational energies. We note that a more general formulation would include allowed energies of electronic transitions, whose energies are coupled to the rotational and vibrational states, and that can have energies corresponding to visible and ultraviolet radiation.

The allowed energy levels of the anharmonic vibrator–rotator, to a first approximation, are given by

$$\omega_v = BJ(J+1) - DJ^2(J+1)^2 + \left(v + \frac{1}{2}\right)\omega_e - \left(v + \frac{1}{2}\right)^2 \omega_e x_e \quad v = 0, 1, 2, \cdots, \tag{4.68}$$

The extra rotational term, which removes the assumption of a rigid rotator and involves the rotational constant, D, arises from the centrifugal force of rotation that increases the internuclear distance between the atoms, and hence affects both rotation and vibration. This is because, in reality, we cannot have a rigid rotator that is also vibrating. However, the effect of the centrifugal force correction is very small as the constant D is much smaller than the rotational constant B. To a very good approximation we can write

$$\omega_v = BJ(J+1) + \left(v + \frac{1}{2}\right)\omega_e - \left(v + \frac{1}{2}\right)^2 \omega_e x_e \quad v = 0, 1, 2, \cdots, \tag{4.69}$$

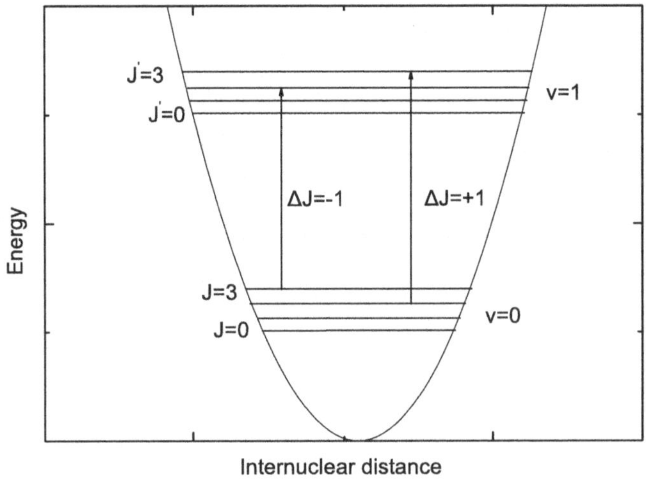

FIG. 4.13. Permitted transitions ($\Delta J = \pm 1$) between rotational levels corresponding to the vibrational ground level $v = 0$ and the first vibrational excited state $v = 1$.

with the selection rules being $\Delta v = \pm 1, \pm 2, \cdots$ and $\Delta J = \pm 1$. These rules correspond to simultaneous vibrational and rotational energy changes. In Fig. 4.13 are shown the allowed vibrational–rotational transitions. For the $v = 0 \rightarrow v = 1$ vibrational transition we can write the energy of the rotational transitions as

$$\omega_{J,v} = \omega_1 + 2Bm \qquad m = \pm 1, \pm 2, \cdots, \tag{4.70}$$

where $m > 0$ corresponds to $\Delta J = +1$, and $m < 0$ corresponds to $\Delta J = -1$. The vibration–rotation spectrum consists of equally spaced ($2B$ cm^{-1}) rotational lines on each side of the vibration–rotation band centre, ω_1. The idealized envelope or contour of rotational line strength about the band centre is shown in Fig. 4.14. Lines at wave numbers below the band centre correspond to $m < 0$ and are referred to as the P branch of the band while lines at wave numbers above the band centre correspond to $m > 0$ and are referred to as the R branch. No transitions are allowed for $m = 0$. The separation of the P and R maxima is given approximately by $2.36\sqrt{BT}$ where B is the rotational constant in cm^{-1} and T is the temperature in K. Thus for HCl with $B = 10.44$ cm^{-1} the separation is 132 cm^{-1} at $T = 300$ K. As the temperature increases the P and R peaks move to higher m values and so more rotational lines contribute to the band absorption. The first and second overtone bands have the same rotational structure but the line intensities are greatly reduced.

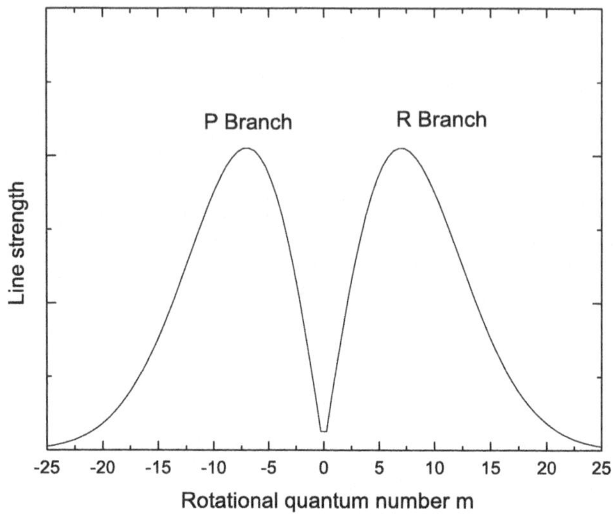

FIG. 4.14. The idealized envelope of line strength exhibiting the P and R branches of a molecular vibrational–rotational infra–red absorption band.

In reality, the P and R branches are not symmetric due to the interaction of vibration and rotation (breakdown of the Born–Oppenheimer approximation) that results in the rotational constant, B, decreasing with increasing vibrational quantum number, v. The result is that the spacing of the rotational lines decreases with increasing m. Furthermore, since in the Boltzmann distribution of the populations of the rotational levels, the rotational quantum number, J, of the lower level of an absorption transition appears in the exponential term, then line intensities are higher for the R branch (originating from lower J values) than for the P branch.

4.4.4 *Absorption bands of polyatomic molecules*

There is an increase in the complexity of the band envelope and vibrational-rotational structure when we consider the polyatomic molecules CO_2, H_2O, CH_4, NH_3 and O_3 that play an important role in the terrestrial greenhouse effect. This band complexity results from the many different mechanisms that affect the selection rules and rotational line intensities that result in deviations from the idealized band envelope of a diatomic molecule shown in Fig. 4.14 (see Herzberg 1945 and 1950, Banwell and McCash 1994).

As a relatively simple example of a polyatomic molecule, let us consider CO_2 that is a linear molecule that can have three modes of vibration. The most important to atmospheric thermal infra–red absorption are the bending mode and the asymmetric stretch mode, while the symmetric stretch mode does not have an electric dipole and so is infra–red inactive. The vibrational quantum numbers of the symmetric stretch, bending and asymmetric stretch modes are represented by v_1, v_2, v_3, respectively. In general, there can be transitions between the vibrational levels of the different modes, and hence vibrational levels, in the simplest way, are represented by $v_1v_2v_3$. Thus, the ground state of the molecule is denoted by 000 and the first excited state of the bending mode is denoted by 010. The most important vibration–rotation band for atmospheric infra–red absorption is the fundamental band of the bending mode that is centred at $\omega_1 = 667.3$ cm^{-1} = 15 μm and corresponds to the transition 000 – 010. As we have seen, the energy separation between the vibrational levels decreases slowly with vibrational quantum number and so transitions between consecutive subordinate levels result in bands also centred near 15 μm. In fact, there are at least 10 such important bands, termed *hot bands.*

An important characteristic of the rotational lines of the fundamental band of CO_2 is that alternate lines are missing, so that the spacing between the lines is $4B$ and not the expected $2B$. Many of its hot bands also have this characteristic that arises due to nuclear spin effects in symmetric molecules (see Herzberg 1945). The CO_2 molecule can lose its symmetry due to isotopic effects. The fundamental bands of the *symmetric isotopes* $O^{16}C^{12}O^{16}$ and $O^{16}C^{13}O^{16}$ have alternate lines missing, whilst those of the *asymmetric isotopes* $O^{18}C^{12}O^{16}$ and $O^{17}C^{12}O^{18}$ do not have alternate lines missing. The fundamental band, the hot bands and the isotopic bands of CO_2 with band centres near 15 μm overlap each other over the wave number interval between approximately 550 and 800 cm^{-1}. This overlap needs to be taken into account in the calculation of atmospheric transmission in this spectral interval. Generally, other molecules, such as H_2O, also have bands that absorb in this spectral interval and so care must be taken to include also this overlap.

4.5 Band absorptance formulations

4.5.1 *Doppler–broadened rotational lines*

The absorptance of a molecular band can be calculated based in an analogous way to line absorptance. At low pressure the rotational lines of a band are Doppler broadened and essentially non-overlapping, as the full line width, $\sim 2b_D$, typically 0.002 cm^{-1}, becomes smaller than the line spacing, d, typically 1.0 cm^{-1}. Ignoring any asymmetries between P and R branches, we can consider, to a first approximation, that the line strength decreases exponentially with increasing rotational quantum number from the band centre and assume that the lines are symmetrical about it. Edwards and Menard (1964) adopted a normalized line strength distribution of the form

$$\kappa_J = \frac{\kappa d}{2A_o} \mathrm{e}^{-Jd/A_o}, \tag{4.71}$$

where κ_J is the rotational line strength for a transition with lower level J, A_o is referred to as the bandwidth parameter, and κ is the total line strength given by

$$\kappa = 2\sum_{J=0}^{\infty} \kappa_J, \tag{4.72}$$

and so is the sum of the line strengths of the individual rotational lines. The band absorptance, A, can then be computed from the individual line absorptances, A_J, given by

$$A_J = y_J \sqrt{\pi}\gamma_{\mathrm{D}} \sum_{n=0}^{\infty} \frac{(-y_J)^n}{(n+1)!(n+1)^{1/2}}, \tag{4.73}$$

where we have

$$y_J = \frac{\kappa_J X}{\gamma_{\mathrm{D}}\sqrt{\pi}}, \tag{4.74}$$

and, to a good approximation, γ_{D}, see eqn (4.11), is evaluated at the band centre wave number for all lines. If we now let $u = \kappa X/A_o$ and $\delta = \sqrt{\pi}\gamma_{\mathrm{D}}/d$, we can then introduce the dimensionless parameter, $w = u/2\delta$ and obtain

$$y_J = w\mathrm{e}^{-Jd/A_o}, \tag{4.75}$$

where the line strength, κ, is in units of cm^{-2}. For most situations, the lines are close together since the line spacing is of the order of 1 cm^{-1} and the band width is of the order of 100 cm^{-1}, we can replace the summation in the band absorptance expression by the integral

$$A = 2\int_o^{\infty} A_J \mathrm{d}J = \frac{2A_o}{d}\int_0^{w} A_J(y_J)\frac{\mathrm{d}y_J}{y_J}, \tag{4.76}$$

where we have set $\mathrm{d}y_J/y_J = -(d/A_o)\mathrm{d}J$. The band absorptance for non-overlapping Doppler–broadened rotational lines can be evaluated from

$$A = 2A_o\delta \sum_{n=1}^{\infty} \frac{(-1)^{n+1}w^n}{n!n^{3/2}}, \tag{4.77}$$

for values of $w \leq 10$, the above series converges after 25 terms with an error less than 1%. For values $w > 10$ the absorptance can be computed, with a maximum error of 3%, from the expression

$$A = 2A_o\delta(0.7523b^{3/2} + 0.6513b^{1/2} + 0.3013b^{-1/2} + 0.1231), \tag{4.78}$$

where $b = \ln w$, based on an asymptotic expansion for single lines given in Struve and Elvey (1934). Thus, given the total line strength of the band, κ, or *band strength*, the absorber amount (in units corresponding to the band strength), the bandwidth parameter, A_o, and the Doppler shift, γ_D, we can calculate the band absorptance, A. If we now define an effective width of the band, $\omega_{\rm eff}$, which defines the wave number range where the rotational line strengths become significant, then the mean band transmission over this spectral interval can be estimated from $t = 1 - A/\omega_{\rm eff}$. For example, for the most important infra–red band of O_3 (the ν_3 band) with band centre $\omega_o = 1042$ cm^{-1} or 9.6 μm, with a band strength 376 cm^{-2} atm^{-1} STP (standard temperature, 273 K, and pressure, 1 atm =1013.25 mbar), a midlatitude ozone amount of 0.345 atm cm STP, we can calculate the absoptance at 300 K using a $\gamma_D = 1.075 \times 10^{-6}\omega_o$ cm^{-1}, mean line spacing, $d = 0.1$ cm^{-1}, and bandwidth parameter, $A_o = 30$ cm^{-1}, we find that $w = 109$ and so $A = 11$ cm^{-1}, assuming Doppler–broadened rotational lines. This is a fair assumption considering that most of the ozone is located in the stratosphere and hence at low pressures. We note that the weak bands of ozone contribute about another 5 cm^{-1} to the absorptance, bringing the total ozone absorptance to 16 cm^{-1} (see Table 5, Vardavas and Carver 1984). If we adopt an effective bandwidth $\omega_{\rm eff} = 100$ cm^{-1}, then we obtain a mean atmospheric transmission in the spectral interval 992–1092 cm^{-1} due to this band of 0.89.

4.5.2 *Collisionally–broadened rotational lines*

We have seen that the line absorptance for collisionally broadened lines has two limits for its variation with absorber amount, X. For small X the variation is linear, while at large X it is a square-root law. We have also seen that for Doppler broadening, the linear variation also holds for small X, while for large X the variation is logarithmic. For collisionally broadened lines, the band absorptance has three limits for its variation with absorber amount, the linear, square root and logarithmic. The linear limit holds for small absorber amounts, as it does for line absorptance and also for bands with Doppler–broadened lines. For large absorber amounts, the band absorptance varies as the logarithm of the amount. For intermediate absorber amounts, certain conditions need to be fulfilled for the variation to be a square–root law.

A band–absorptance formulation that satisfies the above limits is

$$A(u, \beta) = 2A_o \ln\left[1 + \frac{u}{2 + u^{1/2}(1 + 1/\beta)^{1/2}}\right], \tag{4.79}$$

given by Cess and Tiwari (1972). Here, u is the dimensionless path length, or absorber amount, given by

$$u = \frac{\kappa X}{A_o}, \tag{4.80}$$

with κ the band strength, X the absorber amount, and as before A_o the bandwidth parameter. It is common that κ is given in units of cm^{-2} atm^{-1} at some reference temperature, and so X has units atm cm and A_o is in cm^{-1}.

The dimensionless parameter β is known as the line-structure parameter and is given by

$$\beta = \frac{4b_c p_e}{d} \tag{4.81}$$

and depends on the effective broadening pressure, p_e (atm), given by

$$p_e = \sum_m p_m \zeta_m, \tag{4.82}$$

where p_m is the partial pressure of atmospheric constituent molecule, m, and ζ is the broadening coefficient. For CO_2, ζ takes values 1.3, 1.0, 0.81, and 0.78, for collisions between CO_2 molecules and CO_2 (self broadening), N_2, O_2, and argon, respectively. For ozone, we can take $p_e = 1$ and $\zeta = 1$, for ozone–air collisions, as it is a minor constituent of the atmosphere. Care must be taken with the units when the effective broadening pressure is involved in the definition of the dimensionless line-structure parameter, β. If p_e is in atm, then the band mean line half-width, b_c, must be in cm^{-1} atm^{-1} according to eqn (4.5).

4.5.2.1 *Limiting forms* Let us first examine the variation of the band absorptance with absorber amount. First, we note that the line structure parameter, β, is a measure of the overlapping of the rotational lines in the band. When $\beta \ll 1$, the lines are non-overlapping, as the effective band mean line half-width, $b_c p_e$, for collisionally broadened lines, is much smaller then the mean line spacing. When $\beta \gg 1$ the overlapping of the rotational lines is so large that the line structure of the band is smeared out. In Fig. 4.15 is shown the band absorptance for the 9.6 μm bands of ozone as a function of effective broadening pressure for ozone amounts ranging from a typical present atmospheric amount (0.001 g cm^{-2}) to a very low amount (0.00001 g cm^{-2}). We can observe three important limits. As the absorber amount becomes very small the band absorptance goes to the pressure–independent linear limit. As the pressure becomes very small the band absorptance is that of Doppler–broadened non-overlapping lines, and hence the band absorptance is independent of pressure. At sufficiently high pressures, line overlapping due to collisional broadening smears out the line structure and so the band absorptance becomes independent of pressure.

In Fig. 4.16 is shown the variation of the band absorptance for the CO_2 15 μm bands as a function of absorber amount, for a relatively high pressure of 1 atm and a low pressure of 0.001 atm. First, we note the linear limit, $A = A_o u$, as the absorber amount becomes small, $u \ll 1$, independently of pressure. For very high absorber amounts, $u \gg 1$, the band absorptance varies as the logarithm of the absorber amount, $A = A_o \ln u$ for $\beta \gg 1$ and $A = A_o \ln \beta u$ for $\beta \ll 1$, independently of pressure. The square-root limit, $A = A_o\sqrt{\beta u}$, is attained depending on three conditions. First, the lines must be non-overlapping ($\beta \ll 1$). Secondly, the line strengths are large, $u/\beta \gg 1$, and thirdly the path length, or absorber amount, is not too large, $\beta u \ll 1$. We see that the square

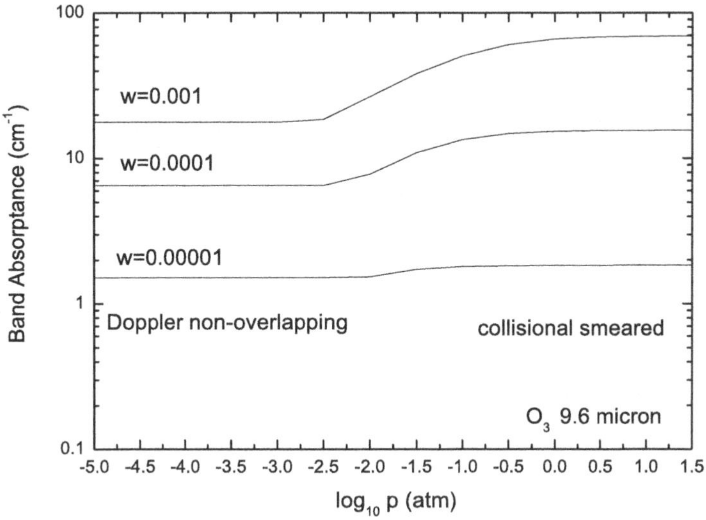

FIG. 4.15. Band absorptance (cm^{-1}) as a function of the effective broadening pressure (atmospheres) for the O_3 9.6 μm bands, depicting the pressure–independent regions, for different ozone amounts (g cm^{-2}), at T=300 K.

root limit is attained for intermediate absorber amounts for the low–pressure case. The criterion for strong lines is satisfied for the CO_2 15 μm band. For the high–pressure case only the linear and logarithmic limits are attained.

4.5.3 *Temperature dependence of absorptance*

The temperature dependence of the band absorptance arises from the temperature dependence of; the band strength, κ, the bandwidth parameter, A_o, and the rotational line half-width, b.

4.5.3.1 *Temperature dependence of band strength* The temperature dependence of the band strength can be understood by considering first the temperature dependence of the line strength of a diatomic molecule. For a single absorption line, m, the line strength, eqn (4.31), can be written in terms of the Einstein coefficient B_{lu} and wave number as

$$\kappa_m = hc\omega_{lu} n_l B_{lu}(1 - e^{-hc\omega_{lu}/kT})/4\pi, \tag{4.83}$$

where ω_{lu} is the wave number of the vibrational–rotational transition, which is essentially the wave number of the band centre, ω_o. Assuming thermodynamic

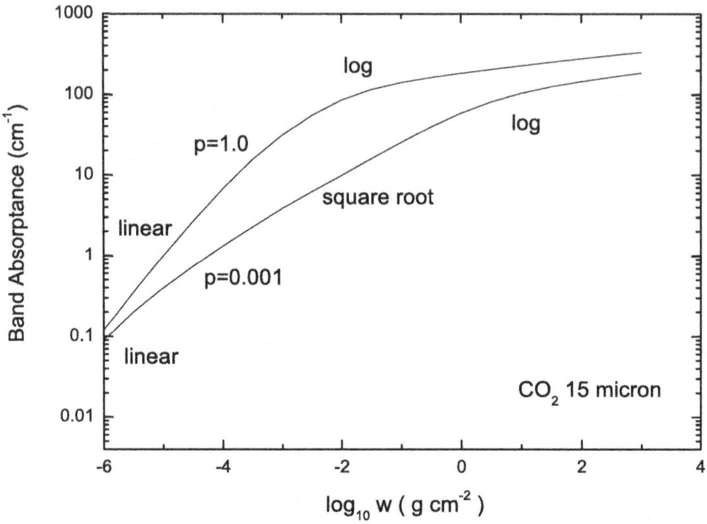

FIG. 4.16. Band absorptance as a function of CO_2 amount (g cm^{-2}) for the 15 μm bands, depicting the linear, square root and logarithmic limits at effective broadening pressures of 1.0 and 0.001 atmospheres, at $T = 300$ K.

equilibrium, we can write the ratio of the population n_l of the lower vibrational–rotational level and the total population, n, of the molecule in terms of the Boltzmann distribution

$$\frac{n_l}{n} = \frac{g_J \exp -(E_J + E_v)/kT}{Q_v Q_r}, \tag{4.84}$$

where, E_v is the energy of the lower vibrational level v, there is no degeneracy associated with v, while g_J and E_J are the degeneracy and energy associated with the lower rotational level, J, of the transition, while Q_v and Q_r are the partition functions for vibration and rotation, respectively. We have assumed that vibration and rotation do not interact, a good approximation for the lower vibrational levels that are of atmospheric interest, and so the total partition function for vibration-rotation, $Q(T) = Q_v Q_r$, the product of the individual partition functions. The rotational partition function for diatomic and linear polyatomic molecules, considered as a rigid rotator, can be written as

$$Q_r = \sum_{J=0}^{\infty} (2J+1) \exp(-J(J+1)\Theta_r/T), \tag{4.85}$$

where $\Theta_r = hcB_r/k$ is a characteristic temperature of rotation and depends on the rotational constant B_r. Thus, we see that since $hc/k = 1.439$ cm K and B_r is typically about 2 cm^{-1}, then Θ_r is about 3 K, and hence for atmospheric temperatures $\Theta_r/T \ll 1$ and hence the sum in the above equation can be replaced by an integral to give

$$Q_r = T/\Theta_r, \tag{4.86}$$

while for polyatomic molecules of atmospheric interest, the rotational partition function varies as $T^{3/2}$ (Herzberg 1945). The partition function for vibration of a diatomic molecule, neglecting anharmonicity effects, to a very good approximation for the lower levels (equally spaced vibrational levels), is

$$Q_v = (1 - e^{-\Theta_v/T})^{-1} \tag{4.87}$$

where $\Theta_v = hc\omega_o/k$ is a characteristic temperature of vibration, typically about 3000 K. Since $\Theta_r \ll \Theta_v$ we can write, ignoring the zero-point energy and the Θ_r value when v=0,

$$\frac{n_l}{n} = \frac{g_J \exp(-v\Theta_v/T)}{Q_v Q_r}. \tag{4.88}$$

We can sum over all the rotational lines, m, in the band to obtain the temperature dependence of the band strength. The value of the band strength κ at some temperature T, compared to its value at some reference temperature T_s, is then given by

$$\kappa(T) = \kappa(T_s) R_r R_v R_b R_{st}, \tag{4.89}$$

where the ratios R are associated with the rotational and vibrational partition functions, the Boltzmann distribution, and the correction for stimulated emission, are given by

$$R_r = Q_r(T_s)/Q_r(T), \tag{4.90}$$

$$R_v = Q_v(T_s)/Q_v(T), \tag{4.91}$$

$$R_b = \frac{\exp(-v\Theta_v/T)}{\exp(-v\Theta_v/T_s)}, \tag{4.92}$$

$$R_{st} = \frac{1 - \exp(-\Theta_v/T)}{1 - \exp(-\Theta_v/T_s)}. \tag{4.93}$$

4.5.3.2 *Temperature-dependence of the bandwidth parameter* The bandwidth parameter, A_o, scales as the separation, $\sqrt{8BkT/hc}$, between the P and R maxima in the band envelope, and so

$$A_o(T) = A_o(T_s)\sqrt{(T/T_s)}. \tag{4.94}$$

4.5.3.3 *Temperature dependence of the line half-width* The Doppler line half-width, b_D, and the collisionally induced line half-width, b_c, scale with temperature according to

$$b_D(T) = b_D(T_s)\sqrt{(T/T_s)}, \tag{4.95}$$

$$b_c(T) = b_c(T_s)(T_s/T)^{\eta}, \tag{4.96}$$

where the parameter η for molecular bands takes a mean value close to 0.7, to a good approximation. From the elementary kinetic theory of gases $\eta = 1/2$ for a collisionally broadened rotational line. This arises from the fact that b_c for a line scales as the frequency of collisions (see §7.3.2) which is proportional to $n\sqrt{T}$, where n is the number density given by $n = p/kT$. Both theoretical work and experimental work indicate that for rotational lines of molecular bands, the parameter η is about 3/4 near the band centre decreasing to 1/2 for lines in the band wings, for atmospheric temperatures (see also the HITRAN database discussed in §4.7 for individual line values for each molecule).

4.6 Important greenhouse gases

The Earth's surface longwave radiation emission peaks at about 10 μm, or 1000 cm^{-1}, and is significant between 5 and 50 μm, as can be seen from Fig. 3.4. Thus, molecular bands that absorb significantly in this spectral region play an important role in atmospheric absorption and emission, and hence control the outgoing longwave radiation to space and hence the Earth's longwave radiation budget. This, in turn, determines the greenhouse effect of the atmospheric molecules as the balance between the net incoming solar radiation and the outgoing longwave radiation primarily determines the Earth's surface temperature.

4.6.1 *Bands in the terrestrial infra-red region*

In Table 4.2 we divide the terrestrial infra-red spectral region into intervals within which there is significant band absorption by the atmospheric molecules. We list spectral regions where there is absorption via rotational bands, vibrational–rotational bands, and continuum absorption. The importance of water vapour is clearly seen.

4.6.2 *Water vapour*

Water vapour is the most important atmospheric greenhouse gas. Its bands cover most of the spectral region of the Earth's surface thermal radiation emission. Water vapour has important rotational bands between 0 and 800 cm^{-1} ($>$ 12 μm), vibrational-rotational bands between 1300 and 2125 cm^{-1} (the 6.3 μm bands), and continuum absorption in the region 769–1250 cm^{-1}. It also has many weak bands in the spectral region between 1 and 10 μm, some of these we

Table 4.2 *The terrestrial infra–red region divided into spectral intervals with centres* ω *cm*$^{-1}$ *and width* $\Delta\omega$ *within which there are important overlapping bands of water vapour, carbon dioxide, methane and ammonia. (Vardavas and Carver 1984)*

	ω_i	$\Delta\omega_i$	Absorbers and Bands
1	20	40	H_2O rot ($\gtrsim$ 12 μm)
2	100	120	H_2O rot ($\gtrsim$ 12 μm)
3	220	120	H_2O rot ($\gtrsim$ 12 μm)
4	330	100	H_2O rot ($\gtrsim$ 12 μm)
5	440	120	H_2O rot ($\gtrsim$ 12 μm)
6	525	50	H_2O rot-cont (16.7–20.8 μm)
7	675	250	H_2O rot-cont, CO_2 (15 μm), NH_3 (10.5 μm)
8	850	100	H_2O rot, H_2O cont (8–13 μm), CO_2 (15 μm) CO_2 (10.4 μm), NH_3 (10.5 μm)
9	950	100	H_2O rot, H_2O cont (8–13 μm), CO_2 (10.4 μm) NH_3 (10.5 μm)
10	1040	80	H_2O cont (8-13 μm), CO_2 (10.4 μm), CO_2 (9.4 μm), O_3 (9.6 μm), NH_3 (10.5 μm)
11	1110	60	H_2O cont (8–13 μm), CO_2 (9.4 μm), NH_3 (10.5 μm)
12	1170	60	H_2O cont (8–13 μm), NH_3 (10.5 μm), CH_4 (7.66 μm)
13	1275	150	H_2O vib-rot (6.3 μm), CH_4 (7.66 μm)
14	1400	100	H_2O vib-rot (6.3 μm),CH_4 (7.66 μm),NH_3 (6.14 μm)
15	1500	100	H_2O vib-rot (6.3 μm), CH_4 (7.66 μm),NH_3 (6.14 μm)
16	1600	100	H_2O vib-rot (6.3 μm), NH_3 (6.14 μm)
17	1700	100	H_2O vib-rot (6.3 μm), NH_3 (6.14 μm)
18	1800	100	H_2O vib-rot (6.3 μm), NH_3 (6.14 μm)
19	1900	100	H_2O vib-rot (6.3 μm), NH_3 (6.14 μm)
20	2000	100	H_2O vib-rot (6.3 μm)
21	2125	150	H_2O vib-rot (6.3 μm)
22	>2200	∞	see Table 6.1

will consider in more detail in Chapter 6, because of their importance to near–infrared (0.85 to 5 μm) solar radiation absorption. There are also very weak bands between 16.7 and 20.8 μm, and between 0.4 and 0.75 μm. Details on how to obtain an estimate of the absorption of the near–infra–red bands can be found in Vardavas and Carver (1984).

4.6.3 *Carbon dioxide*

Carbon dioxide is the second most important greenhouse gas. Carbon dioxide has a strong fundamental band, at least three isotopic bands and at least ten hot bands that are centred near 15 μm, which overlap each other over the spectral interval 550 to 800 cm^{-1}. The fundamental band absorptance for collisionally broadened lines, at 300 K and a pressure of 1 atm, can be estimated from the band–absorptance formulation, eqn (4.79), by setting $A_o = 25$ cm^{-1}, $d = 1.56$ cm^{-1}, $b_c = 0.065p$ cm^{-1}, and $\kappa = 194$ cm^{-2} atm^{-1}. These values give $u = 7.76X$

and $\beta = 0.17p$ where X is the CO_2 amount in atm cm and p is the pressure in atm. One can then use the temperature dependence of κ, β and b_c to evaluate the absorptance at different temperatures, pressures and amounts. The mean transmission of the fundamental band can then be obtained using an effective width of about 235 cm^{-1}. Carbon dioxide has two hot weak bands centred at 10.4 μm and 9.4 μm. The former band overlaps the interval 800–1000 cm^{-1}, while the latter band overlaps the interval 1080–1140 cm^{-1}. A table of total band absorptance for the CO_2 total band absorptance for the CO_2 15 μm bands as a function of amount and pressure is given in Vardavas and Carver (1984), where simple expressions for the transmittance for the weak hot bands are also given.

4.6.4 *Ozone*

Ozone is a strong absorber near 9.6 μm. There are two fundamental bands, a very strong band at 1042.1 cm^{-1} and a weaker one at 1103.16 cm^{-1}, and at least three hot bands and three isotopic bands of the strong band. An estimate of the strong band absorptance for collisionally broadened lines, at 300 K and 1 atm, can also be obtained from eqn (4.79), using $A_o = 30$ cm^{-1}, $b_c = 0.078$ cm^{-1}, $\kappa = 376$ atm^{-1} cm^{-2}, $d = 0.1$ cm^{-1}. These values give $u = 12.5X$, $\beta = 3.12p$, where X is the ozone amount in atm cm and p in atm. The mean transmission can then be obtained using an effective bandwidth of about 100 cm^{-1}. A table of total band absorptance for the O_3 9.6 μm bands as a function of amount and pressure is given in Vardavas and Carver (1984).

4.6.5 *Methane, ammonia and N_2O*

Methane absorbs strongly in the region 1100–1750 cm^{-1} centred at 7.66 μm. From measurements of band absorptance, an estimate of the mean transmission, $\bar{t} = 1 - A/w_{\rm eff} = \exp(-\bar{\tau})$, of this band can be obtained using a mean optical depth, $\bar{\tau} = 8.22X^{0.54}$ with $X = X_o p^\eta$ and X_o the methane amount in g cm^{-2}, p in atmospheres, $\eta = 0.5$ and $w_{\rm eff} = 300$ cm^{-1}. Methane also has a near–infra–red band at 3.31 μm, with $\bar{\tau} = 9.97X^{0.61}$, $\eta = 0.45$ and $w_{\rm eff} = 600$ cm^{-1}.

Ammonia is a very strong absorber in the regions 660–1300 cm^{-1}, the 10.5 μm band, and 1300–2000 cm^{-1}, the 6.14 μm bands. For the 10.5 μm band $\bar{\tau} = 21.2X^{0.604}$ with $\eta = 0.7$ and $w_{\rm eff} = 500$ cm^{-1}, while for the 6.14 μm bands $\bar{\tau} = 38.1X^{0.74}$ with $\eta = 0.6$ and $w_{\rm eff} = 400$ cm^{-1}. Ammonia has also a near–infra–red bands near 3.03 μm with $\bar{\tau} = 32.68X^{0.822}$, $\eta = 0.3$ and $w_{\rm eff} = 400$ cm^{-1}.

Another molecule that contributes to atmospheric absorption is nitrous oxide, N_2O, with bands near 7.8 μm and 17 μm. Line strengths and absorption cross–sections for N_2O are given in the next section for the 7.8 μm band using the HITRAN database.

4.7 The HITRAN database

The HITRAN database, originally developed by the US Air Force, provides a valuable source of spectroscopic parameters for the theoretical calculation of molecular transmittances. Here we shall give parameter symbols, nomenclature and units as close as possible to those given in the HITRAN database, and hence different from those used in the preceding sections of this chapter. For a given molecule, the database contains information for all the known energy transitions, including transitions due to isotopes. The parameters necessary for the calculation of a molecule's absorption coefficient are:

1. The position of each transition, in wave number ω_{lu} (cm^{-1}), that corresponds to the energy of the photon emitted or absorbed by the molecule in the transition between two energy states.

2. The spectral line intensity of the transition, denoted in the database by S_{lu} in units $\mathrm{cm}^{-1}/(\mathrm{molecule\ cm}^{-2})$, [cf. eqn (4.31)]. The line intensity for a rotational-vibrational transition is given in the form

$$S_{\mathrm{lu}} = \frac{h\omega_{\mathrm{lu}}}{c}\frac{n_{\mathrm{l}}}{N}\left(1 - \frac{g_{\mathrm{l}}n_{\mathrm{u}}}{g_{\mathrm{u}}n_{\mathrm{l}}}\right)B_{\mathrm{lu}}, \tag{4.97}$$

where h is Planck's constant, c the speed of light, l and u refer to the lower and upper state, respectively, N is the molecular number density, n_{l} and n_{u} are the populations of each state, g are statistical weights that take into consideration the electronic, vibrational, rotational and nuclear degrees of freedom of the molecule and B_{lu} is the Einstein coefficient for induced absorption in units $\mathrm{cm}^3/(\mathrm{erg\ s}^2)$. For LTE conditions the population of each state is defined by Boltzmann statistics, which means that the line strength is temperature dependent, eqn (4.65).

3. The line pressure-broadened half-width at half-maximum (HWHM), γ_{air} in units $\mathrm{cm}^{-1}/\mathrm{atm}$, which depends on temperature and pressure, eqn (4.5).

The above data allow the calculation of the monochromatic absorption coefficient, κ_ω ($\mathrm{cm}^2\ \mathrm{molecule}^{-1}$), at a specific wave number, ω, due to a single transition, ω_{lu}, given by

$$\kappa_\omega = S_{\mathrm{lu}}\phi(\omega - \omega_{\mathrm{lu}}) \tag{4.98}$$

where $\phi(\omega - \omega_{\mathrm{lu}})$ is the absorption line profile or line shape function in units of cm. The total contribution to wave number ω from all the lines j corresponding to each transition lu with ω_j, will then be

$$k_\omega = \sum_j S_j\phi(\omega - \omega_j). \tag{4.99}$$

The HITRAN data corresponds to reference temperature and pressure conditions ($T_{\mathrm{ref}} = 296$ K and $P_{\mathrm{ref}} = 1$ atm). This means that for different conditions, corrections to the line intensities and half-widths must be done, before the calculation of the absorption coefficient. In order to correct for temperature variations in the line strength, we need to know the temperature dependence of the statistical weight and level populations. Based on the condition of LTE, the Boltzmann distribution provides this information

$$\frac{g_{\mathrm{l}} n_{\mathrm{u}}}{g_{\mathrm{u}} n_{\mathrm{l}}} = \exp(-c_2 \omega_{\mathrm{lu}}/T), \tag{4.100}$$

and

$$\frac{n_{\mathrm{l}}}{N} = \frac{g_{\mathrm{l}} \exp(-c_2 E_{\mathrm{l}}/T)}{Q(T)} \tag{4.101}$$

In the above, $c_2 = hc/k = 1.439$ cm K, with k the Boltzmann constant and E_{l} the energy (cm^{-1}) of the lower state that is also provided in the HITRAN database. $Q(T)$ is the total partition function of the molecule that is given by the usual expression:

$$Q(T) = \sum_{\mathrm{l}} g_{\mathrm{l}} \exp(-c_2 E_{\mathrm{l}}/T). \tag{4.102}$$

Substituting the above in the line strength equation we get

$$S_{\mathrm{lu}} = \left(\frac{h\omega_{\mathrm{lu}}}{c}\right) \frac{g_{\mathrm{l}} \exp(-c_2 E_{\mathrm{l}}/T)}{Q(T)} \left[1 - \exp(-c_2 \omega_{\mathrm{lu}}/T)\right] B_{\mathrm{lu}}. \tag{4.103}$$

From the last result it is clear that the temperature dependence of the line intensity relative to the reference temperature will have the form, eqn (4.89),

$$S_{\mathrm{lu}}(T) = S_{\mathrm{lu}}(T_{\mathrm{ref}}) \frac{Q(T_{\mathrm{ref}})}{Q(T)} \frac{\exp(-c_2 E_{\mathrm{l}}/T)}{\exp(-c_2 E_{\mathrm{l}}/T_{\mathrm{ref}})} \frac{[1 - \exp(-c_2 \omega_{\mathrm{lu}}/T)]}{[1 - \exp(-c_2 \omega_{\mathrm{lu}}/T_{\mathrm{ref}})]}. \tag{4.104}$$

In the last expression all parameters necessary for the calculations are directly provided from the database. Q(T) is calculated using the Gamache *et al.* (1990) parametrization that approximates the partition function as a third order polynomial of temperature

$$Q(T) = a + bT + cT^2 + dT^3, \tag{4.105}$$

where the constants depend on the molecule/isotope under consideration. Typical values of the parameters for sample molecules are given in Table 4.3.

The effects of pressure and temperature on the half-width are described in the form

$$\gamma(p, T) p_{\mathrm{ref}} = \left(\frac{T_{\mathrm{ref}}}{T}\right)^{\eta} \left[(p - p_{\mathrm{s}}) \gamma_{\mathrm{air}}(p_{\mathrm{ref}}, T_{\mathrm{ref}}) + p_{\mathrm{s}} \gamma_{\mathrm{self}}(p_{\mathrm{ref}}, T_{\mathrm{ref}})\right]. \tag{4.106}$$

In the above, p_{s} is the partial pressure of the absorbing molecule, γ_{self} is the self–broadening half–width of the transition line and η is a constant. The last

Table 4.3 *Partition function parameters for sample molecules for the temperature range 70–400 K. (From Gamache et al. 1990)*

Molecule	Parameter			
	a	b	c	d
$^{14}N_2$	7.3548E–01	7.86628E–01	–1.82828E–06	6.8772E–09
$^{12}C^{16}O$	–4.8544	3.4530E–01	5.4835E–05	–6.0682E–08
$^{16}O^{12}C^{16}O$	–2.1995	9.67518E–01	–8.0827E–04	2.80408E–06
$H^{12}C^{14}N$	–9.7107E–01	2.9506	–1.6077E–03	6.11488E–06
$^{14}N^{14}N^{16}O$	–9.5291	1.5719E+01	–1.20638E–02	5.37818E–05
$H^{12}C^{12}CH$	2.5863	1.1921	–7.9281E–04	4.62258E–06

two parameters are also included in the HITRAN database. A typical value for η is 1/2 which comes from collision theory. For most cases $\gamma_{\rm air} \gg \gamma_{\rm self}$ and if the molecule under consideration is a minor species then $p_{\rm s} \ll$ p. Then the above expression takes the simpler form

$$\gamma(p,T) = \frac{p}{p_{\rm ref}} \left(\frac{T_{\rm ref}}{T}\right)^{\eta} \gamma_{\rm air}(p_{\rm ref}, T_{\rm ref}). \tag{4.107}$$

Finally, the position of the line centre for a specific transition can be affected by the pressure. The database provides this information also, in the form of a pressure–shift parameter, δ (cm^{-1}/atm)

$$\omega'_{\rm lu} = \omega_{\rm lu} + p\delta(p_{\rm ref}). \tag{4.108}$$

4.7.1 *Application to N_2O*

As an example of the above formulation, we present an application to N_2O. The HITRAN database contains 25 724 lines for the possible transitions of the N_2O molecule and its isotopes, covering the spectrum between 523 and 5131 cm^{-1}. An example of rotational line strengths is given in Fig. 4.17 for the region of the 7.8 μm band. Figure 4.18 presents the calculated cross–section for the above region at 298 K and 1 atm pressure for a Voigt line profile. In the calculation of the cross–section high resolution must be considered in between the line positions in order to have a correct description of the variation in absorption with wave number.

4.8 Clear–sky fluxes

In Chapter 3, we derived the Schwarzschild–Milne equations, given the source function, for the net upwelling flux at any atmospheric level arising from flux emitted by the surface and by the atmospheric layers below and above that level. For the troposphere and stratosphere, the bulk of the thermally emitting

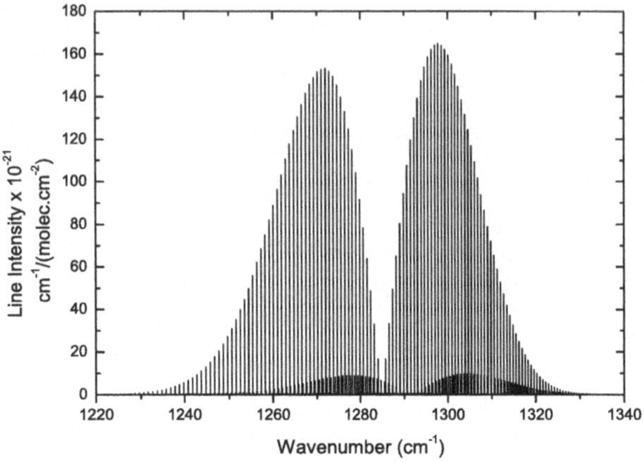

FIG. 4.17. N_2O 7.8 μm band rotational line intensity, including contributions from isotopic bands.

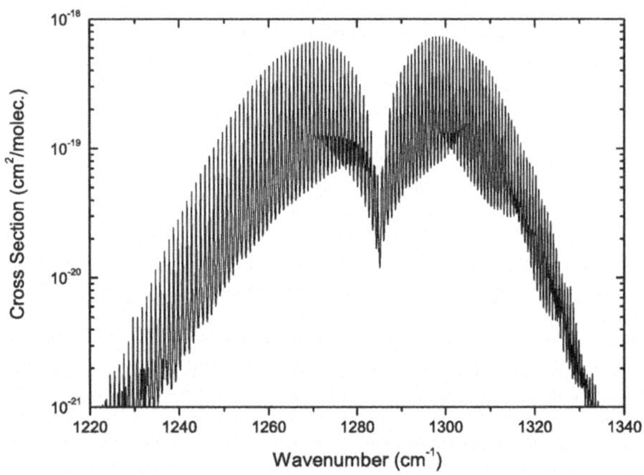

FIG. 4.18. N_2O absorption cross-section in the region of the 7.8 μm band.

atmosphere, collisions determine the populations of the vibrational–rotational states and hence the atmosphere can be taken to be in local thermodynamic equilibrium (LTE). Thus, the source function can be set equal to the Planck function, $S_\lambda = B_\lambda$, which holds below 70 km for most molecules in the Earth's atmosphere (for non–LTE see Lopez–Puertas and Taylor 2001). Equation (3.125) can be simplified by using the diffusivity approximation, discussed in §3.5.5, and the recurrence relation between the exponential integrals E_2 and E_3.

4.8.1 *Upwelling fluxes*

We can rewrite, to a very good approximation, eqn (3.125) for the upwelling component at optical depth τ_λ to obtain

$$f_\lambda^+(\tau_\lambda) = \pi\varepsilon_{\mathrm{g}\lambda}B_\lambda(T_{\mathrm{g}})\mathrm{e}^{-(\tau_{\mathrm{g}\lambda}-\tau_\lambda)/\mu_{\mathrm{c}}} + \int_{\tau_\lambda}^{\tau_{\mathrm{g}\lambda}} \pi B_\lambda(T(\tau_\lambda^{'}))\mathrm{e}^{-|\tau_\lambda^{'}-\tau_\lambda|/\mu_{\mathrm{c}}}\mathrm{d}\tau_\lambda^{'}/\mu_{\mathrm{c}} \tag{4.109}$$

where $\mu_{\mathrm{c}} = 0.6$ and $\tau_{g\lambda}$ is the total optical depth of the atmosphere at wavelength λ, from the top of the atmosphere (TOA) to the surface, and T is the atmospheric temperature. We have also introduced the possibility that the surface does not absorb/emit as an ideal blackbody, at temperature T_{g}, through the emissivity $\varepsilon_{\mathrm{g}\lambda}$, which is unity for an ideal blackbody and takes values ranging between 0 and 1, this effectively means that the surface can reflect infra–red radiation, with thermal infra–red reflectivity $1-\varepsilon_{\mathrm{g}\lambda}$.

Furthermore, in order to perform the integrals we need to divide the atmosphere into sufficiently optically thin layers, $d\tau_\lambda^{'}$, of homogeneous temperature. We see from the above equation that the integral is essentially a summation of contributions of emission $\pi B_\lambda(T(\tau_\lambda^{'}))d\tau_\lambda^{'}/\mu_c$ from the surface of each layer of homogeneous temperature T, since as we saw in eqn (3.84), the emissivity of an optically thin layer of homogeneous temperature is given by $\varepsilon = \mathrm{d}\tau_\lambda^{'}/\mu_{\mathrm{c}}$. If we now define the transmission between levels $\tau_\lambda^{'}$ and τ_λ by

$$t(\tau_\lambda, \tau_\lambda^{'}) = \mathrm{e}^{-|\tau_\lambda^{'}-\tau_\lambda|/\mu_{\mathrm{c}}}, \tag{4.110}$$

then we can write for the *net upwelling flux*

$$f_\lambda(\tau_\lambda) = \pi\varepsilon_{\mathrm{g}\lambda}B_\lambda(T_{\mathrm{g}})t(\tau_{\mathrm{g}\lambda}, \tau_\lambda) - \int_0^{\tau_{\mathrm{g}\lambda}} \pi B_\lambda[T(\tau_\lambda^{'})]\mathrm{d}t(\tau_\lambda^{'}, \tau_\lambda). \tag{4.111}$$

This formulation provides a simpler and smoother way to evaluate the integral in terms of the transmission rather than the optical depth. We note that $\mathrm{dt} > 0$ for atmospheric levels above τ_λ, and vice versa.

4.8.2 *Downwelling fluxes*

The *downwelling flux at the surface* is given by

$$f_\lambda^-(\tau_{\mathrm{g}\lambda}) = \int_0^{\tau_{\mathrm{g}\lambda}} \pi B_\lambda[T(\tau_\lambda^{'})]dt(\tau_\lambda^{'}, \tau_{\mathrm{g}\lambda}), \tag{4.112}$$

while the *net upward flux at the surface* can be written as

$$f_\lambda(\tau_{g\lambda}) = \pi\varepsilon_{g\lambda}B_\lambda(T_g) - \varepsilon_{g\lambda}f_\lambda^-(\tau_{g\lambda}), \qquad (4.113)$$

noting that only the fraction $\varepsilon_{g\lambda}$ of the downwelling flux is absorbed by the surface. The reflected fraction does not contribute to the net upward flux emitted by the surface.

4.8.3 *Outgoing flux at TOA*

We can now write for the *outgoing flux* at the top of the atmosphere

$$\begin{aligned} f_\lambda^+(0) = {} & \pi\varepsilon_{g\lambda}B_\lambda(T_g)t(\tau_{g\lambda},0) + (1-\varepsilon_{g\lambda})f_\lambda^-(\tau_{g\lambda})t(\tau_{g\lambda},0) \\ & - \int_0^{\tau_{g\lambda}} \pi B_\lambda[T(\tau_\lambda')]\mathrm{d}t(\tau_\lambda',0). \end{aligned} \qquad (4.114)$$

The first term represents the contribution from the surface emission with emissivity ε_g, the second represents that from surface reflection of downwelling flux and the third is that from the atmospheric emission directly to space.

4.9 Cloudy–sky fluxes

Clouds are an important determinant of atmospheric radiation as they have lower temperatures than the surface and hence reduce the thermal emission of the planet. Optically thick clouds, like the low–level and middle–level clouds, emit essentially as blackbodies with emissivity close to 1, where $\varepsilon_{c\lambda} = 1 - e^{-\tau_{c\lambda}}$ and $\tau_{c\lambda}$ is the cloud optical depth at wavelength λ. High–level clouds, are usually not ideal blackbodies and can have an emissivity well below unity. In Fig. 4.19 we see that the low–emissivity clouds are mainly located in the tropics. Here, high-level clouds with low emissivity play an important role in the Earth's radiation budget as they allow more of the thermal radiation from the surface to escape to space and usually reflect relatively more of the incoming solar radiation (Chapter 6). The fraction of thermal flux transmitted by the clouds is $1-\varepsilon_{c\lambda}$, assuming no cloud reflection in the infra–red, which in any case would be incorporated in the cloud emission by measurements. This high–level cloud transmission plays a very important role in the Earth's climate since the outgoing thermal flux at TOA has a very significant component of radiation from the surface of the planet that is transmitted to space. Thus, although one would expect that as the emissivity of the high-level clouds increases their contribution to the outgoing flux to increase, they essentially block the larger thermal flux from the warmer Earth's surface from escaping to space. Thus, an increase in high–level cloud emissivity results in reduced outgoing thermal flux at TOA.

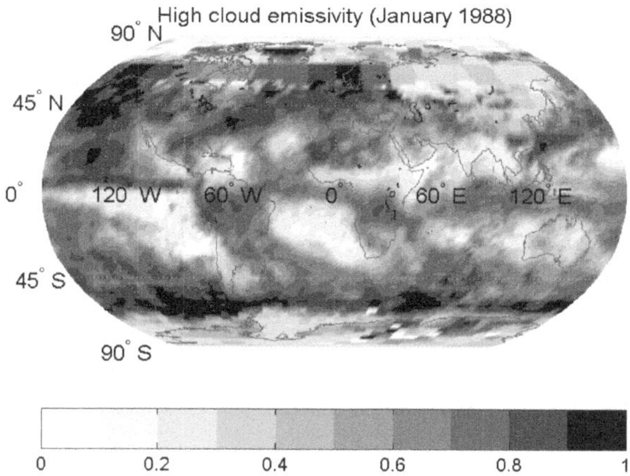

FIG. 4.19. High-level cloud thermal emissivity for January 1988, based on ISCCP data.

4.9.1 *Outgoing flux above a cloud layer*

The *upwelling flux at TOA* above the cloud layer has seven components, as shown in Fig. 4.20, arising from; the cloud-top surface, the atmosphere above the cloud layer directly and via reflection from the Earth's surface, the Earth's surface itself, the atmosphere below the cloud layer directly and via reflection from the Earth's surface, and the cloud–base surface via reflection from the Earth's surface. For optically thick clouds, we can use the expressions for clear–sky fluxes but with the emissivity and surface temperature replaced by that of the cloud surface. For optically thin clouds we need to compute all seven components.

4.9.2 *Downwelling flux at the Earth's surface*

The downwelling flux at the Earth's surface has three components. For a partly cloudy sky we can use the fraction of the sky covered by a particular cloud i, the cloud–cover fraction a_{ci}, to compute the various fluxes from $a_{\mathrm{ci}} f_{\mathrm{c}} + (1 - a_{\mathrm{ci}}) f_{\mathrm{s}}$, where f_{c} corresponds to the flux for a cloudy sky and f_{s} is that for a clear sky. One can extend this idea to many non-overlapping clouds. For overlapping clouds one needs to invoke some cloud–overlapping procedure (see Chapter 7).

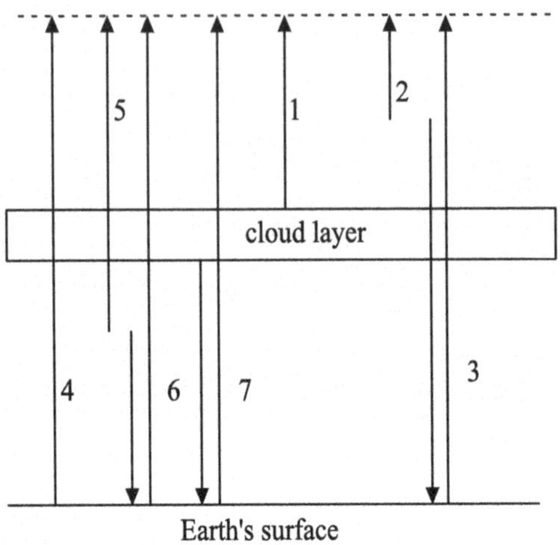

FIG. 4.20. The components of outgoing thermal flux at the top of the atmosphere and downwelling flux at the surface.

4.10 Computation of fluxes

4.10.1 *Data requirements*

To compute the Earth's longwave radiation budget one needs the following atmospheric, Earth surface, and cloud properties:

- Molecular mixing ratio vertical profile
- Molecular band absorption
- Atmospheric vertical temperature and pressure profiles
- Earth's surface temperature and emissivity
- Cloud infrared optical depth
- Cloud–cover fraction
- Cloud–top temperature and pressure
- Cloud–base temperature and pressure

in order to compute, at each wavelength, the thermal fluxes at the Earth's surface, within the atmosphere and at TOA. The next step is to integrate over wavelength to obtain the total outgoing longwave radiation (OLR) and the surface radiation budget (SRB). The difference between the net upward flux at the surface and the outgoing flux at TOA is a measure of the atmospheric greenhouse effect under clear–sky conditions. The integration over wavelength means that the radiative properties of the Earth–atmosphere–clouds system at each wavelength need to be known. This is a formidable task and one needs to define the spectral resolution of one's radiation–transfer model based on the available spectral properties of the system. For molecular absorption one can either attempt a line–by–line transfer (use of HITRAN) or use a broad–spectral–interval approach by dividing the Planck function into intervals with significant molecular band absorption, as shown in Table 4.2. This simpler approach involves the computation of the various thermal infra–red fluxes at wavelengths at the midpoints of each spectral interval.

4.10.2 *Curtis–Godson approximation*

If ω_j is the wave number at the centre of the interval j and $\Delta\omega_j$ is the width of the interval, then we can use the mean band transmission t_{ik} for this interval due to some molecule k. We can then derive a mean optical depth, τ_{jk} for each spectral interval j for each molecule k from

$$\tau_{jk} = -\ln t_{jk}, \tag{4.115}$$

then sum these optical depths for each molecule k to obtain the total optical depth for the interval j, and hence the mean transmission t_j. We can evaluate the mean transmission of the atmosphere by dividing it into pressure levels so that each atmospheric layer is sufficiently optically thin for all j. We can then compute the transmission between each pressure level $t_j(p^{'}, p)$. Note that this transmission is both temperature and pressure dependent since the molecular band absorptances are so dependent. We thus need to obtain a representative temperature, T_{r}, and pressure, p_{r}, between the two atmospheric levels we wish to evaluate the transmission. One approach, the *Curtis–Godson approximation*, is to correct for inhomogeneous atmospheric layers by weighting the pressure or temperature by the absorber amount over the transmission path within the atmosphere so that

$$p_{\mathrm{r}} = \frac{1}{w}\int p_{\mathrm{e}}\eta\frac{\mathrm{d}p}{g}, \tag{4.116}$$

$$T_{\mathrm{r}} = \frac{1}{w}\int T(p)\eta\frac{\mathrm{d}p}{g}, \tag{4.117}$$

where p_{e} is the effective broadening pressure and w is the absorber amount (g cm^{-2}) or column density given by

$$w = \int \eta\frac{\mathrm{d}p}{g}, \tag{4.118}$$

with η the mixing ratio by mass, i.e. g of molecule per g of air.

We can use the flux equations of §4.8 to compute the various fluxes at each wave number ω_j, by replacing πB_ω with $\pi B_j = \pi\int_{\Delta\omega_j} B_\omega \mathrm{d}\omega$ and then sum over all spectral intervals j to obtain the total fluxes. In this way, the clear–sky downwelling flux at the Earth's surface for spectral interval j, would be computed, for example, from

$$f_j^-(p_g) = \int_0^{p_g} \pi B_j[T(p)] \mathrm{d}t_j(p, p_g). \tag{4.119}$$

Note that we need to enforce the normalization condition $\pi \sum_j B_j(T) = \sigma T^4$.

In Chapter 8 we give actual computations for the global distribution of the Earth's thermal infra–red budget.

4.11 Bibliography

4.11.1 *Notes*

For further reading related to band–absorptance formulations see: Goody and Yung and references therein; Cess and Trewari; Tien; and Ramanathan. For the molecular spectroscopy of rotational–vibrational bands see the classic texts by Herzberg, and also Banwell and McCash. Numerical computations of the Voigt function can be found in Humlicek and, for early tabulations, Hummer. The HITRAN (high–resolution transmission molecular absorption) database can be accessed for spectroscopic data.

For a detailed discussion of non-local thermodynamic equilibrium (non–LTE) related to atmospheric radiation see Lopez–Puertas and Taylor.

4.11.2 *References and further reading*

Abramowitz, M. and Stegun, I. A. (1965). *Handbook of mathematical functions.* Dover, New York.

Banwell, C. N. and McCash, E. M. (1994). *Fundamentals of molecular spectroscopy.* McGraw-Hill, London.

Cess, R. D. and Tiwari, S. N. (1972). Infrared radiative energy transfer in gases. *Advances in heat transfer* **8**. Academic Press, New York.

Dirac, P. A. M. (1958). *The principles of quantum mechanics.* Oxford University Press, Oxford.

Edwards, D. K. and Menard, W. A. (1964). Comparison of models for correlation of total band absorption. *Appl. Opt.*, **3**, 621–625.

Gamache, R. R., Hawkins, R. L. and Rothman, L. S. (1990). Total internal partition sums in the temperature range 70–3000 K: Atmospheric linear molecules. *J. Mol. Spectrosc.*, **142**, 205–219.

Goody, R. M. and Yung, Y. L. (1989). *Atmospheric radiation.* Oxford University Press, Oxford.

Herzberg, G. (1945). *Molecular spectra and molecular structure. II. Infrared and raman spectra of polyatomic molecules.* Van Nostrand, New York.

Herzberg, G. (1950). *Molecular spectra and molecular structure. I. Spectra of diatomic molecules.* 2nd edn., Van Nostrand, New York.

Humlicek, J. (1982). Optimized computation of the Voigt and complex probability functions. *J. Quant. Spectrosc. Radiat. Transfer*, **27**, 437–444.

Hummer, D. G. (1965). The Voigt function. *Mem. R. Astr. Soc.*, **70**, 1–32.

López-Puertas, M. and Taylor, F. W. (2001). *Non-LTE radiation transfer in the atmosphere.* World Scientific, Singapore.

Oxenius, J. (1965). Emission and absorption profiles in a scattering atmosphere. *J. Quant. Spectrosc. Radiat. Transfer*, **5**, 771–781.

Ramanathan, V. (1976). Radiative transfer within the Earth's troposphere and stratosphere: A simplified radiative-convective model. *J. Atmos. Sci.*, **33**, 1330–1346.

Struve, O. and Elvey, C. T. (1934). The intensities of stellar absorption lines. *Astrophys. J.*, **79**, 409–440.

Tien, L. (1968). Thermal radiation properties of gases. *Advances in heat transfer* **5**, Academic Press, New York.

Vardavas, I. M. and Carver, J. H. (1984). Solar and terrestrial parameterizations for radiative-convective models. *Planet. Space Sci.*, **32**, 1307–1325.

Vardavas, I. M. (1993). Fast and accurate generation of the curve of growth for the Voigt lineshape. *J. Quant. Spectrosc. Radiat. Transfer*, **49**, 119–127.

5

INCOMING SOLAR RADIATION

5.1 Introduction

The Sun is a dwarf main–sequence star with an age of about 4.6 Byr (billion years) whose luminosity has increased slowly, by about 30%, from the time it joined the main sequence. At the same time, the XUV (1–1200 Å) and ultraviolet emission decreased as the rotation rate of the Sun became slower, and its magnetic activity declined. Thus, the incoming solar radiation (ISR) at the Earth's orbit is determined by solar evolution on timescales of billions of years. Luminosity variations affect climate directly, while variations in the ultraviolet flux affect the molecular composition of the atmosphere, and so indirectly affect climate through the concentrations of greenhouse gases. The ultraviolet flux also has a direct effect on biological evolution that can also indirectly influence climate by controlling land weatherability (§12.2), surface cover and surface reflectivity.

According to analyses of stratigraphic distributions of oxidized and reduced mineral deposits, the atmosphere of the early Earth (4 Byr ago) was anoxic near the surface, with the O_2 mixing ratio less than 10^{-4}, and it is generally accepted that today's oxygen–rich atmosphere is the product of photosynthesis and organic–carbon burial. An anoxic early atmosphere implies no ozone layer, and hence no associated ultraviolet screen to shield early life from the Sun's lethal ultraviolet radiation. Heavy bombardment of the Earth's surface is thought to have occurred between 4.5 and 3.8 Byr ago, based on the lunar–cratering record, so that life became extant about 4 Byr ago, as evidenced by the microfossil record. Thus understanding the evolution of solar ultraviolet flux on timescales of billions of years, which is related to solar activity, is of paramount importance to the biological evolution on Earth.

On timescales of about a million years to thousands of years, changes in the Earth's orbit can produce significant changes in ISR. On human life timescales, there are variations in the luminosity and ultraviolet flux arising from the 11–year solar cycle related to the coverage of the solar surface by sunspots. Solar–cycle luminosity variations are relatively small, but the variation in ultraviolet flux is significant.

The seasonal–latitudinal distribution of the ISR determines the global distribution of the Earth's shortwave radiation budget. This distribution is determined by the tilt of the planet's axis and the eccentricity of its orbit. These in turn determine the duration of the polar winter (or night) and the polar ice caps. The distribution of Earth's total, both shortwave (solar) and longwave (terrestrial), radiation budget determines both atmospheric and oceanic dynamics and hence climate. We shall examine the Earth's radiation budget in Chapter 8 and the evolution of the Earth's climate in §12.2.

5.2 The Sun as a main–sequence star

5.2.1 *Stellar properties*

Main–sequence (dwarf) stars like the Sun are referred to as *late-type* stars and burn H to He. These stars have relatively low effective temperatures and stellar radii. The smallest of these stars are classified as spectral types G, K and M, and have effective temperatures ranging from 5980–5370 K (G0–G9), 5230–4350 K (K0–K5), and below 3840 K for M0–M2 stars. In our galaxy the most numerous are the M stars, then the K stars followed by the G stars.

The colour index B–V of a star is defined as

$$B - V = m_B - m_V, \tag{5.1}$$

where m_B and m_V are the apparent magnitudes of the star in the blue and visual, respectively. To a good approximation the mass of a late-type star can be computed from

$$\log(M/M_\odot) = 0.28 - 0.42(B - V), \tag{5.2}$$

so that the radius of the star can be calculated from

$$R = \sqrt{GM/g}, \tag{5.3}$$

where $G = 6.626 \times 10^{-8}$ dyn cm^2 g^{-2}, is the universal gravitational constant and g the gravitational acceleration at the stellar surface. The effective temperature of a late–type star can be approximated from

$$\log T_{\mathrm{eff}} = 3.908 - 0.234(B - V). \tag{5.4}$$

The Sun is a main–sequence dwarf star of spectral type G2V (V being the luminosity class of the main–sequence stars) according to the effective temperature classification shown in Table 5.1. The present properties of the Sun are given in Table 5.2, where $M_\odot$ is the solar mass, $R_\odot$ the radius, $L_\odot$ the luminosity, $T_\odot$ the effective surface temperature, and $t_\odot$ its age. The present composition of the solar surface is given in Table 5.3.

The effective temperature of the Sun corresponds to a surface temperature of a blackbody that has the equivalent luminosity. The surface of the Sun is referred

Table 5.1 *Thermal parameters for late–type main–sequence stars. (Data from Cram and Kuhi 1989)*

Spectral–type	$M/M_\odot$	$\log g$	$T_{\rm eff}$(K)	$B-V$
G0	1.15	4.32	5980	0.583
G1	1.10	4.34	5900	0.608
G2	1.07	4.35	5800	0.625
G3	1.04	4.37	5710	0.642
G4	1.00	4.38	5690	0.657
G5	0.98	4.40	5620	0.672
G6	0.93	4.42	5570	0.690
G7	0.90	4.44	5500	0.713
G8	0.87	4.46	5450	0.740
G9	0.84	4.48	5370	0.776
K0	0.81	4.49	5230	0.819
K1	0.79	4.50	5080	0.866
K2	0.76	4.52	4920	0.912
K3	0.74	4.53	4810	0.966
K4	0.70	4.54	4640	1.030
K5	0.67	4.55	4350	1.150
M0	0.52	4.63	3840	1.420
M1	0.49	4.66	3710	1.475
M2	0.44	4.70	3620	1.512

Table 5.2 *Present solar mass, radius, luminosity, effective temperature, age and log(g).*

$M_\odot$(g)	$R_\odot$ (cm)	$L_\odot$ (erg/s)	$T_\odot$ (K)	$t_\odot$ (Byr)	log(g)
1.989×10^{33}	6.96×10^{10}	3.845×10^{33}	5777 K	4.6	4.44

Table 5.3 *Per cent abundance by volume (%v) and by mass (%m) of elements in the solar surface.*

Element	%v	%m
H	92.0	73.4
He	7.8	25.0
C	0.02	0.20
N	0.008	0.09
O	0.06	0.8
Ne	0.01	0.16
Mg	0.003	0.06
Si	0.004	0.09
S	0.002	0.05
Fe	0.003	0.14

to as its *photosphere* whose mean temperature is $T_\odot$. The concept of a surface of course is a matter of definition. In Chapter 3, we derived the following variation of temperature with mean all–wave optical depth for an atmosphere in radiative equilibrium

$$T^4 = T_o^4 \left(1 + \frac{3}{2}\tau\right), \tag{5.5}$$

where T_o is the atmospheric skin temperature, corresponding to $\tau = 0$. We also saw that the skin temperature and effective temperature are related by

$$T_o = \left(\frac{1}{2}\right)^{\frac{1}{4}} T_\odot, \tag{5.6}$$

so that $T_o = 0.84 T_\odot$. Thus, for the Sun the atmospheric skin temperature is 4859 K. We also noted that the effective temperature is located at $\tau = \frac{2}{3}$, which defines, essentially, the *photospheric surface.* Above the photosphere we expect the atmospheric temperature to decrease to the skin temperature. In fact for the Sun as we move outwards magnetic–field effects heat the atmosphere so, analogous to the Earth's stratosphere that is heated by solar ultraviolet radiation due to ozone absorption, the Sun's atmospheric temperature begins to rise above the skin temperature in what is called the *solar chromosphere* where the temperature goes through a *temperature minimum* and that has a mean value of about 4600 K. If it were not for the non–radiative heating, associated with the magnetic–field activity of the Sun, the temperature of the outer solar atmosphere would decrease to zero at infinity due to dilution of the radiation field according to the inverse square law. The temperature at 'infinity' is in fact not zero but 2.725 K, the blackbody temperature corresponding to the cosmic microwave background radiation. The temperature in the chromosphere is of the order of 10^4 K, due to non-radiative heating, and rises to about 10^6 K (corresponding to X–ray emission) in the outer atmosphere, called the *solar corona.* The chromospheric temperature is sufficient for atomic hydrogen, the main constituent, to have appreciably populated excited states. This fact is very important to the ultraviolet radiation that reaches the Earth, and hence to the photochemistry of the atmosphere.

5.2.2 *Total solar irradiance*

The mean Sun–Earth distance is $\bar{r} = 1.496 \times 10^{13}$ cm $= 1$ AU, and the mean flux arriving at the Earth's orbit at this mean distance is called the *solar constant* or *total solar irradiance* (TSI) and it is denoted by $S_\odot$. Measurements give a mean value of

$$S_\odot = 1366 \quad \mathrm{W\ m^{-2}}$$

over several solar cycles. Thus, according to the inverse square law of radiation

$$S_\odot = \frac{L_\odot}{4\pi\bar{r}^2}, \tag{5.7}$$

where $L_\odot$ is the solar luminosity. When we consider the relatively long, on human timescales, characteristic time for radiative perturbations in the solar core to reach the solar surface, we can appreciate the historical terminology of *solar constant*.

Free escape of photons from the centre of the Sun would take only $R_\odot/c = 2.3$ *s*. However, the mean free path of photons in their random path outwards is very small. If we let l represent the mean free path then

$$l = 1/\chi = 1/\sigma N \tag{5.8}$$

where χ is the extinction coefficient, and $\sigma = 2 \times 10^{-24}$ cm^2, is the absorption cross–section. Now the number of particles per unit volume, N is given by

$$N = \frac{\rho N_A}{\mu} \tag{5.9}$$

where $N_A = 6.022 \times 10^{23}$ is the Avogadro number, $\mu = 0.6$ is the solar mean molecular weight (§5.4.1), and ρ is the solar mean density equal to 1.409 g cm^{-3}. Thus, $N = 1.4 \times 10^{24}$ and hence $l = 0.36$ cm. Now since each photon absorption/re-emission event takes about $t_{ae} = 10^{-8}$ s, then the travel time to escape is $t_{ae}n_c$, where n_c is the number of such absorption/re–emission events to reach the surface given by $(R_\odot/l)^2$, so that the travel time to escape is of the order of 10 Myr. Thus, perturbations to the photon flux in the solar core take a long time to be reflected in its luminosity. As we shall see, there are radiative perturbations at the solar surface due to photospheric magnetic field activity that result in areas of reduced radiative emission called sunspots that are controlled by the 11–year solar cycle. The total solar irradiance thus varies over the solar cycle.

5.2.3 *The solar cycle*

The 11–year cycle of solar activity manifests itself in significant variation of sunspot (easily observable regions of reduced visible radiation flux) numbers on the solar photosphere. The Sun undergoes a period of very few sunspot numbers, of the order of 10, called the *solar minimum* to a period of sunspot numbers of the order of 100, called the *solar maximum*. The period between the solar minima or between the maxima is about 11 years. Annual mean sunspot numbers since 1730 AD, clearly show the 11–year cycle, as seen in Fig. 5.1 (data given by the NASA Marshall Space Flight Center, based on the International Sunspot Numbers). It is worth noting that between 1645 and 1715 AD, the evidence is that there were very few sunspots, the cycle was unclear during this period, referred to as the Maunder minimum.

The current solar cycle is Cycle 23 that commenced with a minimum in October 1996. Measurements of the total solar irradiance, or solar constant $S_\odot$, over the solar cycle have been performed by various satellites, such as NIMBUS–7,

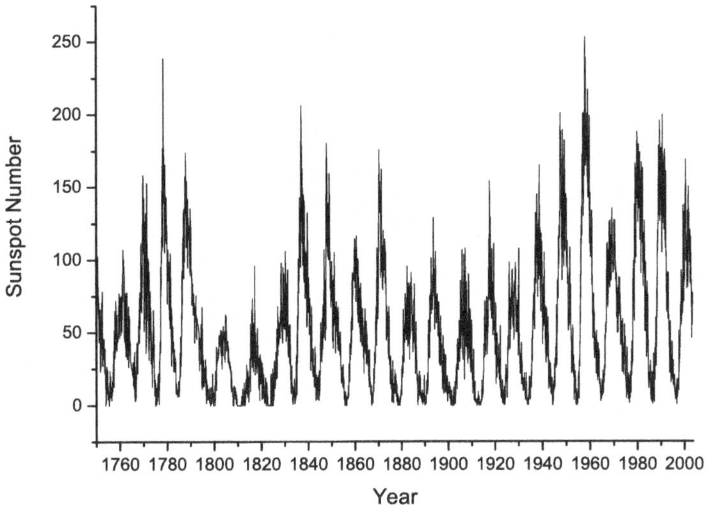

FIG. 5.1. Sunspot number variation since 1750 AD. (Data from NASA)

SMM (Solar Maximum Mission), ERBS (Earth Radiation Budget Satellite), and ACRIM (Active Cavity Radiometer Irradiance Monitor). In Fig. 5.2 is shown the solar constant for the period 1978–2003, based on the ACRIM Composite TSI (Total Solar Irradiance) data, based on ACRIM 1,2,3, Nimbus 7/ERB and VIRGO measurements. The long-term mean value, according to the ACRIM composite is 1366.25 ± 0.71. The SOLCON experiment, which was part of the ATLAS 1 & 2 Missions on the space shuttles Atlantis and Endeavour, measured the solar constant between 9–11 April 1993 and found a mean value of 1366.31 W m^{-2}. There is also a solar–cycle variation, as can be seen in Fig. 5.2, where the minimum in the solar constant in October 1996 corresponds to the minimum in sunspot number. This arises because although sunspots reduce the solar luminosity, they are surrounded by bright regions called *faculae* that increase in number along with sunspots during solar maxima. The overall radiation balance is dominated by the faculae. The variation in solar luminosity over the solar cycle, is about 0.1%, which according to simple climate models would correspond to a < 0.1 K variation in the Earth's mean global surface temperature, a quantity too small to affect climate. However, ultraviolet and energetic–particle flux variations are much larger and play a significant role in upper–atmospheric chemistry.

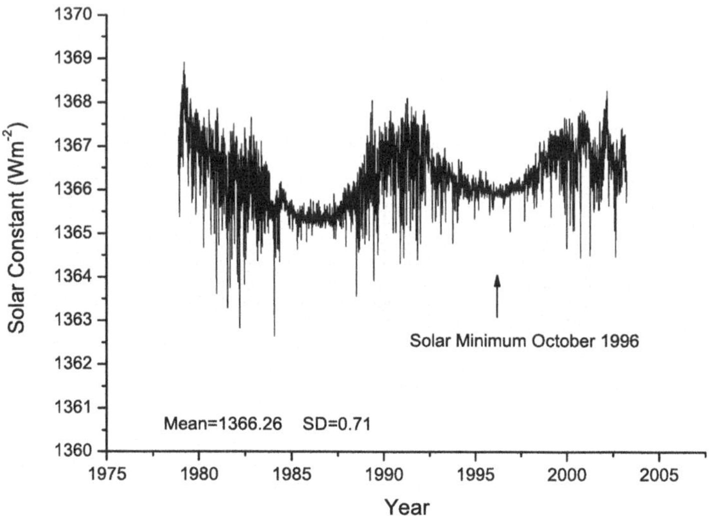

FIG. 5.2. Total solar irradiance (TSI), or solar constant, variation between 1978 and 2003 AD, based on ACRIM Composite data.

5.2.4 *Solar spectral irradiance*

The solar spectral irradiance, $S_{\odot\lambda}$, reaching the Earth is shown in Fig. 5.3 and comprises mostly a Planck function distribution on which are superimposed emission lines that are generated in the solar chromosphere. The most important of these is the Lyman-α of H, which corresponds to the transition between the ground and first excited state, as shown in Fig. 5.4. Transitions originating from the second level are known as the Balmer series, while those originating from the third level are called the Paschen series. For the H atom, the energy difference between any two levels is given by

$$h\nu_{ij} = h\mathrm{R_H}c\left(\frac{1}{n_i^2} - \frac{1}{n_j^2}\right), \tag{5.10}$$

where here $\mathrm{R_H}$ is the Rydberg constant for atomic hydrogen given by

$$\mathrm{R_H} = \frac{2\pi^2 e^4 \bar{m}}{ch^3} \tag{5.11}$$

where e is the electron charge and $\bar{m}$ the reduced mass of the electron–proton system of H, so that $h\mathrm{R_H}c$ is equal to 13.60 eV. Thus, if we set $n_i = 1$ and $n_j = \infty$ then the energy difference is the ionization limit that is equal to 13.60 eV and corresponds to radiation at the wavelength 912 Å. Generally, the ionization

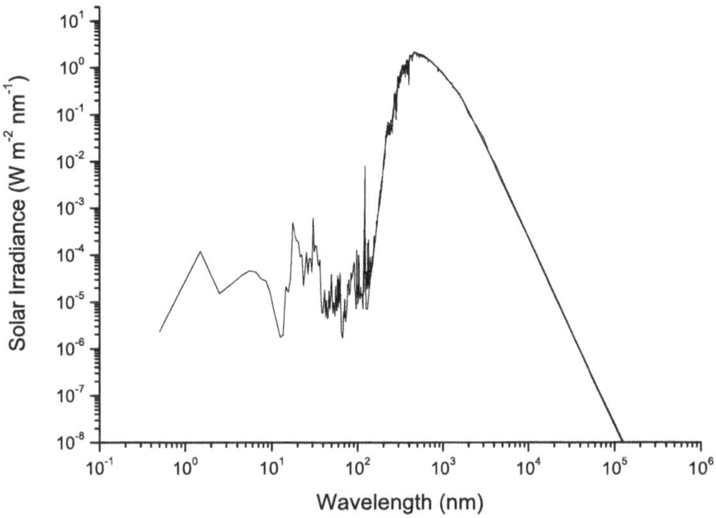

FIG. 5.3. The mean solar spectral irradiance, $S_{\odot\lambda}$ above the atmosphere of the Earth, corresponding to a solar constant 1366 W m^{-2}. (Data from Gueymard 2004)

limit from the n level is equal to $h\mathrm{R_H}c/n^2$. In Table 5.4 are shown the first three transitions and the ionization limit for the Lyman, Balmer and Paschen series.

In thermodynamic equilibrium, we can use the Boltzmann distribution to calculate the ratio of the population of H atoms in the second level and that in the ground state to obtain an idea of the strength of the Lyman–α flux at any given temperature from

$$\frac{n_2}{n_1} = \frac{g_2}{g_1}\mathrm{e}^{-h\nu_{12}/kT}, \tag{5.12}$$

where the statistical weight of each level is given by $(2s+1)(2l+1)$, where $s = \frac{1}{2}$ corresponds to the electron spin quantum number and l the orbital quantum number. For the ground state, $n = 1, l = 0$ and so $g_1 = 2$, whilst for the first excited state $n = 2, l = 0$ or 1, and hence $g_2 = 8$. For the hydrogen atom $g = 2n^2$. Thus, at $T = 6000$ K, $n_2/n_1 = 10^{-8}$, while at $T = 20000$ K the value of the ratio is 0.01. Hence, in the solar chromosphere there is significant Lyman–α emission. As we shall see when we look at the photochemistry of the atmosphere, solar Lyman–α radiation plays a significant role in the dissociation of water vapour and oxygen molecules in the Earth's mesosphere, while Lyman–β plays an important role in the ionization of nitrogen and oxygen in the Earth's thermosphere. In Fig. 5.5 is shown the Lyman–α line profile normalized to a quiet Sun value for the total flux of 3.5×10^{11} photons cm^{-2} s^{-1}.

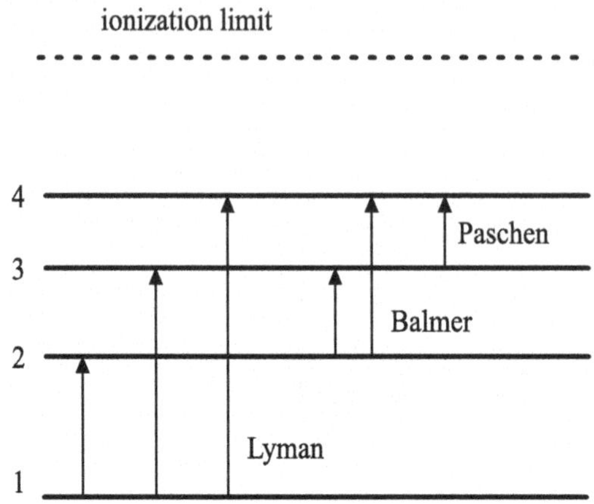

FIG. 5.4. Spectral line series of the H atom.

Table 5.4 *The first three transitions of Lyman, Balmer and Paschen series of the H atom and their ionization limits.*

Series	Transition	Energy (Å)	Label
Lyman	1→ 2	1216	α
(ultraviolet)	1→ 3	1026	β
	1→ 4	972	γ
	1→ ∞	912	limit
Balmer	2→ 3	6563	Hα
(visible)	2→ 4	4861	Hβ
	2→ 5	4340	Hγ
	2→ ∞	3646	limit
Paschen	3→ 4	18751	P4
(infra–red)	3→ 5	12818	P5
	3→ 6	10938	P6
	3→ ∞	8204	limit

5.2.4.1 *Ultraviolet flux variations* We have seen that sunspots produce small variations in solar luminosity, however, the concomitant magnetic activity can produce large variations in the ultraviolet flux. For the present Sun, the solar cycle minimum to maximum variation in the ultraviolet flux above 200 nm is small, typically about 10%, while the variation at Lyman–α can be a factor of about 2. The quiet–sun Lyman–α flux is about 3.0 $\pm 0.1\times 10^{11}$ photons cm^{-2} s^{-1}, while measurements over cycles 21 and 22 give a solar maximum value of 6.75 $\pm 0.25\times 10^{11}$ photons cm^{-2} s^{-1}.

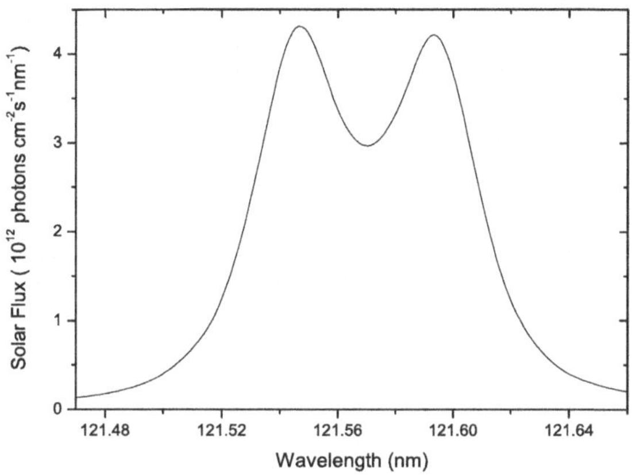

FIG. 5.5. The solar Lyman–α irradiance above the atmosphere, averaged for quiet Sun conditions for solar cycle 23. (Data from Lemaire *et al.* 2005)

5.3 Solar evolution

5.3.1 *Protostar to main sequence*

The Sun started its life as a cloud of very tenuous hydrogen gas that contracted to a protostar in about one million years (Myr). As it contracted its interior temperature increased until its surface temperature rose to over 1000 K and it became an infra–red emitter. A possible next phase is what is called the T–Tauri star stage with an effective temperature of several thousand degrees and a radius of several times the present value, as shown in Table 5.5. The T–Tauri phase of a star is relatively short and the key characteristic of such stars is their *ultraviolet flux excess*, with Lyman–α line flux being typically about 10^4 times the present quiet–Sun value of about 3×10^{11} photons cm^{-2} s^{-1}. T–Tauri stars lose a substantial amount of their mass as a stellar wind. The main sequence (MS) stage is entered when the star's core temperature is high enough ($>10^6$ K) for fusion reactions to occur. This is the most stable phase of solar evolution lasting about 10 Byr. During this phase there is conversion of H to He in the core, which at present extends to about 0.25 $R_\odot$. The primary fusion reaction chain is the proton–proton or *pp* chain whereby four protons are converted to He with the release of energy, Q. Part of this energy, as we shall see later, escapes slowly to the solar surface as radiation. The relatively low–temperature *pp* chain in simplified form can be represented by the following

$$4^1_1\mathrm{H} =^4_2 \mathrm{He} + 2\mathrm{e}^+ + 2\nu \quad \mathrm{Q} = 26.731 \text{ MeV}, \tag{5.13}$$

Table 5.5 *Typical properties of the stars representing possible stages in solar evolution.*

Property	Protostar	T-Tauri	Main sequence	Red giant	White dwarf
$R/R_\odot$	10^6	3–8	1	100	0.01
$L/L_\odot$	10	5	1	1000	0.2
T_{eff}	2000 K	3000 K	6000 K	3000 K	40000 K
Life	1 Myr	10 Myr	11 Byr	500 Myr	1 Byr

where ν represents a neutrino and e^+ a positron. The atomic mass of ${}^1_1$H is 1.00797 amu (atomic mass units) while for ${}^4_2$He it is 4.00260 amu, thus we have a mass defect of $0.029m_p$, where m_p is the proton mass 1.00727 amu. This is equivalent to an energy loss of $0.029m_pc^2$, where c is the speed of light. Thus, for each proton converted we have $0.0073m_pc^2$ energy produced. If, as an approximation, we assume that the Sun's luminosity has not changed (§5.4.2) during its MS life, then the amount of energy released due to the conversion of H to He is $L_\odot t_\odot$, which comes to about 56×10^{50} erg. This is equivalent to a conversion of about 5% $M_\odot$ from H to He and a mass loss of only 0.0073×0.05 $M_\odot$ or $0.00037M_\odot$. Thus, the change in the solar mass due to the conversion of H to He is very small.

5.3.2 *Beyond the main sequence*

When all the hydrogen is converted to helium in the solar core, the Sun will move from the main sequence, and sequentially become a red giant then a white dwarf with a planetary nebula, as the fusion processes in the solar core change. When hydrogen in the core is depleted, the core contracts until hydrogen fusion commences in a shell surrounding the core. This heats the atmosphere that then expands until the sun becomes a red giant, (Table 5.5) with a larger radius but cooler surface, hence the maximum in the blackbody emission moves towards the red. After the hydrogen in the shell has been converted to helium, the core contracts again until the temperature of the core rises to the point where its electron gas pressure becomes independent of temperature (the electrons are said to be degenerate). This leads to further contraction and temperature rise without a rise in pressure to balance the core collapse until the core temperature reaches 100 million K when the helium has enough kinetic energy to fuse to carbon and oxygen. This is the *triple–alpha* or three–helium–nuclei process that is highly temperature dependent. This fusion spreads rapidly within the core in minutes, generating what is termed the *helium flash*. The electron pressure in the core becomes once again temperature dependent at higher temperatures and contraction stops. When the helium in the core is depleted, the core contracts again until helium fusion takes place in a shell surrounding the core. The atmosphere expands once again to an even larger giant star with a very tenuous atmosphere with a strong loss of mass due to an enhanced stellar wind. As the atmosphere loses mass, expansion cooling and contraction heating become more vigorous because of the temperature sensitivity of the triple–alpha process, and the sun then becomes a pulsating star. The instability of these pulsations as mass is lost eventually results in the ejection of the atmosphere to form a planetary nebula leaving a planet–size carbon core or white dwarf that is mainly emitting in the ultraviolet because of its high effective temperature.

5.4 Solar luminosity evolution

5.4.1 *The role of mean molecular weight*

For the Sun we can assign an effective blackbody temperature, $T_\odot$, so that its luminosity is given by

$$L_\odot = 4\pi R_\odot^2 \sigma T_\odot^4, \tag{5.14}$$

where $R_\odot$ is its radius. For a dwarf star like the Sun its luminosity during its MS evolution varies approximately as

$$L \propto \mu^{7.5} M^{5.5} R^{-0.5}, \tag{5.15}$$

where μ is the solar mean relative molecular weight with respect to the proton mass that, as we shall see, controls the luminosity on timescales of billions of years. The mean molecular weight needs to be calculated with some care as it is not simply a weighted mean of the molecular weight of hydrogen and helium, the main solar constituents. In the interior of the Sun these constituents are fully ionized, thus electrons are a component of the particles that determine the mean molecular weight.

The density, ρ, of the solar plasma can be written as

$$\rho = nm = n\mu m_{\mathrm{p}}, \tag{5.16}$$

where n is the number density of the particles and m is the mean mass of the particles, so that the relative mass with regard to the proton mass can be written as

$$\mu = \frac{m}{m_{\mathrm{p}}}, \tag{5.17}$$

where m_{p} is the mass of the proton. The hydrogen density can be written as

$$\rho_{\mathrm{H}} = X_{\mathrm{H}}\rho = n_{\mathrm{H}} m_{\mathrm{H}} \tag{5.18}$$

and similarly for ρ_{He}, with $X_{\mathrm{H}} + X_{\mathrm{He}} = 1$, (ignoring other minor elements) and X represents the fraction of each constituent by mass. The electron number density is then

$$n_{\mathrm{e}} = n_{\mathrm{H}} + 2n_{\mathrm{He}} \tag{5.19}$$

since atomic hydrogen has one electron and helium has two. Thus the total particle number density is

$$n = 2n_{\rm H} + 3n_{\rm He} \tag{5.20}$$

which then can be expressed as

$$n = 2(X_{\rm H}\rho/m_{\rm p}) + 3(X_{\rm He}\rho/4m_{\rm p}) \tag{5.21}$$

on replacing ρ using eqn (5.16), we get

$$\mu = \frac{4}{5X_{\rm H} + 3}. \tag{5.22}$$

5.4.2 *The faint–young–Sun paradox*

If we take a value $X_{\rm H} = 0.80$ for the young Sun, which is an estimate of the initial solar nebula composition based on the H composition by mass of the large planets, Jupiter and Saturn, then $\mu = 0.57$ when the Sun began to convert H to He. If we adopt a present–day value of $X_{\rm H} = 0.734$ for the H composition by mass (Table 5.3), corresponding to $\mu = 0.60$, then we see that the luminosity of the young Sun would have been about 72% that of today's value, assuming that the mass was the same and the radius was about 0.9 of its present value (see next section). This of course has ramifications for the Earth's early climate. Simple climate models give a 1 K change in the global mean surface temperature of the Earth for a 1 % change in the solar luminosity. This implies that the Earth's mean temperature would have been 28 K cooler than today. If we take a present global mean surface temperature of 288 K then the early Earth would have been frozen as its mean surface temperature would have been 260 K. However, there is geological evidence (sedimentary rocks) dating back to 4 Byr ago, suggesting the presence of water, and hence we have what is known as the *faint-young-Sun paradox.* We note that the age of the Earth is estimated to be about 4.5 Byr, and so it took about 100 Myr for the Earth to reach its present mass. The paradox is usually resolved by a CO_2–richer primitive atmosphere than today's (see §12.2).

5.5 Solar ultraviolet flux evolution

5.5.1 *Stellar activity and rotation*

It is well known that stellar chromospheric activity, as measured by the strength of emissions lines such as Ca II HK (3934 Å), Mg II (2800 Å), C IV (1550 Å), N V (1240 Å) and Lyα (1216 Å), decreases with main-sequence age, t. Magnetic activity, which controls chromospheric, transition region and coronal activity, is driven by the interaction of stellar rotation, internal differential rotation and convection (the dynamo action). The primary indicator of MS activity is thus the stellar rotation rate, which varies with MS age essentially as $t^{-1/2}$. In addition, the depth of a star's convection zone is thought to play a role in the dynamo regeneration of magnetic fields. The deeper the convection zone the larger are the

Table 5.6 *Stellar parameters for solar–like G–type stars based on the work of Guinan and Ribas (2002); and Ribas et al. (2005).*

Star	Spectral type	Age (Byr)	P (d)	Age indicator
EK Dra	G0 V	0.1	2.68	Local Association and Li
π^1 UMa	G1.5 V	0.30	4.90	UMa group
χ^1 Ori	G1 V	0.30	5.24	UMa group
9 Cet	G3 V	0.65	7.6	Hyades stream
κ^1 Cet	G5 V	0.65	9.21	P_{rot}-age & L_x
β Com	G0 V	1.6	12.0	P_{rot}-age rel.
15 Sge	G1 V	1.9	13.5	P_{rot}-age rel.
Sun	G2 V	4.6	25.4	Isotopic dating
18 Sco	G2 V	4.9	23.0	Isochrones
β Hyi	G2 IV	6.7	~28	Isochrones
16 Cyg A	G1.5 V	8.5	~35	Isochrones

convection cells and hence what is called the *convective overturn time*, τ_c. Thus both the stellar rotation period, P (days), and depth of the convection zone determine stellar chromospheric activity, and hence ultraviolet flux. Stars less massive than the Sun rotate more slowly at a given MS age, while higher–mass stars rotate faster. The cooler MS stars (increasing B–V) have deeper convection zones so that the convective overturn time increases with B–V. Further, the higher the *metallicity* (proportion of elements heavier than He), the deeper is the convection zone due to increased opacity by these elements. Chromospheric and transition-region activity, which determines radiation emission from Ca HK to Lyα and XUV, has been correlated with the *Rossby number*, $\Re$. This parameter is a measure of the rate of fluid flow (convection speed) against rotational speed and is given by

$$\Re = P/\tau_c. \tag{5.23}$$

We shall give simple expressions to derive $\Re$ also from spectral type (B–V) and age. The stellar Rossby number and effective temperature ($T_{\rm eff}$) can be used to calculate stellar surface XUV and Lyα emission flux from G–type solar–like stars, relative to present solar values. Furthermore, given the spectral type, effective temperature and radius with age, one can calculate the variation of the ultraviolet flux emission from such stars and hence, through photochemical models (see Chapter 7), its effects on the evolution of atmospheres of planets orbiting such stars. The above are also directly applicable to solar ultraviolet flux evolution, and hence to the evolution of the Earth's atmosphere from the Precambrian era (see Chapter 12). We shall give the evolution of the solar XUV and Lyα emission flux with age from 0.1 to 8.5 Byr, based on the model of Girardi *et al.* (2000), for a 1 solar mass star with solar metallicity.

5.5.2 *Rotation with spectral type and age*

For solar–like stars, Ayres (1997) found a rotation period–age relation $P \propto t^{0.6}$, while Barry (1988) found for G stars a spin-down dependence on spectral type according to $P \propto \mathrm{e}^{-A}$, where the parameter A is a function of B–V. Here, we shall combine the rotation–age relation with the rotation period–spectral type dependence to obtain the age evolution of the rotational period for each spectral type according to

$$P(t, \mathrm{B}-\mathrm{V}) = P_{\odot}\mathrm{e}^{-(A-A_{\odot})}\left(\frac{t}{t_{\odot}}\right)^{m}, \tag{5.24}$$

where $P_{\odot}$ is the present solar rotational period of 26.09 days (Messina *et al.* 2003) and the MS age t is measured in Byr. The above expression has been normalized to the present solar rotation period with a present solar age $t_{\odot} = 4.56$ Byr, with $A_{\odot} = 0.77$ corresponding to $(\mathrm{B}-\mathrm{V})_{\odot} = 0.65$ based on the work of Friel (1993). The problem of the solar B–V has been discussed by Taylor (1994) who proposed a value of 0.63, while the solar evolution models of Girardi *et al.* (2000) give a value of 0.67 for the present Sun. The parameter A is tabulated by Barry (1988) but it can be approximately computed from the expression

$$\log(A/A_{\odot}) = 1.48 - 2.28(\mathrm{B}-\mathrm{V}). \tag{5.25}$$

Thus given B–V and age, the rotation period can be calculated from eqn (5.24).

In Table 5.6 are given the rotation periods and estimated ages, for stars with ages ranging from 0.1 to 8.5 Byr, from the *Sun in Time* programme using different age–indicator techniques. The rotation period based on eqn (5.24) was fitted to the observations with a value of $m = 0.52 \pm 0.02$, with the standard error at the 95% confidence level and normalized to the present solar value. The computed rotation period is compared with that based on observations in Fig. 5.6. Also shown is the power–law fit $P = P_{\odot}(t/t_{\odot})^{n}$, where $n = 0.60 \pm 0.04$, in agreement with the rotation period–age relation.

5.5.3 *Rossby number with spectral type and age*

The convective overturn time, τ_{c}, for G, K and M stars can be computed from the expression of Noyes *et al.* (1984)

$$\log(\tau_{\mathrm{c}}) = 1.362 - 0.166x + 0.025x^{2} - 5.323x^{3} \quad x > 0, \tag{5.26}$$

$$= 1.362 - 0.14x \quad x \leq 0, \tag{5.27}$$

where $x = 1 - (\mathrm{B}-\mathrm{V})$. Thus, given the stellar rotation period and B–V the Rossby number can be computed from eqn (5.23). The stellar B–V (available from the SIMBAD database), effective temperature, radius, convective overturn time and Rossby number are given in Table 5.7. The variation of the Rossby

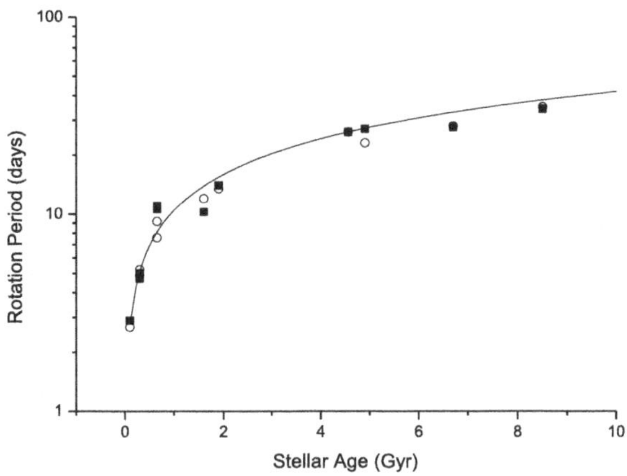

FIG. 5.6. Stellar rotation period versus age for solar–like stars. Empty circles are based on observations while full squares are computed values using eqn (5.24). The solid line is the power–law curve $P = P_{\odot}(t/t_{\odot})^{0.6}$. (Vardavas 2005)

number with stellar age is shown in Fig. 5.7. We note that the Rossby number of 18 Sco does not follow the trend of the others and its rotation period seems low for its estimated age. One can also compute the Rossby number from the stellar age and B–V from eqn (5.24) according to

$$\Re = P_{\odot} \frac{e^{-(A-A_{\odot})}}{\tau_c} \left(\frac{t}{t_{\odot}} \right)^{0.52}, \tag{5.28}$$

as shown in Fig. 5.7. This is useful for studies of the evolution of planetary atmospheres given the evolution of stellar B–V with age. Furthermore, a simple power law fits the present data of form

$$\Re = \Re_{\odot} \left(\frac{t}{t_{\odot}} \right)^{k}, \tag{5.29}$$

where $k = 0.51 \pm 0.03$.

5.5.4 *XUV–Lyα emission and Rossby number*

A parameter that is commonly used to give a measure of stellar chromospheric activity is denoted by $R'_{\rm HK}$. It is defined as the ratio of the emission flux, $F_{\rm HK}$,

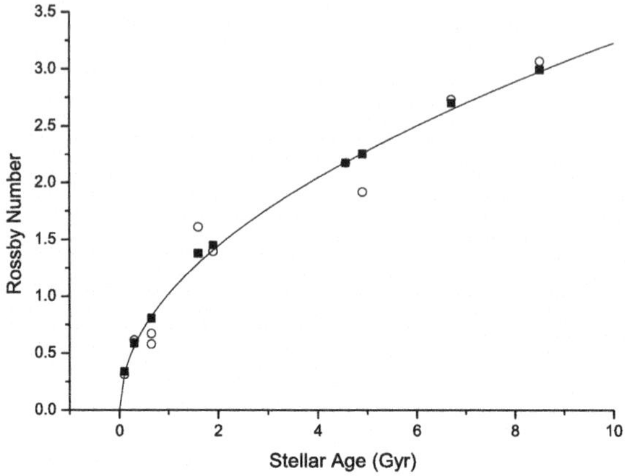

FIG. 5.7. Stellar Rossby number computed using eqn (5.23) (empty circles) versus stellar age. The solid line is the power–law fit $\Re_\odot(t/t_\odot)^{0.5}$. Solid squares are values calculated using eqn (5.28). (Vardavas 2005)

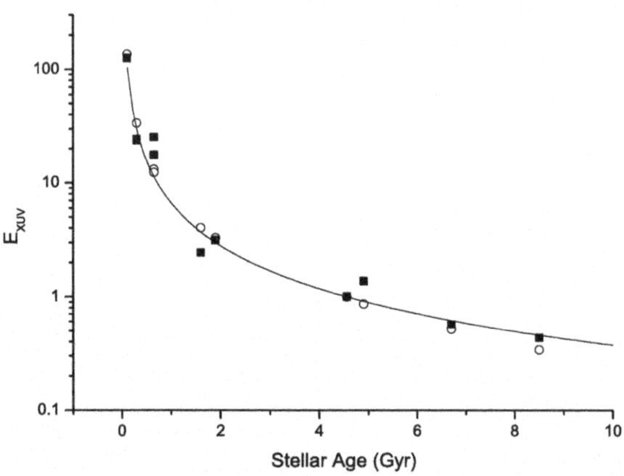

FIG. 5.8. XUV(1–1200 Å) stellar surface emission flux ratios versus age for solar–like stars. Empty circles are based on measurements, while full squares are computed values using the stellar Rossby number and effective temperature in eqn (5.33). The solid line is a $(t/t_\odot)^{-1.25}$ power–law fit. (Vardavas 2005)

Table 5.7 *Stellar B–V, effective temperature, radius, convective–overturn time and Rossby number for solar-like G–type stars.*

Star	B–V	$T_{\rm eff}$	$R/R_\odot$	$\tau_{\rm c}$	$\Re$
EK Dra	0.59	5870	0.94	8.54	0.31
π^1 UMa	0.58	5850	0.97	7.99	0.61
χ^1 Ori	0.59	5890	1.00	8.54	0.61
9 Cet	0.67	5740	1.00	13.14	0.58
κ^1 Cet	0.68	5750	0.93	13.71	0.67
β Com	0.57	6000	1.08	7.45	1.61
15 Sge	0.61	5850	1.09	9.67	1.40
Sun	0.65	5777	1.00	11.99	2.18
18 Sco	0.65	5785	1.02	11.99	1.92
β Hyi	0.62	5774	1.89	10.24	2.73
16 Cyg A	0.64	5790	1.27	11.40	3.07

Table 5.8 *XUV and Lyα emission–flux ratios at 1 AU and scaled to the solar radius at the star's age (proxy solar flux ratios). Also given are the ratios of the corresponding fluxes at the stellar and solar surface.*

Star	XUV (1 AU)	XUV (surface)	Lyα (1 AU)	Lyα (surface)
EK Dra	110.7	135.8	14.6	17.9
π^1 UMa	27.9	33.9	6.8	8.3
χ^1 Ori	27.9	33.9	6.8	8.3
9 Cet	10.3	12.4	4.4	5.3
κ^1 Cet	11.0	13.2	4.7	5.6
β Com	3.5	4.0	2.2	2.5
15 Sge	2.9	3.3	1.9	2.2
Sun	1	1	1	1
18 Sco	0.90	0.87	0.96	0.92
β Hyi	0.63	0.52	0.76	0.63
16 Cyg A	0.50	0.34	0.63	0.43

from the cores of the Ca II H and K lines to the total bolometric emission ($\sigma T_{\rm eff}^4$) of the star

$$R'_{\rm HK} = \frac{F_{\rm HK}}{\sigma T_{\rm eff}^4}, \tag{5.30}$$

and related to the Rossby number by an expression of the form

$$R'_{\rm HK} = c e^{-\alpha \Re}. \tag{5.31}$$

We shall choose the following simpler power–law dependence on Rossby number for the XUV (1–1200 Å) and Lyα activity, defined as the ratio of the total emission flux within these spectral regions to the total bolometric emission of the star

$$R'_{\rm XUV,Ly\alpha} = \Re^{-\beta}. \tag{5.32}$$

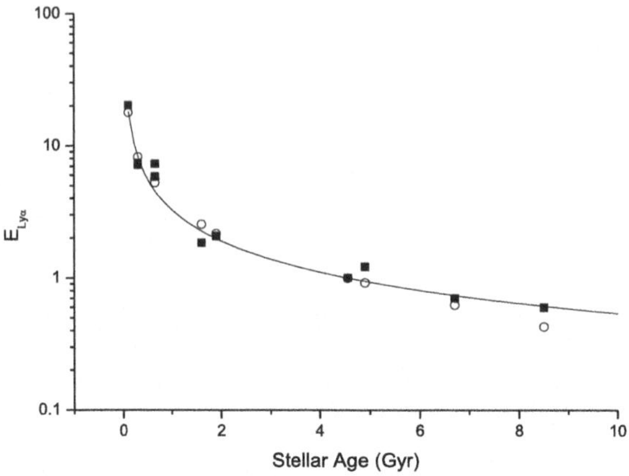

FIG. 5.9. Lyα stellar surface emission flux ratios versus age for solar–like stars. Empty circles are based on measurements while full squares are computed values using the stellar Rossby number and effective temperature in eqn (5.33). The solid line is a $(t/t_{\odot})^{-0.78}$ power–law fit. (Vardavas 2005)

The ratio of the stellar surface (in fact chromospheric layers and coronal layers close to the star) emission flux ($F_{\rm S}$ in erg cm^{-2} s^{-1}) relative to the corresponding present solar mean value ($F_{\rm S\odot}$), defined by $E = F_{\rm S}/F_{\rm S\odot}$, can then be calculated for the XUV and Lyα from

$$E = \left(\frac{\Re_{\odot}}{\Re}\right)^{\beta} \left(\frac{T_{\rm eff}}{T_{\odot}}\right)^{4}. \tag{5.33}$$

From observations by the ASCA, ROSAT EUVE, FUSE and IUE satellites, Lammer et al. (2004) derived, for the stars of Table 5.6, stellar/solar XUV (1–1000Å) emission flux ratios, while Ribas et al. (2005) derived improved XUV (1–1200Å) flux ratios for six stars (EK Dra, π^1 UMa, χ^1 Ori, k^1 Cet, β Com and β Hyi) at 1 AU and scaled to the expected solar radius at the star's age based on the evolution of a 1 solar mass star with solar metallicity. Similarly, from high-resolution spectroscopic observations by the Hubble Space Telescope, Lyα emission flux ratios were derived for the stars of Table 5.6. The measured XUV and Lyα flux ratios at 1 AU are given in Table 5.8, taking into account the improved values for the stars mentioned above. These scaled ratios ($F/F_{\odot}$) at 1 AU are related to the corresponding ratios based on fluxes at the stellar and solar surface via

$$\frac{F}{F_{\odot}} = \frac{F_{\rm S}}{F_{\rm S\odot}} \left(\frac{R_{\odot}(t)}{R_{\odot}}\right)^{2}, \tag{5.34}$$

where $R_\odot(t)$ is the expected solar radius at the star's age based on the evolution of a 1 solar mass star with solar metallicity. In Figures 5.8 and 5.9 the emission flux ratios at the stellar surface are compared with those derived from the stellar Rossby number and effective temperature, using eqn (5.33) and data from Table 5.7, with $\beta = 2.46 \pm 0.12$ for the XUV and $\beta = 1.52 \pm 0.07$ for the Lyα.

If we insert the power–law expression, eqn (5.29), for $\Re$ in eqn (5.33) with the appropriate values for β we obtain a one-parameter fit to the flux ratio values at the stellar surface

$$E_{\mathrm{XUV}} = \left(\frac{t}{t_\odot}\right)^{-1.25} \left(\frac{T_{\mathrm{eff}}}{T_\odot}\right)^4, \tag{5.35}$$

and

$$E_{\mathrm{Ly}\alpha} = \left(\frac{t}{t_\odot}\right)^{-0.78} \left(\frac{T_{\mathrm{eff}}}{T_\odot}\right)^4. \tag{5.36}$$

The variation of the flux ratios is close to those of Ribas et al. who obtained power–law fits of $3.10t^{-0.72}$ and $6.40t^{-1.23}$ for the scaled Lyα and XUV (1–1200 Å) ratios, respectively, at 1 AU. Similarly, Wood *et al.* (2005) found for G V and F V stars a power–law fit of Lyα flux at the stellar surface $\propto P^{-1.09 \pm 0.08}$, which when combined with the age–rotation relationship $P \propto t^{0.6}$ gives Lyα flux $\propto t^{-0.65 \pm 0.12}$.

5.5.5 *Solar XUV and Ly–α emission flux evolution*

The method can be applied to calculate the evolution of the solar XUV and Ly–α emission flux with age. In Fig. 5.10 is given the evolution with age of the solar B–V, radius and effective temperature, based on the model for a 1 solar mass star and metallicity ($Z_\odot = 0.019$) of the sun. The evolution of stellar B–V with age can be used with eqn (5.24) and eqn (5.28) to calculate the evolution of the rotation period and Rossby number with age, respectively. The variation with stellar age of the XUV and Ly–α emission flux at a planetary distance of r measured in AU from a star, relative to present–day mean solar values at 1 AU, can be calculated from

$$\frac{F(r,t)}{F_\odot} = \left(\frac{\Re_\odot}{\Re}\right)^\beta \left(\frac{T_{\mathrm{eff}}}{T_\odot}\right)^4 \left(\frac{R(t)}{rR_\odot}\right)^2. \tag{5.37}$$

Solar and proxy solar (stars of Table 5.6) results for $r = 1$ AU are shown in Figures 5.11 and 5.12. At an age of 100 Myr the solar XUV and Lyα emission flux ratios are 104.0 and 15.6, decreasing to 5.7 and 2.7, respectively, at an age of 1 Byr (3.56 Byr ago). However, beyond the present age, for the next several billion years the solar emission flux ratios decrease very slowly up to 8 Byr. Although the Rossby number increases over this time interval its effect is compensated by a corresponding increase in solar radius so that the ultraviolet

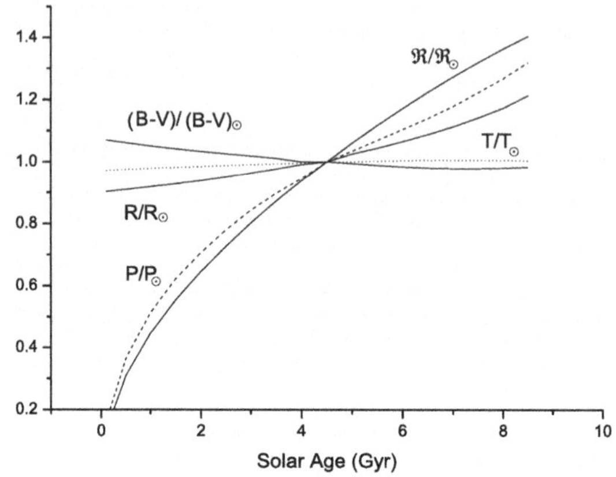

FIG. 5.10. Evolution of solar radius, effective temperature and B–V with age, based on the model of Girardi *et al.* (2000) for a star with 1 solar mass and solar metallicity. Also shown is the evolution of the rotation period, P from eqn (5.24), and Rossby number, $\mathfrak{R}$ from eqn (5.23). (Vardavas 2005)

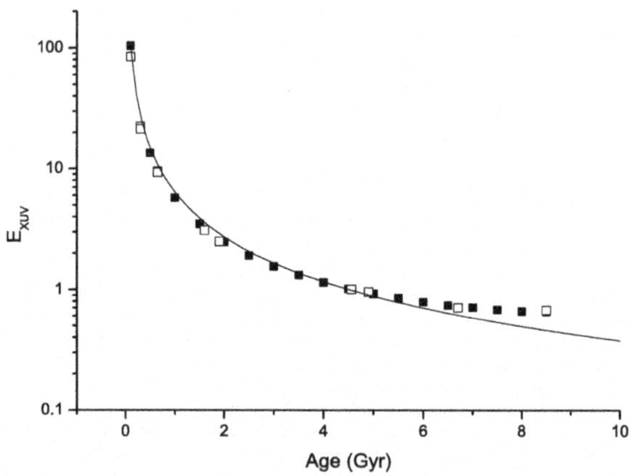

FIG. 5.11. Evolution of XUV(1–1200 Å) emission flux ratios with age at 1 AU using eqn (5.37) for the sun (full squares) and proxy solar (empty squares) *Sun in Time Programme* stars (Table 5.6), with radius scaled to the solar value at the age of the star. The solid line is the power–law fit $6.4t^{-1.23}$. (Vardavas 2005)

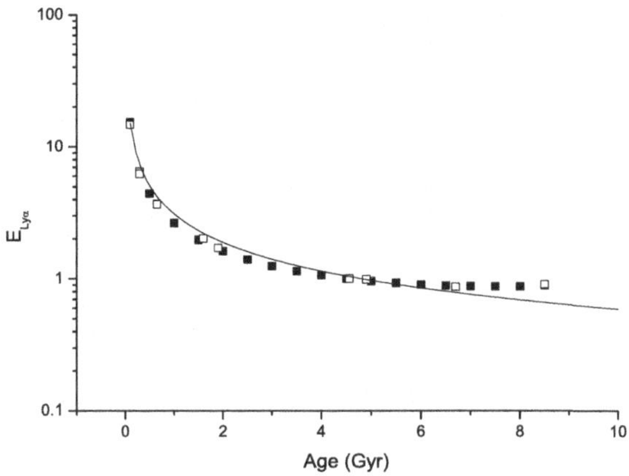

FIG. 5.12. Evolution of Ly–α emission flux ratios with age at 1 AU using eqn (5.37) for the sun (full squares) and proxy solar (empty squares) *Sun in Time Programme* stars (Table 5.6) with radius scaled to the solar value at the age of the star. The solid line is the power–law fit $3.1t^{-0.72}$. (Vardavas 2005)

emission flux ratios decrease very slowly. Finally, we note that for a star like β Hyi whose actual radius is $1.89R_\odot$ at age 6.7 Byr, the emission flux ratios at 1 AU would be about 3 times those of the corresponding proxy solar star at the same age with radius of $1.1R_\odot$ (Fig. 5.10).

The stellar Rossby number increases with stellar age essentially according to $t^{0.5}$ and the period of rotation according to $t^{0.6}$, for G–type solar-like stars. Given the stellar rotation period and B–V we can compute the Rossby number and knowing the effective temperature compute the stellar surface XUV and Ly–α emission flux relative to the present sun. For stellar evolution studies, we can use the variation of the stellar B-V to derive the Rossby number with age. From the evolution of both the effective temperature and the stellar radius we can calculate the stellar XUV and Ly–α emission flux ratios with age, at any planetary distance from the star. These emission flux ratios can then be used in photochemical models (see Chapter 7) to examine the evolution of the composition of planetary atmospheres orbiting such stars.

5.5.6 *Solar photospheric irradiance evolution*

From Fig. 5.10, we can see that the effective temperature of the Sun at zero main sequence age was about 0.97 of its present value of 5777 K and that its radius was 0.9 times its present value. These translate to a luminosity that was 0.72 its present value some 4.6 Byr ago. Thus, its effective temperature would

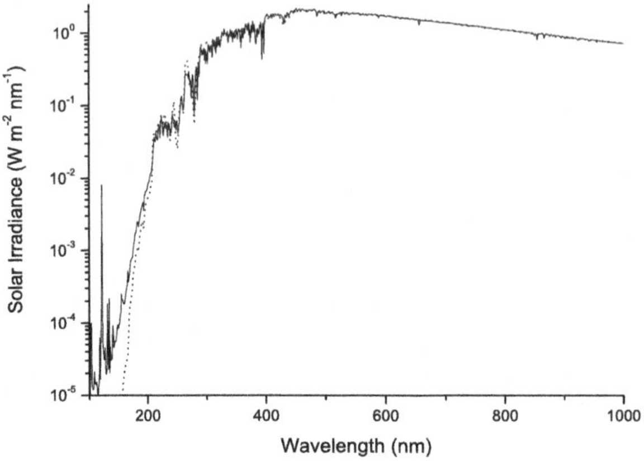

FIG. 5.13. Photospheric irradiance at 1 AU for the Sun from the library of Lejeune *et al.* (1997) is compared with the solar cycle mean solar irradiance of Gueymard (2004).

have been about 5600 K, which according to the Wien displacement law (§3.3.4) and the Planck function would have meant that the spectral irradiance in the visible and near–infra–red would have been slightly lower and its peak shifted from 5016 Å to 5175 Å. In contrast, the ultraviolet and XUV flux would have been appreciably higher, as we saw in the previous section. Further, if we assume that its mass has not changed appreciably (§5.3.1) since 4.6 Byr ago, then its gravity would have been 1.23 its present value with $\log(g) = 4.53$, instead of its present value of 4.44.

Stellar photospheric irradiance F_λ is tabulated in standard stellar libraries (see Buser and Kurucz 1992, Lejeune *et al.* 1997) given the stellar effective temperature, $T_{\rm eff}$, gravity, $\log(g)$ and metallicity, Z of the star. In Fig. 5.13 we compare the photospheric irradiance at 1 AU for the Sun taken from the Lejeune *et al.* (1997) standard stellar library with the solar cycle mean solar irradiance of Gueymard (2004). Down to 200 nm there is good agreement between the two profiles, given that the solar irradiance includes contributions from the chromosphere. The differences can be seen below 200 nm where the ultraviolet flux arises from the chromosphere and corona, and that varies with solar activity. As the Sun aged, its photospheric irradiance increased with time and solar activity decreased. The result was more visible and near–infra–red flux and less ultraviolet flux below 200 nm. The solar ultraviolet flux was probably enhanced 3.5 Byr ago (when life was well established on Earth), by a factor of about 3 at Lyman–α, as shown in Fig. 5.12, with little enhancement at wavelengths greater than 200 nm.

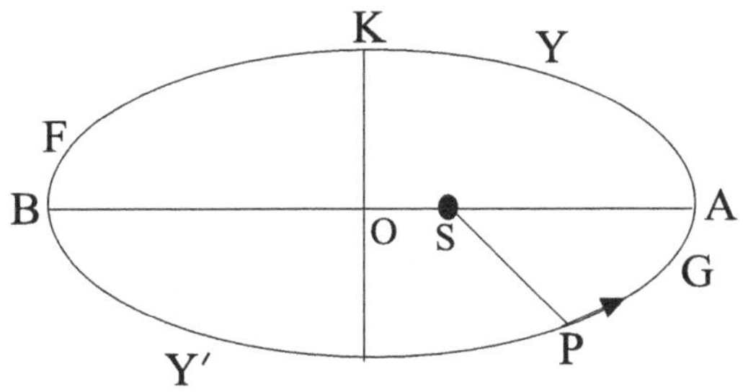

FIG. 5.14. Elements of the Earth's elliptical orbit.

5.6 Solar flux at the Earth's orbit

5.6.1 *The Earth's elliptical orbit*

According to Kepler's first law, the orbit of a planet around the Sun is an ellipse, with the Sun one focus. The Sun–Earth distance, r, thus varies with time as the Earth's orbit around the Sun is elliptical. The elements of this elliptical orbit are shown in Fig. 5.14 where: S is one focus corresponding to the Sun (the second focus is its reflection on the segment BO across the KO plane); A is perihelion; B is aphelion; a = OA is the semimajor axis; b = OK is the semiminor axis; e = OS/OA is the eccentricity. The eccentricity varies over timescales of hundreds of thousands of years with a present mean value of 0.0167 (Astronomical Almanac 1992). If P is the position of the planet then SP is the radius vector, of length equal to r. For an ellipse the following relation holds between a and b

$$b^2 = a^2(1 - e^2). \tag{5.38}$$

Figure 5.14 shows the orbit of the Earth in a heliocentric system. For the Northern Hemisphere, the point Y is the vernal equinox, or spring equinox, and Y′ is its autumnal counterpart. G is the winter solstice and F is the summer solstice.

5.6.2 *The plane of the ecliptic*

We can also represent the apparent orbit of the Sun in a geocentric system as shown in Fig. 5.15. P is the pole of the celestial sphere, while the circle YEY′C is the celestial equator. The plane YGY′F is the *ecliptic plane* that defines the Sun's apparent orbit in an ascending mode from A, now called perigee (corresponding to perihelion for the Earth's orbit), to Y. The ecliptic plane makes an angle ε, corresponding to the tilt in the Earth's axis of rotation, with the celestial equator arc GE, and is known as the *obliquity of the ecliptic* that had a value of $\varepsilon_O = 23.439° = 0.13022\pi$ radians in the year 2000 AD. The obliquity has varied very slowly on timescales of tens of thousands of years and its current trend can be estimated (in radians) from

$$\varepsilon(\text{year}) = \varepsilon_O - 7.22 \times 10^{-7}\pi(\text{year} - 2000), \tag{5.39}$$

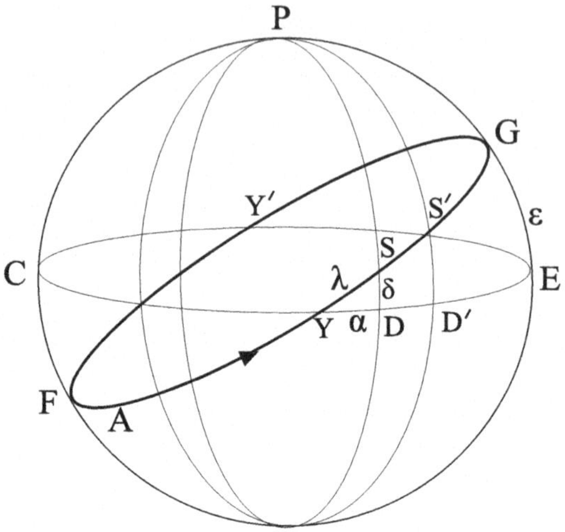

FIG. 5.15. The celestial sphere and the plane of the ecliptic.

where *year* is years AD. The arc SD $= \delta$ is the solar *declination* with a maximum value equal to ε. The arc YD $= \alpha$ is called the *right ascension* of the Sun, and corresponds to the solar longitude as measured on the celestial equator, measured from the vernal equinox where $\alpha = 0$. Thus, α and δ define the coordinates of the Sun, point S, on the celestial sphere. The arc λ is the *ecliptic longitude* that is zero at the vernal equinox and measured in the ascending mode. Thus the vernal equinox, or first point of Aries, Y, is the point of reference for the Sun's apparent motion.

5.6.3 *Sun–Earth distance and solar longitude*

The equation for the Sun–Earth distance is given by

$$r = \frac{a(1 - e^2)}{1 + e \cos v}, \tag{5.40}$$

where v is arc AS, and is measured from perigee and is called the *true anomaly*, given by

$$v = \lambda - \omega, \tag{5.41}$$

and ω is the solar longitude of perigee that is equal to $\varpi + \pi$, where ϖ is the Earth's *longitude of perihelion*, the direction of perihelion (as viewed from the Sun) is opposite to the direction of perigee (as viewed from the Earth). Thus, $v = 0$ at perigee or perihelion, on about 3 January, and $v = \pi$ at apogee or aphelion, about 3 July. Thus, $r_{\mathrm{P}} = a(1 - e)$ at perihelion, shortest Earth–Sun distance, and $r_{\mathrm{a}} = a(1 + e)$ at aphelion, largest Sun–Earth distance, and the mean Sun–Earth distance $\bar{r} = 0.5(r_{\mathrm{P}} + r_{\mathrm{a}}) = a$. In Table 5.9 are given the values

Table 5.9 *The solar longitude and the seasonal cycle.*

Y	$\lambda=0$	vernal equinox	20 March
G	$\lambda=\pi/2$	summer solstice	21 June
B	$\lambda=\varpi$	aphelion	3 July
Y'	$\lambda=\pi$	autumnal equinox	22 September
F	$\lambda=3\pi/2$	winter solstice	21 December
A	$\lambda=\pi+\varpi$	perihelion	3 January

of λ for the various points along the Earth's orbit corresponding to times of the seasonal cycle.

5.6.4 *The equation of time*

The Sun's apparent elliptical motion around the Earth is not uniform, however, according to Kepler's Second Law, the radius vector SP sweeps out equal areas in equal times. An element of area corresponding to an element of angle, $\delta\theta$ swept out on an ellipse can be written as $r(r+\delta r)\delta\theta/2$. Neglecting the $\delta r\delta\theta$ term we have

$$\frac{\mathrm{dArea}}{\mathrm{d}t_{\mathrm{d}}}=\frac{1}{2}r^2\frac{\mathrm{d}\theta}{\mathrm{d}t_{\mathrm{d}}}=\text{constant}. \tag{5.42}$$

Time, t_{d} in days from 1st January as measured by a clock, is based on a fictitious mean Sun moving on the celestial equator at a uniform rate, i.e. circular motion, which takes one year to complete its orbit. Thus, the mean Sun sweeps out an angle of 2π over one clock year. We can thus define a *mean longitude*, L, of the mean Sun in its apparent motion around the Earth on the celestial equator, and neglect the time dependence of r, hence, we can write in terms of the mean solar longitude L

$$\frac{\mathrm{dArea}}{\mathrm{d}t_{\mathrm{d}}}=\frac{1}{2}\overline{r}^2\frac{\mathrm{d}L}{\mathrm{d}t_{\mathrm{d}}}=\text{constant} \tag{5.43}$$

so that the mean Sun sweeps out an area πa^2 in a clock year. Thus

$$\frac{\mathrm{d}L}{\mathrm{d}t_{\mathrm{d}}}=\text{constant} \tag{5.44}$$

and so

$$L(t)=L_{\mathrm{o}}+\frac{2\pi t_{\mathrm{d}}}{P}, \tag{5.45}$$

where P is the period of the Earth's orbit, 365.256 days, and $L_o = 1.5581\pi$ (Astronomical Almanac 1992) at $t_d = 0$ midnight on the 1st January. Thus, as can be seen in Fig. 5.15, the mean Sun would be at some right ascension given by α + arc DD′, instead of at its true right ascension of α. For the rotating Earth at the centre of the celestial sphere, an observer located on the celestial meridian PGE, sees the Sun at an *hour angle* GPS, where the hour angle of the observer is defined by

$$h = (t - 12)\pi/12, \tag{5.46}$$

where now t is the clock time in hours, so that h is equal to $-\pi$ radians at midnight when $t = 0$, zero at noon $t = 12$, and π radians at midnight again. Thus, the mean sun is located at an hour angle of GPS′ instead of at the true hour angle GPS. Thus, we can define an *equation of time*, E, which is the difference between the true hour angle, h, of the Sun and the hour angle, h_m, of the mean Sun

$$E = h - h_{\rm m} = arc({\rm DD'}) = L - \alpha. \tag{5.47}$$

Now, every 15° of hour angle or of mean longitude corresponds to one hour of clock time, so that the correction that needs to be applied to the clock time to obtain the sundial time is $E/15$ h, with E in degrees. The true hour angle is then given by

$$h(t) = (t - 12 + E/15)\pi/12. \tag{5.48}$$

The equation of time varies from about –14 min in February to +16 min in November (Norton's Star Atlas). If a reference observer is at the Greenwich meridian (longitude $\phi = 0$) then the clock time t is called Universal Time (UT) beginning at midnight, and so for another observer located at longitude ϕ in degrees, the local hour angle is given by

$$h(t) = (t - 12 + \phi/15 + E/15)\pi/12. \tag{5.49}$$

The mean Sun anomaly, g, is different to the true Sun anomaly v, and is given by

$$g = L - \omega, \tag{5.50}$$

through a small correction according to

$$v = g + 2e\sin g + 5e^2 \sin 2g/4 \tag{5.51}$$

keeping second–order terms in the eccentricity e. The mean anomaly can be calculated from

$$g = 1.986\pi + 2\pi t_{\rm d}/P, \tag{5.52}$$

where $t_{\rm d} = n-1+t/24$, n is the day of the year with $n = 1$ on January 1, and t is UT in hours. The above expressions give the true anomaly to a precision of 0.01° between 1950 and 2050. We can now calculate the mean anomaly and hence the true anomaly and calculate the Sun–Earth distance r. Thus, at perigee, when $t_{\rm d} = 2.625$, $v = g = 0$, and the solar longitude is 283.05° and $\varpi_\circ = 103.05° = 0.5725\pi$. The Earth's longitude of perihelion varies from 0 to 360° in about 21 000 years so that

$$\varpi = \varpi_o + 9.44 \times 10^{-5}\pi(\text{year} - 2000), \tag{5.53}$$

to a good approximation.

Table 5.10 *Orbital variation of incoming solar flux.*

Day	Date	Time of season	$S(t)/S_\odot$
3	3 January	perihelion	1.03426
80	20 March	vernal equinox	1.00794
173	21 June	summer solstice	0.96820
185	3 July	aphelion	0.96742
266	22 September	autumnal equinox	0.99297
356	21 December	winter solstice	1.03339

5.6.5 *Incoming radiation at TOA*

We shall now derive the equations for computing the incoming solar radiation at the top of the atmosphere (TOA), for any day of the year and latitudinal location.

The solar flux reaching the Earth's orbit varies with the time of year according to

$$S(t) = S_\odot \left(\frac{\bar{r}}{r(t)} \right)^2 , \qquad (5.54)$$

this variation is given for times of the seasonal cycle for 2000 AD in Table 5.10.

The incoming solar flux, $F_\odot^\downarrow$, above the Earth's atmosphere, at a particular location on Earth (given by the longitude, ϕ, and the latitude, θ) and at a particular time (given by the time of day, t in UT, the day of the year, n, and the year), is given by

$$F_\odot^\downarrow(t) = S_\odot \mu \left(\frac{\bar{r}}{r(t)} \right)^2 , \qquad (5.55)$$

with $\mu = \cos z$, z is the solar zenith angle at any time t during the day and can be calculated from

$$\mu(t) = A + B \cos h(t), \qquad (5.56)$$

with

$$A = \sin\theta \sin\delta, \qquad (5.57)$$

$$B = \cos\theta \cos\delta, \qquad (5.58)$$

where δ and h are the Sun's declination and hour angle, in radians, respectively. The solar declination is given by

$$\delta = \arcsin(\sin\varepsilon \sin\lambda), \qquad (5.59)$$

and we note that at the equinoxes it is zero. The true solar longitude can be calculated from

$$\lambda = L + 2e \sin g + (5e^2/4) \sin 2g. \qquad (5.60)$$

To calculate the true hour angle h we need to calculate the equation of time E, eqn (5.47), from L and the solar right ascension α, which can be calculated from

$$\alpha = \lambda - u \sin 2\lambda + (u^2/2) \sin 4\lambda, \tag{5.61}$$

where

$$u = \tan^2(\varepsilon/2). \tag{5.62}$$

We can also calculate the day length or duration of daytime from the times of sunrise and sunset by setting $\mu = 0$. This gives the corresponding hour angles $-H$ and H for sunrise and sunset, where

$$H = \arccos(-\tan\theta \tan\delta). \tag{5.63}$$

The day length, as a fraction of one day, is then $d = 24H/\pi$. Furthermore, we can calculate the mean value μ_n between sunrise and sunset on day n and obtain

$$\mu_n = A_n + \frac{B_n \sin H_n}{H_n}. \tag{5.64}$$

Thus the mean daily solar flux (W m^{-2}) on day n can be calculated from

$$F^{\downarrow}_{\odot n} = S_{\odot}\mu_n d_n \left(\frac{\bar{r}}{r_n}\right)^2. \tag{5.65}$$

We note that at the equinoxes $\delta = 0$ and so $H = \pi/2$ so that $d = 1/2$. In the polar regions it is possible for H to be zero, $d = 0$, i.e. polar night (or winter) that can last up to six months depending on latitude. At this time the other pole has $H = \pi$, $d = 1$ polar day (or summer). The polar night can happen for latitudes that satisfy $\tan\theta = -\cot\delta$ for $\delta \neq 0$. This occurs for $\theta > 90° - \delta = 66.56°$. During the polar day $H = \pi$ and $\mu > 0$ for 24 hours and the sun simply remains above the horizon circling the sky in an upward spiral motion, beginning at the spring equinox, with an elevation δ above the horizon reaching a maximum value of ε at the summer solstice for the Pole ($\theta = 90°$).

5.6.6 *Global distribution of incoming radiation*

5.6.6.1 *Zonal–seasonal variation* The zonal–seasonal variation of the daily (24-h) average in W m^{-2}, mean monthly incoming solar radiation is given in Fig. 5.16. We note that the seasonal variability of ISR increases as we move towards the poles.

5.6.6.2 *Latitudinal variation* The long–term (1984–1997) latitudinal variation of mean annual TOA incoming solar radiation is given in Fig. 5.17 for a solar constant of 1367 W m^{-2}. Mean monthly 10-degree latitude zonal fluxes were computed by averaging along 2.5-degree latitude zones, taking into account the

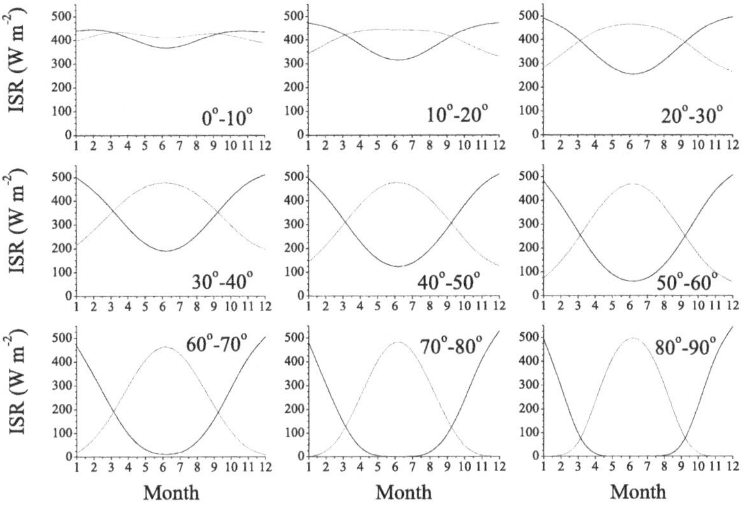

FIG. 5.16. Long-term (1984-1997) zonal-seasonal variation of incoming solar radiation (W m^{-2}) at the top of the atmosphere (TOA) for the Northern Hemisphere (dotted lines) and Southern Hemisphere (solid lines).

fraction of surface area contained in each 2.5-degree zone. The fraction of surface area in each zone can be taken, to a very good approximation, to be $E_j = \sin j\delta\theta - \sin(j-1)\delta\theta$ where $\delta\theta$ is the width of each latitudinal zone, for $j = 1, 2, ..., 36$ zones in the case of 2.5-degree zones spanning from equator to pole, assuming the Earth is a perfect sphere. Subsequently, annual mean quantities can be computed by summing the corresponding monthly means for each 10-degree latitudinal zone over the 12 months of the year.

5.6.6.3 *Seasonal variation* The yearly averaged incoming solar radiation has a maximum value of about 425 W m^{-2} along the equator, decreasing rapidly to about 175 W m^{-2} towards the poles, due to the latitudinal variation of solar zenith angle. However, note that the equator-to-pole decrease in ISR is associated with a corresponding increase in ISR seasonal variability. The mean hemispherical ISR fluxes have opposite seasonality (Fig. 5.18 with values ranging between about 200 and 475 W m^{-2}, resulting in a mean global ISR ranging from 330 to 352 W m^{-2} throughout the year. The ISR seasonal range for the Southern Hemisphere is slightly larger than for the Northern Hemisphere as perihelion occurs in January, corresponding to a slightly larger incoming solar flux during summer in the Southern Hemisphere.

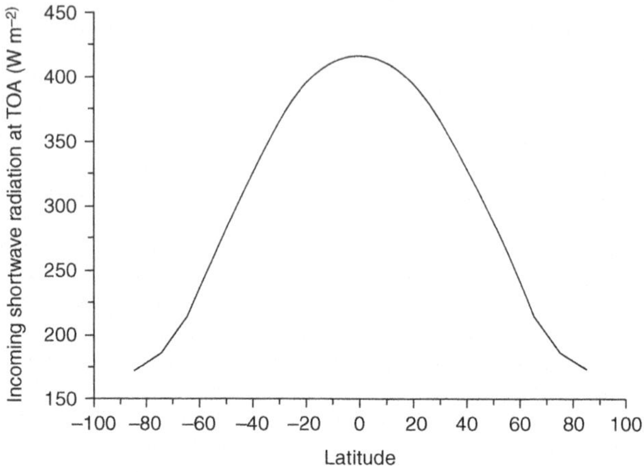

FIG. 5.17. Long–term (1984–1997) mean annual latitudinal variation of computed incoming shortwave radiation (W m^{-2}) at the top of the atmosphere. (Hatzianastassiou *et al.* 2004)

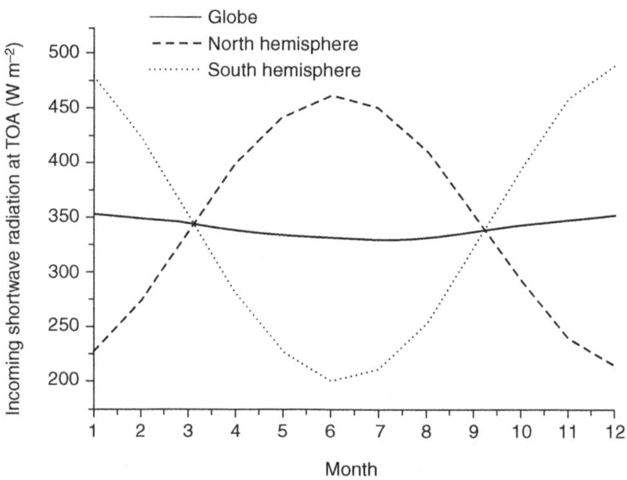

FIG. 5.18. Long–term (1984–1997) latitudinal average seasonal variation of incoming shortwave radiation (W m^{-2}) at the top of the atmosphere. (Hatzianastassiou *et al.* 2004)

5.7 Bibliography

5.7.1 *Notes*

For a more detail description of the *pp* chain see Sturrock; and Tayler.

A more detailed discussion on photon escape from the centre of stars may be found in the book by Böhm–Vitense. For more details on the solar cycle see the early work of Eddy.

For the faint–young–Sun paradox and for conditions on the early Earth see §12.2. The article by Walter and Barry in the *Sun in Time* text by Sonnet *et al.* and the text itself provide details on solar flux evolution; see also Zahnle and Walker, and Chyba and Sagan.

For solar cycle ultraviolet variations see Lean *et al.*; Woods and Rottman; and Rottman et al.

For an alternative theory to that of a fainter young Sun see Sackman and Boothroyd, who propose that the Sun was initially more massive (by 1–7%) so that its luminosity was sufficiently high to maintain warm conditions on the primitive Earth.

On solar rotation and age see Skumanich, for stellar activity and rotation see the early work of Willson and Skumanich, for stellar activity and Rossby number see Noyes et al. and the monograph by Cram and Kuhi and references therein.

For the spectral variation of the solar irradiance see also the profile of Thekaekara and Drummond, which can be renormalized to a solar constant of 1366 W m^{-2}.

A classical text on spherical astronomy is that of Smart.

5.7.2 *References and further reading*

Astronomical Almanac (1992). US Government Printing Office, Washington, D.C.

Ayres, T. R. (1997). Evolution of the solar ionizing flux. *J. Geophys. Res.*, **102**, 1641–1651.

Barry, D. C. (1988). The chromospheric age dependence of the birthrate, composition, motions, and rotation of late F and G dwarfs within 25 parsecs of the Sun. *Astrophys. J.*, **334**, 436–448.

Böhm-Vitense, E. (1992). *Introduction to stellar astrophysics. Stellar structure and evolution.* Cambridge University Press, Cambridge.

Buser, R. and Kurucz, R. L. (1992). A library of theoretical flux spectra I - Synthetic UBVRI photometry and the metallicity scale for F- to K-type stars. *Astron. Astrophys.*, **264**, 557–591.

Chyba, C. and Sagan, C. (1992). Endogenous production, exogenous delivery and impact–shock synthesis of organic molecules: an inventory for the origins of life. *Nature*, **355**, 125–132.

Cram, L. E. and Kuhi, L. V. (1989). *FGK stars and T Tauri stars.* NASA SP-502, Washington DC.

Eddy, J. A. (1977). Historical evidence for the existence of the solar cycle. In *The solar output and its variation.* ed. O. R. White, Colorado Assoc. Univ. Press, Boulder.

Friel E., Cayrel de Strobel G., Chmielewski Y., Spite M., Lèbre A., Bentolila C. (1993). In search of real solar twins. III. *Astron. Astrophys.*, **274**, 825–837.

Gueymard, C. A. (2004). The Suns total and spectral irradiance for solar energy applications and solar radiation models. *Solar Energy*, **76**, 423–453.

Girardi, L., Bressan, A., Bertelli, G. and Chiosi, C. (2000). Evolutionary tracks and isochrones for low- and intermediate-mass stars: From 0.15 to 7 $M_\odot$, and from $Z = 0.0004$ to 0.03. *Astron. Astrophys. Suppl. Ser.*, **141**, 371–383.

Guinan E. F. and Ribas I. (2002). Our changing Sun: The role of solar nuclear evolution and magnetic activity on Earth's atmosphere and climate. In B. Montesinos, A. Giménez, Guinan E.F., ed., ASP Conf. Ser., 269, *The evolving Sun and its influence on planetary environments*, San Francisco.

Hatzianastassiou, N., Fotiadi, A., Matsoukas, C., Drakakis, E., Pavlakis, K. G., Hatzidimitriou and Vardavas, I. M. (2004). Long-term global distribution of Earth's shortwave radiation budget at the top of atmosphere, *Atmos. Chem. Phys.*, **4**, 1217–1235.

Lammer, H., Ribas, I., Grießmeier, J.-M., Penz T., Hanslmeier A. and Biernat, H. K. (2004). A brief history of the solar radiation and particle flux evolution. *Hvar. Obs. Bull.*, **28**, 139–155.

Lean, J. L., Rottman, G. J., Kyle, H. L., Woods, T. N., Hickey, J. R. and Puga, L. (1997). Detection and parameterization of variations in solar mid- and near-ultraviolet radiation (200-400 nm). *J. Geophys. Res.*, **102**, 29939–29956.

Lejeune, Th., Cuisinier, F. and Buser, R. (1997). A standard stellar library for evolutionary synthesis I. Calibration of theoretical spectra. *Astron. Astrophys. Suppl. Ser.*, **125**, 229–246.

Lemaire, P., Emerich, C., Vial, J.-C., Curdt, W., Schüle, U. and Wilhelm, K., (2005). Variation of the full Sun hydrogen Lyman profiles through solar cycle 23. Adv. Space Res., **35**, 384–387.

Messina, S. and Guinan, E. F. (2003). Magnetic activity of six young solar analogues II. Surface differential rotation from long-term photometry. *Astron. Astrophys.*, **409**, 1017–1030.

Montesinos B. and Jordan, C. (1993). On magnetic-fields, stellar coronae and dynamo action in late-type dwarfs. *Mon. Not. R. Astron. Soc.*, **264**, 900–918.

Norton's *Star atlas and reference handbook* (1998). Prentice Hall, London.

Noyes R. W., Hartmann L. W., Baliunas S. L., Duncan D. K. and Vaughan A. H. (1984). Rotation, convection, and magnetic activity in lower main-sequence stars. *Astrophys. J.*, **279**, 763–777.

Ribas, I., Guinan, E. F., Güdel, M. and Audard, M. (2005). Evolution of the solar activity over time and effects on planetary atmospheres: I. High-energy irradiances (1-1700 Å). *Astrophys. J.*, **622**, 680–700.

Rottman, G. L., Floyd, L., Viereck, R. (2004). Measurement of solar ultraviolet irradiance. In *Solar variability and its effect on the Earth's atmosphere and climate system.* ed. J. Pap, and coauthors. AGU Monograph Series, Washington DC.

Sagan, C. and Mullen, C. (1972). Earth and Mars: Evolution of atmospheres and surface temperatures. *Science*, **177**, 52–56.

Sagan, C. and Chyba, C. (1997). The early faint sun paradox: Organic shielding of ultraviolet-labile greenhouse gases. *Science*, **276**, 1217–1221.

Sackmann, I.-J. and Boothroyd, A. I. (2003). Our sun. V. A bright young sun consistent with helioseismology and warm temperatures on ancient Earth and Mars. *Astrophys. J.*, **583**, 1024–1039.

Skumanich A., (1972). Time scales for Ca II emission decay, rotational braking, and lithium depletion. *Astrophys. J.*, **171**, 565–567.

Smart, W. M. (1977). *Text book on spherical astronomy.* Cambridge University Press, Cambridge.

Sonnet, C. P., Giampapa, M. S. and Matthews, M. S. (1991). The Sun in time. The University of Arizona Press, Tucson.

Sturrock, P. A. (ed.) (1985). *Physics of the Sun. Vol. 1: The solar interior.* D. Reidel, Dordrecht.

Tayler, R. J. (1994). *Stars: Their structure and evolution.* Cambridge University Press, Cambridge.

Taylor, B. J. (1994). Analyses of archival data for cool dwarfs. I. The solar B-V problem reconsidered. *Pub. Astron. Soc. Pacif.*, **106**, 444–451.

Thekaekara, M. P. and Drummond, A. J. (1971). Standard values for the solar constant and its spectral components. *Nature Phys. Sci.*, **229**, 6–9.

Vardavas, I. M. (2005). The dependence of Rossby number and XUV-Lyα emission with age for solar-like G-type stars. *Mon. Not. R. Astron. Soc. Lett.*, **363**, L51–L55.

Walter, F. M. and Barry D. C. (1991). Pre- and main–sequence evolution of solar activity. In Sonnet, C.P., Giampapa, M.S., Matthews, M.S., ed., *The Sun in Time*, University of Arizona Press, Tucson.

Willson, O. C., Skumanich, A. (1964). Dependence of chromospheric activity upon age in main-sequence stars: Additional evidence. *Astrophys. J.*, **140**, 1401–1408.

Wood, B. E., Redfield S., Linsky J. L., Müller H-R. and Zank, G. P. (2005). Stellar Ly alpha emission lines in the Hubble Space Telescope archive: Intrinsic line fluxes and absorption from the heliosphere and astrospheres. *Astrophys. J. Suppl. Ser.*, **159**, 118–140.

Woods, T. N. and Rottman G. J. (1997). Solar Lyman-α irradiance measurements during two solar cycles. *J. Geophys. Res.*, **102**, 8769–8779.

Zahnle, K. J. and Walker, J. C. G. (1982). The evolution of solar ultraviolet luminosity. *Rev. Geophys. Space Phys.*, **20**, 280–292.

6

SOLAR RADIATION TRANSFER IN THE ATMOSPHERE

6.1 Introduction

The absorption and scattering of the incoming solar radiation depends on the molecular composition of the atmosphere, the presence of clouds and aerosols in the atmosphere, and on the radiative properties of the Earth's surface with water, snow and ice playing an important role. It is only weakly dependent on the vertical thermal structure. Incoming solar radiation follows a downward path through the atmosphere to the surface and through reflection an upward path through the atmosphere out to space. Incoming direct solar radiation, as we have seen in Chapter 5, is characterized by a direction given by the solar zenith angle z, or $\mu = \cos z$, which depends on latitude and time of year. Radiation that has undergone scattering or isotropic reflection, i.e. diffuse radiation, can be characterized by a fixed value $\mu = 3/5$, the standard diffusivity approximation for converting diffuse transmission to an equivalent direct transmission, as was discussed in §3.5.5. In reality, the radiation undergoes multiple scattering between the various atmospheric layers, Fig. 6.1, and the Earth's surface, especially over highly reflecting surfaces such as ice or snow. The effect of the multiple reflections between the ground and clouds or aerosols is to increase atmospheric and ground absorption. The incoming solar flux can be divided into two broad spectral regions comprising 60% mainly UV–visible radiation ($0.2 \leq \lambda \leq 0.85$ μm) and 40% near–infra–red radiation ($0.85 \leq \lambda \leq 5.0$ μm), and their transfer through each atmospheric layer needs to be treated separately. The UV-visible component in

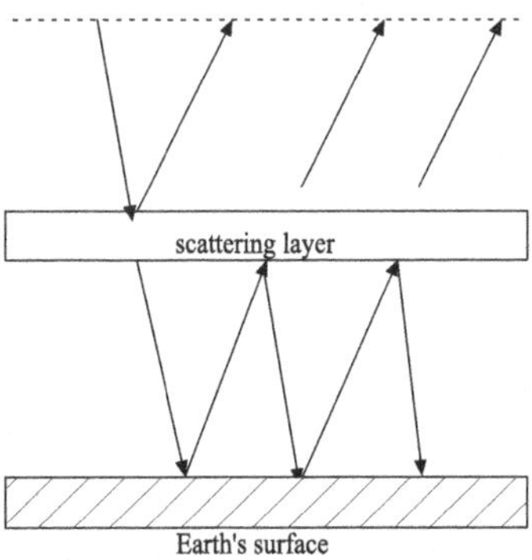

FIG. 6.1. Multiple reflections between scattering layer and the Earth's surface.

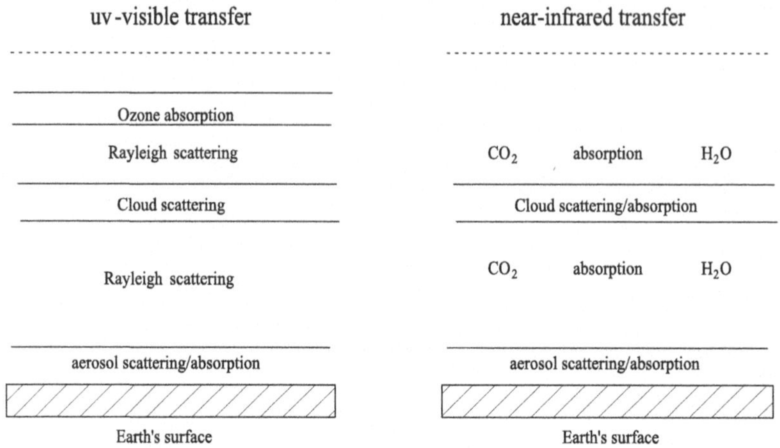

FIG. 6.2. Scattering and absorption of UV–visible and near–infra–red solar radiation by components of the atmosphere.

its transfer through the atmosphere (Fig. 6.2) undergoes cloud scattering, absorption mainly by ozone, Rayleigh scattering by air molecules, absorption and scattering by aerosols, and some very weak absorption by water vapour. The near–infra–red component undergoes cloud and aerosol scattering and absorption, absorption by water vapour, carbon dioxide, methane and ammonia. Near and below 0.2 μm, O_2 plays a very significant role in ultraviolet absorption, but this mainly affects the Earth's photochemistry, as will be discussed in Chapter 7. Both components of the solar radiation undergo reflection and absorption by the Earth's surface.

6.2 Atmospheric molecular absorption

In Fig. 6.3 is shown the solar irradiance, at low resolution (see §5.2.4), and the atmospheric molecules that absorb it, mainly in the near–infra–red. In Table 6.1 are shown the main absorbing molecules in each spectral interval, including Rayleigh scattering, from the ultraviolet to the near–infra–red.

6.2.1 *Ultraviolet–visible absorption*

The most important molecular absorption relevant to the 0.20–0.85 μm range of solar radiation is that of O_3. The absorption is temperature–dependent in the range 0.20–0.35 μm (Hartley–Huggins bands), corresponding to about 4.5% of the incoming flux. The ozone layer also absorbs in the visible between 0.45 and 0.85 μm (Chappuis bands), corresponding to about 55.4% of the incoming solar

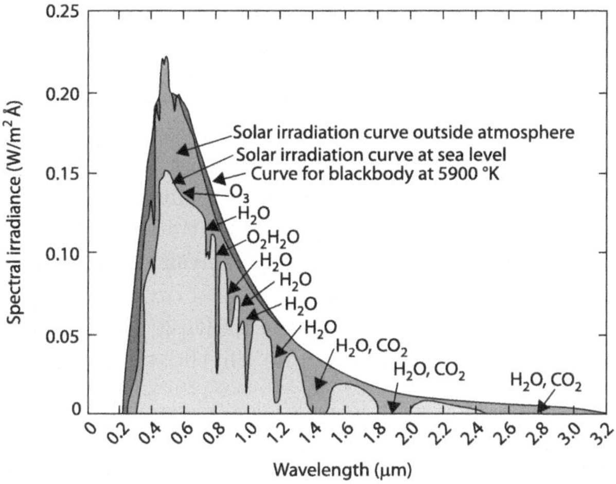

FIG. 6.3. Low–resolution solar irradiance and atmospheric absorption. (Source: CSR University of Texas, based on the original from Valley 1965, Air Force Cambridge Research Laboratories)

Table 6.1 *The distribution of the incoming solar flux in the spectral range 0.1–10.0 μm, and the molecules and Rayleigh scattering in each spectral interval. (Vardavas and Carver 1984)*

Interval	Wavelength (μm)	% Solar flux	Molecules/scattering
1	0.10–0.20	0.00644	NH_3, CH_4, CO_2, H_2O, O_2, Rayleigh
2	0.20–0.35	4.46	O_3, Rayleigh
3	0.35–0.45	10.63	Rayleigh
4	0.45–0.85	44.77	O_3, H_2O, Rayleigh
5	0.85–1.00	9.48	H_2O, Rayleigh
6	1.00–1.04	2.12	H_2O
7	1.04–1.22	7.55	H_2O
8	1.22–1.58	9.19	H_2O, CO_2
9	1.58–1.95	4.81	H_2O
10	1.95–2.12	1.24	H_2O, CO_2
11	2.12–2.61	2.41	H_2O
12	2.61–2.92	0.92	H_2O, CO_2
13	2.92–3.88	1.34	H_2O, CH_4, NH_3
14	3.88–4.60	0.40	CO_2
15	4.60–9.50	0.64	H_2O, CH_4, NH_3

radiation. Absorption cross-section data in the Chappuis bands (0.40–0.85 μm) are shown in Chapter 7. Detailed ultraviolet–visible cross–section data are available from the Jet Propulsion Laboratory (JPL 2006) and there is also available the MPI–Mainz–UV–VIS Spectral Atlas that is a comprehensive collection of absorption cross–sections for 630 molecular species. Both databases can be accessed via the web. More details are given in Chapter 7 in relation to atmospheric photochemistry.

In the UV–visible spectral range, the ozone cross–sections are sufficiently smooth to allow the wavelength range to be finely subdivided (e.g. 200 spectral intervals) into monochromatic absorption optical depths, from Lyman-α at 0.1216 μm out to 0.85 μm, not including the highly structured Schumann–Runge (S–R) bands (0.18–0.20 μm) of O_2, which require special treatment. More details are given regarding the S–R bands in relation to the photolysis of molecules in Chapter 7. Having the fine–resolution spectral distribution of ozone UV–visible absorption cross–sections one can incorporate these in a multiple scattering and absorption code (§6.7) to include the effects of ozone on the transfer of solar radiation through the atmosphere.

For ozone in relation to the Earth's radiation budget, one can make a quick estimate of the flux–mean atmospheric transmission, from the expressions given below. The transmitted fractions of the total incoming solar radiation due to the absorption of visible and UV radiation by ozone are given, respectively, by

$$t'_{\mathrm{vis}} = 1 - 0.023\,(W_{\mathrm{O_3}}/\mu)^{0.18}\,, \tag{6.1}$$

and

$$t'_{\mathrm{uv}} = 1 - 0.023 W_{\mathrm{O_3}}/\mu, \tag{6.2}$$

where $W_{\mathrm{O_3}}$ represents the total atmospheric ozone amount in atm–cm STP. The minimum value of t'_{UV} is thus 0.955, while for t'_{vis} it is 0.446.

6.2.2 *Near–infra–red absorption*

The transmission, t, for each near–infra–red interval (Table 6.1) where rotational lines of molecular vibrational–rotational bands become important, can be computed using the HITRAN database by subdividing the solar near–infra–red spectral intervals into higher–resolution subintervals and then computing the transmission within each subinterval as a function of pressure, temperature and absorber amount for each molecule (§4.7). This approach is now computationally feasible and preferred. However, for climate models (e.g. general circulation models, §11.4) such an approach may not be feasible and a simpler approach is needed. We note that when the optical depth varies non–linearly with absorber amount we cannot simply divide the atmosphere into layers and then multiply their transmissivities. Methods that are employed to overcome this problem are the *exponential–sumfit*, *k–distribution* and the *correlated–k approximation*, according to which the average transmissivity within a spectral interval is obtained from

$$t = \sum_n \alpha_n \exp(-k_n y), \tag{6.3}$$

for an absorber amount y, in terms of effective monochromatic absorption coefficients, k_n, and associated discrete probability distribution weights α_n, with the normalization

$$\sum_n \alpha_n = 1. \tag{6.4}$$

The sets of k_n and α_n for each spectral interval are not unique and can be obtained using, for example, the Fibonacci minimization technique (Vardavas 1989). For inhomogeneous layers, the correlated–k approximation can be used to allow k_n to be temperature and pressure dependent and the α_n are fixed and correspond usually to Gaussian-Legendre quadrature points.

The monochromatic radiation–transfer equation is then solved for each optical depth, $\tau_n = k_n y$ for incoming solar flux $\alpha_n F_i$, where F_i is the flux contained in each spectral interval i. For overlapping bands of two different molecules, with absorber amounts y and v, the transmission can be taken to be

$$t_{nm} = \alpha_n \beta_m \exp(-k_n y) \exp(-l_m v), \tag{6.5}$$

with the normalization condition

$$\sum_{n,m} \alpha_n \beta_m = 1, \tag{6.6}$$

and the transfer equation is solved for $\alpha_n \beta_m F_i$ with optical depth $\tau_{nm} = k_n y + l_m v$ for each combination of n and m. This approach can be extended to overlapping bands of more than two types of molecules. One can then use a monochromatic multiple–scattering radiation–transfer code to solve the radiation–transfer equation, as discussed in §6.7. Coefficients and weights for the near–infra–red intervals of Table 6.1 for water vapour, carbon dioxide, methane and ammonia are given in Vardavas and Carver (1984) based on laboratory measurements at 1 atm and 300 K.

6.2.2.1 *Water vapour* has significant absorption bands in the near–infra–red (0.85–5.0 μm), which corresponds to about 40.1% of the incoming solar radiation. The fraction of the total incoming solar flux absorbed by a H_2O layer can be quickly estimated from

$$t'_w = 1 - 0.106 \left(W_{H_2O}/\mu\right)^{0.31}, \tag{6.7}$$

where W_{H_2O} is the amount of H_2O, g cm^{-2}, in the layer. The above simple expression is based on the data of Wyatt *et al.* (1964).

6.2.2.2 *Carbon dioxide* absorption of solar radiation takes place throughout the atmosphere. We can neglect the UV absorption of CO_2 that occurs below 0.2 μm where the solar flux is less than 0.01% of the incoming flux but we need to include the near–infra–red absorption (Stull *et al.* 1964). The fraction of the total incoming solar flux transmitted by a CO_2 layer can be quickly estimated from

$$t'_c = 1 - 0.015\,(W_{CO_2}/\mu)^{0.263}\,, \qquad (6.8)$$

where W_{CO_2} is the total amount of CO_2, g cm^{-2}, in the layer. For a clear sky we have one CO_2 layer below the ozone layer, while for a cloudy sky we have two, one above and one below. If p_o represents the surface pressure at sea level then the amount of CO_2 above a cloud layer with cloud–top pressure p_{cla} is $W_{CO_2}(p_{cla}/p_o)$ while in the layer below the cloud layer the amount is $W_{CO_2}(1 - p_{clb}/p_o)$, where p_{clb} is the cloud–base pressure.

6.2.2.3 *Ammonia* absorbs in the near–infra–red intervals 1300–2000 cm^{-1} (6.14 μm) and 3100–3500 cm^{-1} (3.03 μm). The mean transmission of each band can be obtained from the following simple analytical fits to the laboratory data (France and Williams 1966) in the form

$$t = \exp(-w\sigma(w)). \qquad (6.9)$$

For the ammonia 6.14 μm band $\sigma(w) = 38.09w^{-0.26}$ with a corrected absorber amount $w = w_0(p_e/p_o)^C$ with $C = 0.6$ and $p_e = p_{N2} + (B-1)p_{NH3}$ and $B = 5.77$. For the 3.03 μm band of ammonia we have $\sigma(w) = 32.68w^{-0.178}$ with $C = 0.3$ and $B = 6.1$.

6.2.2.4 *Methane* absorbs in the intervals 1100–1750 cm^{-1} (7.66 μm) and 2700–3300 cm^{-1} (3.31 μm). For the 7.66 μm methane band we have, from fits to laboratory measurements (Burch and Williams 1962), $\sigma(w) = 8.22w^{-0.46}$ with $C = 0.5$ and $B = 1.38$, while for the 3.31 μm band $\sigma(w) = 9.97w^{-0.39}$ with $C = 0.45$ and $B = 1.3$.

6.3 Particle absorption and scattering

Aerosols and cloud droplets scatter radiation in different directions, coherently in frequency, and absorb radiation, depending on their shape, chemical composition, and size. In the case of simple spherical and homogenous scatterers, Mie theory provides a complete solution to the problem (Mie 1908, Debye 1909) for light scattering by small particles. Analytical forms are provided for the total extinction cross–section, σ_{ext} (cm^2), the scattering cross–section, σ_{scat} and through energy conservation, the absorption cross–section, σ_{abs}, where

$$\sigma_{ext} = \sigma_{scat} + \sigma_{abs}, \qquad (6.10)$$

and also the scattering phase function, $p(\cos\theta)$, where θ is the angle between the incoming and outgoing radiation, as discussed in Chapter 3. From the phase function we can also derive the asymmetry factor g, defined by eqn (3.58). Knowing the cross–sections we can compute the corresponding *optical depths*, the *single*

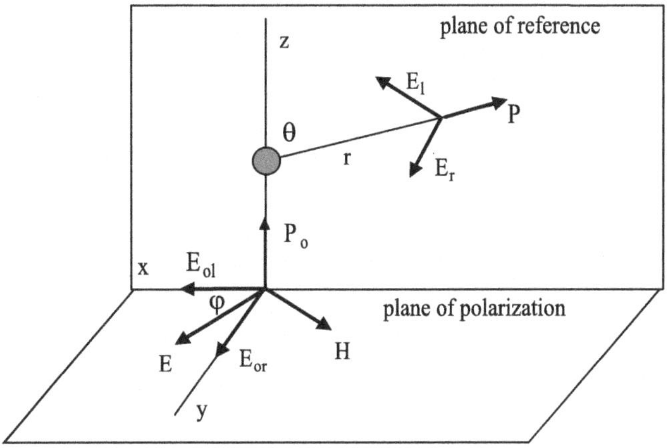

FIG. 6.4. Scattering of electric vectors of a plane wave by a homogeneous spherical scatterer. Plane of reference taken through incident and scattered wave. Plane of polarization contains the orthogonal electric field, **E** and magnetic field **H** vectors. The Poynting vector **P**, is in the direction of wave propagation.

scattering albedo, ω, defined by eqn (3.36), and knowing the *asymmetry factor* we can solve the multiple–scattering problem within an atmosphere, as we shall see in §6.4.

6.3.1 *Mie scattering*

Details on the electromagnetic theory and the solution of Maxwell's equations to derive the Mie solution are given in the literature (e.g. van de Hulst 1981). Here, we will give the relevant equations and definitions to allow one to compute the above particle radiative properties with a physical understanding of how these quantities are determined.

In Fig. 6.4, we depict the scattering of the electric field **E** of an electromagnetic wave by a homogeneous spherical particle. The propagation of the incident wave is orthogonal to both the electric field and the magnetic field, **H**. The electric field has two components, $E_{\rm ol}$ parallel to the *plane of reference* (containing the directions of the incident and scattered wave) and $E_{\rm or}$ perpendicular to it. The electric–field components lie on the *plane of polarization* and the two generally unequal components are said to describe elliptically polarized radiation since the variation of the electric field on this plane describes an ellipse. The radiation flux (W m^{-2}), flowing through an element of area in the direction of the incident wave is given by the *Poynting vector* where

$$\mathbf{P} = \mathbf{E} \times \mathbf{H}. \tag{6.11}$$

The scattered wave has electric–field components $E_{\rm l}$ and $E_{\rm r}$ and the Poynting vector is **P**. The angle between the incident and scattered radiation is θ, and ϕ describes the azimuthal angle on the plane of polarization such that the components of the incident wave are $E_{\rm ol} = E_{\rm o}\cos\phi$ and $E_{\rm or} = E_{\rm o}\sin\phi$, where $E_{\rm o}$ is the amplitude or scalar magnitude of the incident electric field. For an EM wave the magnitude of the magnetic field, H is related to the magnitude of the electric field, E through

$$H = \epsilon_0 cE, \tag{6.12}$$

where ϵ_0 is the permittivity of free space and c is the speed of light in vacuum. The magnitude of the Poynting vector at any instant can be written as

$$P = \epsilon_0 cE^2. \tag{6.13}$$

The average rate of energy flow of a sinusoidal incident wave is then

$$\overline{P}_{\rm o} = \frac{1}{2}\epsilon_0 cE_{\rm o}^2. \tag{6.14}$$

In mks units, $\epsilon_0 c$ is equal to 2.66×10^{-3} W V^{-2}, and the magnitude of the electric field then has units of V m^{-1}. Thus, the magnitude of the Poynting vector has units of W m^{-2}. We also recall that $c = (\epsilon_0\mu_0)^{-\frac{1}{2}}$, where μ_0 is the permeability of free space.

In a medium other than vacuum, the speed of light is reduced according to

$$v = c/n, \tag{6.15}$$

where n is unity for vacuum and greater than unity for a scattering medium, such as the atmosphere or droplets of clouds. The factor n is know as the real part of the *refractive index*, m, which is a complex number $m = n - {\rm i}n'$, where n', as we shall see, gives the absorption of the wave, and ${\rm i} = \sqrt{-1}$, so that ${\rm i}^2 = -1$. The frequency of a wave remains the same in vacuum and the medium, so that only the wavelength changes according to $\lambda' = \lambda/m$, where λ is the wavelength in the surrounding space, usually taken to be vacuum. The wave number in the medium is $k' = 2\pi/\lambda'$ where $k' = km$, and k is the wave number in vacuum corresponding to the angular frequency $\omega = 2\pi\nu = kc$.

In Mie theory, formal solutions for the scattered electric–field components, and hence Poynting vector, are derived based on the scattering of a plane sinusoidal wave by a spherical scatterer. We can represent the electric–field components of an incident plane sinusoidal wave travelling in the direction z in vacuum by

$$E_{\rm l} = E_{\rm o}\cos\phi {\rm e}^{{\rm i}(\omega t - kz)}, \tag{6.16}$$

$$E_{\rm r} = E_{\rm o}\sin\phi {\rm e}^{{\rm i}(\omega t - kz)}, \tag{6.17}$$

with amplitude $E_{\rm o}$. This complex number notation is for mathematical expediency since

$${\rm e}^{{\rm i}(\omega t - kz)} = \cos\omega(t - z/u) + {\rm i}\sin\omega(t - z/u), \tag{6.18}$$

where $u = \omega/k$ is the phase velocity, $\omega(t - z/u)$ is the phase, and $t - z/u$ is the phase angle. We recall that any complex number $x + {\rm i}y$ can also be written as

$r\exp(\mathrm{i}\vartheta)$, where $r^2 = x^2 + y^2$, and represents a vector of distance r from the origin to the point (x, y), making an angle $\vartheta = \omega t$ with the X–axis. Thus, as t varies, $r\exp(\mathrm{i}\omega t)$, represents a rotating vector at an angular frequency ω. The distance r projected on the X–axis represents its amplitude, which thus varies sinusoidally.

The above complex form of the electric–field components satisfies the wave equation for a wave travelling in the Z–direction with no attenuation. If the medium surrounding the spherical scatterer has $m = n - \mathrm{i}n'$, then we can write

$$E_\mathrm{l} = E_\mathrm{o}\cos\phi \mathrm{e}^{\mathrm{i}(\omega t - k(n - \mathrm{i}n')z)}, \tag{6.19}$$

$$E_\mathrm{r} = E_\mathrm{o}\sin\phi \mathrm{e}^{\mathrm{i}(\omega t - k(n - \mathrm{i}n')z)}, \tag{6.20}$$

or

$$E_\mathrm{l} = E_\mathrm{o}\cos\phi \mathrm{e}^{-kn'z}\mathrm{e}^{\mathrm{i}(\omega t - knz)}, \tag{6.21}$$

$$E_\mathrm{r} = E_\mathrm{o}\sin\phi \mathrm{e}^{-kn'z}\mathrm{e}^{\mathrm{i}(\omega t - knz)}, \tag{6.22}$$

so that the magnitude of the Poynting vector is reduced according to

$$P = \epsilon_0 c E^2 \mathrm{e}^{-2kn'z}, \tag{6.23}$$

where the optical depth for absorption in the surrounding medium is $\tau_\mathrm{abs} = 2kn'z$, after travelling a distance z. The factor $2kn'$ has units of inverse length and can be thought of as the absorption coefficient of the surrounding medium.

For simplicity, let us assume that the refractive index of the surrounding medium is that of vacuum, and ignore attenuation. After the plane wave undergoes scattering with the spherical particle, considered as a point source, the outgoing wave is a spherical wave with origin the particle. For a spherical wave, the wave equation can be written as

$$\frac{\partial^2(r\psi)}{\partial r^2} - \frac{1}{c^2}\frac{\partial^2(r\psi)}{\partial t^2} = 0, \tag{6.24}$$

where

$$E_\mathrm{l} = E_\mathrm{o}\cos\phi\psi(t - r/c), \tag{6.25}$$

$$E_\mathrm{r} = E_\mathrm{o}\sin\phi\psi(t - r/c), \tag{6.26}$$

with

$$\psi(t - r/c) = \frac{1}{r}\mathrm{e}^{\mathrm{i}\omega(t - r/c)}. \tag{6.27}$$

Thus, the amplitude of a spherical wave decreases with the radius r from the source. However, because of refraction (scattering plus absorption) within the scattering sphere, the Mie solution involves a 2×2 scattering matrix, with diagonal components S_1 and S_2 with offdiagonal components zero for spherical particles, which gives the distribution of energy that is scattered at the different angles θ, between incident and scattered wave, as shown in Fig. 6.4, so that the scattered components of the electric field are

$$E_{\mathrm{l}} = E_{\mathrm{o}} \cos\phi\psi(t - r/c)[-\frac{\mathrm{i}}{k}S_2(\theta)], \tag{6.28}$$

$$E_{\mathrm{r}} = E_{\mathrm{o}} \sin\phi\psi(t - r/c)[-\frac{\mathrm{i}}{k}S_1(\theta)], \tag{6.29}$$

where S_1 and S_2 are called the Mie scattering functions (§6.3.2).

The time–average magnitude of the Poynting vector of the scattered wave can be written as

$$\overline{P} = \overline{P}_{\mathrm{o}} \frac{F(\theta, \phi)}{k^2 r^2}, \tag{6.30}$$

where $F(\theta, \phi)$ is given by

$$F(\theta, \phi) = S_1^2 \sin^2\phi + S_2^2 \cos^2\phi. \tag{6.31}$$

The total rate of scattered energy, dL (in W), flowing through an element of area dA is then

$$\mathrm{d}L = \mathbf{P} \cdot \mathbf{n}\mathrm{d}A, \tag{6.32}$$

where $\mathbf{n}$ is a unit vector normal to the element of surface. If we now replace $\mathrm{d}A/r^2$ by $\mathrm{d}\Omega = \sin\theta\mathrm{d}\theta\mathrm{d}\phi$, an element of solid angle, then the total rate of energy, L, scattered into a sphere around the scattering particle is

$$L = \overline{P}_{\mathrm{o}} \int_0^{\pi} \int_0^{2\pi} \frac{F(\theta, \phi)}{k^2} \sin\theta\mathrm{d}\theta\mathrm{d}\phi \tag{6.33}$$

$$= \overline{P}_{\mathrm{o}}\sigma_{\mathrm{scat}}, \tag{6.34}$$

where σ_{scat} is the cross–section for scattering. If we integrate over the azimuthal angle we then have

$$\sigma_{\mathrm{scat}} = \frac{\pi}{k^2} \int_0^{\pi} \left[S_1^2(\theta) + S_2^2(\theta)\right] \sin\theta\mathrm{d}\theta. \tag{6.35}$$

The phase function for Mie scattering is then given by

$$\frac{p(\theta)}{4\pi} = \frac{[S_1^2(\theta) + S_2^2(\theta)]}{\int_0^{\pi} [S_1^2(\theta) + S_2^2(\theta)] \sin\theta\mathrm{d}\theta} \tag{6.36}$$

with the normalization condition

$$\int_0^{\pi} \frac{p(\theta)}{4\pi} \sin\theta\mathrm{d}\theta = 1. \tag{6.37}$$

The polarization of the scattered wave with respect to the scattering angle is given by

$$\frac{i_1(\theta) - i_2(\theta)}{i_1(\theta) + i_2(\theta)}, \tag{6.38}$$

where $i_1(\theta) = |S_1(\theta)|^2$ and $i_2(\theta) = |S_2(\theta)|^2$. The asymmetry factor can be computed from

$$g = \frac{\int_0^\pi p(\theta) \cos\theta \sin\theta \mathrm{d}\theta}{\int_0^\pi p(\theta) \sin\theta \mathrm{d}\theta} \tag{6.39}$$

$$= \frac{\int_0^\pi [S_1^2(\theta) + S_2^2(\theta)] \cos\theta \sin\theta \mathrm{d}\theta}{\int_0^\pi [S_1^2(\theta) + S_2^2(\theta)] \sin\theta \mathrm{d}\theta}. \tag{6.40}$$

The extinction cross–section per particle, σ_{ext}, computed by Mie theory needs some care. The extinction (scattering plus absorption) of an EM wave by an infinitesimal layer, $\mathrm{d}z$, of a medium is given, according to Mie theory (see van de Hulst), by the reduction in the magnitude of the electric–field components

$$E_{\mathrm{l}} = E_{\mathrm{o}} \cos\phi(1 - \beta_{\mathrm{ext}}), \tag{6.41}$$

$$E_{\mathrm{r}} = E_{\mathrm{o}} \sin\phi(1 - \beta_{\mathrm{ext}}), \tag{6.42}$$

where the fraction removed is

$$\beta_{\mathrm{ext}} = 2\pi k^{-2} N \mathrm{d}z \mathrm{Re}[S(0)], \tag{6.43}$$

with N the number density of the scatterers, all assumed to be of the same radius and refractive index. We have taken the real part of the complex scattering functions, as we are dealing with amplitudes, setting $\theta = 0$ as we are concerned with extinction of the incident wave, and $S(0) = S_1(0) = S_2(0)$. Thus, the energy of the incident wave that has been attenuated by forward scattering and absorption within the infinitesimal layer is

$$\overline{P} = \overline{P}_{\mathrm{o}}(1 - \beta_{\mathrm{ext}})^2. \tag{6.44}$$

However, since β is small, we can neglect the β_{ext}^2 term and write

$$\overline{P} = \overline{P}_o(1 - 2\beta_{ext}), \tag{6.45}$$

hence the fraction of the incident energy that is removed is

$$2\beta_{\mathrm{ext}} = 4\pi k^{-2} \mathrm{Re}[S(0)](N\mathrm{d}z), \tag{6.46}$$

where $N\mathrm{d}z$ has units of inverse area and so we can write the extinction cross–section for a single particle as

$$\sigma_{\mathrm{ext}} = 4\pi k^{-2} \mathrm{Re}[S(0)]. \tag{6.47}$$

Knowing the extinction and scattering cross–sections for a single particle we can compute the absorption cross–section of the scattering particle. For a medium consisting of scattering spheres of different radii, a, we can weight a particle cross section by the number density distribution of sizes

$$\sigma = \int_0^\infty \sigma(a)N(a)\mathrm{d}a, \tag{6.48}$$

and we note that the functions S_1 and S_2 are functions of the particle radius and the particle refractive index.

6.3.2 *The Mie scattering functions*

The functions $S_1(\theta)$ and $S_2(\theta)$ can be computed from

$$S_1(\theta) = \sum_{n=1}^{\infty} \frac{2n+1}{n(n+1)}[a_n\pi_n(\cos\theta) + b_n\tau_n(\cos\theta)], \tag{6.49}$$

$$S_2(\theta) = \sum_{n=1}^{\infty} \frac{2n+1}{n(n+1)}[b_n\pi_n(\cos\theta) + a_n\tau_n(\cos\theta)], \tag{6.50}$$

where

$$a_n = \frac{\psi_n'(y)\psi_n(x) - m\psi_n(y)\psi_n'(x)}{\psi_n'(y)\xi_n(x) - m\psi_n(y)\xi_n'(x)}, \tag{6.51}$$

$$b_n = \frac{m\psi_n'(y)\psi_n(x) - \psi_n(y)\psi_n'(x)}{m\psi_n'(y)\xi_n(x) - \psi_n(y)\xi_n'(x)}, \tag{6.52}$$

with the prime superscript denoting a first derivative with respect to the argument of a function, $x = kr$ and $y = mx$, r is the radius of the particle, m its refractive index. In the above Mie solution, the medium surrounding the scattering particle was taken to be vacuum (m=1), however, if the surrounding medium has a real refractive index m_2 (no absorption), then we obtain the solution by replacing m by m/m_2 and λ (in vacuum) by λ/m_2. Also

$$\pi_n(\cos\theta) = \frac{1}{\sin\theta}P_n^1(\cos\theta), \tag{6.53}$$

$$\tau_n(\cos\theta) = \frac{d}{d\theta}P_n^1(\cos\theta). \tag{6.54}$$

In the above, $P_n^1(\cos\theta)$ is the associated Legendre polynomial, ψ_n and ξ_n are the Ricatti–Bessel functions that are related to the Bessel functions through

$$\psi_n(z) = z \cdot j_n(z) = \sqrt{\frac{\pi z}{2}} J_{n+\frac{1}{2}}(z), \tag{6.55}$$

$$\xi_n(z) = z \cdot h_n^{(2)}(z) = \sqrt{\frac{\pi z}{2}} H_{n+\frac{1}{2}}^{(2)}(z), \tag{6.56}$$

where $j_n(z)$ and $h_n^{(2)}(z)$ are the spherical Bessel functions, while $J_{n+\frac{1}{2}}(z)$ and $H_{n+\frac{1}{2}}^{(2)}(z)$ are the fractional–order Bessel functions, and z is a complex number. The scattering function for forward scattering can be computed from

$$S(0) = \frac{1}{2}\sum_{n=1}^{\infty}(2n+1)(a_n + b_n). \tag{6.57}$$

We can define a dimensionless *efficiency factor* for extinction by $Q_{\text{ext}} = \sigma_{\text{ext}}/\pi r^2$, and similarly for scattering and absorption. We can then compute Q_{ext} from

$$Q_{\text{ext}} = \frac{2}{x^2}\sum_{n=1}^{\infty}(2n+1)\text{Re}(a_n + b_n), \tag{6.58}$$

and the corresponding efficiency factor for scattering from

$$Q_{\text{sca}} = \frac{2}{x^2}\sum_{n=1}^{\infty}(2n+1)(|a_n|^2 + |b_n|^2). \tag{6.59}$$

We can compute, for example, the scattering cross–section for a distribution of particle sizes from

$$\sigma_{\text{scat}} = \int_0^{\infty}\pi r^2 Q_{\text{scat}}(r)N(r)\text{d}r. \tag{6.60}$$

The asymmetry factor, g, can be computed from

$$g = \frac{4}{x^2 Q_{\text{scat}}}\sum_{n=1}^{\infty}\frac{n(n+2)}{n+1}\text{Re}[a_n a_{n+1}^* + b_n b_{n+1}^*] \tag{6.61}$$

$$+ \frac{4}{x^2 Q_{\text{scat}}}\sum_{n=1}^{\infty}\frac{2n+1}{n(n+1)}\text{Re}[a_n b_n^*], \tag{6.62}$$

where a^* denotes the complex conjugate of a. We recall that the complex conjugate of a complex number is obtained by replacing i by $-$i, wherever it occurs (implicitly or explicitly), and that the product of a complex number and its conjugate is equal to the square of its magnitude.

In Fig. 6.5, we present the variation of Q_{scat} as a function of the size parameter x, for $n = 1.4$. We see that if the scattering particle has no absorption ($n' = 0$), Q_{scat} has a sinusoidal form with damping, arising from the interference of light that is refracted through the particle and diffracted around the particle. The irregularities superimposed on the form, correspond to edge phenomena around the scattering sphere. In this case, $Q_{\text{ext}} = Q_{\text{scat}}$ and for large x, $Q_{\text{scat}} \to 2$. As the absorption by the particle increases (increasing values of the imaginary part n' of the refractive index), the damping becomes stronger until there is

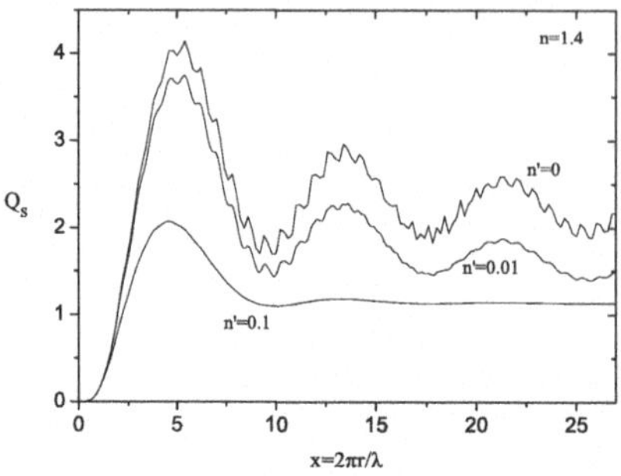

FIG. 6.5. Q_{scat} versus size parameter x, for $n = 1.4$, for four values of the imaginary part of the refractive index.

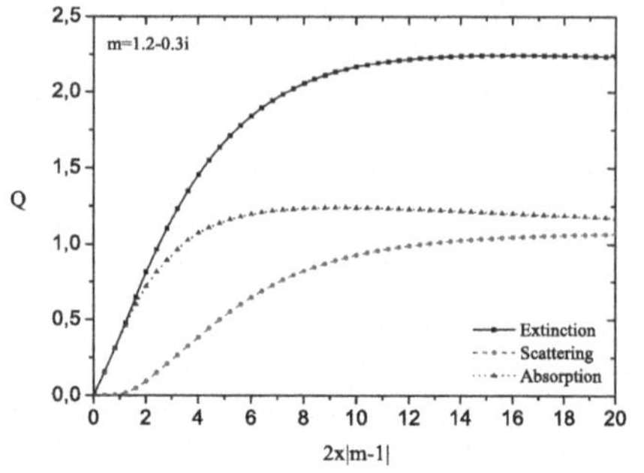

FIG. 6.6. Q_{ext}, Q_{scat} and Q_{abs} versus the phase–lag parameter $2x|m-1|$.

saturation beyond $x = 10$. For large particle sizes and small wavelengths, x becomes very large and the propagation of the EM wave can be described in terms of geometric optics. In Fig. 6.6, we demonstrate the effect of absorption on Q_{scat} and Q_{abs}, for the case of $m = 1.2 - 0.3i$, as functions of the phase–lag parameter $2x|m - 1|$ describing the phase lag between the wave passing through the sphere and outside the sphere. As the phase–lag parameter increases $Q_{\text{scat}} \to Q_{\text{abs}} \to 1$. In Fig. 6.7 we show the various Q values for the case of a factor ten less absorption, $n^{'} = 0.03$. In Fig. 6.8 we show the variation of the asymmetry factor g for $n = 1.33$ for different absorption as described by $n^{'}$. In the case of zero absorption $n^{'} = 0$ we see that g approaches about 0.87 at large particle sizes.

In Fig. 6.9 we show the shape of the phase function with scattering angle θ, for different particle sizes for a refractive index $m = 1.371 - 0.272\text{i}$, which is typical for water at 3 μm. In the natural atmosphere the size of the particles varies within a large range of values. Usually, a Gamma function distribution is used to describe the probability of finding a particle of specific size r

$$p(r) = \frac{\alpha^{\gamma-1}e^{-\alpha}}{\beta\Gamma(\gamma)}, \tag{6.63}$$

where $\alpha = (r - r_{\text{min}})/\beta$, r_{min} is the smallest radius considered, γ is the shape parameter of the distribution that defines the relative population of particles smaller than the most probable size, β is the scale parameter that controls the width of the distribution and Γ is the Gamma function. As an example, Fig. 6.10 represents the resulting average phase function distribution over the range of values covered by the Gamma function distribution shown in Fig. 6.11.

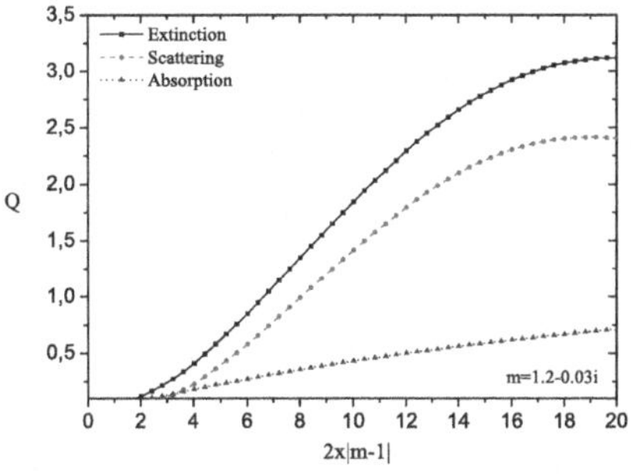

FIG. 6.7. The variation of Q_{ext}, Q_{scat} and Q_{abs} versus the phase–lag parameter $2x|m-1|$, for the case of small absorption.

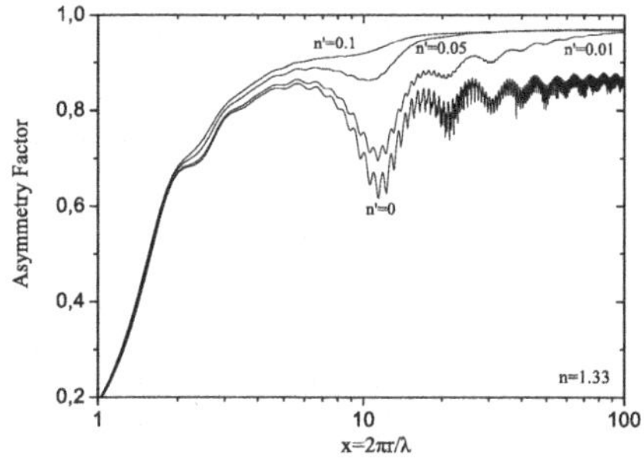

FIG. 6.8. The variation of the asymmetry factor g for $n = 1.33$ for different absorption as described by n'.

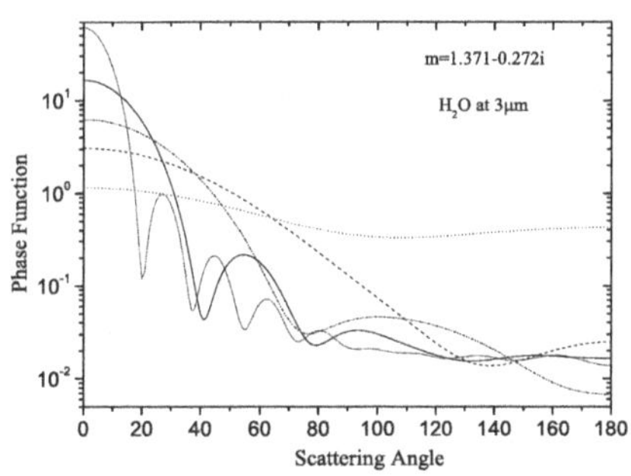

FIG. 6.9. The shape of the phase function for different particle sizes, x, for a refractive index m typical of water at 3 μm.

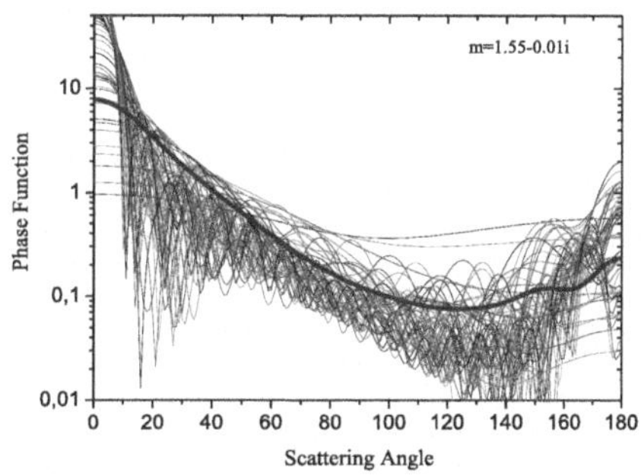

FIG. 6.10. Phase function with refractive index $m = 1.55 - 0.01$i, for a Gamma distribution in sizes (shown below). Phase functions for the different particle sizes of the distribution are also shown.

6.3.3 *Rayleigh scattering*

Rayleigh scattering by atmospheric molecules is important for wavelengths up to 1.0 μm, as the Rayleigh scattering cross–section decreases rapidly with wave-

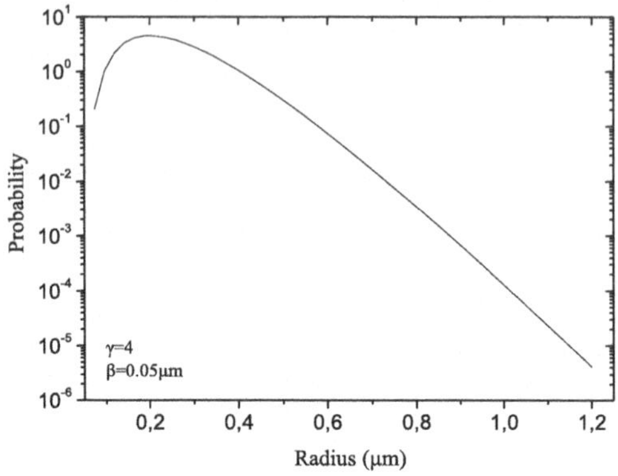

FIG. 6.11. A Gamma function distribution of particle sizes. The shape parameter is γ and the scale parameter is β.

length. The real part of the refractive index of an atmosphere consisting of scattering molecules that are far apart is given by

$$n = 1 + 2\pi\alpha N, \tag{6.64}$$

where N is the number density of the molecules and α is called the polarizability of the molecule (see van de Hulst). The Rayleigh scattering cross–section for a molecule is given by

$$\sigma_{\mathrm{R}} = \frac{8\pi}{3}k^4\alpha^2\delta, \tag{6.65}$$

where $k = 2\pi/\lambda$, and λ is the wavelength in vacuum. The original result of Rayleigh was for spherical top molecules, so there is a small but significant correction δ for nonspherical molecules. The refractive indices of molecular gases at STP are tabulated, for example, in Allen (1976) and have the form

$$n_{\mathrm{STP}} - 1 = A\left(1 + \frac{B}{\lambda^2}\right), \tag{6.66}$$

where λ is in μm. Thus, we can write

$$\alpha = \frac{A}{2\pi N_{\mathrm{STP}}}\left(1 + \frac{B}{\lambda^2}\right). \tag{6.67}$$

If we replace k in the expression for the cross–section we have

$$\sigma_{\mathrm{R}} = \frac{128\pi^5\delta}{3\lambda^4}\alpha^2, \tag{6.68}$$

Table 6.2 *Coefficients for computing the Rayleigh scattering cross–section.*

Molecule	A($\times 10^{-5}$)	B($\times 10^{-3}$)	Δ	δ
N_2	29.06	7.70	0.0305	1.05
O_2	26.63	5.07	0.0540	1.10
Ar	29.06	7.70	0.0305	1.00
air	28.71	5.67	0.0350	1.06
CO_2	43.9	6.4	0.0805	1.15
CO	32.7	8.1	0.080	1.15
H_2	13.58	7.52	0.022	1.038
NH_3	37.0	12.0	0.001	1.002

and so the cross–section (cm^2) can be computed from

$$\sigma_{\mathrm{R}} = 4.577 \times 10^{-21} \frac{\delta}{\lambda^4} \left[A \left(1 + B/\lambda^2 \right) \right]^2 , \tag{6.69}$$

where λ is in μm, the coefficients A and B are tabulated in Allen (1976) for various molecules, while $\delta = (6 + 3\Delta)/(6 - 7\Delta)$, where Δ is the *depolarization factor*, whose values are given in Table 6.2, together with the coefficients A and B for various molecules.

For a mixture of atmospheric gases we obtain the Rayleigh cross-section from

$$\sigma_{\mathrm{R}} = \sum_m \sigma_{\mathrm{R}m} n_m \tag{6.70}$$

where n_m is the mixing ratio by volume, of molecules m.

The Rayleigh reflectivity of an atmospheric layer can be quickly estimated using the two–stream approximation (§3.5.7) from

$$R_{\mathrm{r}} = \frac{(1 - g_{\mathrm{r}})\tau_{\mathrm{r}}/\mu}{2 + (1 - g_{\mathrm{r}})\tau_{\mathrm{r}}/\mu} \tag{6.71}$$

where the transmissivity $t_{\mathrm{r}} = 1 - R_{\mathrm{r}}$, g_{r} is the asymmetry factor set to zero (symmetric scattering). For the atmosphere the flux–mean optical depth over the UV–visible range of the incoming solar radiation is 0.187. The flux–mean Rayleigh scattering optical depth can be computed from the incoming solar flux spectral distribution, normalized to a solar constant of 1366 W m^{-2}. For a solar zenith angle corresponding to $\mu = 0.5$ the Rayleigh reflectivity of the atmosphere is about 16%, according to the above simple expression.

6.4 Clouds absorption and scattering

6.4.1 *Cloud types*

The total fraction of the sky covered by clouds can be represented by A_{c} and taken to comprise three non–overlapping: low–level cumulus (Cu), stratocumu-

Table 6.3 *Typical values of the visible (0.6 μm) scattering optical depth, τ_s, and corresponding approximate albedo, for cloud types belonging to low–level, middle–level and high–level clouds, based on the ISCCP classification.*

Cloud type	Cloud-top level	τ_s	R_{swc}
Low	>680 mbar, <3.2 km	>22.63	>0.77
Middle	440–680 mbar, 3.2–6.5 km	3.55–22.63	0.35–0.77
High	$\leq$ 440 mbar, >6.5 km	$\leq$ 3.55	$\leq$ 0.35

lus (Sc), stratus (St); middle-level altostratus (As), altocumulus (Ac), nimbostratus (Ns); and high-level cirrus (Ci), cirrostratus (Cs), deep convective (Dc); with cloud components, A_{cl}, A_{cm} and A_{ch}, respectively. Cloud radiative property climatologies, spanning decades, (3–hourly on 2.5°×2.5° latitude-longitude resolution) for the globe are given by the International Cloud Climatology Project (ISCCP), accounting for 15 different cloud types, including both liquid and ice phase (§8.2).

6.4.2 *Visible scattering*

Typical scattering optical depths in the visible (0.6 μm) are given in Table 6.3 for each cloud type. The cloud–top pressure and visible scattering optical depth is used to classify the various cloud types.

The reflectivity, R_{swc}, and transmissivity, t_{swc}, of each cloud type can be analytically estimated based on expressions derived using the modified two–stream approximation (Sagan and Pollack 1967, Irvine 1968). For the solar UV–visible radiation we use the result for pure scattering

$$R_{swc} = \frac{(1 - g_c)\tau_s/\mu}{2 + (1 - g_c)\tau_s/\mu}, \tag{6.72}$$

and $t_{swc} = 1 - R_{swc}$, where τ_s is the scattering optical depth and g_c the cloud scattering asymmetry factor with a typical value of 0.85. Note that the scattering optical depth of clouds (Mie scattering) is approximately independent of wavelength over the solar spectrum as they have a particle–size distribution of sufficiently large radii compared with the radiation wavelength (Fig. 6.12). In Table 6.3 an approximate reflectivity is given for the different cloud types based on an incident angle corresponding to $\mu = 0.5$ using the two-stream approximation. As the scattering optical depth increases, the reflectivity of the cloud approaches unity. Given that Mie scattering is primarily in the forward direction this might appear paradoxical. However, the explanation is that some of the incident photons on the cloud are scattered backwards out of the cloud as they have a smaller optical depth (larger mean free path) to traverse and so escape backwards with a smaller probability of undergoing a scattering event. Those travelling forwards have a higher probability of being scattered and are gradually all reflected backwards and so the reflectivity rises to unity.

6.4.3 *Near–infra–red absorption and scattering*

For the solar near–infra–red radiation we also include cloud absorption and so

$$R_{\mathrm{irc}} = \frac{(u+1)(u-1)(\mathrm{e}^{v} - \mathrm{e}^{-v})}{(u+1)^2\mathrm{e}^{v} - (u-1)^2\mathrm{e}^{-v}}, \tag{6.73}$$

$$t_{\mathrm{irc}} = \frac{4u}{(u+1)^2\mathrm{e}^{v} - (u-1)^2\mathrm{e}^{-v}}, \tag{6.74}$$

where

$$u = \left[\frac{1-\omega g_{\mathrm{c}}}{1-\omega}\right]^{\frac{1}{2}}, \tag{6.75}$$

$$v = [(1-\omega)(1-\omega g_{\mathrm{c}})]^{\frac{1}{2}} (\tau_{\mathrm{s}} + \tau_{\mathrm{a}})/\mu \tag{6.76}$$

and the cloud single scattering albedo is

$$\omega = \frac{\tau_{\mathrm{s}}}{\tau_{\mathrm{s}} + \tau_{\mathrm{a}}} \tag{6.77}$$

with τ_{s} and τ_{a} being the scattering and near–infra–red absorption cloud optical depths, respectively, while the asymmetry factor g_{c} is set to a lower value of about 0.8. The cloud absorption is then given by $a_{\mathrm{irc}} = 1 - R_{\mathrm{irc}} - t_{\mathrm{irc}}$.

6.5 Aerosol absorption and scattering

A major difference between aerosols and greenhouse gases is that aerosols have a much shorter atmospheric lifetime (from 10^{-4} days for natural, to tens of days for anthropogenic aerosols) compared with the important greenhouse gases (decades to centuries). This, together with microphysical and mixing processes, results in larger spatial and temporal aerosol variability. Thus, quantification of aerosol radiative effects is complex, as this requires comprehensive global micro-physical, microchemical, and radiative aerosol properties. Characteristics such as size distribution, composition, and radiative properties have to be adequately determined on a global scale.

6.5.1 *Aerosol radiative properties*

Given the high spatial and temporal variability of atmospheric aerosols, unique aerosol types never occur in the atmosphere; in fact, aerosol size–distribution spectra are present as internal plus external mixtures of various aerosol components. Such mixtures, along with complex aerosol properties, can only be simulated and represented by global aerosol models.

Table 6.4 *Long-term mean annual global aerosol optical thickness (τ_{aers}), aerosol single scattering albedo (ω_{aer}), and aerosol asymmetry parameter (g_{aer}) derived from the Global Aerosol Data Set (GADS) for actual atmospheric and surface conditions at wavelengths 0.25, 0.5, and 0.8 μm. (Hatzianastassiou et al. 2004)*

Wavelength (μm)	τ_{aers}	ω_{aer}	g_{aer}
0.25	0.12	0.91	0.76
0.50	0.08	0.96	0.73
0.80	0.06	0.97	0.73

The Global Aerosol Data Set (GADS), (Koepke *et al.* 1997) is a starting point by providing long–term mean aerosol radiative properties. These properties are highly dependent on the atmospheric relative humidity, which controls the water content of the aerosols and hence their scattering properties. The aerosol radiative properties are very sensitive to the ambient relative humidity, because it determines, along with the particle composition and the ambient temperature, the rate of condensation of water vapour on the particle's surface, thereby determining the growth rate of the aerosol particle, and hence the particle size and refractive index. In Table 6.4 are given the global mean spectral radiative properties of aerosols derived from the GADS data and NCEP/NCAR humidity data for the globe.

6.5.2 *Particle size and Ångstrom parameter*

The aerosol extinction coefficient, σ_λ in cm^2 (combined scattering and absorption) is generally a function of wavelength, depending on the size of the particle. For large particles we saw that we can employ Mie theory to compute the extinction properties of particles with Rayleigh scattering being the small–size limit. We may approximate the wavelength dependence of the extinction coefficient by the form

$$\sigma_\lambda = \sigma_{\text{o}} \left(\frac{\lambda}{\lambda_o} \right)^{-\alpha}, \tag{6.78}$$

where the extinction optical depth at wavelength λ_{o} (also known as the Ångstrom turbidity coefficient) is $\tau_{\text{o}} = \sigma_{\text{o}} W$ and W is the absorber amount in particles per cm^2 (the number column density). The wavelength exponent α is known as the Ångstrom parameter, based on the observations of Ångström (1929) in the wavelength range 0.35–1.00 μm. This parameter can be written as

$$\alpha = -\frac{\partial \ln \sigma_\lambda}{\partial \ln \lambda}, \tag{6.79}$$

and thus approximately can be calculated from

$$\alpha = \frac{\ln(\sigma_1/\sigma_2)}{\ln(\lambda_1/\lambda_2)}, \tag{6.80}$$

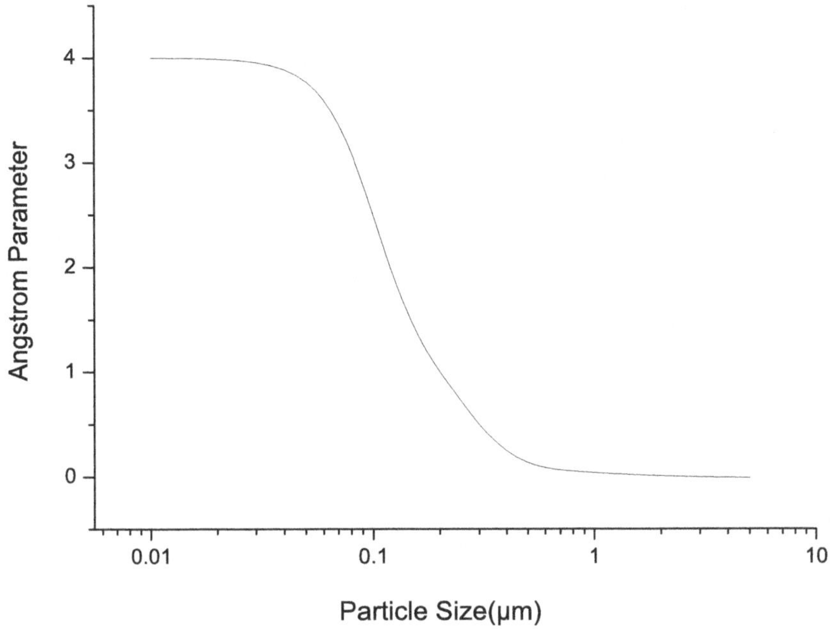

FIG. 6.12. Ångstrom parameter as a function of particle radius for a single particle with refractive index 1.33–0.5i, calculated from the scattering efficiency at wavelengths of 440 nm and 870 nm.

$$= \frac{\ln(\tau_1/\tau_2)}{\ln(\lambda_1/\lambda_2)}. \tag{6.81}$$

Thus, if the aerosol extinction optical depth is known at two wavelengths (usually one taken in the visible and the other in the near–infra–red) then the Ångstrom parameter can be estimated. The variation of the Ångstrom parameter with particle radius for a single particle with refractive index 1.33–0.5i, calculated from the scattering efficiency at wavelengths 440 nm and 870 nm is shown in Fig 6.12. We see that as the particle radius becomes small we approach the Rayleigh scattering limit and $\alpha \to 4$. At the other limit of very large particles the extinction cross–section becomes wavelength independent. This variation of the Ångstrom parameter is very useful for distinguishing between large and fine aerosol particles. Values of α above unity indicate fine particles ($r < 0.1$ μm), while below 0.5 we have coarse particles ($r > 0.5$ μm). In Fig. 6.13 is shown a scatter diagram of the Ångstrom parameter, taking $\lambda_1 = 440$ nm and $\lambda_2 = 870$ nm, as a function of τ_{870}, from actual aerosol measurements by the AERONET site in Crete. The measurements in the region of low α and the higher τ_{870} values correspond to major dust events transported from the Sahara, whilst the fine–aerosol measurements correspond to industrial pollution events from Europe. Small values of α (< 0.5) and the larger values of τ_{870} (> 0.20) correspond to

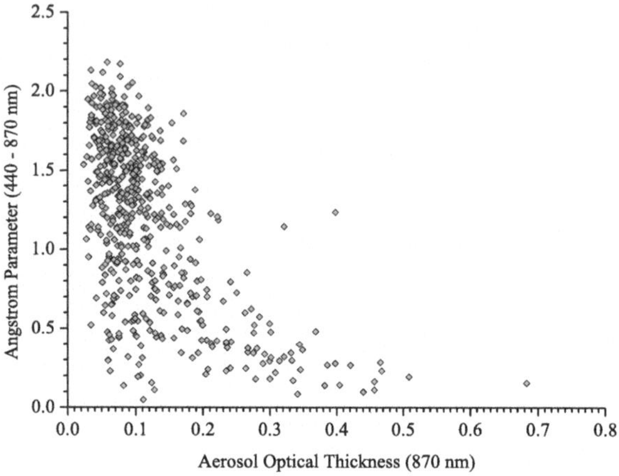

FIG. 6.13. Scatter plot of Ångstrom parameter as a function of aerosol optical depth at 870 nm, measured by the AERONET site in Crete. (Fotiadi *et al.* 2006)

coarse particles, associated with dust outbreaks in North African deserts during spring and late summer to autumn. Points corresponding to small values of α (< 0.5) and small τ_{870} (0.10–0.15) correspond to transported dust during winter, whereas the small τ_{870} values are due to frequent and efficient wet removal processes during the Mediterranean winter. These points can be also associated with transport of maritime aerosols (sea–spray). There is a wide range of α values (~0.10–2.0) for $\tau_{870} < 0.2$, which is due to the presence of various types of aerosol for background conditions, arising from transport and mixing processes. Such mixtures can involve background sulphate maritime aerosols, along with mineral dust, soot, and other continental natural or anthropogenic particles. The contribution of each of these particles varies with time, but coexistence of many types as internal and/or external mixture, is often possible. The bimodal nature of Mediterranean aerosols is clearly seen in Fig. 6.14, where the size distribution is shown for the four seasons. The fine mode (composed of the smaller particles) is principally of anthropogenic origin, with radius between 0.05 and 0.5 μm, whilst the coarse mode consists of larger particles of basically natural origin, with radius ranging from 0.5 to 15 μm.

6.5.3 *Aerosol fine and coarse modes*

The presence of aerosols of different size distributions correspond to different chemical compositions, and these result in different aerosol extinction optical depths. In Fig. 6.15 are shown monthly mean values of aerosol extinction optical thickness (AOT) at 340, 500, 870 and 1020 nm, and Ångstrom parameter α mea-

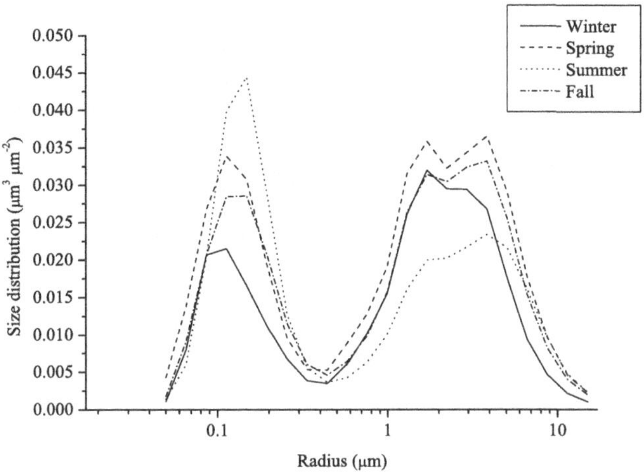

FIG. 6.14. Seasonal variation of aerosol columnar volume size distribution measured by the AERONET site in Crete. (Fotiadi *et al.* 2006)

sured by the AERONET station in Crete. The AOT values at 340, 500, 870 and 1020 nm indicate a significant spectral dependence of aerosol optical thickness, with decreasing values from 0.34 $\pm$0.14 at 340 nm down to 0.11 $\pm$0.09 at 1020 nm, implying an AOT reduction of 68% from the near–UV to near–IR wavelengths. The higher spring and autumn AOT values are associated with strong dust episodes taking place in these seasons when dust particles are transported from North African deserts. This is indicated by decreased values of Ångstrom parameter during spring and autumn. Spectral differences in terms of AOT indicate a heterogeneous aerosol population, which is composed of mineral dust, marine biogenic particles and anthropogenic aerosols at different layers. Table 6.5 presents computed parameters such as the aerosol effective radius ($r_{\rm eff}$) and the columnar volume, V_c, of particles per unit cross–section of atmospheric column $\mu m^3/\mu m^2$ where

$$V_{\rm c} = \int_0^\infty V(\ln r) {\rm d} \ln r, \tag{6.82}$$

for each mode, where $V(\ln r)$ is a log–normal aerosol columnar volume size distribution defined by

$$V(\ln r) = \frac{V_{\rm c}}{\sqrt{2\pi} \ln \sigma_{\rm g}} \exp\left[-\frac{\ln^2(r/r_{\rm g})}{2 \ln^2 \sigma_{\rm g}} \right], \tag{6.83}$$

where r is the particle radius, $r_{\rm g}$ is the *geometric mean particle radius* (50% of the particles have a radius below this value) and $\sigma_{\rm g}$ is the *geometric standard*

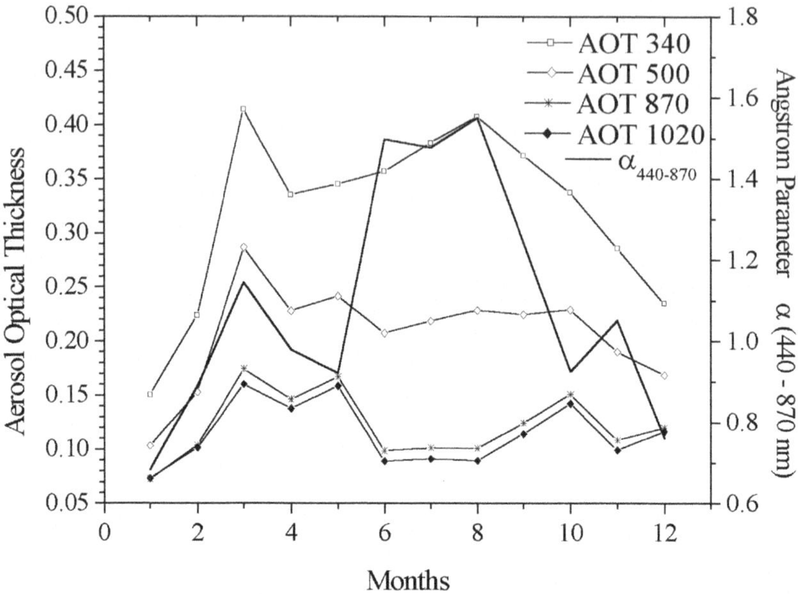

FIG. 6.15. Intra–annual variation of aerosol optical thickness (AOT) at 340, 500, 870 and 1020 nm. The Ångstrom parameter $\alpha_{440-870}$, derived from the AOT at 440 and 870 nm is also shown. Measurements by the AERONET site in Crete. (*Fotiadi et al.* 2006)

deviation. These are defined for particles with columnar population n_i of radius r_i by

$$\ln r_{\mathrm{g}} = \frac{1}{N}\sum_i n_i \ln r_i, \tag{6.84}$$

$$\ln^2 \sigma_{\mathrm{g}} = \frac{1}{N}\sum_i n_i \ln^2(r_{\mathrm{g}}/r_i), \tag{6.85}$$

where the total columnar number of particles (particles per unit cross-section of atmospheric column or column number density) of all sizes is $N = \sum n_i$. We can also define an effective aerosol radius for the whole aerosol column based on the columnar volume size distribution

$$r_{\mathrm{eff}} = \frac{\int V(\ln r) r^3 \mathrm{d}\ln r}{\int V(\ln r) r^2 \mathrm{d}\ln r}. \tag{6.86}$$

The effective radius for the fine mode shown in Fig.6.14 is about 0.13 μm throughout the year, while that of the coarse mode varies within the range 2.00–2.32 μm.

Table 6.5 *Seasonal averages of aerosol volume size distribution parameters at the AERONET station in Crete; r_{eff} is the effective radius (in μm) and V_{c} is the columnar volume of particles per unit cross section of atmospheric column ($\mu m^3/\mu m^2$). (Fotiadi et al. 2006)*

	Fine mode		Coarse mode		Fine + coarse
	r_{eff}	V_{c}	r_{eff}	V_{c}	V_{c}
Winter	0.131	0.025	2.004	0.057	0.082
Spring	0.128	0.042	2.026	0.075	0.117
Summer	0.132	0.048	2.322	0.046	0.094
Autumn	0.135	0.037	2.141	0.066	0.103
Annual	0.132	0.038	2.123	0.061	0.099

It should be noted that the r_{eff} of the coarse mode is slightly larger in summer (the dry season) than in the other seasons due to the absence of wet removal processes of dust. The columnar volume of aerosols, V_c, ranges from 0.025 to 0.048 for the fine mode and from 0.046 to 0.075 for the coarse mode. The maximum columnar volume of fine mode aerosols takes place in summer, owing to the significant contribution of anthropogenic and natural fine pollution particles. The coarse mode has higher columnar volumes in spring and in autumn, when there is strong transport of dust. In spring and autumn, when the frequency of occurrence of dust episodes is highest, the V_c of the coarse mode is larger than that of the fine mode by up to 1.8 times. The very large coarse–to–fine ratio of columnar volume in winter may be attributed to strong processing of aerosols by winter clouds and also to the presence of mineral dust and maritime particles.

6.6 Surface reflection

In the calculation of the surface reflectivity we can consider four types of surface; land, ocean, snow, and ice (frozen ocean). The surface albedo can be computed from:

$$R_{\mathrm{g}} = f_{\mathrm{land}} R_{\mathrm{l}} + f_{\mathrm{ocean}} R_{\mathrm{o}} + f_{\mathrm{snow}} R_{\mathrm{s}} + f_{\mathrm{ice}} R_{\mathrm{i}} \tag{6.87}$$

where f is the fraction of the Earth's surface covered by each type of reflecting surface.

6.6.1 *Snell and Fresnel laws*

The ocean reflectivity, R_{o}, can be computed using Fresnel reflection of natural (unpolarized) light, corrected for a nonsmooth surface for an incidence angle $i = \cos^{-1} \mu$ from

$$R_{\mathrm{o}} = 0.5 \left[\left(\frac{\sin a}{\sin b} \right)^2 + \left(\frac{\tan a}{\tan b} \right)^2 \right], \tag{6.88}$$

with

$$a = i - r, \tag{6.89}$$

$$b = i + r, \tag{6.90}$$

and r the *angle of refraction*, as shown in Fig. 6.16. According to Snell's law r is related to the angle of incidence via

$$\frac{\sin i}{\sin r} = \frac{n_{\text{sea}}}{n_{\text{air}}}, \tag{6.91}$$

where $n_{\text{sea}} = 1.34$ is the refractive index of vacuum (ratio of speed of light in sea water to speed of light in vacuum), and n_{air} is that of air, given at STP (0 °C, 1 atm) by

$$n_{\text{air}} = 1 + A(1 + B/\lambda^2), \tag{6.92}$$

where A and B are given in Table 6.2, and so $n_{\text{air}} = 1.0002918$ at 0.59 μm (the wavelength of the sodium D lines).

For very small i, that is essentially normally incident direct solar radiation onto a perfectly smooth ocean surface,

$$R_{\text{o}} \approx (a/b)^2 = \left(\frac{i - r}{i + r}\right)^2, \tag{6.93}$$

and on using Snell's law for small i and r we have $i/r \approx 1.34$ we obtain a reflectivity of 0.021. For an ocean we can correct the Fresnel reflectivity to take into account surface roughness

$$R_{\text{o}}^{'} = R_{\text{o}} + 0.016 \cos i, \tag{6.94}$$

which gives a higher reflectivity of about 0.04 for incident radiation normal to the surface (Kontratyev 1973). The correction factor goes to zero when the radiation is parallel to the ocean surface to maintain a Fresnel reflectivity of unity in this case. The radiation is taken to be isotropically reflected from the non-smooth

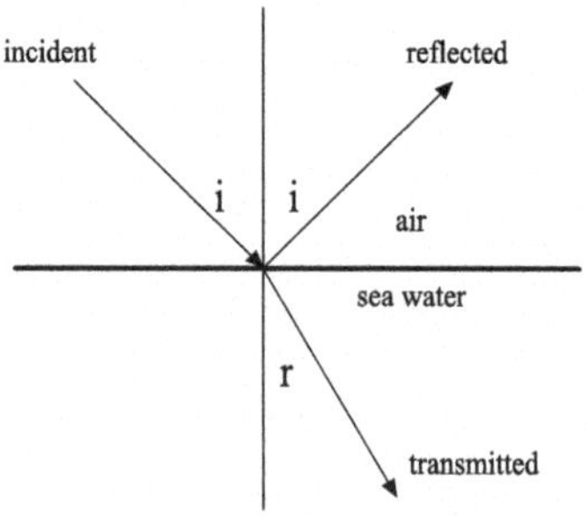

FIG. 6.16. The incident radiation at angle i to the sea surface is reflected and transmitted by the sea water. The angle of transmission is the angle of refraction r.

ocean surface. There are measurements that indicate that the surface albedo of a rough water surface may not approach the Fresnel limit at low solar elevation angles but can be nearer to 0.3 (Henderson–Sellers and Wilson 1983).

For incident diffuse solar radiation, a mean angle of incidence corresponding to $\mu = 3/5$ can be used, which gives an ocean reflectivity for diffuse radiation of 0.05, consistent with the range 0.03–0.10 given for an ocean by Sellers (1965) for cloudy–sky conditions.

Furthermore, we need to include the condition that if the Fresnel reflectivity is greater than the ice or snow reflectivity, which occurs at low solar elevations, then the reflectivities of both snow and ice are set equal to the Fresnel reflectivity. Kuhn (1989) gives observations that clearly show this behaviour for snow cover in Antarctica.

Beyond the dependence of surface reflection on the angle of incidence of the solar radiation, the reflectivity of natural surfaces can be strongly dependent on wavelength, with large differences as we go from the visible to the near–infra–red. Fresh snow has a peak reflectivity of about 0.80 near 0.6 μm falling to below 0.10 by 2.0 μm. Old snow follows a similar decrease but the reflectivity is generally lower, about 0.7 at 0.6 μm. Sand and loam have a reflectivity of about 0.2 near 0.6 μm rising to 0.4 near 2.0 μm. Vegetation has a reflectivity of about 0.1 below 0.7 μm rising rapidly to 0.4 near 0.8 μm (see Fig. 10.8 in §10.6).

6.7 Multiple scattering solution for inhomogeneous layers

For a multiple anisotropic scattering solution of the transfer equation, a set of monochromatic radiative flux equations can be solved rapidly for an absorbing–scattering inhomogeneous atmosphere using the delta–Eddington method (Joseph *et al.* 1976). The delta–Eddington method is an improvement of the Eddington method (Shettle and Weinmann 1970). The latter workers introduced a simplified approach for inclusion of anisotropic scattering in the Eddington approximation using a truncated-phase function, and gave analytical solutions for the two–boundary valued problem for homogeneous atmospheric layers, that is for ω and g constant within each layer.

Here we shall derive the radiation diffusion equation for anisotropic scattering, following their truncated–phase function approach, but for inhomogeneous layers. This allows the inclusion of molecular, aerosol and cloud scattering to be more easily accomplished in each layer. We then give the appropriate simple transformations of the standard radiation transfer parameters g (asymmetry factor), τ (extinction optical depth), and ω (single scattering albedo) for the delta–Eddington improvement to the solution. The radiation diffusion equation is a second-order ordinary differential equation that can be numerically solved rapidly via the Thomas algorithm. We first re–examine its form for isotropic scattering based on the Eddington approximation discussed in §3.5.6.

6.7.1 *Isotropic scattering solution*

For coherent and isotropic scattering we saw in Chapter 3 that the Eddington approximation results in a second–order differential equation for the mean radiance J_λ in the form of a diffusion equation

$$\frac{1}{3}\frac{\partial^2 J_\lambda(\tau_\lambda)}{\partial \tau_\lambda^2} = J_\lambda(\tau_\lambda) - S_\lambda(\tau_\lambda), \tag{6.95}$$

with the angle–independent source function S_λ given by

$$S_\lambda = (1-\omega_\lambda)B_\lambda + \omega_\lambda J_\lambda, \tag{6.96}$$

where ω_λ is the single–scattering albedo and τ_λ is the extinction optical depth. When the expression for the source function is inserted into the above diffusion equation we get

$$\frac{1}{3}\frac{\partial^2 J_\lambda(\tau_\lambda)}{\partial \tau_\lambda^2} = (1-\omega_\lambda)(J_\lambda(\tau_\lambda) - B_\lambda(\tau_\lambda)). \tag{6.97}$$

Given appropriate boundary conditions, one can rapidly solve this boundary–valued problem using a numerical technique such as the Thomas algorithm, described below, if one replaces the second derivative by a three–term finite–differences expression.

6.7.2 *Thomas algorithm*

The Thomas algorithm is Gauss elimination applied to the inversion of a tridiagonal matrix (see Williams 1972). If the first and second derivatives involved in the radiation diffusion equation are replaced by finite differences and keeping only three terms then the Thomas algorithm can be applied and is very rapid. We may replace first and second derivatives by expressions of the form

$$J_i^{''} = \beta_{1i}J_{i+1} + \beta_{2i}J_i + \beta_{3i}J_{i-1}, \tag{6.98}$$

so that the radiation diffusion equation at atmospheric level i takes the form

$$c_i J_{i+1} + a_i J_i + b_i J_{i-1} = v_i. \tag{6.99}$$

We then solve the linear system of n equations

$$\mathbf{AJ} = \mathbf{V}, \tag{6.100}$$

where $\mathbf{J}$ and $\mathbf{V}$ are vectors and $\mathbf{A}$ is a matrix of form

$$\begin{pmatrix} a_1 & c_1 & 0 & \cdots & \cdots & 0 \\ b_2 & a_2 & c_2 & 0 & \cdots & 0 \\ \cdots & \cdots & \cdots & \cdots & \cdots & \cdots \\ \cdots & \cdots & \cdots & \cdots & \cdots & \cdots \\ 0 & \cdots & 0 & b_{n-1} & a_{n-1} & c_{n-1} \\ 0 & \cdots & \cdots & 0 & b_n & a_n \end{pmatrix}$$

with imposed upper and lower boundary conditions to provide information on J_0 and J_{n+1}, required by the finite–difference expressions for the first and second derivatives.

6.7.2.1 *Solution procedure* For $i = 1$ we set

$$\alpha_1 = a_1 \qquad \gamma_1 = c_1/\alpha_1 \qquad u_1 = v_1/\alpha_1,$$

for $i = 2, 3, ...n$ we set

$$\alpha_i = a_i - b_i\gamma_{i-1}, \tag{6.101}$$

$$u_i = (v_i - b_i u_{i-1})/\alpha_i, \tag{6.102}$$

whilst for $i = 2, 3, ...n - 1$ we have

$$\gamma_i = c_i/\alpha_i, \tag{6.103}$$

with the back–substitution procedure for $i = n - 1, n - 2, ...1$

$$J_n = u_n, \tag{6.104}$$

$$J_i = u_i - \gamma_i J_{i+1}. \tag{6.105}$$

Note that matrix inversion operation counts (and hence computer time) depend on n^3, whilst the Thomas algorithm operation counts vary as n. This makes the Thomas algorithm relatively fast, especially if one needs to solve for many atmospheric layers. Furthermore, one can apply the method to a tridiagonal matrix where the matrix elements are themselves matrices, an example is the solution of the continuity equation for many species concentrations in atmospheric photochemistry (see Chapter 7, also Lavvas *et al.* 2007). In this case divisions by scalar factors are replaced by matrix inversions in the Thomas solution procedure. Generally, one can speed up matrix inversions by partitioning the matrix into four submatrices and so reduce the size by 2 and the time for each inversion by 8, at the expense of more matrix multiplications.

6.7.3 *Anisotropic scattering solution*

6.7.3.1 *Truncated phase–function approximation* Shettle and Weinman (1970) expressed the scattering phase function, $p(\vartheta)$, as a two–term truncated series of associated Legendre polynomials in the manner of Chandrasekhar (1960) to obtain

$$p(\vartheta) = 1 + 3g\cos\vartheta, \tag{6.106}$$

where ϑ is the scattering angle. This form arises by applying two conditions, that this phase function is normalized and that it satisfies eqn (3.58). For an azimuthally averaged radiation field, we can average also the phase function over the incoming azimuthal direction by replacing $\cos\vartheta$ given by eqn (3.69) to obtain

$$p(\mu,\mu^{'}) = 1 + 3g\mu\mu^{'}, \tag{6.107}$$

where $\mu^{'}$ is the cosine of the incoming zenith direction and μ that of the outgoing zenith direction. The angle-dependent source function can now be expressed as

$$S_\lambda(\mu) = S_{\mathrm{e}\lambda} + S_{\mathrm{s}\lambda} + S_{\mathrm{i}\lambda} \tag{6.108}$$

where the thermal emission source function is

$$S_{\mathrm{e}\lambda} = (1-\omega_\lambda)B_\lambda, \tag{6.109}$$

the scattering source function is

$$S_{\mathrm{s}\lambda}(\mu) = \frac{\omega_\lambda}{2}\int_{-1}^{1} p(\mu,\mu^{'})I_\lambda(\mu^{'})\mathrm{d}\mu^{'}, \tag{6.110}$$

and the source function for the incoming direct solar radiation is

$$S_{\mathrm{i}\lambda}(\mu) = \frac{\omega_\lambda}{4\pi}p(\mu,\mu_{\mathrm{o}})F^{\downarrow}_{\mathrm{d}\lambda} \tag{6.111}$$

where

$$F^{\downarrow}_{\mathrm{d}\lambda} = S_{\odot\lambda}\mathrm{e}^{-\tau_\lambda/\mu_{\mathrm{o}}} \tag{6.112}$$

is the direct solar flux, at optical depth τ_λ, normal to the direction of incidence defined by μ_{o}, the cosine of the solar zenith angle. $S_{\odot\lambda}$ is the incoming solar spectral flux (erg cm^{-2} s^{-1} μm^{-1}) at the Earth's orbit. We note that the phase function is divided by 4π sr so that $S_{i\lambda}(\mu)$ has units of radiance (erg cm^{-2} s^{-1} sr $^{-1}$ μm^{-1}.

If we take μ_o positive in the downwards direction (increasing optical depth), integrate over the incoming direction and use the Eddington approximation $I_\lambda = J_\lambda + 3\mu H_\lambda$, eqn (3.89), we get

$$S_\lambda(\mu) = (1-\omega_\lambda)B_\lambda + \omega_\lambda(J_\lambda + 3g\mu H_\lambda) + \frac{\omega_\lambda}{4\pi}(1+3g\mu\mu_o)F^{\downarrow}_{d\lambda}, \tag{6.113}$$

where H_λ is the second moment of the radiance, I_λ, given by eqn (3.93), but now defined by

$$\frac{\partial H_\lambda}{\partial \tau_\lambda} = -J_\lambda + S_\lambda, \tag{6.114}$$

where the angle–averaged source function (*first moment of the source function*) is given by

$$S_\lambda = \frac{1}{2}\int_{-1}^{1} S_\lambda(\mu)\mathrm{d}\mu \tag{6.115}$$

$$= (1 - \omega_\lambda)B_\lambda + \omega_\lambda J_\lambda + \frac{\omega_\lambda}{4\pi}F^{\downarrow}_{\mathrm{d}\lambda}. \tag{6.116}$$

The equation of the third moment of the radiance, given by eqn (3.94), can now be written

$$\frac{\partial K_\lambda}{\partial \tau_\lambda} = -H_\lambda + Q_\lambda, \tag{6.117}$$

where Q_λ is here defined as the *second moment of the source function* defined by

$$Q_\lambda = \frac{1}{2}\int_{-1}^{1} \mu S_\lambda(\mu)\mathrm{d}\mu \tag{6.118}$$

$$= \omega_\lambda g H_\lambda + \frac{\omega_\lambda}{4\pi} g\mu_o F^{\downarrow}_{\mathrm{d}\lambda}. \tag{6.119}$$

Since $K = J/3$, eqn (3.100), we can replace K in terms of J in eqn (6.117) to obtain

$$\frac{1}{3}\frac{\partial J_\lambda(\tau_\lambda)}{\partial \tau_\lambda} = -\alpha H_\lambda(\tau_\lambda) + \beta, \tag{6.120}$$

where

$$\alpha = 1 - \omega_\lambda g, \tag{6.121}$$

$$\beta = \frac{\omega_\lambda}{4\pi} g\mu_o F^{\downarrow}_{\mathrm{d}\lambda}. \tag{6.122}$$

We can divide the equation by α and differentiate a second time to obtain the radiation diffusion equation, cf. eqn (6.95), on replacing the derivative of H in terms of J from eqn (6.114)

$$\frac{\partial}{\partial \tau_\lambda}\left(a(\tau_\lambda)\frac{\partial J_\lambda(\tau_\lambda)}{\partial \tau_\lambda}\right) = J(\tau_\lambda) - S(\tau_\lambda) + \frac{\partial b(\tau_\lambda)}{\partial \tau_\lambda}, \tag{6.123}$$

where $a = 1/3\alpha$ and $b = \beta/\alpha$. For the case of isotropic scattering, $g = 0$, and for no incoming solar flux $a = 1/3$ and $b = 0$, and so we obtain the radiation diffusion equation given by eqn (6.97).

If we now substitute for the source function S_λ from eqn (6.116) we get a second-order differential equation

$$c_1\frac{\partial^2 J_\lambda}{\partial \tau_\lambda^2} + c_2\frac{\partial J_\lambda}{\partial \tau_\lambda} + c_3 J_\lambda + c_4 = 0, \tag{6.124}$$

where

$$c_1 = a(\tau_\lambda), \tag{6.125}$$

$$c_2 = \frac{\partial a_\lambda}{\partial \tau_\lambda}, \tag{6.126}$$

$$c_3 = \omega_\lambda - 1, \tag{6.127}$$

$$c_4 = -c_3 B_\lambda - \frac{\partial b_\lambda}{\partial \tau_\lambda} + \frac{\omega_\lambda}{4\pi} F^{\downarrow}_{\mathrm{d}\lambda}, \tag{6.128}$$

that we can solve rapidly to obtain J_λ as a function of optical depth for each wavelength, given the boundary conditions at the top of the atmosphere and at the surface. The upwards and downwards diffuse spectral fluxes at each atmospheric level can then be calculated from

$$F^{\uparrow}_\lambda = \pi(J_\lambda - 2H_\lambda), \tag{6.129}$$

$$F^{\downarrow}_\lambda = \pi(J_\lambda + 2H_\lambda), \tag{6.130}$$

so that the net downwards diffusive flux is $4\pi H_\lambda$, while the total flux available for photolysis is $4\pi J_\lambda + \mu_\mathrm{o} F^{\downarrow}_{\mathrm{d}\lambda}$, so that the enhancement factor for photolysis is

$$f = \frac{4\pi J_\lambda + \mu_\mathrm{o} F^{\downarrow}_{\mathrm{d}\lambda}}{\mu_\mathrm{o} S_{\odot\lambda}}. \tag{6.131}$$

H can be computed from

$$H_\lambda = -a \frac{\partial J_\lambda(\tau_\lambda)}{\partial \tau_\lambda} + b. \tag{6.132}$$

The total downwards flux, $F^{\downarrow}_{\mathrm{T}\lambda}$, comprises the diffuse and direct components

$$F^{\downarrow}_{\mathrm{T}\lambda} = F^{\downarrow}_\lambda + \mu_\mathrm{o} F^{\downarrow}_{\mathrm{d}\lambda}, \tag{6.133}$$

and the net downwards flux is

$$F^{\downarrow}_{\mathrm{net}\lambda} = F^{\downarrow}_{\mathrm{T}\lambda} + F^{\uparrow}_{\mathrm{d}\lambda}. \tag{6.134}$$

The surface boundary condition is then

$$F^{\uparrow}_\lambda = R_{\mathrm{s}_\lambda} F^{\downarrow}_{\mathrm{T}\lambda}, \tag{6.135}$$

where R_{s_λ} is the surface spectral reflectance and both diffuse and direct solar radiation is diffusely reflected upwards at the surface. Thus the total upwards flux is equal to the diffuse component, $F^{\uparrow}_{\mathrm{T}\lambda} = F^{\uparrow}_\lambda$. At the top of the atmosphere the boundary condition is that there is no incoming diffuse component so $F^{\downarrow}_\lambda = 0$. The boundary conditions in terms of J_λ then become

$$d_1 \frac{\partial J_\lambda(\tau_\lambda)}{\partial \tau_\lambda} + d_2 J_\lambda + d_3 = 0, \tag{6.136}$$

where for the surface boundary condition

$$d_1 = 2\pi a(1 + R_{\mathrm{s}\lambda}), \tag{6.137}$$

$$d_2 = \pi(1 - R_{s\lambda}), \tag{6.138}$$

$$d_3 = -\pi 2b(1 + R_{s\lambda}) - R_{s\lambda}\mu_o F^{\downarrow}_{d\lambda}, \tag{6.139}$$

while for the upper boundary condition

$$d_1 = -2a, \tag{6.140}$$

$$d_2 = 1, \tag{6.141}$$

$$d_3 = 2b. \tag{6.142}$$

6.7.3.2 *Enhancement factor for photolysis* The enhancement factor for photolysis is given by

$$f = \frac{4\pi J_\lambda + \mu_o F^{\downarrow}_{d\lambda}}{\mu_o S_{\odot\lambda}}, \tag{6.143}$$

where $\mu_o S_{\odot\lambda}$ is equal to the incoming radiation at the top of the atmosphere (TOA). At TOA the boundary condition gives $J_\lambda = -2H_\lambda$ so that the upwards diffusive flux is then equal to $2\pi J_\lambda$. The maximum value for the enhancement factor for photolysis at TOA is then obtained for a purely scattering atmosphere that overlays a surface with reflectance $R_{s\lambda} = 1$. Under these conditions the upwards diffusive flux is equal to the incoming radiation so that $2\pi J_\lambda = \mu_o S_{\odot\lambda}$ and so the maximum value of f at TOA is 3.

Under the above scattering conditions, J_λ becomes constant deep in the atmosphere for large scattering optical depths, where the direct component vanishes, and is given by

$$J_{\text{deep}\lambda} = \mu_o S_{\odot\lambda}\left(\frac{1}{2\pi} + \frac{3\mu_o}{4\pi}\right), \tag{6.144}$$

so that the maximum value for f is then attained for $\mu_o = 1$ and is equal to 5. We note that the ratio of $J_{\text{TOA}\lambda}$ to $J_{\text{deep}\lambda}$ is then

$$\frac{J_{\text{TOA}\lambda}}{J_{\text{deep}\lambda}} = \frac{2}{2 + 3\mu_o}, \tag{6.145}$$

and hence, when $\mu_o = 1$ the above ratio is $\frac{2}{5}$ (cf. the limb–darkening law we discussed in §3.5.10).

6.7.3.3 *Delta–Eddington approximation* Joseph *et al.* (1976) improved the truncated phase function approach by including a strong forward scattering component in the phase function

$$p(\vartheta) = 2\text{f}(1 - \cos\vartheta) + (1 - \text{f})(1 + 3g\cos\vartheta), \tag{6.146}$$

where f is the fraction of scattered photons in the forward direction. Assuming that the original phase function that they were approximating was the Henyey–Greenstein phase function, they obtained f$= g^2$. They then found simple transformations that significantly improved the truncated–phase function approach. These were

$$\tau' = (1 - \omega \mathrm{f})\,\tau, \tag{6.147}$$

$$\omega' = \frac{(1 - \mathrm{f})\,\omega}{1 - \omega \mathrm{f}}, \tag{6.148}$$

$$g' = \frac{g}{1 + g}. \tag{6.149}$$

6.7.4 *Atmospheres with clouds and aerosols*

The atmosphere can be divided into layers whose individual optical depth is evaluated, according to its properties, from

$$\tau = \tau_{\mathrm{cs}} + \tau_{\mathrm{ca}} + \tau_{\mathrm{aers}} + \tau_{\mathrm{aera}} + \tau_{\mathrm{ma}} + \tau_{\mathrm{R}}, \tag{6.150}$$

where τ_{cs} is the cloud–scattering optical depth, τ_{ca} is the cloud–absorption optical depth, τ_{aers} is the aerosol–scattering optical depth, τ_{aera} is the aerosol–absorption optical depth, τ_{ma} is that for molecular absorption, and τ_{R} is that for Rayleigh or molecular scattering. The single scattering albedo for each layer is

$$\omega = \omega_{\mathrm{c}} + \omega_{\mathrm{aer}} + \omega_{\mathrm{R}}, \tag{6.151}$$

$$\omega_{\mathrm{c}} = \tau_{\mathrm{cs}}/\tau, \tag{6.152}$$

$$\omega_{\mathrm{aer}} = \tau_{\mathrm{aers}}/\tau, \tag{6.153}$$

$$\omega_{\mathrm{R}} = \tau_{\mathrm{R}}/\tau, \tag{6.154}$$

where

$$g\omega = g_{\mathrm{c}}\omega_{\mathrm{c}} + g_{\mathrm{aer}}\omega_{\mathrm{aer}} + g_{\mathrm{R}}\omega_{\mathrm{R}}, \tag{6.155}$$

with g_{c} and g_{aer} being the cloud and aerosol asymmetry factors, respectively, with the Rayleigh asymmetry factor $g_{\mathrm{R}} = 0$.

To proceed, we need to divide the atmosphere into clear– and cloudy–sky components. The cloudy-sky component is subdivided into non–overlapping components covered by low–level, middle–level and high-level clouds. There are methods for taking into account cloud overlap (Chapter 8). The delta–Eddington method computes the fraction of the incoming solar flux that is reflected by the planet to space, R_{p}, the fraction absorbed by the atmosphere, a_{pa}, and that absorbed by the Earth's surface, a_{pg}, where the planetary absorptivity $a_{\mathrm{p}} = a_{\mathrm{pa}} + a_{\mathrm{pg}}$, and so $R_{\mathrm{p}} + a_{\mathrm{pa}} + a_{\mathrm{pg}} = 1$. These fractions define the shortwave radiation budget of the planet.

The planetary absorptivity has a clear–sky component a_{s} and three cloudy–sky components a_{ci} and is expressed as

$$a_{\mathrm{p}} = (1 - A_{\mathrm{c}})a_{\mathrm{s}} + \sum_i A_{\mathrm{ci}}a_{\mathrm{ci}}, \tag{6.156}$$

where the cloud–cover fractions A_{ci} correspond to each of the individual cloud–cover components.

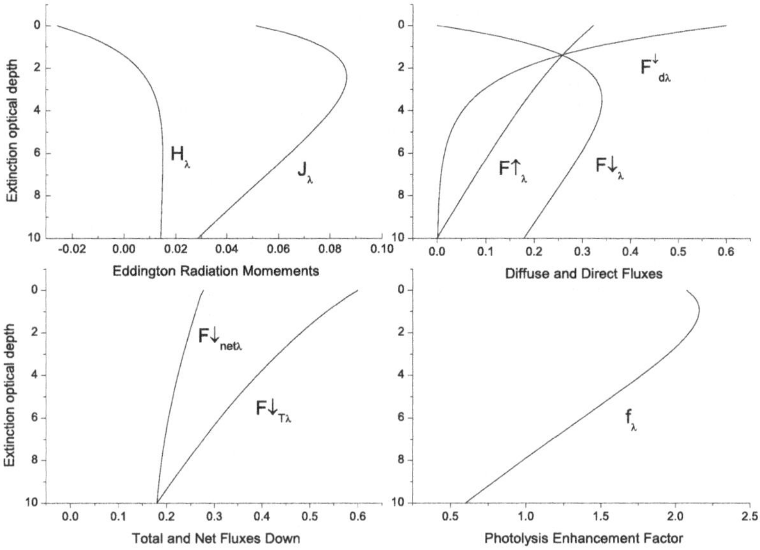

FIG. 6.17. Variation of the Eddington radiation moments J_λ and H_λ, and fluxes; diffuse up $F_\lambda^\uparrow$, diffuse down $F_\lambda^\downarrow$, direct down $F_{\mathrm{d}\lambda}^\downarrow$, net down $F_{\mathrm{net}\lambda}^\downarrow$, total down $F_{\mathrm{T}\lambda}^\downarrow$, and photolysis enhancement factor f_λ. The atmosphere has total extinction optical depth $\tau_\lambda = 10$, single scattering albedo $\omega_\lambda = 0.99$, asymmetry factor $g = 0.8$, incoming solar flux $S_{\odot\lambda} = 1$, solar zenith angle $\mu_\odot = 0.6$, surface albedo $R_{\mathrm{s}\lambda} = 0$, and no thermal emission.

6.7.5 *Sample computations*

Here, we present sample computations using the delta–Eddington method and solve numerically, via the Thomas algorithm, the radiation diffusion equation as a second–order ordinary differential equation, as described above, using 40 optical depth levels. We consider an atmosphere with total monochromatic extinction optical depth $\tau_\lambda = 10$, single scattering albedo $\omega_\lambda = 0.99$ and asymmetry factor $g_\lambda = 0.8$ at all levels, and no thermal emission. We set the incoming solar flux $S_{\odot\lambda} = 1$ and the solar zenith angle by $\mu_\odot = 0.6$. The resulting Eddington radiation moments $J_\lambda(\tau_\lambda)$ and $H_\lambda(\tau_\lambda)$ are shown in Fig. 6.17. The various radiation fluxes are also shown together with the photolysis enhancement factor. The fraction of the radiation reflected to space, i.e. the planetary albedo at the top of the atmosphere ($\tau_\lambda = 0$) is 0.538, the fraction absorbed by the surface is 0.301 and the fraction absorbed by the atmosphere is 0.161 and sum up to unity.

6.8 Bibliography

6.8.1 *Notes*

For more details regarding the exponential–sumfit, k–distribution and correlated–k methods for solar radiation transfer see, for example, Kato *et al.*, and for an application see Irwin *et al.*

For ISCCP cloud climatologies see Rossow *et al.* and Rossow and Schiffer.

For more details on scattering of light by particles see the text by van de Hulst.

For tables of refractive index for standard air and Rayleigh scattering coefficients see the early work of Penndorf.

For wavelength variation of the asymmetry factor for water clouds see the report by Stephens.

6.8.2 *References and further reading*

Allen, C. W. (1976). *Astrophysical quantities.* Athlone Press, London.

Ångström, A. (1929). On the atmospheric transmission of sun radiation and on dust in the air. *Geograf. Ann. Deut.*, **11**, 156–166.

Burch, D. E. and Williams, D. (1962). Total absorptance of carbon monoxide and methane in the infrared. *Appl. Opt.*, **1**, 587–594.

Chandrasekhar, S. (1960). *Radiative transfer.* Dover Publications Inc., New York.

Debye, P. (1909). Der Lichtdruk auf Kugeln von beliebigem Material. *Ann. Physik. (Leipzig)*, **30**, 57–136.

Fotiadi, A., Drakakis, E., Hatzianastassiou, N., Matsoukas, C., Pavlakis, K. G., Hatzidimitriou, D., Gerasopoulos, E., Mihalopoulos, N., and Vardavas, I. M., (2006). Aerosol physical and optical properties in the eastern Mediterranean basin, Crete, from Aerosol Robotic Network data. *Atmos. Chem. Phys.*, **6**, 5399–5412.

France, W. L. and Williams, D. (1966). Total absorptance of ammonia in the infrared. *J. Opt. Soc. Amer.*, **56**, 70–74.

Hatzianastassiou, N., Katsoulis, B. and Vardavas, I. M. (2004). Global distribution of aerosol direct radiative forcing in the ultraviolet and visible arising under clear skies. *Tellus*, **56B**, 51–71.

Henderson-Sellers, A. and Wilson, M. F. (1983). Surface albedo data for climate modelling. *Rev. Geophys. Space Phys.*, **21**, 1743–1778.

Irvine, W. M. (1968). Multiple scattering by large particles II. Optically thick layers, *Astrophys. J.*, **152**, 823–834.

Irwin, P.G., Calcutt, S. B., Taylor, F. W. and Weir, A. L. (1996). Calculated k distribution coefficients for hydrogen– and self–broadened methane in the range 2000–9500 cm^{-1} from exponential sum fitting to band–modelled spectra. *J. Geophys. Res.*, **101**, 26137–26154.

Joseph, J. H., Wiscombe, W. J. and Weinman, J. A. (1976). The delta-Eddington approximation for radiative flux transfer. *J. Atmos. Sci.*, **33**, 2452–2459.

JPL 2006 : *Chemical Kinetics and Photochemical Data for Use in Atmospheric Studies Evaluation Number 15.* NASA Panel for Data Evaluation: S. P. Sander, R. R. Friedl, A. R. Ravishankara, D. M. Golden, C. E. Kolb, M. J. Kurylo, J. Molina, G. K. Moortgat, H. Keller-Rudek, B. J.Finlayson-Pitts, P.H. Wine, R. E. Huie, and V. L. Orkin. Jet Propulsion Laboratory Publication 06-2, California Institute of Technology, Pasadena.

Kato, S., Ackerman, T. P., Mather, J. H., Clothiaux, E. E. (1999). The k-distribution method and correlated–k approximation for a shortwave radiative transfer model. *J. Quant. Spectrosc. Radiat. Trans.*, **62**, 109–121.

Koepke, P., Hess, M., Schult, I. and Shettle, E. P. (1997). *Global Aerosol Data Set.* Report No 243, Max-Planck Institut für Meteorologie, Hamburg.

Kontratyev, K. Ya. (1973). *Radiation characteristics of the atmosphere and the Earth's surface.* Amerind. Pub. Co., New Delhi.

Kuhn, M. H. (1989). The role of land ice and snow in climate. In *Understanding climate change*, eds. A. Berger, R.E. Dickinson, and J.W. Kidson, Geophys. Monogr., 52, IUGG 7, AGU, Washington, DC.

Lavvas, P., Coustenis, A. and Vardavas, I. M. (2007). Coupling photochemistry with haze formation in Titan's atmosphere. Part I. Model description. *Planet. Space Sci.*, in press.

Mie, G. (1908). *Contributions to the optics of turpid media, especially colloidal metal solutions. Ann. Physik. (Leipzig)*, **25**, 377–445.

Penndorf, R. (1957). Tables of refractive index for standard air and the Rayleigh scattering coefficient for the spectral region between 0.2 and 20.0 μm and their application to atmospheric optics. *J. Opt. Soc. Am.*, **47**, 176–182.

Rossow, W. B., Walker, A. W., Beuschel, D. E. and Roiter, M. D. (1996). *International Satellite Cloud Climatology Project (ISCCP). Documentation of new cloud datasets*, 115 pp., World Meteorological Organisation, Geneva.

Rossow, W. B. and Schiffer, R. A. (1999). Advances in understanding clouds from ISCCP. *Bull. Am. Meteorol. Soc.*, **80**, 2261–2287.

Sagan, C. and Pollack, J. B. (1967). Anisotropic non-conservative scattering and the clouds of Venus. *J. Geophys. Res.*, **72**, 469–477.

Sellers, W. D. (1965). *Physical climatology.* University of Chicago Press, Chicago.

Shettle, E. P. and Weinmann, J. A. (1970). The transfer of solar irradiance through inhomogeneous turbid atmospheres evaluated by Eddington's approximation. *J. Atmos. Sci.*, **27**, 1048–1055.

Stull, V. R., Wyatt, P. and Plass, G. N. (1964). The infrared transmittance of carbon dioxide. *Appl. Opt.*, **3**, 241–254.

Stephens, G. L. (1979). Optical properties of eight water cloud types. *CSIRO Div. of Atmos. Phys. Tech. Pap. 36*, Melbourne.

Valley, S. L. (ed.) (1965). *Handbook of geophysics and space environments.* Air Force Cambridge Research Laboratories. McGraw-Hill, New York.

van de Hulst, H. C. (1981). *Light scattering by small particles.* Dover, New York.

Vardavas, I. M. and Carver, J. H. (1984). Solar and terrestrial parameterizations for radiative-convective models. *Planet. Space Sc.*, **32**, 1307–1325.

Vardavas, I. M. (1989). A Fibonacci search technique for model parameter selection. *Ecol. Model.*, **48**, 65–81.

Williams, P. W. (1972). *Numerical computation.* Nelson, London.

Wyatt, P. J., Stull, V. R. and Plass, G. N. (1964). The infrared transmittance of water vapour. *Appl. Opt.*, **3**, 229–241.

7

ATMOSPHERIC PHOTOCHEMISTRY

7.1 Introduction

Solar ultraviolet radiation plays a key role in the determination of a planet's climate through photolysis reactions involving the molecules that constitute their atmospheres. In the Earth's atmosphere, radiation below about 300 nm plays a key role as important molecules such as H_2O, CO_2, O_2 and O_3, are easily photolyzed in this spectral region. Most of the photolysis occurs in the upper atmosphere, and via diffusion turbulence and advection the photolysis products are transferred to the lower atmosphere where they take part in the chemistry of the atmospheric constituents. The atmospheric composition in turn controls both the solar and terrestrial radiation fields and the thermal structure. There is thus a strong coupling between atmospheric photochemistry, composition and climate.

The Earth's surface also plays a crucial role in the control of species concentrations via emission (both anthropogenic and natural) and deposition processes (both dry and wet removal), representing species sources and sinks. Loss to space via diffusion and thermal (Jeans) escape is important for the lighter species, such as H and H_2. Inflow from space of molecules, such as H_2O and CO_2, via meteorites can also be important to the long–term evolution of planetary atmospheres. Thus whatever species is emitted into the atmosphere of a planet, there must be ultimately a loss mechanism to the surface or space, otherwise there will be an accumulation of the emitted species or its photochemical products.

The concentration of molecular species at some level within the atmosphere is controlled by a number of physico–chemical processes that include

1. Emission from the Earth's surface
2. Ejection and infall into the atmosphere
3. Dry deposition at the Earth's surface
4. Wet deposition or rainout within the troposphere
5. Transport by turbulence, diffusion, and advection
6. Collisionally driven chemical production and destruction
7. Radiation–driven production and destruction.

7.2 The continuity equation

The number density, n_i, of species i, is determined at each atmospheric level by the continuity equation that describes the rate of change of the number density due to production and loss mechanisms arising from photochemical reactions, emission, surface deposition and transport. The general form of the continuity equation for a molecular species i is

$$\frac{\partial n_i}{\partial t} + \nabla \cdot (n_i \mathbf{u}) = C_i + P_i + S_{\mathrm{e}i} + S_{\mathrm{d}i}, \tag{7.1}$$

where $\mathbf{u}$ is the three–dimensional velocity vector, $\phi_i = n_i \mathbf{u}$ is the total flux of particles arising from diffusion and advection, C_i is the net production rate arising from collisionally induced reactions, P_i is the net production rate from photolysis, $S_{\mathrm{e}i}$ is the net production rate arising from surface emission processes and $S_{\mathrm{d}i}$ is the net loss rate from deposition processes at the Earth's surface.

The 1D continuity equation for species i at altitude z can be written as

$$\frac{\partial n_i}{\partial t} = -\frac{\partial \phi_i}{\partial z} + C_i + P_i + S_{\mathrm{e}i} + S_{\mathrm{d}i} \tag{7.2}$$

where t is time, z is altitude, and ϕ_i is the total vertical diffusive flux. The net production term C_i from collisionally driven reactions is determined from the expression

$$C_i = \sum_j \psi_{ij} k_j \prod_l n_l^{\nu_{lj}}, \tag{7.3}$$

where ν_{lj} is the reactant *stoichiometric coefficient* of species l involved in each collisional reaction j with forward reaction rate $k_j(T,p)$. We recall that the general elementary reaction can be written as

$$\sum_l \nu_{lj} X_l \rightleftharpoons \sum_l \nu'_{lj} X_l, \tag{7.4}$$

where X_l is species l taking part in reaction j, ν'_{lj} is the product stoichiometric coefficient and ψ_{lj} is the net stoichiometric coefficient for species l, given by

$$\psi_{lj} = \nu'_{lj} - \nu_{lj}, \tag{7.5}$$

so that $\psi_{lj} < 0$ if species l is a reactant species and $\psi_{lj} > 0$ if it is a product species. As an example, in the bimolecular collisional reaction

$$\mathrm{N_2 + O_2 \rightleftharpoons 2NO} \tag{7.6}$$

for species NO, we have $\nu = 0$, $\nu' = 2$, and $\psi = 2$.

The net photolysis production term P_i of species i is determined from the expression

$$P_i = \sum_k \psi_{ik} j_k n_i \tag{7.7}$$

for each photolysis reaction k with rate j_k and

$$\psi_{ik} = \nu'_{ik} - \nu_{ik}, \tag{7.8}$$

so that $\psi_{ik} < 0$ if species i is a photolyzed species and $\psi_{ik} > 0$ if it is a photolysis product species. As an example, in the photolysis reaction

$$\mathrm{H_2O} + \mathrm{h}\nu \rightarrow \mathrm{OH} + \mathrm{H} \tag{7.9}$$

for species OH, we have $\nu = 0$, $\nu' = 1$, and $\psi = 1$.

7.3 Brownian and turbulent diffusion

7.3.1 *Fick's law*

Let us define l as the mean free path between collisions between like molecules undergoing random thermal motion (Brownian). If $\bar{v}$ is the mean velocity of the molecules then the frequency of collision is $f = \bar{v}/l$ (s^{-1}). For an isotropic velocity distribution one third of the molecules will move in the X–direction and of these half will go forward. Thus, the number of collisions in the forward direction will be $\bar{v}n/6l$, where n is the particle density in molecules cm^{-3}. For molecules moving along the X–axis towards the origin $x = 0$, the probability of reaching the origin after colliding at x without further collisions is $\mathrm{e}^{-x/l}$. Thus, the change in the net flux ϕ in the positive X–direction about $x = 0$ is

$$\mathrm{d}\phi = \frac{\bar{v}}{6l} n(x) \mathrm{e}^{-x/l} \mathrm{d}x, \tag{7.10}$$

on integrating, we get the flux at the origin

$$\phi_0 = \frac{\bar{v}}{6l} \left[\int_{-\infty}^{0} n(x) \mathrm{e}^{x/l} \mathrm{d}x - \int_{0}^{\infty} n(x) \mathrm{e}^{-x/l} \mathrm{d}x \right]. \tag{7.11}$$

Now, for atmospheres the mean free path is very small compared to distances over which the density is changing (see next section), thus we can consider small gradients in the number density about $x = 0$ and so through a Taylor expansion we obtain

$$n(x) = n_0 + x \left(\frac{\mathrm{d}n}{\mathrm{d}x} \right)_0 + ..., \tag{7.12}$$

so that

$$\phi_0 = \frac{\bar{v}}{6l} \left(\frac{\mathrm{d}n}{\mathrm{d}x} \right)_0 \left[\int_{-\infty}^{0} x \mathrm{e}^{x/l} \mathrm{d}x - \int_{0}^{\infty} x \mathrm{e}^{-x/l} \mathrm{d}x \right], \tag{7.13}$$

and so we can write

$$\phi_0 = -D\left(\frac{\mathrm{d}n}{\mathrm{d}x}\right)_0 \tag{7.14}$$

where $D = \bar{v}l/3$ is the molecular diffusion coefficient in $\mathrm{cm^2\ s^{-1}}$ and the above expression is known as *Fick's law.*

7.3.2 *Bimolecular diffusion*

If we are considering the diffusion of molecule A in a bath molecule B, then the frequency of collision is given by $f_{\mathrm{AB}} = \bar{v}_{\mathrm{AB}}/l_{\mathrm{AB}}$ where the mean molecular speed, assuming a Maxwellian distribution of velocities is given by

$$\bar{v}_{\mathrm{AB}} = \left(\frac{8kT}{\pi m_{\mathrm{AB}}\delta}\right)^{\frac{1}{2}}, \tag{7.15}$$

where $m_{\mathrm{AB}} = m_{\mathrm{A}}m_{\mathrm{B}}/(m_{\mathrm{A}} + m_{\mathrm{B}})$ is the reduced mass of the system and δ is a symmetry factor equal to 1 for unlike molecules and 2 for like molecules. The mean free path is

$$l_{\mathrm{AB}} = \frac{1}{\pi n_{\mathrm{B}} d_{\mathrm{AB}}^2 \sqrt{1 + m_{\mathrm{A}}/m_{\mathrm{B}}}}, \tag{7.16}$$

where $d_{\mathrm{AB}} = (d_{\mathrm{A}} + d_{\mathrm{B}})/2$ is the mean diameter of the colliding molecules. For like particles the reduced mass becomes $m/2$ and for molecular diameter d we have

$$l = \frac{1}{\sqrt{2}n\pi d^2}, \tag{7.17}$$

and

$$\bar{v} = \left(\frac{8kT}{\pi m}\right)^{\frac{1}{2}}. \tag{7.18}$$

For example, for N_2 at a pressure of 1 atm, at 300 K the number density is 2.4×10^{19} molecules $\mathrm{cm^{-3}}$ and for a typical molecular diameter of 2×10^{-8} cm we get $l = 3\times10^{-5}$ cm. From Avogadro's number, the mass is equal to $28/(6\times10^{23})$ g, and so the mean speed is 5×10^4 cm $\mathrm{s^{-1}}$. Thus, the diffusion coefficient is 0.5 $\mathrm{cm^2\ s^{-1}}$ and the collision frequency is $5\times10^9\ \mathrm{s^{-1}}$.

In the atmosphere we consider diffusion of a molecule i in the vertical direction and define the flux to be positive in the direction of the decreasing atmospheric density with altitude z. We can then write

$$\phi_{ij} = -D_{ij}\frac{\partial n_i}{\partial z}, \tag{7.19}$$

where D_{ij} ($\mathrm{cm^2\ s^{-1}}$) is the molecular diffusion coefficient of molecule i in a bath molecule j, which can be expressed as $D_{ij} = b_{ij}/n$, where b_{ij} is the binary collision parameter ($\mathrm{cm^{-1}\ s^{-1}}$) and n is the total number density. For a gas

Table 7.1 *Bimolecular diffusion coefficients, at $T_o = 273$ K and $p_o = 1$ atm. (see Notes)*

System	D_0	m
$H–CO_2$	0.96	1.70
$H–N_2$	1.23	1.70
$He–CO_2$	0.494	1.80
$He–N_2$	0.621	1.73
$H_2–CO_2$	0.575	1.76
$H_2–N_2$	0.689	1.72
$H_2O–CO_2$	0.146	1.84
$H_2O–N_2$	0.260	1.84
$O–N_2$	0.248	1.77
$N–N_2$	0.251	1.77

mixture the binary collision parameter will correspond to the two types of colliding molecules. As we saw from simple kinetic theory, the bimolecular diffusion coefficient has a pressure and temperature dependence of the form

$$D_{ij} = D_{o,ij}\left(\frac{p_o}{p}\right)\left(\frac{T}{T_o}\right)^{m_{ij}}, \tag{7.20}$$

where $D_{o,ij}$ is the diffusion coefficient of molecule i in a bath molecule j at standard conditions $T_o = 273$ K, $p_o = 1$ atm, with values usually in the range 0.1 and 1.0 cm^2 s^{-1}, and m_{ij} is usually about 1.70 as shown in Table 7.1.

Similarly, we can define a turbulent or eddy diffusion coefficient, K, for the gas mixture, which needs to be determined empirically (usually based on inert or tracer species in the atmosphere). In the troposphere $K(z)$ has a constant value of about 1.0×10^5, as shown in Fig. 7.1. From the pressure dependence of the molecular diffusion coefficient, we see that the eddy diffusion coefficient is larger for pressures below about $p_o/p < 10^6$, that is below about 100 km altitude.

7.3.3 *Diffusive flux*

In a 1D stratified atmosphere where the density decreases with altitude z, the diffusive flux equation needs to be corrected for *apparent diffusion* due to the density gradient that arises from gravitational stratification in the number density (see Chapter 2). In the absence of any diffusion, we would still have an apparent component of Brownian diffusive flux implicitly included through the stratification density gradient, even if the atmosphere was in diffusive equilibrium. The corrected Brownian diffusive flux is then

$$\phi_i = -D_i\frac{\partial n_i}{\partial z} + D_i\left(\frac{\partial n_i}{\partial z}\right)^*, \tag{7.21}$$

where for the correction term

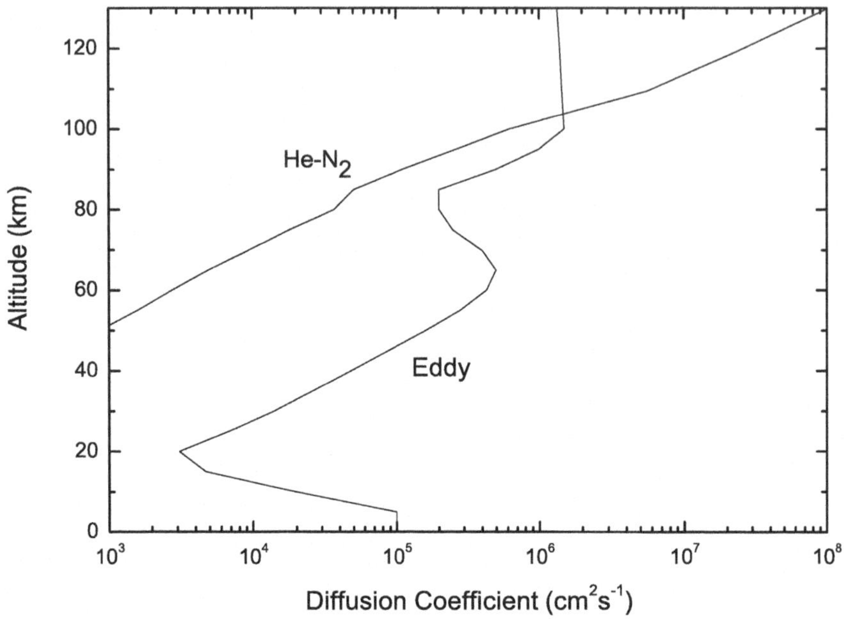

FIG. 7.1. Semiempirical eddy–diffusion coefficient profile and the bimolecular diffusion coefficient for helium in nitrogen. (Based on data from Brasseur and Solomon 1986, Strobel 1989)

$$\left(\frac{\partial n_i}{\partial z}\right)^* = -\frac{n_i}{H_i^*}, \tag{7.22}$$

and the density stratification scale height H_i^* for species i can be derived from hydrostatic equilibrium (see Chapter 2) and is given by

$$\frac{1}{H_i^*} = \frac{M_i g}{RT} + \frac{1}{T}\frac{\mathrm{d}T}{\mathrm{d}z}, \tag{7.23}$$

where M_i is the molecular weight, g is gravitational acceleration, T is atmospheric temperature, and R is the universal gas constant. We see that when the molecular density structure follows that given by the stratification structure we have zero Brownian diffusive flux.

Similarly, the atmospheric scale height when we have complete mixing is given by

$$\frac{1}{H} = \frac{Mg}{RT} + \frac{1}{T}\frac{\mathrm{d}T}{\mathrm{d}z}, \tag{7.24}$$

where M is the atmospheric mean molecular weight and

$$\frac{\partial n}{\partial z} = -\frac{n}{H}. \tag{7.25}$$

We can now express the Brownian diffusive flux in terms of the mixing ratio $f_i = n_i/n$ by

$$\phi_{i\mathrm{B}} = b_i f_i \left(\frac{1}{H} - \frac{1}{H_i^*}\right) - D_i n \frac{\partial f_i}{\partial z}. \tag{7.26}$$

For turbulent or eddy diffusion we need to correct the diffusive flux to remove the apparent flux of molecules i that arises from the stratification in the total number density

$$\phi_{i\mathrm{E}} = -K\frac{\partial n_i}{\partial z} + K\left(\frac{\partial n_i}{\partial z}\right)^*, \tag{7.27}$$

where the second term is the fraction of the apparent total turbulent flux that corresponds to molecules i. For a well-mixed atmosphere we can now write

$$\left(\frac{\partial n_i}{\partial z}\right)^* = -\frac{n_i}{H}. \tag{7.28}$$

The eddy diffusive flux then reduces to

$$\phi_{i\mathrm{E}} = -Kn\frac{\partial f_i}{\partial z}. \tag{7.29}$$

The total diffusive flux can then be written in terms of the mixing ratio f_i as

$$\phi_i = b_i f_i \beta_i - (D_i + K_i)\, n\frac{\partial f_i}{\partial z}, \tag{7.30}$$

where the buoyancy term β_i is given by

$$\beta_i = \gamma_i \left(\frac{1}{H} - \frac{1}{H_i^*}\right), \tag{7.31}$$

and we have also introduced a correction factor γ_i. In the lower atmosphere, where eddy diffusion maintains a well–mixed atmosphere, the buoyancy term needs to be corrected as the stratification of light species is destroyed by turbulent mixing with the result that the scale height H_i^* needs to be replaced by H. Thus, the correction factor γ_i is equal to zero near the Earth's surface increasing to 1 in the upper atmosphere. A possible variation of γ for the present atmosphere is

$$\gamma(z) = 1 - \mathrm{e}^{-z/10} \tag{7.32}$$

where the altitude z is in km so that by 100 km $\gamma = 1$.

Another component of diffusion is thermal diffusion with the flux given by

$$\phi_{i\mathrm{T}} = -D_i n_i \frac{\alpha_i}{\mathrm{T}} \frac{\mathrm{d}T}{\mathrm{d}z}, \tag{7.33}$$

where α_i is the thermal diffusion factor. When this is included we get

$$\beta_i = \gamma_i \frac{(M - M_i)\, g}{RT} - \frac{\alpha_i}{T} \frac{\mathrm{d}T}{\mathrm{d}z}. \tag{7.34}$$

We now see that for a light molecule, with M_i much less than the ambient or mean atmospheric molecular weight M we have a strong buoyancy force for diffusion to space. When the molecular mixing ratio is constant, $\partial f_i/\partial z = 0$, we are left with the buoyancy term that is commonly referred to as the *limiting flux*.

7.4 Surface deposition

7.4.1 *Surface deposition loss rate*

The surface deposition loss rate term S_{di} (molecules cm^{-3} s^{-1}) depends on the atmospheric number density of the particles that are being deposited onto the surface. It can be written as

$$S_{\mathrm{d}i} = \left(\frac{\partial n_i}{\partial t}\right)_{\mathrm{d}} = \frac{\partial n_i v_i}{\partial z}, \tag{7.35}$$

where n_i is the number density, and v_i is the velocity of particle i measured positive in the downwards direction. Expanding the above derivative into its two components we can write

$$S_{\mathrm{d}i} = v_i \frac{\partial n_i}{\partial z} + n_i \frac{\partial v_i}{\partial z}. \tag{7.36}$$

If we are considering deposition onto a surface from an atmospheric layer just above the surface, then the deposition velocity at the surface boundary, z_1, is equal to the dry deposition velocity, v_{di}, while at the upper boundary, z_2, of the layer it is zero. From eqn (7.28) we can approximately write for the loss rate at the lower boundary of the layer

$$S_{1\mathrm{d}i} = -n_{1i} \frac{v_{\mathrm{d}i}}{H} - n_{1i} \frac{v_{\mathrm{d}i}}{z_2 - z_1}, \tag{7.37}$$

where n_i is the atmospheric number density of molecule i at the surface and H is the atmospheric number density scale height. We can derive a surface deposition flux $F_{\mathrm{d}i}$ (molecules cm^{-2} s^{-1}) from the surface deposition velocity (cm/s) according to

$$F_{\mathrm{d}i} = v_{\mathrm{d}i} n_{1i}, \tag{7.38}$$

Table 7.2 *Total surface resistance (s m^{-1}) of a deciduous forest at 0 and 25 °C for various values of the incident solar flux W m^{-2}. (From Wesely 1988)*

Molecule	T(°C)	800	500	300	100	0
SO_2	0	980	980	980	990	1000
SO_2	25	120	130	160	370	1000
O_3	0	550	610	700	1000	3000
O_3	25	100	110	130	320	950
NO_2	0	3800	3800	3900	4300	9500
NO_2	25	120	130	160	480	2800
H_2O_2	0	390	420	460	610	980
H_2O_2	25	80	90	110	250	640
N_2O	0	980	980	980	990	1000
N_2O	25	100	110	140	340	1000
NH_3	0	1500	1500	1500	1500	1500
NH_3	25	70	80	100	310	2600

and define a first–order, in n_i, loss rate (s^{-1}) at the surface from

$$\frac{1}{\tau_{1di}} = \frac{v_{di}}{H} + \frac{v_{di}}{z_2 - z_1}, \tag{7.39}$$

where τ_{di} is the mean lifetime of the molecule with respect to deposition.

7.4.2 *Dry deposition velocities*

The deposition velocity of gases is usually derived from a formula analogous to Ohm's law in electrical circuits

$$v_{id} = (R_a + R_b + R_c)^{-1}, \tag{7.40}$$

where R_a is the aerodynamic resistance determined by turbulence, R_b is referred to as the quasilaminar boundary layer resistance, often significantly smaller than R_a (typical values 50 s m^{-1}), and R_c is the total surface resistance arising from uptake by vegetation, soil, water or ice. Considerable effort has been placed in evaluating R_c from

$$R_c = R_{veg} + R_{soil} + R_{water/ice}. \tag{7.41}$$

The vegetation resistance (stomatal resistance) is controlled by the solar flux (W m^{-2}) incident on the vegetation and the surface air temperature. Thus, there is a feedback mechanism between molecules that control the radiation field and the indirect effect of radiation on atmospheric composition through vegetation uptake. In Table 7.2 are given typical values for the total surface resistance of a deciduous forest as a function of downwelling solar flux for various molecules.

In Table 7.3 are given the resistances to soil uptake and deposition onto water/ice surfaces. For large bodies of water, dry deposition (as opposed to removal by

Table 7.3 *Typical resistances (s m^{-1}) of various molecules to soil uptake and to deposition onto water/ice surfaces.*

Molecule	Soil	Water/ice
SO_2	125	227
O_3	400	2000
NO_2	600	10^5
HNO_3	10	10
NO_2	600	10^5
NO	10^5	10^5
CO	2000	2.0×10^4
H_2	2650	3×10^4
HCHO	280	110
H_2O_2	110	100

rain) of gases is primarily dependent on water solubility. Uptake of gases by soils is primarily dependent on biological activity in the soil, although other factors such as temperature, acidity and moisture are also important. Gases that react with the surface (O_3, NO_2), and gases that are highly soluble in water (HNO_3, SO_2, H_2O_2) are better sequestered by the surface, while inert gases (CO) have relatively low rates of surface deposition.

If we have a surface deposition rate L_{id} of a particular molecule i in Tg (10^{12} g) per year, then we can express this in terms of a mean deposition velocity and vice versa through

$$v_{\mathrm{id}} = \frac{L_{\mathrm{id}} 10^{12} N_{\mathrm{A}}}{3.15 \times 10^7 n f_i M_i A_{\mathrm{d}}} \tag{7.42}$$

where n is the atmospheric number density (equal to the Loschmidt number 2.687×10^{19} cm^{-3} at STP), f_i the mixing ratio of molecule i, $N_{\mathrm{A}} = 6.0224 \times 10^{23}$ is Avogadro's number, M_i is the molecular weight, and A_{d} is the total deposition surface area. As an example, for H_2 with a mean global soil deposition rate of 56 Tg $year^{-1}$, with global soil cover taken as 70% of global land surface area $A_{\mathrm{d}} = 0.70A_{\mathrm{l}}$, where the Earth's land surface area is $A_{\mathrm{l}} = 1.5 \times 10^{18}$ cm^2, $n = 2.55 \times 10^{19}$ molecules cm^{-3} (at 288 K), and $f_i = 530$ ppbv we obtain a deposition velocity of 0.038 cm s^{-1} corresponding to a soil resistance of 2650 s m^{-1} or a soil deposition flux $F_{\mathrm{d}i}$ of 5.1×10^{11} molecules cm^{-2} s^{-1}. Globally, this corresponds to a mean surface flux of 1.06×10^{11} molecules cm^{-2} s^{-1}, a deposition velocity of 0.0078 cm s^{-1}, which corresponds to a loss rate of 7.8×10^{-8} s^{-1} for a 1–km atmospheric layer.

Tropospheric gases that are highly soluble in water can be removed from the atmosphere through scavenging by rain drops. Their removal rate depends on cloud formation and rainfall. Gases like HNO_3, HNO_4, HCHO, CH_3OOH and H_2O_2 have relatively short residence times against *rainout* of the order of 10 days,

Table 7.4 *Estimates of total global surface emission rates for various gases, arising from natural and anthropogenic direct sources as at 1998 AD.*

Molecule	Tg year^{-1}	molecules cm^{-2} s^{-1}
H_2	51	9.72×10^{10}
CH_4	600	1.44×10^{11}
CO	2800	3.78×10^{11}
N_2O	25	2.17×10^{9}
CO_2	26	2.25×10^{9}
SO_2	150	8.93×10^{9}

compared with say O_2 with residence times of the order of 10^5 years. Thus, for these soluble gases a rainout rate of 10^{-6} s^{-1} is typical.

7.5 Surface emission

There are both natural and anthropogenic emission sources of molecules at the Earth's surface. Water vapour enters the atmosphere through evaporation from the oceans, evapotranspiration from plants and ejection from volcanoes. Carbon dioxide, methane, N_2O, hydrogen, ammonia, for example, are produced by many natural processes related to biological cycles and also by industrial activities.

The surface emission term $S_{\mathrm{e}i}$ can be written in terms of the surface emission flux, $F_{\mathrm{e}i}$,

$$S_{\mathrm{e}i} = \left(\frac{\partial n_i}{\partial t}\right)_{\mathrm{e}} = -\frac{\partial F_{\mathrm{e}i}}{\partial z}. \tag{7.43}$$

If we are considering emission from a surface into an atmospheric layer we can approximate the derivative of the flux in the layer by

$$S_{\mathrm{e}i} = -\frac{\mathrm{flux}_2 - \mathrm{flux}_1}{z_2 - z_1}, \tag{7.44}$$

where z_2 is the upper boundary of the layer and z_1 is the lower. For surface emission we can set $\mathrm{flux}_1 = F_{\mathrm{e}i}$, where $F_{\mathrm{e}i}$ is the surface flux and $\mathrm{flux}_2 = 0$, hence the production rate within the layer is determined by the surface emission term

$$S_{\mathrm{e}i} = \frac{F_{\mathrm{e}i}}{z_2 - z_1}, \tag{7.45}$$

$$= \frac{\text{surface flux}}{\text{width of layer}}, \tag{7.46}$$

$$= \frac{\text{molecules emitted per second}}{\text{volume of layer}}. \tag{7.47}$$

The atmospheric production rate is independent of the choice of the width of the layer, since for example doubling the layer width simply halves the production rate but production is distributed over twice as much of the atmosphere.

If we have a surface emission rate L_{ei} of a particular molecule i in Tg (10^{12} g) per year, then we can express this as a surface flux, F_{ei} of molecules cm^{-2} s^{-1} via

$$F_{ei} = \frac{L_{ei} 10^{12} N_{A}}{3.15 \times 10^{7} M_{i} A_{e}}, \tag{7.48}$$

where N_A is Avogadro's number, M_i is the molecular weight, and A_e is the total emitting surface area. In Table 7.4 are given estimates of total global emission rates for various gases.

7.6 Photolysis

Photolysis or photodissociation of a molecule is the dissociation of the molecule by radiation. Photodissociation is the process by which a molecule absorbs a photon of sufficient energy to overcome the bond holding the constituent atomic nuclei together. The bond arises from the shared electrons of the atoms that provide an electric field that overcomes the repulsive forces of the positively charged nuclei. Photoionization of a molecule or atom is the process by which a molecule or atom absorbs photons of sufficiently high energy to remove bound electrons and thus form positively charge ions.

7.6.1 *Photolysis rate*

The photolysis rate per second, j, at altitude z at time t is the inverse of the lifetime of a molecule before photolysis occurs, and is given by

$$j(z,t) = \int_{\lambda_a}^{\lambda_b} f_{\lambda}(z,t)\mu_{o}(t)S_{\odot\lambda}\sigma_{\lambda}(z)\Phi_{\lambda}(z)d\lambda, \tag{7.49}$$

where $S_{\odot\lambda}$ is the total solar irradiance or solar constant, f_λ is the enhancement factor, $\mu_o(t)$ is the cosine of the solar zenith angle. The absorption cross–section, σ_λ can be temperature dependent as can be the quantum yield Φ_λ, which represents the probability that if a molecule absorbs a photon it will be photodissociated.

If there is no Earth–atmosphere reflectance of radiation to space at a particular wavelength, because for example the absorption is so strong, then there is only the downwards photon flux and so $f_\lambda = 1$ at the top of the atmosphere. If we have no atmospheric absorption or scattering but only surface reflection, by a surface with an albedo of 0.5, then $f_\lambda = 1.5$ at all altitudes. Finally, if we include multiple scattering between this reflecting surface and the atmosphere above, then the enhancement factor can rise above 2.0. Thus, multiple scattering and surface reflection can enhance the photolysis rate of an atmospheric molecule, while absorption of solar radiation by other molecules reduces the enhancement factor. We note that the maximum value for the enhancement factor at the top

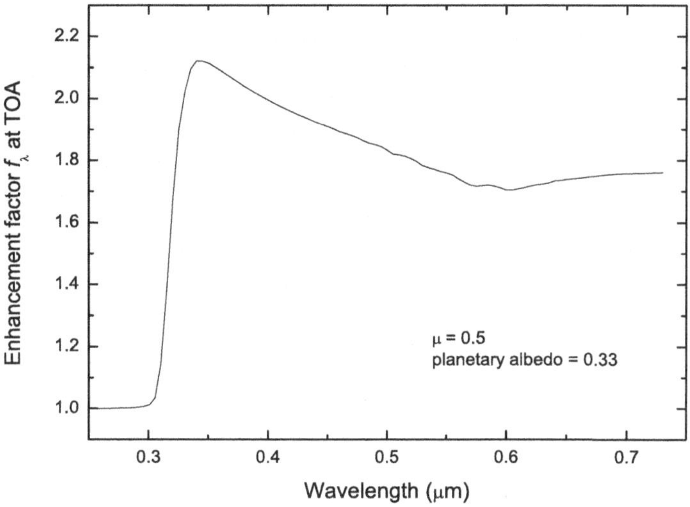

FIG. 7.2. The photolysis enhancement factor f_λ at the top of the atmosphere for global mean Earth conditions.

of the atmosphere is 3.0 (§6.7). This arises because we are summing photons from all directions, both diffuse and direct as shown by eqn (6.135), that are incident on an element of volume containing the molecule.

In Fig. 7.2 is shown the enhancement factor f_λ at the top of the atmosphere for solar ultraviolet–visible radiation. The enhancement factor was calculated for global mean conditions with a solar zenith angle χ, given by $\mu = \cos(\chi) = 0.5$ and a planetary albedo of 0.33, based on a radiative–convective 1D model (Vardavas and Carver 1984) that includes cloud and Rayleigh scattering, molecular absorption, and a mean surface albedo of 0.1.

7.6.2 *Quantum yield*

The *quantum yield* is defined in various ways depending on its use. Some of the definitions are listed below:

$$\Phi = \frac{\text{number of molecules reacting per second}}{\text{number of quanta absorbed per second}},$$

or

$$\Phi = \frac{\text{number of product molecules}}{\text{number of quanta absorbed}}.$$

If the quantum yield is less than one it indicates that de-excitation of the molecule occurs by other processes (e.g. collisions) before photolysis takes place. If the quantum yield is greater than one it indicates a chain reaction.

Depending on the photon wavelength, a molecule can dissociate into different products. In this case we can add the various photolysis rates, corresponding to different spectral regions, to obtain the total photolysis rate. If the molecule can dissociate to different products for the same spectral interval, then we can define a probability for each possible outcome of products, or *branching ratio*, so that the sum of the branching ratios is unity.

In Appendix B (Tables B.2 and B.3) are given the spectral intervals of significant photolysis for various molecules, together with the total photolysis rate, and photolysis products and rates for the photolysis channels. Also given are the ionization rates for the main constituents of the upper atmosphere, and ionization products (Table B.1).

7.6.3 *O_2 photolysis*

Absorption of visible and ultraviolet radiation by molecules occurs via transitions between vibrational–rotational sublevels of electronic states. Idealized potential–energy curves for some of the electronic states of O_2 are shown in Fig. 7.3. Photodissociation of O_2 can occur via transitions from the electronic ground state $X^3\Sigma_g^-$ to the excited states $A^3\Sigma_u^+$ and $B^3\Sigma_u^-$. The notation is similar to electronic states of atoms with S, P, D, $\cdots$, replaced by $\Sigma, \Pi, \Delta, \cdots$ (see Herzberg 1950). Photodissociation can occur via transitions from a bound level to the continuous region of an excited electronic state. The dissociation limit is then exceeded and the molecule breaks up. Photodissociation can also occur from bound–bound electronic transitions if there is a repulsive electronic state (dissociation continuum) that interacts with the vibrational-rotational energy levels of the upper electronic state of the transition. This is referred to as *predissociation* and it involves a nonradiative transition from the vibrational–rotational levels of the upper bound electronic state to the dissociation continuum of the repulsive state. The probability of predissociation increases if the potential energy of the vibrational–rotational level of the upper electronic state of the transition is higher than the crossing point of the potential curves of the repulsive and upper electronic states of the transition. Transitions from $X^3\Sigma_g^-$ to the dissociation continuum of the state $A^3\Sigma_u^+$ (the forbidden *Herzberg system*) result in the dissociation of molecular oxygen via the channel

$$O_2 + h\nu \rightarrow O(^3P) + O(^3P), \tag{7.50}$$

with atomic oxygen in the ground state. This Herzberg dissociation continuum is weak, extends from 185 nm to 242 nm, and is the main source of O atoms for altitudes below 60 km in the Earth's atmosphere. The absorption cross–section of O_2 as a function of wavelength for the Herzberg continuum is given in Fig. 7.4.

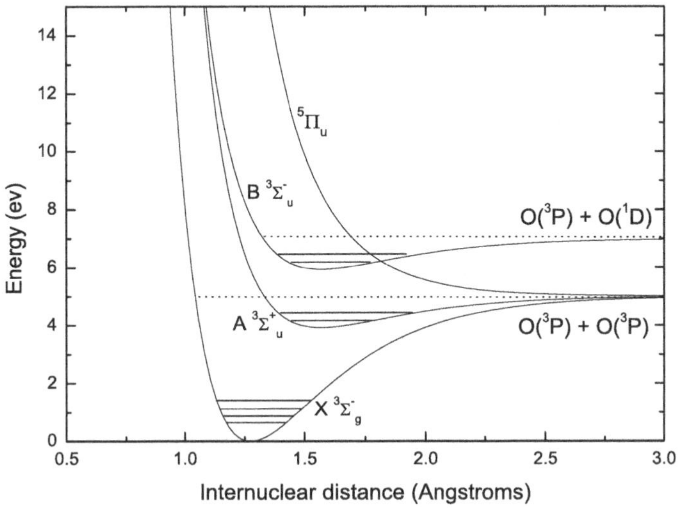

FIG. 7.3. Idealized potential energy curves for O_2 and vibrational–rotational energy levels, with dissociation limits at 7.047 and 5.080 eV.

Bound–bound transitions from $X^3\Sigma_g^-$ to $B^3\Sigma_u^-$ involve nonradiative transitions to the unstable repulsive states $^5\Pi_u$, $^3\Pi_u$, $^1\Pi_u$ and $2^3\Sigma_u^+$ leading to predissociation. This is the system of allowed strong Schumann–Runge bands with a sharp band structure from 175 to 200 nm that results in the dissociation channel

$$O_2 + h\nu \rightarrow O(^3P) + O(^1D), \tag{7.51}$$

with one of the oxygen atoms in the excited state $O(^1D)$. The absorption–cross–section has a detailed vibrational–rotational band structure. In Fig. 7.4 are given averaged values for the cross–section. Dissociation of O_2 by absorption in the Schumann–Runge bands is dominant between 60 and 90 km.

Bound–continuum transitions from $X^3\Sigma_g^-$ to the dissociation continuum of $B^3\Sigma_u^-$ also result in atomic oxygen products as for the Schumann–Runge bands. Such transitions result in the intense Schumann–Runge dissociation continuum between 137 and 175 nm. The cross–section peaks at 142.5 nm, as shown in Fig. 7.4, with three diffuse bands (interaction between discrete states and the continuum) at 129.4, 133.4, and 135.4 nm. Absorption in the Schumann–Runge continuum is the primary O_2 dissociation process above 90 km.

Small portions of the two continua, Herzberg and Schumann–Runge, underlie the Schumann–Runge bands. Nicolet and Kennes (1986) give the formula for the

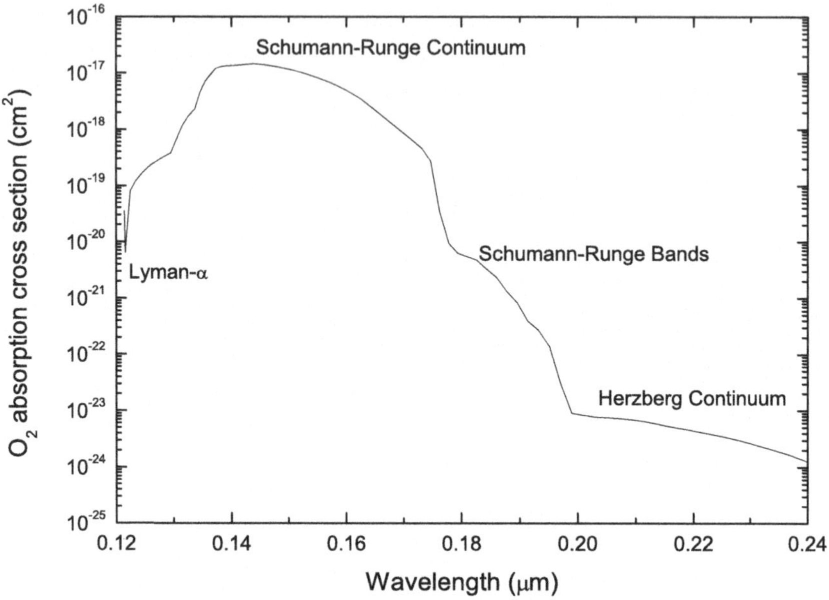

FIG. 7.4. The absorption cross–section of O_2 at 273 K (band structure has been averaged out).

Herzberg continuum cross–section

$$\sigma(x) = 7.5 \times 10^{-24} x \exp[-50(\ln x)^2], \tag{7.52}$$

where $x = 199/\lambda$, with λ in nm, and recommended for use above 210 nm. The cross–section in the Schumann–Runge bands is highly variable with wavelength owing to the thousands of overlapping rotational lines that are broadened by predissociation and by Doppler broadening, resulting in Voigt rotational line profiles. The band cross–sections are temperature dependent as the temperature affects the populations in the vibrational and rotational levels of the ground electronic state and it also affects the broadening of the individual rotational lines. It is thus difficult to give average cross–sections over the wavelength intervals used to model atmospheric photochemistry. Photons that have wavelengths between band peaks can penetrate deeper into the atmosphere, as the cross–sections can vary by two orders of magnitude over an interval of 20 cm^{-1}, or of the order of 10^{-5} nm. Murtagh (1988) used high spectral resolution cross–sections to derive average cross–sections in seventeen broad spectral intervals, from 175.4 to 205.8 nm, tabulated as functions of the O_2 column number density (molecules cm^{-2}). Allen and Frederick (1982) give temperature–dependent expressions for the cross–section in seventeen broad spectral intervals, for use in photochemical modelling studies.

Below 137 nm, the solar Lyman–α line becomes important for the photolysis of O_2. The cross–section is temperature dependent and varies significantly, by almost one order of magnitude, over the Lyman–α line (Fig. 7.4). Lewis *et al.* (1983) tabulate the cross–section for the wavelength interval 121.40 to 121.90, in intervals of 0.01 nm, for the temperatures 84, 203, 288 and 366 K. We recall that the solar Lyman–α flux varies by a factor of 2 over the solar cycle.

The photoionization limit of O_2 is 102.78 nm. Photoionization yield is unity below 60 nm and 0.65 at 102.57 nm, corresponding to Lyman–β . The cross–section between 137 and 105 is highly structured, varying between 10^{-20} and 10^{-17} cm^2, while below 100 nm there is an ionization continuum. For photochemical modelling it is useful to have mean cross–sections over broad intervals and for this purpose Banks and Kockarts (1973) tabulate the absorption and photoionization cross–sections at 102.57, 97.7 (corresponding to the solar C(III) doubly ionized carbon line), and for seven spectral intervals from 91 to 8 nm. Below 71.9 nm, dissociative photoionization is possible

$$O_2(^3\Sigma_g^-) + h\nu \rightarrow O^+ + O(^3P) + e, \tag{7.53}$$

$$O_2(^3\Sigma_g^-) + h\nu \rightarrow O^+ + O(^1D) + e, \tag{7.54}$$

where the second channel is possible below 60 nm.

7.6.4 *O_3 photolysis*

Ozone absorbs strongly in the wavelength region between 200 and 300 nm. From 300 to 350 nm it absorbs in the relatively weak and temperature–dependent Huggins bands. It also absorbs in the visible region between 410 and 850 nm in the Chappuis bands. The photodissociation threshold of ozone is 1.14 μm. Ozone is the main atmospheric absorber of solar ultraviolet and visible radiation above the cloud deck so that the variation of the enhancement factor f_λ, shown in Fig. 7.2, follows inversely the ozone absorption cross–section, shown in Fig. 7.5, between 0.3 and 0.75 μm. Below 0.3 μm the very strong absorption of the Hartley band results in no reflected radiation to space and so $f_\lambda = 1$. The minimum in f_λ at 0.6 μm corresponds to the peak in the absorption cross–section of the visible Chappuis bands of ozone. Ozone can photodissociate to different products depending on the photon wavelength (see Appendix B, Table B.3)

$$O_3 + h\nu \rightarrow O + O_2, \tag{7.55}$$

$$O_3 + h\nu \rightarrow O(^1D) + O_2. \tag{7.56}$$

The quantum yield for the production of $O(^1D)$ is negligible for $\lambda > 330$ nm but rises rapidly near 310 nm to 0.9 for $\lambda < 306$ nm, independently of temperature (JPL 2006).

Absolute measurements in the Hartley band were reported by Hearn (1961) at wavelengths of several mercury lines 253.7, 289.4, 296.7 and 302.1 nm. In particular, the 253.65 nm cross–section of Hearn (1.147×10^{-17} cm^2 molecule^{-1}) is

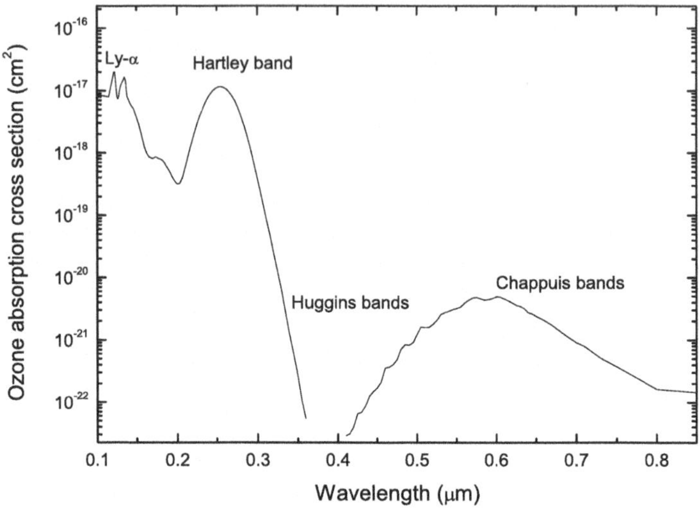

FIG. 7.5. The absorption cross–section of ozone from Lyman–α to the visible, at 273 K (band structure has been averaged out).

the primary standard for international agencies. These results have served as the basis for normalizing the relative measurements that cover the entire spectrum. Absorption cross–sections for the interval 175.4 to 362.5 nm are tabulated in JPL (2006) for 273 K. An expression for the temperature–dependent quantum yield for $O(^1D)$ production between 305 and 320 nm is also given. Bass and Paur (1981) measured the temperature dependence of the cross–section and found that it was significant for the region 263 to 350 nm, mainly corresponding to the Huggins bands. The original measurements for the Chappuis bands were made by Vigroux (1953) and later verified by Griggs (1968).

7.6.5 *H_2O photolysis*

The absorption cross–section for water vapour is shown in Fig. 7.6. Water vapour can be photodissociated by photons with wavelength below 246 nm. There are three possible channels below 145 nm

$$H_2O + h\nu \rightarrow H_2 + O(^1D), \tag{7.57}$$

$$H_2O + h\nu \rightarrow 2H + O, \tag{7.58}$$

$$H_2O + h\nu \rightarrow H + OH, \tag{7.59}$$

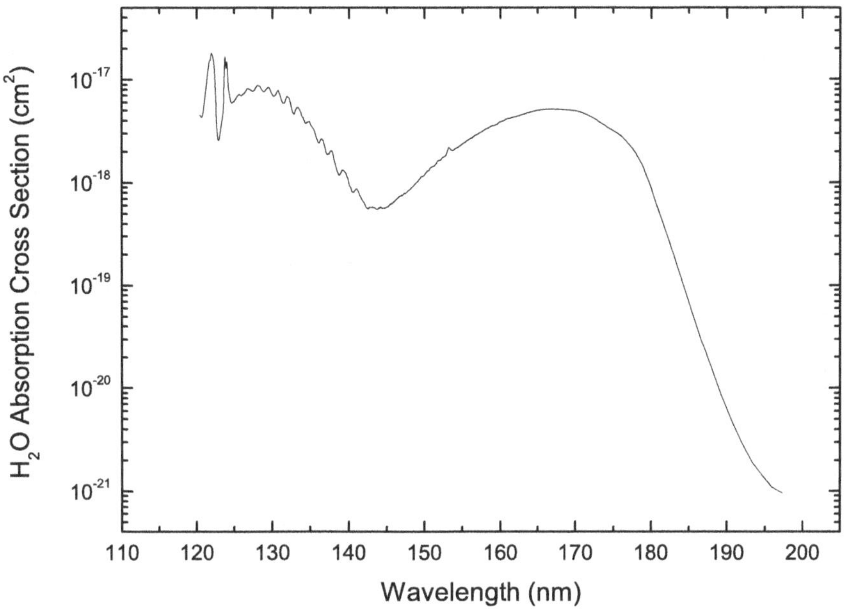

FIG. 7.6. The absorption cross–section of H_2O at 296 K.

where the branching ratio $\alpha = 0.11$ for the first and second channels, and $\alpha = 0.78$ for the third. These channels take place primarily by Lyman–α photons. The mean absorption cross–section over the Lyman–α interval 121.48–121.64 nm is 1.55×10^{-17} cm^2 for $T = 292$ K. Above 145 nm, water vapour photolyzes according to the third channel. The absorption cross–section of H_2O at 296 K is shown in Fig. 7.6, based on the measurements of Yoshino *et al.* (1996) and Parkinson and Yoshino (2003).

7.6.6 *CO_2 photolysis*

In Fig. 7.7 we show the temperature-dependent absorption cross–section of CO_2 from Lyman–α to about 0.2 μm based on the high–resolution (0.005 nm) measurements of Lewis and Carver (1983). The temperature dependence has also been demonstrated by Jensen *et al.* (1997), who give temperature–dependent absorption cross–sections at high temperatures (above 1000 K) in the spectral range 0.19 μm to 0.32 μm. Between 0.200 and 0.206 μm, Karaiskou *et al.* (2004) give absorption cross–section measurements at 295 K and at 373 K.

CO_2 photodissociates below 0.2275 μm according to the two channels

$$CO_2 + h\nu \rightarrow CO + O(^3P) \quad 0.2275 > \lambda > 0.1672 \quad \mu m, \tag{7.60}$$

$$CO_2 + h\nu \rightarrow CO + O(^1D) \quad \lambda < 0.1672 \quad \mu m. \tag{7.61}$$

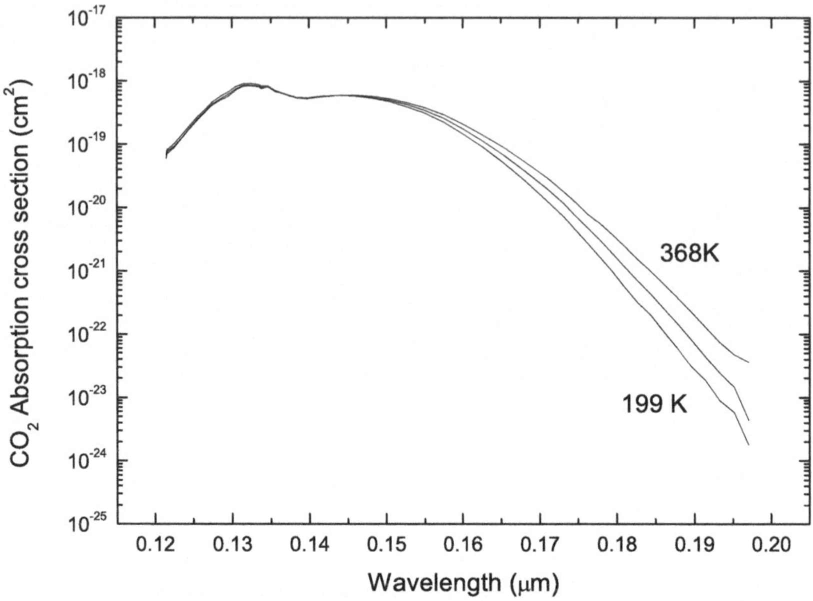

FIG. 7.7. The absorption cross-section of CO_2 at 199 K, 294 K and 368 K (band structure at low temperature has been averaged out.

The quantum yield for the first channel is unity up to about 0.21 μm (Okabe 1978). The most important region is between 0.14–0.18 μm, which corresponds to the absorption region of the molecular oxygen Schumann–Runge bands and continuum.

7.6.7 *CH_4 photolysis*

Methane absorbs strongly at wavelengths less than 145 nm, and photodissociates according to the following important channels

$$CH_4 + h\nu \rightarrow CH_2 + H_2, \tag{7.62}$$

$$CH_4 + h\nu \rightarrow CH_2 + 2H, \tag{7.63}$$

$$CH_4 + h\nu \rightarrow CH_3 + H. \tag{7.64}$$

For Lyman–α, the branching ratio for the first channel is 0.24, for the second it is 0.25 and for the third it is 0.51. Above Lyman–α the branching ratio is unity for the third channel. In Fig. 7.8 is shown the absorption cross-section of methane taken from various sources. For the region 37.43–91.11 nm the data are from Samson *et al.* (1989), for 38.0–18.0 nm from Lee *et al.* (1977), at Lyman–α Brownsword *et al.* (1997) give a mean value of $(2.0 \pm 0.1) \times 10^{-17}$ cm^2, while

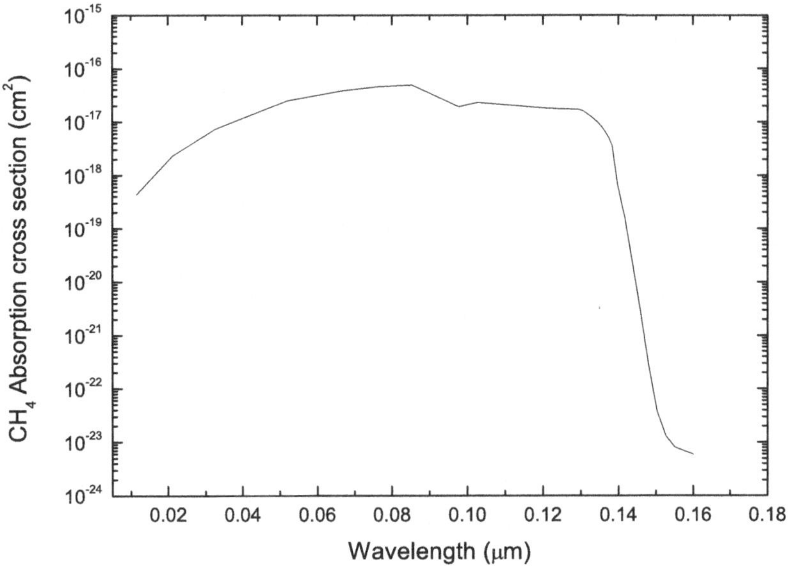

FIG. 7.8. The absorption cross-section of CH_4 at room temperature.

for 137–160 nm the data are from Mount *et al.* (1977). Further references for other cross-section data, photodissociation pathways and their corresponding branching ratios, are given in Lavvas *et al.* (2007).

7.6.8 *Atmospheric photolysis rates*

In Appendix B, Tables B.1 to B.3, are given typical values for the total photodissociation and photoionization rates of atmospheric species for quiet-Sun irradiance in the ultraviolet and visible spectral regions at the top of the atmosphere, for mean global conditions. In Fig. 7.9 are shown the photodissociation rates of various molecules with altitude for global mean atmospheric conditions.

The photolysis rate of ozone with altitude is given for the two channels, for the mean global conditions mentioned earlier in relation to Fig. 7.2. The second channel is significant above 40 km with a photolysis rate of 7.1×10^{-3} s^{-1} at TOA. The first channel is significant below 40 km and has a rate of 1.5×10^{-3} s^{-1} at TOA. In the upper atmosphere the CO_2 second channel is important, while in the stratosphere the first channel is important. Due to the temperature dependence of the CO_2 cross-section below 0.167 μm, the photolysis rate of the second channel increases to 7.45×10^{-7} s^{-1} at TOA. Water vapour is primarily dissociated by the solar Lyman–α line above 70 km altitude in the atmosphere and by photons of wavelength greater than Lyman–α below 70 km, as shown in

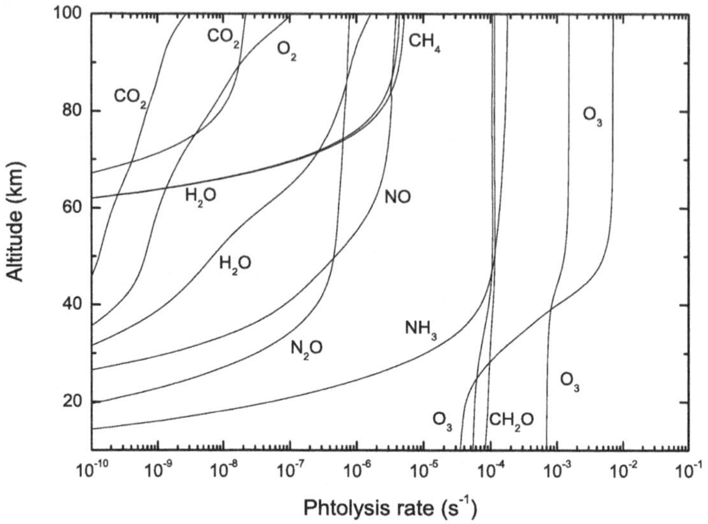

FIG. 7.9. Photolysis rate with altitude for various important molecules, for global mean atmospheric conditions.

Fig. 7.9.

7.7 Collisionally induced reactions

7.7.1 *Types of reactions*

Let us consider a diatomic molecule consisting of two atoms A and B and that the molecule has been excited to some higher energy state, AB* by collisions. The molecule may then undergo spontaneous decomposition into its constituent atoms via the *unimolecular* reaction

$$AB^* \rightarrow A + B. \tag{7.65}$$

The molecule AB may dissociate into its constituent atoms by collision with a second or *catalytic molecule* M via the *bimolecular* reaction

$$AB + M \rightarrow A + B + M. \tag{7.66}$$

Alternatively, two atoms A and B can undergo an association reaction to form the molecule AB in a bimolecular reaction

$$A + B \rightarrow AB. \tag{7.67}$$

A collision between the constituent atoms and a catalytic molecule M may result in the *termolecular recombination reaction*

$$\mathrm{A + B + M \rightarrow AB + M}. \tag{7.68}$$

Collisions involving four or more molecules are very infrequent.

7.7.2 *Bimolecular reactions*

For a bimolecular reaction driven by collisions, we can define a bimolecular collision rate between molecules per unit volume per second according to simple collision theory

$$Z_{\mathrm{AB}} = Z_{\mathrm{o}} n_{\mathrm{A}} n_{\mathrm{B}} \tag{7.69}$$

where

$$Z_{\mathrm{o}} = \pi d_{\mathrm{AB}}^2 \bar{v}_{\mathrm{AB}} \tag{7.70}$$

which has units of $\mathrm{cm^3\ molecule^{-1}\ s^{-1}}$, with the average molecular speed given by eqn (7.15), and d_{AB}, is the mean diameter of the colliding molecules, as in eqn(7.16). The collision rate, Z_{AB} has units of molecules $\mathrm{cm^{-3}\ s^{-1}}$, for a population of colliding molecules. It varies as $\sqrt{T}$, in disagreement with experiment. The reason is that the rate at which molecules react is much less than the rate at which they collide. For a reaction to proceed, the reactants, A and B, must form a transition state AB* that is at a higher potential energy than the reacting molecules or reactants, or the products, C and D, as in

$$\mathrm{A + B \rightleftharpoons AB^* \rightleftharpoons C + D}. \tag{7.71}$$

For the reaction to proceed in the forward direction, the reactants require an activation energy, E_{r}, to form the transition state. Similarly, the products require an activation energy, E_{p}, to form the transition state, for the backward reaction.

The rate of destruction of molecules A per $\mathrm{cm^3}$ per s is proportional to the product of the number densities of the reacting molecules. This is the *Law of Mass Action*, which can be written mathematically as the *rate equation*

$$\frac{\partial n_{\mathrm{A}}}{\partial t} = -k_{\mathrm{f}} n_{\mathrm{A}} n_{\mathrm{B}}, \tag{7.72}$$

where the constant of proportionality, k_{f} in $\mathrm{cm^3\ molecules^{-1}\ s^{-1}}$, is the *forward rate constant*, with a similar equation for the destruction of molecule B. The forward rate constant of the reaction, and the rate constant for the backward reaction, k_{b}, are related through the *equilibrium constant*, K, defined by

$$K = \frac{n_{\mathrm{C}} n_{\mathrm{D}}}{n_{\mathrm{A}} n_{\mathrm{B}}} = \frac{k_{\mathrm{f}}}{k_{\mathrm{b}}}, \tag{7.73}$$

where n is the number density of each molecule. The equilibrium constant is given by the thermodynamic expression

$$\Delta G^{\circ} = -RT \ln K, \tag{7.74}$$

where R is the universal gas constant, and ΔG° the change in the molar Gibb's free energy of the system given by

$$\Delta G^{\circ} = \sum G_f^{\circ}(\text{products}) - \sum \mathrm{G}_f^{\circ}(\text{reactants}), \tag{7.75}$$

with G_f° the free energy of formation of the species involved in the reaction in their standard state, usually at a pressure of 1 atm and room temperature of 298 K. If ΔG° is negative, K will be greater than unity and the forward reaction rate will be greater than the backward reaction rate, eqn (7.73).

The change in free energy is related to the change in enthalpy of the system, H, and the change in the entropy, S, according to

$$\Delta G^{\circ} = \Delta H^{\circ} - T\Delta S^{\circ}, \tag{7.76}$$

where the changes are evaluated from the enthalpy and entropy of formation for each reactant and product species, and from eqn (7.76) we have

$$\ln K = -\frac{\Delta H^{\circ}}{RT} + \frac{\Delta S^{\circ}}{R}. \tag{7.77}$$

Now since ΔH° and ΔS° are weakly temperature dependent, we have the van't Hoff equation

$$\frac{\mathrm{d}\ln K}{\mathrm{d}T} = \frac{\Delta H^{\circ}}{RT^2}. \tag{7.78}$$

For an endothermic reaction $\Delta H^{\circ} = E_{\mathrm{f}} - E_{\mathrm{b}}$ is positive and K increases with temperature, where E_{f} and E_{b} are the activation energies for the forward and backward reactions, respectively. For an exothermic reaction ΔH° is negative and K decreases as the temperature is increased. The *heat of reaction* is $-\Delta H^{\circ}$ and is thus positive for an exothermic reaction, and energy is released in the forward reaction. The forward reaction rate constant can then be written

$$\frac{\mathrm{d}\ln k_f}{\mathrm{d}T} = \frac{E_f}{RT^2}, \tag{7.79}$$

and on integrating we get the form of the *Arrhenius equation* that was first obtained based on experiment

$$k_{\mathrm{f}} = A_{\mathrm{o}} e^{-E_{\mathrm{f}}/RT}, \tag{7.80}$$

where A_{o} is the integration constant. More generally, in the Arrhenius expression, A_{o} can be weakly dependent on temperature. We can see this from simple collision theory. The probability that two molecules A and B collide with a combined energy of at least E_f is given by the exponential term in the above equation. This is equal to the fraction of effective collisions, where the total number of colliding molecules per cm^3 per s is given by $Z_o n_A n_B$. The rate of change in the number density of molecules A is equal to the number of effective collisions between the two types of molecules and so can be written as

$$\frac{\partial n_A}{\partial t} = -n_A n_B Z_o e^{-E_f/RT} = -n_A n_B k_f \tag{7.81}$$

thus $A_o = Z_o$, which varies as $\sqrt{T}$, and hence the main temperature dependence of k_f comes from the exponential term. For large molecules their orientation on collision becomes important and a correction to A_o is required. This is usually introduced as the *steric factor* or probability factor P, so that $A_o = PZ_o$, with $P \leq 1$. The steric factor requires a more complex treatment than that from simple collision theory and this is manifested within the *transition–state model* that involves a transition state or activated complex AB*.

7.7.3 *Termolecular reactions*

Recombination reactions involving a third body are termed termolecular reactions. The probability of a three–body collision at atmospheric conditions is very small (< 1%). The termolecular reaction involves two steps. First the two reacting bodies, A and B collide to form an activated body C*

$$\mathrm{A + B \rightarrow C^*}. \tag{7.82}$$

If the activated body does not suffer a collision it will decompose to its constituent reactants. If it undergoes a deactivating collision with a catalytic body (usually N_2 and O_2 in the atmosphere) then it will become a stable body C

$$\mathrm{C^* + M \rightarrow C + M}. \tag{7.83}$$

The overall reaction is the *termolecular* recombination reaction

$$\mathrm{A + B + M \rightarrow C + M}. \tag{7.84}$$

Termolecular recombination reactions are thus also pressure dependent as the rate of deactivation of the activated body depends on the collision frequency of the three bodies. The general form of the rate constant (see JPL 2006) is

$$k_f([M], T) = \left(\frac{a}{1+b}\right) f^d, \tag{7.85}$$

where

$$a = k_o(T)[M], \tag{7.86}$$

$$b = a/k_\infty(T), \tag{7.87}$$
$$c = \log b, \tag{7.88}$$
$$d = \left(1 + c^2\right)^{-1}, \tag{7.89}$$

and f is a constant that for atmospheric conditions a value of about 0.6 adequately fits the reaction data. There are simpler limiting forms for k_f depending whether or not we have low or high pressure. At low pressures

$$k_\mathrm{f} \rightarrow k_\mathrm{o}(T) = k_\mathrm{o}(300)\left(\frac{T}{300}\right)^{-n}, \tag{7.90}$$

while at high pressures

$$k_\mathrm{f} \rightarrow k_\infty(T) = k_\infty(300)\left(\frac{T}{300}\right)^{-m}. \tag{7.91}$$

If the number density $[M]$ of the catalytic third body is in units of molecules cm^{-3} then the units of k_o are cm^6 $\mathrm{molecules}^{-2}$ s^{-1} while the units of k_∞ are cm^3 $\mathrm{molecules}^{-1}$ s^{-1}. Usually, $k_\mathrm{o}(300)$, $k_\infty(300)$, n and m are tabulated (JPL 2006, IUPAC 2006, NIST 2006).

7.8 Ozone photochemistry

7.8.1 *The Chapman mechanism*

In 1930 Chapman proposed a mechanism for the production of stratospheric ozone. The construction of the photoelectric spectrophotometer for measuring the amount (column density) of atmospheric ozone by Dobson (1931) and Chapman's mechanism played a key role in initiating research on stratospheric ozone. Chapman proposed that an ozone layer should form in the stratosphere through the photolysis of O_2 resulting in the production of atomic oxygen both in the ground state $O(^3P)$ and in the excited state $O(^1D)$, eqn (7.53) and eqn (7.54), which could recombine with molecular oxygen to produce ozone. As we have seen in §7.6.3, $O(^1D)$ is formed above 60 km altitude while O forms also below 60 km with diffusion transporting these atoms to lower altitudes, as can be seen in Fig. 7.10. Above about 110 km atomic oxygen becomes the second most abundant species after molecular nitrogen. The reaction scheme

$$\mathrm{O_2} + h\nu \rightarrow 2\mathrm{O}, \tag{7.92}$$
$$\mathrm{O_3} + h\nu \rightarrow \mathrm{O} + \mathrm{O_2}, \tag{7.93}$$
$$\mathrm{O_3} + h\nu \rightarrow \mathrm{O(^1D)} + \mathrm{O_2}, \tag{7.94}$$
$$\mathrm{O} + \mathrm{O} + \mathrm{M} \rightarrow \mathrm{O_2} + \mathrm{M}, \tag{7.95}$$
$$\mathrm{O} + \mathrm{O_2} + \mathrm{M} \rightarrow \mathrm{O_3} + \mathrm{M}, \tag{7.96}$$

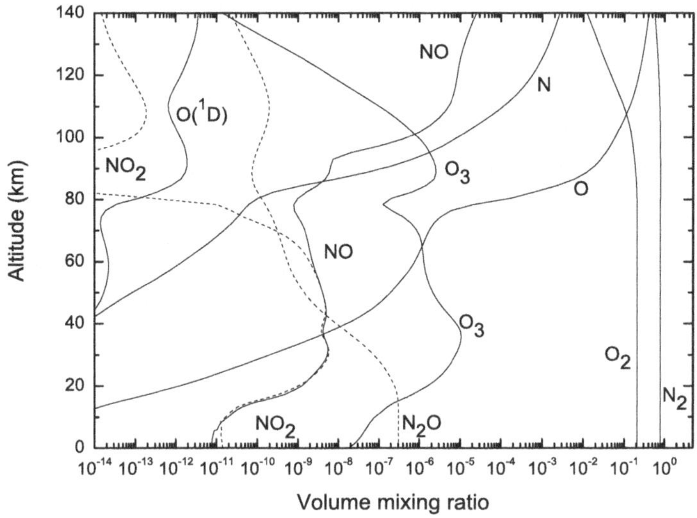

FIG. 7.10. Global mean vertical profiles of oxygen and nitrogen species involved in O_3 photochemistry.

$$O(^1D) + M \to O + M, \tag{7.97}$$

$$O_3 + O \to 2O_2, \tag{7.98}$$

$$O_3 + O(^1D) \to 2O_2, \tag{7.99}$$

$$O_3 + O(^1D) \to O_2 + 2O, \tag{7.100}$$

involves only oxygen species with the deactivation of $O(^1D)$ primarily by molecular nitrogen and oxygen (Appendix B, Table B.4). These reactions produce an ozone layer between 20 and 40 km with a peak concentration at 30 km of about 1.5×10^{13} molecules cm^{-3} that is a factor 4 greater than the global mean observed stratospheric peak value of about 3.5×10^{12} at 25 km. Note that the peak in ozone mixing ratio occurs at about 35 km with a value close to 10 ppmv, as shown in Fig. 7.10. The discrepancy arises from the absence of important catalytical destruction cycles involving NO and OH, and related species NO_2 and HO_2.

7.8.2 *N_2O and NO_x photochemistry*

The oxides of nitrogen, especially NO and NO_2 referred to as NO_x, play an important role in reducing stratospheric ozone levels and explain the difference between the ozone levels predicted by the Chapman mechanism and observations

(Crutzen 1970, 1971, McElroy and McConnell 1971). The destruction mechanism is the following catalytic cycle

$$NO + O_3 \rightarrow NO_2 + O_2, \tag{7.101}$$
$$NO_2 + O \rightarrow NO + O_2, \tag{7.102}$$
$$net: O_3 + O \rightarrow 2O_2. \tag{7.103}$$

The source of NO is primarily nitrous oxide N_2O from biological processes in soil and in the oceans with an annual emission rate of about 25 Tg (Table 7.4). Nitrous oxide is a very stable molecule within the troposphere with a residence time of 120 years but is oxidized by $O(^1D)$ to NO in the stratosphere, and photolyzed to N_2. NO is also produced by oxidation of N_2 by the high temperatures of jet engines.

In Fig. 7.10 is shown the vertical distribution of N_2O where it can be seen that its mixing ratio is constant at 314 ppbv until about 20 km altitude. Other important reactions involving nitrogen-containing species are given in Appendix B, Tables B.4 and B.6, amongst which are NO_3, HNO_3, HNO_4, N_2O_5, referred to as *nitrogen reservoir species* as they do not directly destroy ozone. The NO_x species plus the reservoir species together are referred to as the NO_y chemical family.

Daytime–nighttime effects become important for the NO_y family as NO_3 and N_2O_5 are formed at night since NO_3 is rapidly destroyed by sunlight. Also, HNO_3 is formed by reaction between OH and NO_2 during nighttime as OH is formed in the night by the reaction of $O(^1D)$ with H_2O up to an altitude of 60 km. We note that HNO_3 (nitric acid) is highly soluble and is removed from the atmosphere by rain.

Another significant source of atmospheric NO arises from the reaction of excited nitrogen atoms, $N(^2D)$ with molecular oxygen above 100 km altitude. Nitric oxide is produced primarily via the following chain of reactions (Appendix B, Table B.5)

$$N(^2D) + O_2 \rightarrow NO + O, \tag{7.104}$$
$$NO^+ + e \rightarrow N(^2D) + O, \tag{7.105}$$
$$O^+ + N_2 \rightarrow NO^+ + N, \tag{7.106}$$
$$N^+ + O_2 \rightarrow NO^+ + O, \tag{7.107}$$
$$N_2 + h\nu \rightarrow N(^4S) + N^+ + e, \tag{7.108}$$
$$O + h\nu \rightarrow O^+ + e, \tag{7.109}$$
$$O_2 + h\nu \rightarrow 2O. \tag{7.110}$$

Thus, the above chain of reactions is initiated by the photolysis of molecular oxygen and the dissociative ionization of molecular nitrogen below 37.5 nm. Another

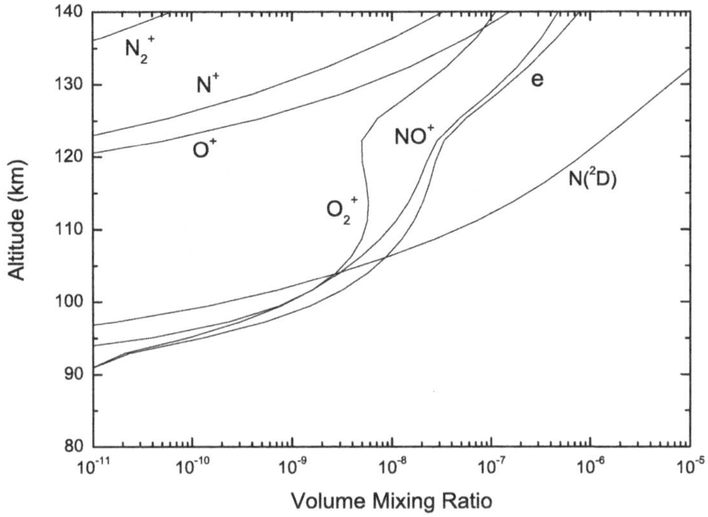

FIG. 7.11. Global mean vertical profiles of ionic species that control mesospheric NO^+, together with $N(^2D)$.

important production mechanism for $N(^2D)$ is the dissociation of molecular nitrogen by energetic electrons (about 10 eV) that are produced by solar soft X–rays (2–7 nm) and extreme ultraviolet radiation (EUV) in the spectral region 7–20 nm. We note that the solar flux in these spectral regions varies significantly over the solar cycle (see §5.2.3) and so does thermospheric NO. Electron precipitation in the auroral polar region is an additional source of energetic electrons. The NO plays an important role in the energy balance of the thermosphere via cooling to space and if transported to the mesosphere it can contribute to the catalytic destruction of ozone. Global mean vertical profiles of ionic species, which control mesospheric NO^+ together with $N(^2D)$, are shown in Fig. 7.11. The dissociation of molecular nitrogen, oxygen, and the hydrogen flux control the variation of the atmospheric molecular weight with altitude in the thermosphere and lead to the formation of the heterosphere, where the atmospheric composition is no longer uniform, as shown by the rapid decrease in the atmospheric molecular weight above about 100 km in Fig. 7.12.

7.8.3 *Water vapour and HO_x photochemistry*

The water–vapour mixing ratio is determined by ocean evaporation (see §8.9), the tropospheric air temperature and by diffusion and transport to the stratosphere.

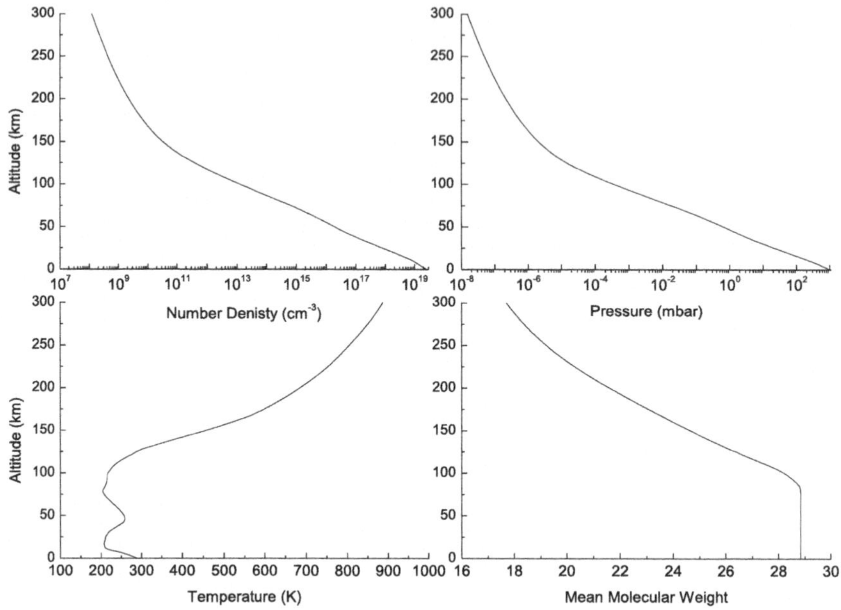

FIG. 7.12. Model vertical profiles of number density, pressure, temperature and mean molecular weight.

It is also produced by methane oxidation in the stratosphere and its mixing ratio there is fairly constant, ranging between 3 and 6 ppmv (see Fig. 7.13). In the stratosphere it is destroyed by $O(^1D)$ during nighttime via the reaction

$$H_2O + O(^1D) \rightarrow 2OH, \tag{7.111}$$

while above 70 km H_2O is readily destroyed by solar Lyman–α radiation (see Fig. 7.9 and Lewis *et al.* 1983). Photolysis of water vapour by Lyman–α can result in the production of active and inactive hydrogen species, with respect to ozone destruction. As we saw earlier, the third channel is the dominant with branching ratio $\alpha = 0.78$ (§7.6.5) so the direct production of H_2 from the photodissociation of water vapour is a small fraction of the process of Lyman–α photodissociation. The role of the HO_x species (H, OH and HO_2) in destroying ozone was investigated in detail by Allen *et al.* (1984) who stressed the key role played by reactions that convert the active-hydrogen HO_x species, to the inactive–hydrogen species H_2 and H_2O.

The hydroxyl radical can efficiently destroy ozone in the stratosphere and mesosphere via the catalytic cycle

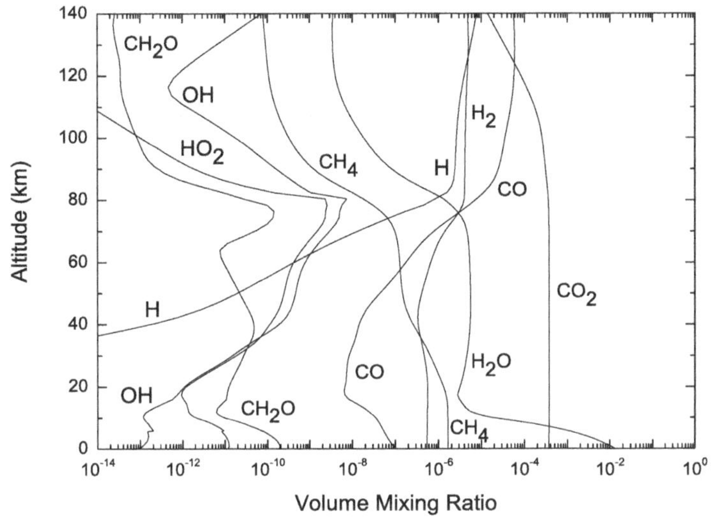

FIG. 7.13. Global mean vertical profiles of key species involved in water vapour and CO_2–initiated photochemistry.

$$OH + O_3 \rightarrow HO_2 + O_2, \tag{7.112}$$
$$HO_2 + O_3 \rightarrow OH + 2O_2, \tag{7.113}$$
$$net : 2O_3 \rightarrow 3O_2. \tag{7.114}$$

In the mesosphere, atomic hydrogen plays an important role in the destruction of ozone via the reaction

$$O_3 + H \rightarrow O_2 + OH, \tag{7.115}$$

resulting in a deep minimum in mesopause ozone. The depth of this minimum can range from a mixing ratio of 0.1 ppmv to 0.01 ppmv. This depends on the reaction of H with HO_2 and on the branching ratio α for H_2O photolysis. The reaction of H with HO_2 has three reaction channels (Appendix B, Table B.6)

$$H + HO_2 \rightarrow OH + OH \qquad a = 0.90 \pm 0.04, \tag{7.116}$$
$$H + HO_2 \rightarrow O + H_2O \qquad b = 0.02 \pm 0.02, \tag{7.117}$$
$$H + HO_2 \rightarrow H_2 + O_2 \qquad c = 0.08 \pm 0.04, \tag{7.118}$$

where channel a that leads to the production of the active hydrogen species OH is dominant (branching ratio 0.9). It is the strength of channel c that limits the role of water vapour photolysis in ozone destruction through the conversion of

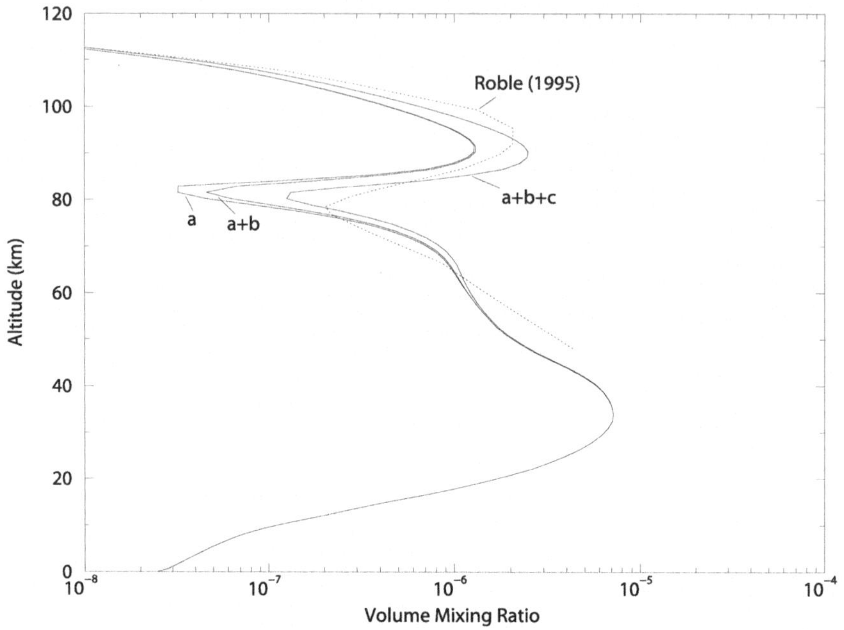

FIG. 7.14. Model global mean diurnally averaged ozone profiles with and without channels a, b, and c of the reaction $H + HO_2$, with $\alpha = 1$. (Vardavas *et al.* 1998)

atomic hydrogen to inactive molecular hydrogen. In Fig. 7.14 is shown the effect of including sequentially channels a, b and c, for a photolysis branching ratio $\alpha = 1$.

7.8.4 *Chlorine and ClO_x photochemistry*

Chlorine is naturally emitted into the atmosphere from the oceans and biomass burning in the form of methyl chloride, CH_3Cl. Significant anthropogenic sources are carbon tetrachloride, CCl_4, methylchloroform, CH_3CCl_3 and chlorofluorocarbons such as $CFCl_3$ (CFC–11) and CF_2Cl_2 (CFC–12), amongst others. CFCs were introduced into the atmosphere after 1960 through their industrial use, refrigeration and in sprays.

In 1974 Molina and Rowland warned of the potential role of CFCs in the destruction of stratospheric ozone associated with their increasing atmospheric concentrations. As can be seen in Table 7.5, the total Cl mixing ratio trebled in about 40 years from about 1 ppbv pre–1960. Molina and Rowland showed that the catalytic destruction of ozone by ClO_x (Cl and ClO) would have a major effect on stratospheric ozone. This and the discovery of a substantial reduction

Table 7.5 *Approximate surface chlorine species mixing ratios (ppbv) and total Cl from pre-1960 to 1998. (WMO 1999, IPCC 2001)*

Species	pre-1960	1980	1990	1998
$CFCl_3$(CFC–11)	0.00	0.17	0.24	0.27
CF_2Cl_2(CFC–12)	0.00	0.29	0.50	0.53
CH_3Cl	0.62	0.62	0.62	0.62
CH_3CCl_3	0.00	0.10	0.10	0.07
CCl_4	0.10	0.10	0.10	0.10
Total Cl	1.02	2.41	3.04	3.10

in the ozone column density in Antarctica (*ozone hole*) led to the Montreal Protocol (1987) that has resulted in the control and reduction in CFC emissions (IPCC 2001). Their suggested replacements by the ozone–inert hydrochlorofluorocarbons (HCFCs) and hydrofluorocarbons (HFCs), however, have introduced a new potential source of greenhouse warming.

CFC molecules are stable in the troposphere but undergo photolysis in the stratosphere by solar ultraviolet radiation (see Appendix B) releasing chlorine atoms

$$CF_2Cl_2 + h\nu \rightarrow CF_2Cl + Cl, \tag{7.119}$$
$$CFCl_3 + h\nu \rightarrow CFCl_2 + Cl, \tag{7.120}$$
$$CH_3Cl + h\nu \rightarrow CH_3 + Cl, \tag{7.121}$$
$$CH_3CCl_3 + h\nu \rightarrow CH_3CCl_2 + Cl, \tag{7.122}$$
$$CCl_4 + h\nu \rightarrow CCl_3 + Cl. \tag{7.123}$$

The chlorine source species diffuse slowly into the stratosphere (Fig. 7.15) where they are destroyed and have tropospheric lifetimes ranging between 2 and 100 years. The ClO_x family can destroy ozone via the catalytic cycle

$$Cl + O_3 \rightarrow ClO + O_2, \tag{7.124}$$
$$ClO + O \rightarrow Cl + O_2, \tag{7.125}$$
$$net : O_3 \rightarrow 2O_2. \tag{7.126}$$

The catalytic cycle is limited by the conversion of ClO_x to *chlorine reservoir species* such as HCl and $ClONO_2$ via the reactions (Appendix B, Table B.7)

$$Cl + CH_4 \rightarrow HCl + CH_3, \tag{7.127}$$
$$ClO + NO_2 + M \rightarrow ClONO_2 + M, \tag{7.128}$$

so there is coupling between chlorine and methane chemistry, and chlorine and NO_x chemistry. The reservoir species HCl is soluble in water and so can be removed from the troposphere by rain, while $ClONO_2$ can be converted to Cl through heterogeneous chemistry (gas and solid/liquid phase reactions) taking place on polar stratospheric clouds (PSCs).

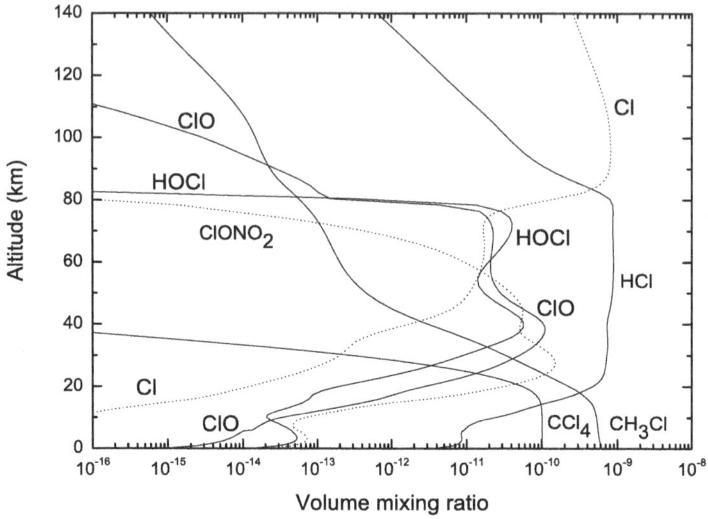

FIG. 7.15. Global mean vertical profiles of chlorine species based on pre–1960 sources of chlorine.

7.8.5 *Polar stratospheric clouds*

The lifetime of HCl is weeks and that of $ClONO_2$ a day. HCl is destroyed by OH, which we saw is a nighttime species in the stratosphere, while $ClONO_2$ is destroyed by photolysis. Hence, during nighttime conditions the $ClONO_2$ species are protected from decomposition back to ozone-active components. Thus, within the polar winter there should be a buildup of these reservoir species. However, the very low temperatures (typically 195 K) of the lower stratosphere during the Antarctic winter allows stratospheric clouds made of ice crystals to form at such low pressures. On ice crystals (containing water and HCl), the following heterogeneous reactions take place that decompose the chlorine reservoir species

$$\mathrm{ClONO_2(gas) + HCl(solid) \rightarrow Cl_2(gas) + HNO_3(solid)}, \quad (7.129)$$

$$\mathrm{ClONO_2(gas) + H_2O(solid) \rightarrow HOCl(gas) + HNO_3(solid)}. \quad (7.130)$$

Molecular chlorine and HOCl are easily photolysed to produce Cl during the polar spring (daytime) and hence significant ozone depletion. The HNO_3 can be removed from the atmosphere through coagulation and particle sedimentation with the result that NO_2 decreases and formation of the reservoir species $ClONO_2$ is limited.

The discovery of the decrease in the total ozone column in Antarctica by the British Antarctic Survey (WMO 1995), by about a factor of 2 between 1955 and 1985, led eventually to the correlation of the presence of high ClO concentrations with Antarctic ozone depletion. This led to the understanding of the role of the ClO dimer (ClOOCl) and BrO in ozone depletion when ClO concentrations are maintained high by the Antarctic vortex (wind circulation about the pole) that limits flow into the region during winter.

7.9 Methane and hydrogen photochemistry

Methane is an important greenhouse gas that is emitted at the Earth's surface through biological activity in swamps, marshes, rice paddies, lakes and oceans. It is also a product in emissions from agriculture, animal husbandry, mining, landfills and volcanoes. Tropospheric methane concentrations have increased from about 0.70 ppmv pre–1800 to 1.75 ppmv in 1998 (IPCC 2001). About half of the current emissions are of anthropogenic origin. Methane is removed from the atmosphere primarily through oxidation by tropospheric OH and subsequent deposition of its products at the Earth's surface, and in the case of its constituent hydrogen (both atomic and molecular forms) by escape to space at the top of the atmosphere. Methane's photochemistry is thus important as it affects stratospheric water–vapour levels, molecular hydrogen and CO levels.

Methane oxidation is initiated by its reaction with OH and follows a sequence of reactions that lead to the production of formaldehyde, CH_2O, that is photolysed to CO and molecular hydrogen. One important path is

$$CH_4 + OH \to CH_3 + H_2O, \tag{7.131}$$

$$CH_3 + O_2 + M \to CH_3O_2 + M, \tag{7.132}$$

$$CH_3O_2 + HO_2 \to CH_3OOH + O_2, \tag{7.133}$$

$$CH_3OOH + OH \to CH_2O + OH + H_2O, \tag{7.134}$$

$$CH_2O + h\nu \to CO + H_2. \tag{7.135}$$

Other possible paths are given in Appendix B, Table B.8. Reaction of CH_3O_2 with NO produces NO_2 that photolyses to produce atomic oxygen as a net effect and is thus an important source of ozone in the lower stratosphere (Crutzen 1971). Methane is thus destroyed in the lower atmosphere by oxidation, as shown in Fig. 7.13. It then undergoes a second major decrease above 80 km due to photolysis and reaction with $O(^1D)$.

7.9.1 *H_2 in the atmosphere*

The atmospheric H_2 has been examined for many years with detailed measurements made by many workers. The measurements reported by Novelli *et al.* (1999) give a global mean H_2 mixing ratio within about 1% of 530 ppbv, with an

Table 7.6 *Model global mean surface emission and deposition rates together with global mean mixing ratios for CH_4, CO and H_2.*

Rate	CH_4	CO	H_2
Mass emission (Tg year^{-1})	1038	3295	47
Molecular emission (cm^{-2} s^{-1})	2.5×10^{11}	4.5×10^{11}	9.0×10^{10}
Mass deposition (Tg year^{-1})	30	198	51
Deposition rate (s^{-1})	2.0×10^{-9}	1.0×10^{-7}	9.0×10^{-8}
Soil resistance (s m^{-1})	4.9×10^{4}	2000	2500
Mixing ratio (ppmv)	1.70	0.10	0.54

estimated global emission rate of 51 Tg year^{-1}. Estimates for the various sources and sinks of tropospheric hydrogen have varied over the years, with the emission rate varying within a range of 30–70 Tg year^{-1}. Hydrogen sources, beyond methane oxidation, include oxidation of non–methane hydrocarbons, fossil–fuel combustion and biomass burning. Additional to the soil sink of H_2 is oxidation by OH and reaction with HO_2, $O(^1D)$, O and Cl. A significant potential source of hydrogen in the troposphere, is the reaction of CO with OH, however, in the present oxygen–rich atmosphere this is rapidly converted to hydrogen peroxide, H_2O_2, via the reactions

$$CO + OH \rightarrow CO_2 + H, \tag{7.136}$$

$$H + O_2 + M \rightarrow HO_2 + M, \tag{7.137}$$

$$HO_2 + HO_2 \rightarrow H_2O_2 + O_2. \tag{7.138}$$

Hydrogen peroxide is highly water soluble and is removed by rain from the atmosphere. In an anoxic atmosphere, one that has little oxygen, the production of H, from the reaction of CO and OH, may become a significant source of atmospheric hydrogen.

In Table 7.6 are shown the emission and deposition rates of CH_4, CO and H_2 that give near–surface tropospheric mixing ratios close to the observed values, based on 1D radiative/convective–photochemical model calculations (see Chapter 11). The soil resistance corresponds to a soil cover fraction 0.21 of the Earth's total surface area, so the global mean is reduced by this fraction.

The vertical variation of molecular hydrogen is shown in Fig. 7.13, where we see that it remains fairly constant near 540 ppbv within the troposphere but gradually rises to 6000 ppbv near 140 km. The escape rate to space is 2.2×10^8 molecules cm^{-2} s^{-1}. Atomic hydrogen becomes significant above 80 km.

7.9.2 *Effects of increasing CH_4 emission*

Molecular hydrogen, like carbon monoxide and methane, is oxidized by atmospheric OH

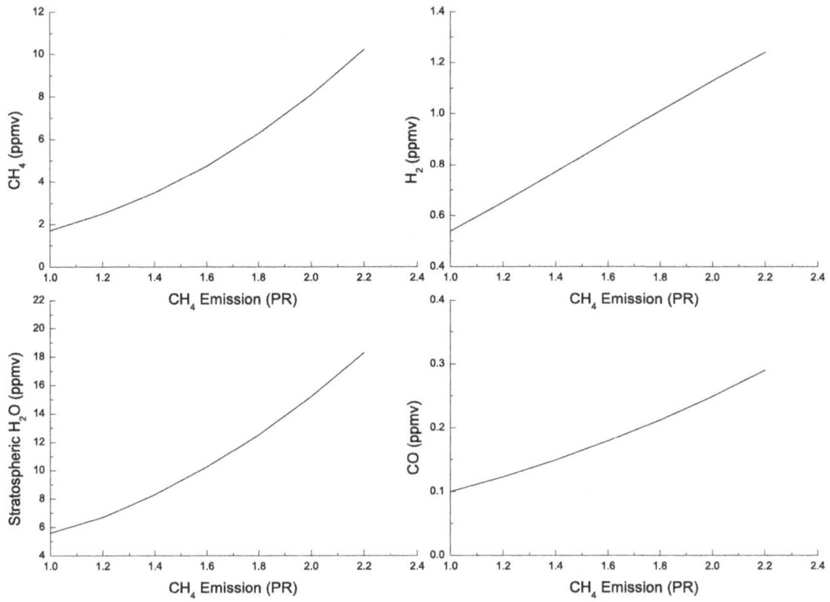

FIG. 7.16. The variation of methane, H_2 and CO surface mixing ratios with methane surface emission (1 PR = 2.5×10^{11} molecules cm^{-2} s^{-1}). Also shown is the rise in global mean stratospheric H_2O mixing ratio.

$$H_2 + OH \rightarrow H_2O + H \tag{7.139}$$

and its atmospheric mixing ratio is thus dependent on the efficiency of methane and CO oxidation by OH. As methane emission increases the OH that is available to oxidize it is reduced resulting in a non–linear rise in methane surface mixing ratio with increasing methane surface emission. In Fig. 7.16 is shown the non-linear rise in the tropospheric methane mixing ratio with methane surface emission measured in units of a model present rate (PR) of 1 PR = 2.5×10^{11} molecules cm^{-2} s^{-1}) equivalent to 1080 Tg year^{-1}. Also shown are the rate of increase in the H_2 and CO surface mixing ratios with increasing methane surface emission. Also shown is the resultant stratospheric peak H_2O mixing ratio. The rise in the global mean surface temperature corresponding to double methane surface emission is about 1.0 K (see §11.3.5).

7.9.3 *Effects of increasing CO_2 levels*

In Table 7.7 are shown the effects of increasing CO_2 from its present atmospheric level (1 PAL = 365 ppmv) on various climatic parameters of the atmosphere.

Table 7.7 *Effects of increasing atmospheric CO_2 levels (1, 2, 4 and 8 PAL) on methane, carbon monoxide and hydrogen levels (ppmv). Water–vapour column W_{H_2O} is in g cm^{-2}. Stratospheric peak water vapour mixing ratio H_2O St is in ppmv.*

Parameter	1	2	4	8
CH_4	1.70	1.46	1.22	0.99
CO	0.101	0.093	0.085	0.076
H_2	0.539	0.496	0.447	0.389
T_s	288.15	289.43	290.90	292.50
W_{H_2O}	2.09	2.35	2.65	3.02
H_2O St	5.75	5.43	4.79	3.19

A radiative–convective model (see §11.3) was run in conjunction with a photochemistry model where the surface emission rates of methane, CO and H_2 are fixed to their present levels. As the carbon–dioxide level increases, the surface temperature increases and so does the water vapour amount in the atmosphere. The effect is that the oxidation of CH_4, CO and H_2 by OH increases and so the atmospheric content of these three gases decreases. Thus methane levels in the atmosphere are linked to CO_2 levels through the water–vapour content. The stratospheric water vapour mixing ratio decreases because of the cooling of the tropospheric–stratospheric region (§11.3) and also by the decrease in atmospheric methane.

7.10 Bibliography

7.10.1 *Notes*

Tables of bimolecular diffusion coefficients can be found in Vargaftik, and in Mason and Marrero. For more details regarding molecular diffusion in the atmosphere see Banks and Kockarts; and for the mathematical theory of diffusion see Crank.

Atmospheric chemistry texts include: McEwan and Phillips; Brasseur and Solomon; Hobbs; Seinfeld and Pandis; Jacob; Warnek.

For more details on dry surface deposition of molecules see; Wesely; Wesely and Hicks; Ganzeveld and Lelieveld; for wet and dry deposition see Warnek.

For surface emission rates of gases see IPCC 2001; for molecular hydrogen see Novelli *et al.*

For more details on predissociation in molecules see; Lefebvre-Brion and Field. For absorption in the the Schumman–Runge bands of oxygen see Allen and Frederick; Lewis *et al.*; and Murtagh. For the Herzberg continuum see Nicolet and Kennes. For photoionization of oxygen see Watanabe *et al.*; and Banks and Kockarts.

For further details regarding ozone absorption see Orphal and Chance for the HITRAN database. For water–vapour absorption see Lewis *et al.*; and the work of Yoshino *et al.*

Branching ratios for methane photolysis are given by; Mordaunt *et al.*; Mount *et al.*; and Toublanc *et al.*

For bimolecular collision theory consult; Vincenti and Kruger; Stevens; and Rosenberg.

Chemical reaction databases that can be easily accessed include: JPL, IUPAC, NIST, and UDfA.

For the role of water vapour on the destruction of mesospheric ozone see Allen *et al.*; Roble; and Vardavas *et al.* For nighttime/daytime ratios of ozone and OH see Vardavas *et al.*; for the diurnal averaging of nighttime/daytime effects see Turco and Whitten.

Mechanisms for the formation of thermospheric NO are discussed in: Barth; Bailey *et al.*; and Barth and Bailey.

Branching ratios for the three reaction channels for the reaction of H with HO_2 are given in: Keyser 1986; and JPL 2006.

For the numerical solution of a system of coupled chemical rate equations see Vardavas, and Lavvas *et al.*

7.10.2 *References and further reading*

Allen, M. and Frederick, J. E. (1982). Effective photodissociation cross–sections for molecular oxygen and nitric oxide in the Schumann-Runge bands. *J. Atmos. Sci.*, **39**, 2066–2075.

Allen, M., Lunine, J. I. and Yung, Y. L. (1984). The vertical distribution of ozone in the mesosphere and lower atmosphere. *J. Geophys. Res.*, **89**, 4841–4872.

Bailey, S. M., Barth, C. A. and Solomon, S. C. (2002). A model of nitric oxide in the lower thermosphere. *J. Geophys. Res.*, **107**, doi:1029/2001JA000258.

Banks, P. M. and Kockarts, G. (1973). *Aeronomy: Parts A and B.* Academic Press, New York.

Barth, C. A. (1992). Nitric oxide in the lower thermosphere. *Planet. Space Sci.*, **40**, 315–336.

Barth, C. A. and Bailey, S. M. (2004). Comparison of a thermospheric photochemical model with Student Nitric Oxide Explorer (SNOE) observations of nitric oxide. *J. Geophys. Res.*, **109**, A03304.

Bass, A. M. and Paur, R. J. (1981). UV absorption cross–sections for ozone: The temperature dependence. *J. Photochem.*, **17**, 141–141.

Brasseur, G. M. and Solomon, S. (1984). *Aeronomy of the middle atmosphere.* D. Reidel, Dordrecht.

Brownsword, R. A., Hillenkamp, M., Laurent, T., Vatsa, R. K., Volpp, H.-R. and Wolfrum, J. (1997). Quantum yield for H atom formation in the methane dissociation after photoexcitation at the Lyman–α (121.6 nm) wavelength. *Chem. Phys. Lett.*, **266**, 259–266.

Chapman, S. (1930). On ozone and atomic oxygen in the upper atmosphere. *Philos. Mag.*, **10**, 369–383.

Crank, J. (1995). *The mathematics of diffusion.* Oxford University Press, Oxford.

Crutzen, P. (1970). The influence of nitrogen oxides on the atmospheric ozone content. *Quart. J. Roy. Met. Soc.*, **96**, 320–325.

Crutzen P. (1971). Ozone production rates in an oxygen-hydrogen-nitrogen oxide atmosphere. *J. Geophys. Res.*, **76**, 7311–7327.

Dobson, G. M. B. (1931). A photoelectric spectrophotometer for measuring the amount of atmospheric ozone. *Proc. Phys. Soc.*, **43**, 324–339.

Ganzeveld, L. and Lelieveld, J. (1995). Dry deposition parameterization in a chemistry general circulation model and its influence on the distribution of reactive trace gases. *J. Geophys. Res.*, **100**, 20999–21012.

Griggs, M. (1968). Absorption coefficients of ozone in the ultraviolet and visible regions. *J. Chem. Phys.*, **49**, 857-859.

Hearn, A. G. (1961). The absorption of ozone in the ultraviolet and visible regions of the spectrum. *Proc. Phys. Soc.*, **78**, 932–940.

Herzberg, G. (1950). *Molecular spectra and molecular structure: I. Spectra of diatomic molecules.* 2nd ed., Van Nostrand, New York.

Hobbs, P. V. (2000). *Introduction to atmospheric chemistry.* Cambridge University Press, Cambridge.

IPCC, 2001: *Climate change 2001: The scientific basis. Contribution of working group I to the Third Assessment Report of the Intergovernmental Panel on Climate Change.* Houghton, J. T., Ding, Y., Griggs, D. J., Noguer, M., van der Linden, P. J., Dai, X., Maskell, K. and Johnson, C. A. (ed.), Cambridge University Press, Cambridge, United Kingdom and New York, NY, USA.

IUPAC, 2006: *Summary of evaluated kinetic and photochemical data for atmospheric chemistry.* Atkinson, R., Baulch, D. L., Cox, R. A., Crowley, J. N., Hampson Jr, R. F., Hynes, R. G., Jenkin, M. E., Kerr, J. A., Rossi, M. J.

and Troe, J. (ed.), Centre for Atmospheric Science, University of Cambridge. [http://www.iupac-kinetic.ch.cam.ac.uk/]

Jacob, D. J. (1999). *Introduction to atmospheric chemistry.* Princeton University Press, Princeton.

Jensen, R. J., Guettler, R. D., Lyman, J. L. (1997). The ultraviolet absorption spectrum of hot carbon dioxide. *Chem. Phys. Lett.*, **277**, 356–360.

JPL 2006 : *Chemical kinetics and photochemical data for use in atmospheric studies evaluation number 15.* NASA Panel for Data Evaluation: S. P. Sander, R. R. Friedl, A. R. Ravishankara, D. M. Golden, C. E. Kolb, M. J. Kurylo, J. Molina, G. K. Moortgat, H. Keller-Rudek, B. J.Finlayson-Pitts, P. H. Wine, R. E. Huie, and V. L. Orkin. Jet Propulsion Laboratory Publication 06-2, California Institute of Technology, Pasadena. [http://jpldataeval.jpl.nasa.gov/]

Karaiskou, A., Vallance, C., Papadakis, V., Vardavas, I. M., and Rakitzis, P. (2004). Absolute absorption cross–section measurements of CO_2 in the ultraviolet from 200 to 206 nm at 295 K and 373 K. *Chem. Phys. Lett.*, **400**, 30–34.

Keyser, L. F. (1986). Absolute rate constant and branching fractions for the H + HO_2 reaction from 240–300 K. *J. Phys. Chem.*, **90**, 2994–3003.

Lavvas, P. P., Coustenis, A. and Vardavas, I. M. (2007). Coupling photochemistry with haze formation in Titan's atmosphere. Part I: Model description. *Planet. Space Sci.*, in press.

Lefebvre-Brion, H. and Field, R. W. (1986). *Perturbations in the spectra of diatomic molecules.* Academic Press, New York.

Lee, L. C., Philips, E. and Judge, D. L. (1977). Photoabsorption cross–sections for CH_4, CF_4, CF_3Cl and C_2F_6. *J. Chem. Phys.*, **67**, 1237–1246.

Lewis, B. R. L., Vardavas, I. M. and Carver, J. H. (1983). The aeronomic dissociation of water vapor by solar H Lyman–α radiation. *J. Geophys. Res.*, **88**, 4935–4940.

Lewis, B. R. L. and Carver, J. H. (1983). Temperature dependence of the carbon dioxide photoabsorption cross–section between 1200 and 1970 Å. *J. Quant. Spectrosc. Radiat. Transfer*, **30**, 415–470.

Lewis, B. R. L., Gibson, S. T. and Dooley, P. M. (1994). Fine–structure dependence of predissociation linewidth in the Schumann–Runge bands of molecular oxygen. *J. Chem. Phys.*, **100**, 7012–7035.

Mason, E. A. and Marrero, T. R. (1970). The diffusion of atoms and molecules, in: Bates, D. R., Esterman, I. (ed.), Advances in atomic and molecular physics. Academic Press, San Diego.

McElroy, M. B. and McConnell, J. C. (1971). Nitrous oxide: A natural source of stratospheric NO. *J. Atmos. Sci.*, **28**, 1095–1098.

McEwan, M. J. and Phillips, L. F. (1975). *Chemistry of the atmosphere.* Edward Arnold, London.

Molina, M. and Rowland, F. S. (1974). Stratospheric sink for chlorofluoromethanes: chlorine atom-catalyzed destruction of ozone. *Nature*, **249**, 810–812.

Mordaunt, D. H., Lambert, I. R., Morley, G. P., Ashfold, N. R. Dixon, R. N., Western, C. M., Schnieder, L. and Welge, K. H. (1993). Primary product channels in the photodissociation of methane at 121.6 nm, *J. Chem. Phys.*, **98**, 2054–2065.

Mount, G. H., Warden, E. S. and Moos, H. W. (1977). Photoabsorption cross–section of methane from 1400 to 1850 Å. *Astrophys. J. Lett.*, **214**, L47–L49.

Murtagh, D. P. (1988). The O_2 Schuman-Runge system - new calculations of photodissociation cross–sections. *Planet. Space Sci.*, **36**, 819–828.

Nicolet, M. and Kennes, R. (1986). Aeronomic problems of the molecular oxygen photodissociation-I: The O_2 Herzberg continuum. *Planet. Space Sci.*, **34**, 1043–1059.

NIST, 2006: *Chemical Kinetics Database on the Web.* Standard Reference Database 17, Version 7.0, Release 1.4. A compilation of kinetics data on gas-phase reactions. [http://kinetics.nist.gov/kinetics/index.jsp]

Novelli, P. C., Lang, P. M., Masarie, K. A., Hurst, D. F., Myers, R. and Elkins, J. W. (1999). Molecular hydrogen in the troposphere: Global distribution and budget. *J. Geophys. Res.*, **104**, 30427–30444.

Okabe, H. (1978). *Photochemistry of small molecules.* John Wiley and Sons Inc., New York.

Orphal, J. and Chance, K. (2003). Ultraviolet and visible absorption cross–sections for HITRAN. *J. Quant. Spec. Rad. Trans.*, **82**, 491–504.

Parkinson, W. H. and Yoshino, K. (2003). Absorption cross–section measurements of water vapor in the wavelength region 181-199 nm. *Chem. Phys.*, **294**, 31–35.

Roble, R. G. (1995). Energetics of the mesosphere and thermosphere. In *The upper mesosphere and lower thermosphere: A review of experiment and theory.* (ed.) R. M. Johnson, and T. L. Killeen, Geophysical Monograph 87, American Geophysical Union, Washington DC.

Rosenberg, R. M. (1977). *Principles of physical chemistry.* Oxford University Press, Oxford.

Samson, J. R., Haddad, G. N., Masuoka, T., Pareek, P. N. and Kilcoyne, D. A.

L. (1989). Ionization yields, total absorption, and dissociative photoionization cross–sections of CH_4 from 110 to 950 Å. *J. Chem. Phys.*, **90**, 6925–6932.

Seinfeld, J. H. and Pandis, S. N. (1998). *Atmospheric chemistry and physics: From air pollution to climate change.* John Wiley & Sons Inc., New York.

Stevens, B. (1970). *Chemical kinetics for general students of chemistry.* Chapman and Hall, London.

Strobel, D. F. (1989). Constraints on gravity-wave induced diffusion in the middle atmosphere. *PAGEOPH*, **130**, 533–546.

Turco, R. P. and Whitten, R. C. (1978). A note on the diurnal averaging of aeronomical models. *J. Atmos. Terr. Phys.*, **40**, 13–20.

Toublanc, D., Parisot, J. P., Brillet, J., Gautier, D., Raulin, F. and McKay, C. P. (1995). Photochemical modeling of Titan's atmosphere. *Icarus*, **113**, 2–26.

UDfA 2005: The UMIST database for Astrochemistry. see Le Teuff, Y.H. Millar, T. J. and Markwick, A. J. (2000). *Astron. Astrophys. Suppl. Ser.*, **146**, 157–168. also Woodall, J., Agundez, M., Markwick-Kemper, A. J., and Millar, T. J. (2007). *Astron. Astrophys.*, **466**, 1197-1204. [http://www.udfa.net/]

Vardavas, I. M. (1984). Modelling reactive gas flows within shock tunnels. *Aust. J. Phys.*, **37**, 1–21.

Vardavas, I. M. and Carver, J. H. (1984). Solar and terrestrial parameterizations for radiative-convective models. *Planet. Space Sci.*, **32**, 1307–1325.

Vardavas, I. M., Carver, J. H. and Taylor, F. W. (1998). The role of water-vapour photodissociation on the formation of a deep minimum in mesopause ozone. *Ann. Geophysicae*, **16**, 189–196.

Vargaftik, N. B. (1975). *Tables on the thermophysical properties of liquids and gases.* John Wiley and Sons Inc, New York.

Vigroux, E. (1953). Contribution a l'étude éxperimentale de l'absorption l'ozone. *Ann. Phys.*, **8**, 709–762.

Vincenti, W. G. and Kruger, C. H. (1977). *Introduction to physical gas dynamics.* Krieger, New York.

Warnek, P. (1988). *Chemistry of the natural atmosphere.* Academic Press, New York.

Watanabe, K. E., Inn, C. Y. and Zelikoff, M. (1953). Absorption coefficients of oxygen in the vacuum ultraviolet. *J. Chem. Phys.*, **21**, 1026–1030.

Wesely, M. L. and Hicks, B. B. (2000). A review of the current status of knowledge on dry deposition. *Atmos. Env.*, **34**, 2261–2282.

Wesely, M. L. (1989). Parameterization of surface resistances to gaseous dry deposition in regional–scale numerical models. *Atmos. Env.*, **23**, 1293–1304.

WMO 1995 and 1999. *Scientific assessment of ozone depletion.* Global Ozone Research and Monitoring Project, World Meteorological Organisation, Reports 37 and 44, Geneva.

Yoshino, K., Esmond, J. R., Parkinson, W. H., Ito, K. and Matsui, T. (1996). Absorption cross–section measurements of water vapour in the wavelength region 120 to 188 nm. *Chem. Phys.*, **211**, 387–391.

8

THE EARTH'S RADIATION BUDGET

8.1 Introduction

The Sun is the source of energy that sustains life on the Earth. Solar energy enters and is redistributed within the Earth–atmosphere system, which subsequently emits radiation at longer wavelengths. The *Earth's radiation budget* (ERB) drives the general circulation of the atmosphere and determines the climate of the Earth–atmosphere system, being an indicator of possible climatic changes. The ocean–surface energy budget ultimately influences the major features of ocean circulation, and hence regional climate. Thus, it is very important to be able to determine as accurately as possible its components at the top of the atmosphere (TOA), within the atmosphere and at the Earth's surface for both shortwave (SW) and longwave (LW) radiative fluxes. At TOA, the net energy input is determined by the incident SW radiation from the Sun minus the reflected SW energy. The difference defines the net SW radiative flux at TOA. To balance this inflow of SW energy, the Earth–atmosphere system emits LW radiation to space. For the planet to be in radiative equilibrium, the emitted LW radiation must equal the net inflow of SW radiation at TOA. Radiative processes that perturb this equilibrium can produce global climatic change.

In this chapter we shall examine the role of atmospheric molecules, aerosols, clouds and properties of the Earth's surface in the Earth's radiation budget. The net solar and terrestrial longwave irradiances at the Earth's surface, i.e. the *surface radiation budget*, (SRB), and at the top of the atmosphere (TOARB), determine the distribution of energy in the Earth–atmosphere–ocean system and hence climate. Any externally imposed perturbation on the Earth's energy budget arising from changes in the concentration of infrared radiatively active molecules or greenhouse gases, H_2O, CO_2, CH_4, N_2O, clouds and aerosols, incoming solar irradiance, or changes in surface reflection is called *radiative forcing*. Such perturbations can lead to changes in climatic parameters, resulting in a new equilibrium state of the climate system (IPCC 2001). Thus, the radiation budget of the Earth–atmosphere system plays a fundamental role in determining the thermal conditions and the circulation of the atmosphere and the ocean, shaping the main characteristics of the Earth's climate.

The ultimate source of energy that drives the climate system is radiation from the Sun. The mean global daily incoming solar flux at the top of the Earth's atmosphere is 342 W m^{-2}. About a third of that is reflected directly back to space

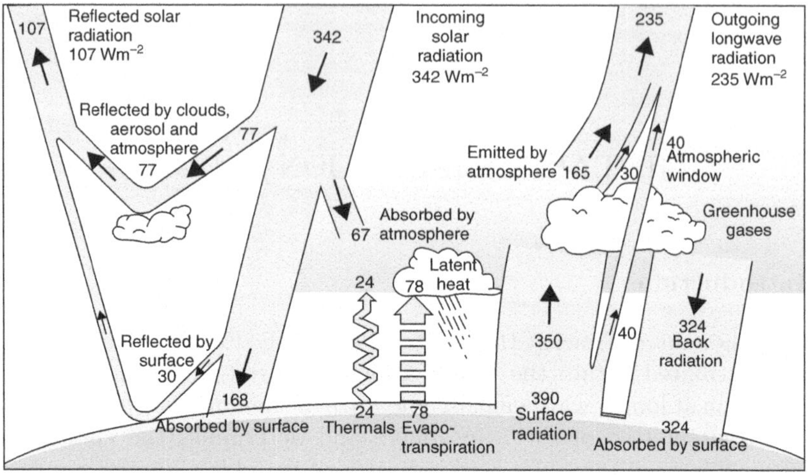

FIG. 8.1. Schematic representation for the partitioning of the Earth's global mean energy balance. For updated values of the various components see text. (Kiehl and Trenberth 1997)

by clouds, by the atmosphere and by the surface of the Earth. The remaining radiation is partly absorbed by the atmosphere and partly by the surface, thus warming it up. This heat is returned to the atmosphere as thermal infrared radiation, sensible heat and latent heat, Fig. 8.1. For a stable climate, the amount of net incoming solar radiation must equal the amount of net thermal infrared radiation emitted back to space. The radiation budget represents this balance of energy.

Attempts to construct a global annual mean ERB date back to the beginning of the twentieth century. There has been a long history of early studies that, however, were limited by the lack of knowledge of the planetary albedo. Satellite observations, such as those from Nimbus-7 and the Earth Radiation Budget Experiment (ERBE) have greatly improved our knowledge of TOARB. Nevertheless, satellites do not measure directly radiative fluxes at the Earth's surface. Therefore, knowledge of SRB is far less advanced than that of TOARB. The SRB, however, is a major component of the energy exchanges between the atmosphere and the land/ocean surface, and hence affects temperature fields, atmospheric and oceanic circulation, and the hydrological cycle. For example, about half of the solar energy absorbed at the surface is used to evaporate water, which eventually forms clouds. SRB data are also valuable for initializing and testing simple climate models and the more complex general circulation models (GCMs).

The estimation of SRB represents one of the most significant objectives of the World Climate Research Programme (WCRP) as demonstrated by its Global Energy and Water Cycle Experiment (GEWEX), and in particular the GEWEX

SRB project. Due to the inability of directly measuring radiation at the Earth's surface with satellites, the SRB has to be derived, as surface observations cannot solve the problem, because of the sparsity and inhomogeneity of meteorological stations, especially over oceans. Thus, modelling remains as the only alternative, provided that the model results are validated against quality measurements.

One modelling approach is to establish correlations between TOA and surface radiative fluxes to derive surface fluxes directly from satellite measurements of TOA fluxes. There is, however, uncertainty regarding the reliability of this method. Another modelling approach, is to develop algorithms and models to compute SRB fluxes on a global scale on the basis of global observations of atmospheric, cloud, and surface properties, preferably from operational satellite sources, since they provide global coverage.

The amount and quality of satellite data have substantially improved with time, especially in terms of cloud properties. The International Satellite Cloud Climatology Project (ISCCP) has provided a complete and comprehensive global cloud climatology. Thus, in the last two decades, a large number of studies that estimate SRB have been published. These were based on the C-series data. The new D-series include significant improvements, especially in terms of cloud structure and cloud detection, over highly reflecting surfaces as well as covering a longer time period, extending from 1983 to 2004. The ISCCP–D2 data are monthly averaged, on a 2.5×2.5 degree (latitude–longitude) resolution.

Early estimates of the planetary albedo ranged from 40% to 50%. Satellite observations from Nimbus–7 and ERBE have greatly improved estimates of the planetary albedo, giving values near 30%. Although the determination of radiative fluxes at TOA is more advanced than at the surface, due to the recent advances in space–borne measurements that provide good knowledge of the net solar energy absorbed by the global climate system, there is a need to model accurately the ERB at TOA. Model validation at TOA is a pre–requisite for obtaining reliable model computations of ERB at the surface, where direct satellite measurements are not possible and surface measurements cannot provide complete global coverage. Satellite observations of ERB have limitations in their accuracy and do not provide continuous data. An outgoing SW radiation (OSR) data set has been produced from the ERB measurements taken aboard Nimbus satellites. The Nimbus ERB scanner data cover the period from January 1979 through May 1980, while non-scanner data extend from November 1978 till December 1993. The ERBE scanner OSR data are considered to be of very good quality in terms of spatial and temporal coverage and instrument accuracy, but they cover only the 5–year period from November 1984 to February 1990. Therefore, although the ERB data obtained from the Nimbus–7 measurements along with those from ERBE, span over a decade (1978–1990), they do not provide continuity. Beyond this, there is a gap in the time series between those and more recent accurate measurements, such as the Scanner for Radiation Budget (ScaRaB) and the Clouds and the Earth's Radiant Energy System (CERES), especially during the 1990s. More sophisticated space–borne instruments such as the Moderate Resolution Imaging Spectroradiometer (MODIS) on board the Terra and Aqua Satellites and the Geostationary Earth Radiation Budget (GERB) instrument on EUMETSAT's Meteosat Second Generation (MSG) satellite have started high–accuracy and high–resolution measurements since early 2000. Furthermore, apart from the fact that satellite measurements have their own errors, especially the SW retrievals, they do not provide complete spatial (global) coverage, as they often miss areas poleward of 70–degree latitude, where model results can be used as first estimates, even though not very accurate. Reliable model ERB computations at TOA, are very important to the study of the temporal variation of the Earth's radiation budget and help us to understand the physical processes that determine the TOA radiation budget.

8.2 Model input data

In order to calculate the longwave and shortwave radiation budgets using the models described in Chapters 4 and 6, various atmospheric, cloud and surface properties are required. These are listed in Table 8.1.

Table 8.1 *Input data required to model the Earth's radiation budget. For clouds the data required are for each cloud type (low, middle and high–level).*

Cloud properties	Atmospheric properties	Surface properties
Cloud amounts	Temperature	Pressure (topography)
Scattering optical depth	Specific humidity	Surface temperature
Absorption optical depth	CO_2, O_3, CH_4, N_2O	Albedo
Cloud-top pressure	Aerosol optical depth,	Emissivity
Cloud-top temperature	Single scattering albedo,	Land/water/coast cover
Cloud-base temperature	Asymmetry parameter	Ice/snow cover
Geometrical thickness		

8.2.1 *Cloud radiative properties*

Clouds also play a very important role in determining the Earth's radiation budget, as they cover on average about 60% of the Earth's surface. Most clouds are so optically thick in the infra–red that they emit as blackbodies. As they are located at high altitudes within the troposphere, their effective blackbody temperature is much lower than that of the Earth's surface. Hence, they absorb infra–red radiation from below, that is emitted at higher temperatures, and re–radiate it to space as cooler blackbodies with the result that they warm the planet. The low–level and middle–level clouds are sufficiently optically thick in the infra–red to be regarded usually as blackbodies. High–level clouds can be optically thin and hence they absorb and emit as non–ideal blackbodies.

Scattering and absorption of the incoming SW radiation at TOA, depends strongly on the presence and type of clouds in the atmosphere, the composition of the atmosphere (gases plus particulate matter) and the reflectivity of the Earth's surface. It is weakly dependent on the thermal structure of the atmosphere. The cloudy–sky component is subdivided into components covered by low–, middle–, and high–level clouds.

8.2.2 *Cloud data sets*

All of the cloud meteorological data listed in Table 8.1, except for the cloud-base temperature, are available from the International Satellite Cloud Climatology Project (ISCCP), established in 1982 as part of the World Climate Research Programme to collect and analyse satellite radiance measurements to infer the global distribution of clouds, their properties, and their diurnal, seasonal and interannual variations. Data collection began in July 1983 and continues to date. ISCCP currently provides the most extensive and comprehensive cloud climatological database that quantifies the variations of cloud properties at the global scale for over more than 20 years.

The ISCCP–D2 data set, which contains monthly data on a 2.5–degree resolution are derived from the ISCCP–DX data, which are 3–hourly and with a resolution of 30 km. There are also the ISCCP–D1 data, which are 3–hourly also on a 2.5–degree grid. Missing data in specific gridboxes may be replaced with values derived by linear interpolation between the values of the neighbouring gridboxes. Cloud-base temperature is necessary for the estimation of the longwave downwelling flux at the surface but it is not provided by satellite data. This parameter can be estimated from the cloud–top pressure and the cloud physical thickness.

ISCCP–D2 cloud properties are provided for nine cloud types that are grouped into three categories: low–level clouds having top pressures greater than or equal to 680 mbar, high–level clouds with top pressures less than 440 mbar and middle–level clouds in between. Cumulus (Cu), stratus (St) and stratocumulus (Sc) clouds are considered low–level clouds. The middle–level clouds include altocumulus (Ac), altostratus (As) and nimbostratus (Ns) clouds, while cirrus (Ci), cirrostratus (Cs) and deep–convective clouds are considered as high–level clouds. Figure 8.2 summarizes the ISCCP radiometric cloud classification. High–level clouds are treated as ice clouds, while each of the low and middle clouds are subdivided into liquid– or ice–phase clouds, resulting in 15 cloud types for which properties such as cloud cover, cloud optical thickness, cloud–top temperature and pressure, as well as liquid or ice water path are provided by ISCCP–D2. The values for the 15 independent cloud types can be appropriately averaged to yield data for low, middle and high–level clouds. LW and SW radiation transfer models

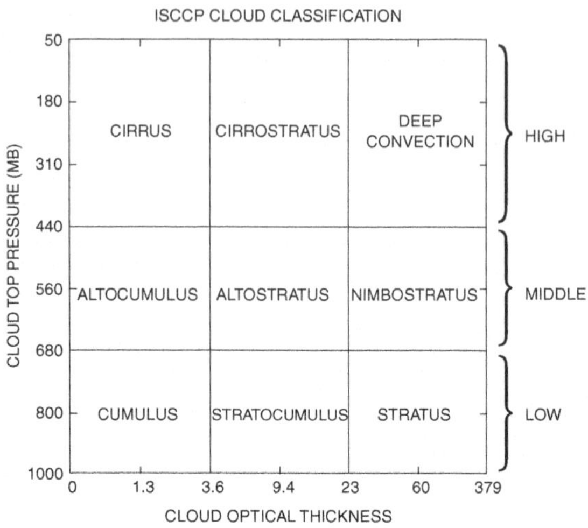

FIG. 8.2. The ISCCP radiometric cloud classification. Cloud–top pressure in mbar.

can use sub-type cloud properties derived from a combination of visual, infrared and near–infra–red measurements. Cloud–cover fractions for low–, middle– and high–level clouds are calculated from the sum of cover fractions for independent ISCCP–D2 cloud types, A_{ci}, whereas the total cloud–cover fraction, A_c, is calculated from the sum of fractions for low, middle and high clouds. Figure 8.3 shows the global distribution of long–term average total cloud cover for January.

The ISCCP–D2 provides cloud scattering optical depth, τ_c^s, only for the visible wavelength of $\lambda = 0.6$ μm and at the infra–red wavelength of 11 μm through a conversion. For the computation of the cloud SW transmissivity and reflectivity in the near–infrared, a radiation transfer model also requires cloud absorption optical depth, τ_c^a and τ_c^s data in the near–infra–red. The near–infra–red values can be derived from the ratios τ_c^s(near–infra–red)/τ_c^s(UV–visible) and τ_c^a(near-infra–red)/τ_c^s(UV–visible) that result either from Mie computations (Chapter 6) using the standard ISCCP liquid–droplet spectrum, with a specific gamma distribution function having an effective radius, r_{eff} of 10 μm and an effective variance of 0.15 μm, or from existing parameterizations for liquid and ice clouds. The standard ISCCP–D2 ice–cloud model assumes a random fractal crystal shape and a 2–power–law size distribution from 20 to 50 μm, with an effective radius of 30 μm and an effective variance of 0.1 μm. Typically, values of τ_c^a(near–infra–red)=0.08τ_c^s(UV–visible) and τ_c^a(near–infra–red)=0.03τ_c^s(UV–visible) can be used in a model for the liquid and ice clouds, respectively.

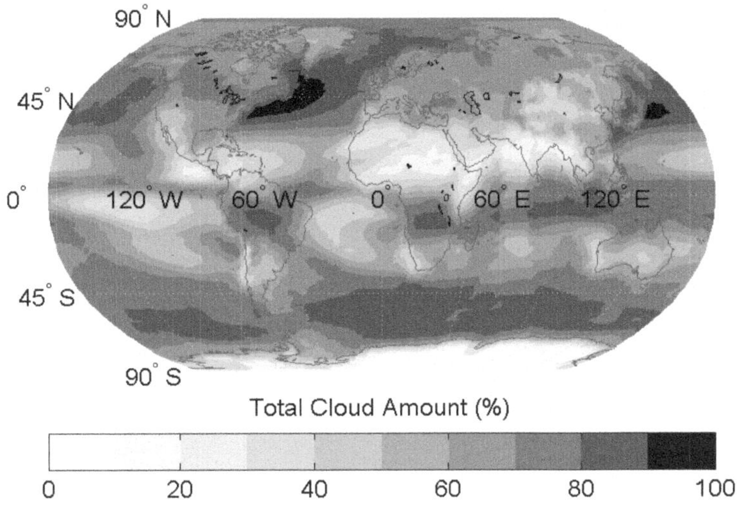

FIG. 8.3. Long–term average total cloud cover for January based on ISCCP–D2 data.

8.2.3 *Water vapour and temperature profiles*

The vertical temperature and humidity profiles (including surface pressure) are available from three different reanalyses projects: (i) NCEP/NCAR (National Center for Environmental Prediction/National Center for Atmospheric Research) Reanalysis project; (ii) ECMWF (European Centre for Medium–Range Weather Forecasts Reanalysis: ERA 15 and 40); and (iii) GEOS–1 Reanalysis (Goddard Earth Observing System). Each data set can be used separately, after averaging and remapping them to match the spatial resolution required. The different data sets for the temperature and humidity profiles of the atmosphere have significant differences, which are subsequently reflected in model fluxes.

Air temperature plays an important role particularly in determining the long-wave fluxes. For example, a comparison of the global distribution of the difference between the mean temperature of the lowest 100 mbar of the atmosphere given by GEOS and that given by NCEP/NCAR, for the month of January, shows that the largest differences, reaching values of 6 K (with NCEP/NCAR giving the higher values), occur over land, particularly in extended regions of the Northern Hemisphere in winter (North America, Siberia, Antarctica). Over oceans the differences are smaller, of the order of 1 K. The discrepancies be-

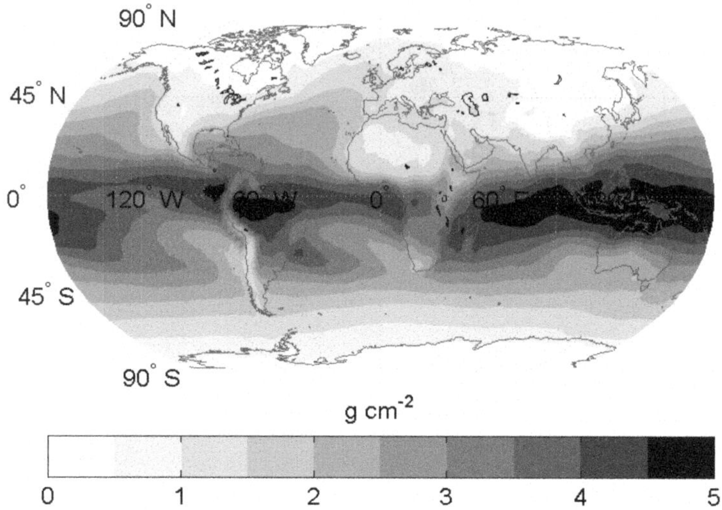

FIG. 8.4. Long–term (1984–2004) average global distribution of precipitable water vapour (g cm^{-2}) for January, based on NCEP/NCAR data.

tween NCEP/NCAR and ECMWF are generally less pronounced. Again, best agreement is found over oceans, while differences of up to 4 K can be found over land, especially in high–altitude regions (Andes, Greenland, Tibetan plateau, Antarctica). An intercomparison of the water–vapour content in the lower 100 mbar of the atmosphere as given by the different data sets give differences that are typically within 25% over much of the globe, although there are extended regions in the Northern Hemisphere, mostly above 40° (North America and Asia), where there are much larger differences, that exceed 60%. In Fig. 8.4 is shown the global distribution of precipitable water vapour (g cm^{-2}) for January 1988, based on NCEP/NCAR data.

8.2.4 *Other greenhouse gases*

Monthly 2.5–degree latitude–longitude pixel data of the total O_3 column abundance (in Dobson units), are available from the Television Infrared Observational Satellite (TIROS) Operational Vertical Sounder (TOVS), archived in the ISCCP–D2 package. For CO_2 a fixed total atmospheric amount can be taken, equal to 0.57 g cm^{-2}, corresponding to 365 (1998 value) parts per million by

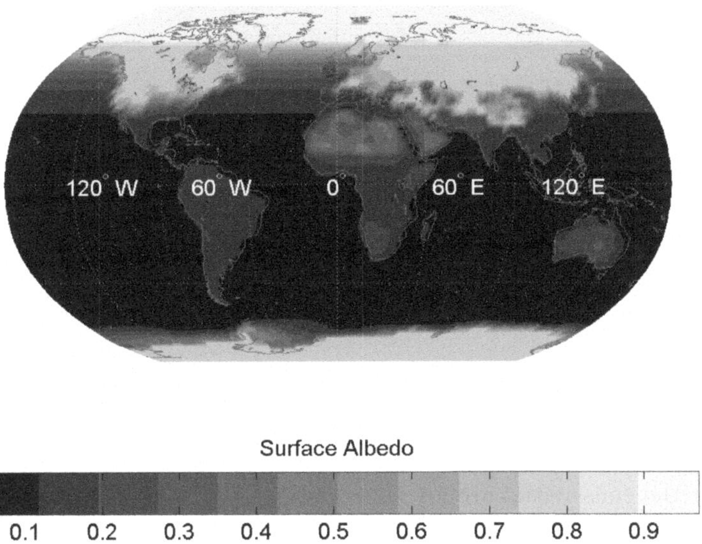

FIG. 8.5. Long–term (1984–1997) global distribution of surface albedo for January. The effects of Fresnel reflection at low solar elevation angles are shown by zonal bands for the Northern Hemisphere winter, where the Arctic has polar night.

volume (ppmv). Similarly, one can set the amounts for CH_4, and N_2O equal to 1.7 and 0.3 ppmv, corresponding to 9.0 and 4.6×10^{-4} g cm^{-2}, respectively. We note that the methane and nitrous oxide mixing ratio is not constant up to high altitudes as for CO_2, so that the amounts in g cm^{-2} are slightly lower than expected for a constant mixing ratio in the atmosphere.

8.2.5 *Surface properties*

ISCCP–D2 provides surface–type fractions such as land, water, ice and snow cover. Land–surface–type information is also provided by the International Satellite Land Surface Climatology Project (ISLSCP). Fifteen different surface types are included in this data set. Reflectances for water and ice can be computed as described in Chapter 6.

8.2.5.1 *Surface albedo* The land–surface reflectance, R_g can be computed based on surface–type information and corresponding reflectances. Figure 8.5 shows the global distribution of shortwave albedo.

8.2.5.2 *Surface emissivity* Computation of the thermal infra–red fluxes requires the total surface emissivity, ϵ_g. In the calculation of ϵ_g we need to consider the following four types of surface: land, ocean, snow, and ice (frozen ocean). The

Table 8.2 *Thermal infra–red emissivity for various surfaces. (Houghton 1985)*

Surface	Emissivity
Water	0.92–0.96
Snow	0.82–0.99
Ice	0.96
Sand	0.89–0.90
Grass	0.90
Forest	0.90

surface emissivity can then be computed from

$$\epsilon_{\rm g} = f_{\rm land}\epsilon_{\rm l} + f_{\rm ocean}\epsilon_{\rm o} + f_{\rm snow}\epsilon_{\rm s} + f_{\rm ice}\epsilon_{\rm i}, \tag{8.1}$$

where f is the fraction of the Earth's surface covered by each type of emitting surface. The various fractions f change with the season and location. Typical values for the emissivities are given in Table 8.2 for various surfaces.

8.2.5.3 *Surface topography* A topography scheme is needed in a model, and this can be obtained using surface–pressure data given either by the NCEP or from ECMWF Global Reanalysis Projects gridded in 2.5× 2.5 degree pixels. Consideration of topography is important for regions with high altitudes, such as the Tibetan plateau, Antarctica, Rocky mountains or the Andes mountains, for the correct computation of layer and total atmospheric amounts of the gases considered, as well as the correct extent of the Rayleigh–scattering layer and the appropriate computation of the mean humidity of the aerosol layer.

8.2.6 *Aerosol particles*

The aerosol particles perturb the radiation field sufficiently, especially SW radiation, to warrant their consideration in the estimation of the SW radiation budget. The radiative effect of aerosols at TOA, within the atmosphere, and at the surface, can be computed by a SW model, using a modified two–stream approximation or the modified Delta–Eddington method (see §6.7) allowing for scattering and absorption in the UV-visible and near–infra–red. Such models require aerosol optical properties such as aerosol extinction optical thickness (AOT), single scattering albedo (ω), and asymmetry parameter (g). Climatological aerosol data can be derived from the Global Aerosol Data Set (GADS). In GADS, aerosols are described as internal plus external mixtures of 10 main aerosol components, which are representative of the atmosphere and characterized through their size–distribution and refractive index, depending on wavelength. Data of AOT, $\omega_{\rm aer}$, and $g_{\rm aer}$, are provided by GADS at 61 wavelengths from 0.25 to 40 μm, 27 of which lie in the SW range, for 8 values of the relative humidity (0, 50, 70, 80, 90, 95, 98, and 99%). Given the strong dependence of

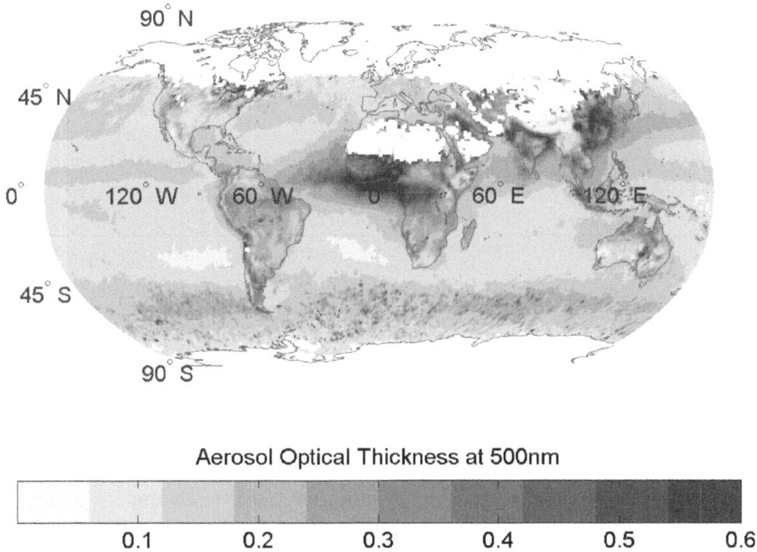

FIG. 8.6. The global mean (2000–2005) distribution of the visible (0.50 μm) aerosol optical thickness for the month of January, from MODIS based on the Terra satellite data. The main aerosol source regions are clearly seen, white areas correspond to either polar night or missing data.

aerosol optical properties on ambient relative humidity, the original GADS properties need to be computed for actual relative humidity values within the aerosol layer. Subsequently, the computed values of aerosol optical properties can be averaged within the UV–visible and near–infra–red spectral intervals, weighted by the spectral distribution of the incoming solar flux.

Real–time satellite data are becoming available, for example there are available the MODIS data from the Terra and Aqua satellites. In Fig. 8.6, we show the global distribution of the visible (0.5 μm) aerosol optical thickness for the month of January, from the MODIS (2000–2005) data set, based on the Terra satellite data.

8.3 Validation data

Validation of model results both for the downward fluxes reaching the surface of the Earth and the outgoing fluxes at TOA is only possible through comparison with ground–based measurements at specific sites for the downward fluxes, and with satellite measurements for the outward fluxes. Table 8.3 summarizes a sample of validation data sets that are available.

Table 8.3 *Validation data sets. Fluxes at the top of the atmosphere (TOA) are outgoing to space. Fluxes are both longwave and shortwave.*

Dataset	Description	Years	Time resolution	Spatial resolution	Coverage
BSRN	Surface fluxes	92–01	minute-daily	Point location	Global
	Synoptic data				
GEBA	Surface fluxes	84–97	mean monthly	Point location	Global
ERBE	TOA Fluxes	85–89	mean monthly	2.5°×2.5°	Global
ERBS-NS	TOA Fluxes	85–99	mean monthly	5°×5°	60°S-60°N

Table 8.4 *Global Energy Balance Archive surface fluxes. Volume corresponds to the amount of data stored as at 2002.*

Code	Name	Instrument	Volume
2	Direct solar radiation	pyrrheliometer	2700
3	Diffuse solar radiation	pyranometer with shading device	10000
1	Global solar radiation	pyranometer up–facing	180000
5	Reflected solar	pyranometer down–facing	3300
16	Absorbed solar radiation	difference of 1 and 5	5000
4	Albedo	quotient of 5 and 1; albedometer	3800
19	Circumglobal radiation	spherical pyranometer	500
6	Incoming longwave radiation	pyrgeometer up–facing	1900
7	Outgoing longwave radiation	pyrradiometer down–facing	1000
8	Longwave net radiation	difference of 6 and 7	800
9	Radiation balance	net radiometer	21000
17	Total outgoing radiation	sum of 5 and 7	30
14	UV radiation	broadband UV instrument	300
10	Sensible heat flux	aerodynamic method	300
11	Latent heat flux	aerodynamic method or lysimeter	800
18	Sensible and latent heat flux	sum of 10 and 11 difference of 9 and 12	20
12	Subsurface heat flux	thermistors or heat plates	300
13	Latent heat of melt	measurement of water–equivalent of snow before melt or of ablation of glacier ice	60

8.3.1 *Global energy balance archive*

The Global Energy Balance Archive (GEBA) contains quality checked monthly means of surface energy fluxes. Data have been extracted from periodicals, monographs, data reports, and unpublished manuscripts. GEBA sites are distributed all over the world. Table 8.4 lists the surface energy flux data that are stored in the GEBA database.

Table 8.5 *Basic radiation measurements of BSRN stations.*

Quantity	Accuracy
Direct SW	1% (2 W m^{-2})
Diffuse radiation	4% (5 W m^{-2})
Global radiation	2% (5 W m^{-2})
Downwelling LW radiation	5% (10 W m^{-2})
Upwelling LW radiation	5% (10 W m^{-2})

8.3.2 *Baseline radiation network*

BSRN has records of solar and atmospheric longwave radiation, along with surface and upper–atmospheric observations. BSRN stations are located in a variety of climates. The stations have to follow specific observational and calibration procedures, resulting in standardized data of very high accuracy. The data provided also have very high temporal resolution (minutes). The basic measurements that exist for all BSRN certified stations include: global (solar) radiation, direct solar radiation, diffuse sky irradiance and downward longwave radiation (see Table 8.5).

Some stations also have measurements that include: direct solar spectral radiation at the WMO specified λ and $\delta\lambda$, synoptic measurements (wind, temperature, dew point, cloud amount and type determined by experienced observers), radiosonde data (vertical distribution of temperature and water vapour), cloud base height, water–vapor vertical distribution and total ozone amount. They also give height for low plus middle clouds (when there is collocation with synoptic stations). There are currently 36 operational BSRN stations around the world, 9 of which are located in Europe. Radiosonde data exist for 16 of the stations, eight of which have data since 1992. Table 8.6 lists the names and locations of the 36 BSRN stations. Sample long–term data from the Sahara station of Tamanrasset are given in Table 8.7.

8.3.3 *ERBE data*

The Earth Radiation Budget Experiment (ERBE) data provide sufficient spatial, temporal and diurnal sampling to measure the radiation budget of the Earth. The Goddard Space Flight Center built the Earth Radiation Budget Satellite (ERBS) on which the first ERBE instruments were launched by the Space Shuttle Challenger in 1984. ERBE instruments were also launched on two National Oceanic and Atmospheric Administration weather monitoring satellites, NOAA 9 and NOAA 10 in 1984 and 1986. There are two types of instruments: (a) Scanner – comprising a set of three coplanar detectors (longwave, shortwave and total energy), all of which scan from one limb of the Earth to the other, across the satellite track (in its normal operational mode); (b) Non–scanner – comprising a

Table 8.6 *BSRN operational stations*

	Station name	Sponsor	Latitude	Longitude
1	Ny Elesund, Spitsbergen (N)	Germany/Norway	78° 56' N	11° 57' E
2	Barrow, Alaska	USA	71° 19' N	156° 36' W
3	Lerwick, Shetland Islands	Great Britain	60° 08' N	1° 11' W
4	Toravere	Estonia	58° 16' N	26° 28' E
5	Lindenberg	Germany	52° 13' N	14° 07' E
6	Regina	Canada	50° 12' N	104° 43' W
7	Camborne	Great Britain	50° 13' N	5° 19'W
8	Palaiseau Cedex	France	48° 43' N	2° 12' E
9	Rock Springs SURFRAD	USA	40° 43' N	77° 56' W
10	S. Great Plains ARM Ext. Facil. 13	USA	36° 36' N	97° 30' W
11	Fort Peck, SURFRAD, CO, Montana	USA	48° 19' N	105° 06' W
12	Payerne	Switzerland	46° 49' N	6° 57' E
13	Carpentras	France	44° 03' N	5° 02' E
14	Albany New York	USA	42° 42' N	73° 50' W
15	Boulder SURFRAD, Co.	USA	40° 08' N	105° 14' W
16	Bondville, Illinois	USA	40° 04' N	88° 22' W
17	Boulder, Colorado	USA	40° 03' N	105° 00' W
18	Chesapeake Lt. Station	Virginia, USA	36° 54' N	75° 43' W
19	Desert Rock SURFRAD, PA	USA	36° 39' N	116° 1' W
20	Billings, ARM/CART, OK	USA	36° 36' N	97° 31' W
21	Tateno	Japan	36° 03' N	140° 08' E
22	Goodwin Creek, Mississippi	USA	34° 15' N	89° 52' W
23	Bermuda	USA	32° 18' N	64° 46' W
24	Sede Boqer	Israel	30° 52' N	34° 46' E
25	Solar Village, Riyadh	Saudi Arabia	24° 55' N	46° 25' E
26	Tamanrasset	Algeria	22° 47' N	5° 31' E
27	Kwajalein, Marshall Islands	USA	8° 43' N	167° 44' E
28	Ilorin	Nigeria/USA	8° 32' N	4° 34' E
29	Momote, Manus Is. ARM PNG	USA	2° 03' S	147° 26' E
30	Alice Springs	Australia	23° 48' S	133° 53' E
31	Florianopolis	Brazil	27° 32' S	48° 31' W
32	De Aar	South Africa	30° 40' S	24° 00' E
33	Lauder	New Zealand	45° 00' S	169° 41' E
34	Syowa, Antarctica	Japan	69° 00' S	39° 35' E
35	Georg von Neumayer, Ant.	Germany	70° 39' S	8° 15' W
36	South Pole, Antarctica	USA	90° S	-

Table 8.7 *Sample long-term (1991–1995) BSRN station solar radiation data from the Sahara station of Tamanrasset, Algeria (Source: World Radiation Data Center)*

Tamanrasset 22° 47' N, 5° 31' E 1377m(a.s.l.)									Source: WRDC				
	J	F	M	A	M	J	J	A	S	O	N	D	Annual
Global	177	214	237	259	256	253	272	247	225	209	184	156	224
Diffuse	30	40	60	85	101	97	83	83	80	53	37	36	65

set of five detectors, one measuring the total energy from the Sun, two measuring the shortwave and total energy from the entire Earth disk and two measuring the shortwave and total energy from a medium resolution area beneath the satellite.

The ERBE products include instantaneous time-sequenced instrument measurements, as well as monthly averaged regional, zonal, and global estimates of radiation budget parameters (S–2 Solar Incidence; S–4 Regional, Zonal, and Global Averages; S–4G Regional, Zonal, and Global Gridded Averages; S–7 Medium-Wide Field of View Non-scanner Data; S–8 Instantaneous Scanner and Non-scanner Data; S–9 Scanner Earth Radiant Flux and Albedo; S–10 Non-scanner Earth Radiant Flux and Albedo). From these products, one can use the scanner S4 satellite data (1984–1989) to validate model radiative fluxes at TOA, and the S–10N (WFOV NF edition 2) non-scanner data (1984–1999) to compare against TOA flux anomaly time series derived from LW and SW models. We note that there are new satellite data becoming available for the TOA fluxes from the Meteosat Second Generation (MSG) GERB and SEVIRI instruments, CERES and the Terra and Aqua platforms (see also Chapter 10).

8.4 Outgoing solar radiation at TOA

We present global distribution results of the outgoing solar radiation at TOA based on a deterministic radiative–transfer model on a mean monthly and 2.5° × 2.5° longitude–latitude resolution, spanning the 14–year period from January 1984 through December 1997. The model uses data from the ISCCP D2–series supplemented by water–vapour and temperature data taken from NCEP/NCAR. Model input data were also taken from other global databases, such as TIROS-TOVS, ISLSCP, and GADS. The model computations were validated at the pixel level, against 5–year accurate scanner data from ERBE, which ensures the quality of the long–term model results. The incoming solar radiation was computed as described in Chapter 5, where the zonal–seasonal distribution is also given.

8.4.1 *Planetary albedo*

The net incoming solar radiation at TOA (planetary absorption), $F^{\downarrow}_{\mathrm{net,TOA}}$, the outgoing SW radiation (OSR) at TOA, $F^{\uparrow}_{\mathrm{TOA}}$, and the planetary albedo R_{p}, are related through the following expressions

$$F^{\downarrow}_{\mathrm{net,TOA}} = \alpha_{\mathrm{p}} F^{\downarrow}_{\mathrm{TOA}}, \tag{8.2}$$

$$F^{\uparrow}_{\mathrm{TOA}} = (1 - \alpha_{\mathrm{p}})\, F^{\downarrow}_{\mathrm{TOA}}, \tag{8.3}$$

$$R_{\mathrm{p}} = \frac{F^{\uparrow}_{\mathrm{TOA}}}{F^{\downarrow}_{\mathrm{TOA}}}, \tag{8.4}$$

where α_{p} is the planetary absorptivity. The scattering and absorption of the incoming SW radiation at TOA, depends strongly on the presence and type of

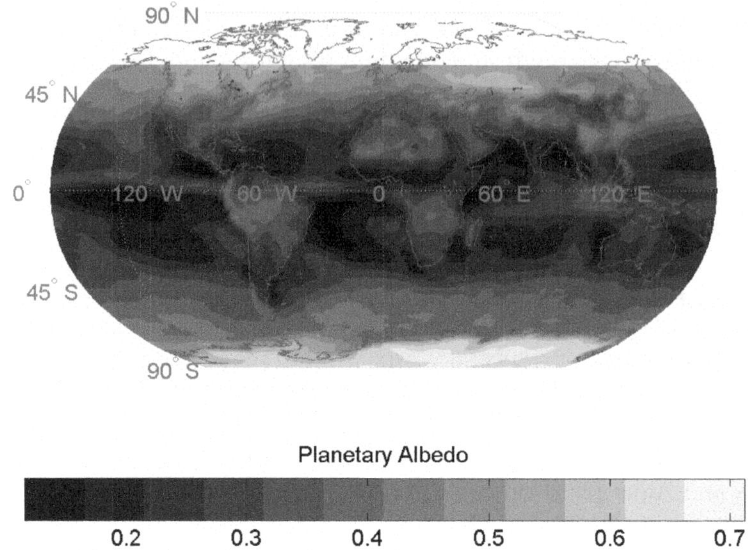

FIG. 8.7. Long–term (1984–1997) mean planetary albedo for January. (Hatzianastassiou *et al.* 2004b)

clouds in the atmosphere, the composition of the atmosphere (gases plus aerosols) and the reflectivity of the Earth's surface. It is weakly dependent on the thermal structure of the atmosphere. The cloudy–sky component can be subdivided into three components covered by low–, middle–, and high–level clouds, according to ISCCP. The planetary absorptivity, and hence the OSR, has a clear–sky component, $\alpha_{\rm s}$, and three cloudy–sky components, $\alpha_{{\rm c}i}$, and is expressed as

$$\alpha_{\rm p} = (1 - A_{\rm c})\,\alpha_{\rm s} + \sum_i A_{{\rm c}i}\alpha_{{\rm c}i}. \tag{8.5}$$

In Fig. 8.7 we show the long–term (1984–1997) mean planetary albedo for January. The planetary albedo, $R_{\rm p}$, is highest over polar and high–altitude areas, which are characterized by large surface albedos, throughout the year. There, $\rm R_p$ values are as high as 75%. Secondary maxima of planetary albedo values occur over cloudy tropical and sub–tropical areas and over highly reflecting surfaces such as the Sahara. The lowest $R_{\rm p}$ values are found over tropical and sub–tropical oceanic areas (with surface albedo lower than 10%), especially those with small cloudiness. When clouds are present above such oceanic areas, e.g. marine stratus clouds off the coasts of western United States, South Africa and South America, then $R_{\rm p}$ increases up to 40%. The systematically large values of $R_{\rm p}$ over mid-to-high latitude and especially polar regions, indicates the importance of these regions in terms of their sensitivity to possible climatic changes. In Fig. 8.8 is shown the high–level cloud cover albedo for January 1988. We see that tropical

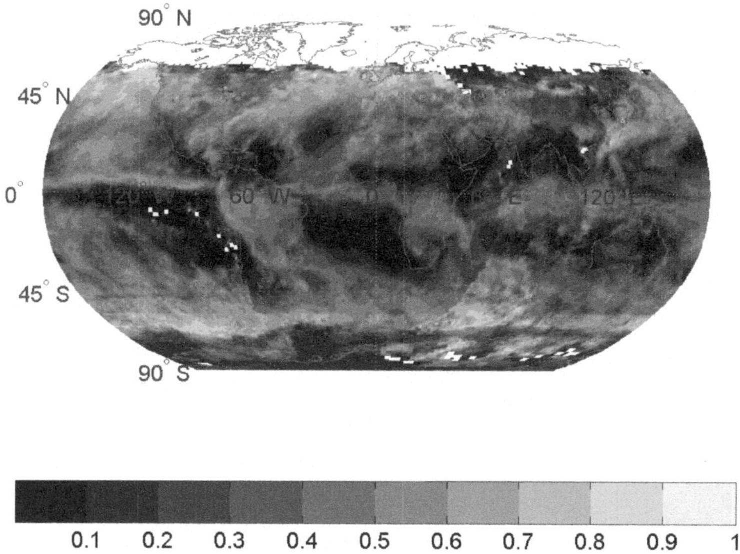

FIG. 8.8. High–level cloud albedo for January 1988. White squares and lines are missing data whilst the white area of the Arctic has nighttime. (Data from NASA–Langley)

high–level clouds over oceans tend to have low albedo, especially in the Eastern Pacific and off the west coast of Africa.

8.4.2 *Global distribution*

The outgoing solar radiation (OSR) fluxes are determined by the incoming solar flux (ISR) and the planetary albedo. The latitudinal gradient (Fig. 8.9) is determined by the incoming solar flux, while the main longitudinal patterns are primarily due to features of the global distribution of surface albedo and cloud cover. Thus, maximum OSR values reaching 350 W m^{-2} are found in polar regions during local summer, i.e. in January and July in the Southern and Northern hemispheres, respectively, when the incoming solar radiation reaches maximum values of 500 W m^{-2}, especially during perihelion. The OSR decreases gradually from the summer towards the winter pole, in January and July, where it switches off poleward of mid-to-high latitudes. Such a strong latitudinal OSR gradient is not found during spring and autumn. Large OSR fluxes are found over areas characterized by large surface albedo. For example, as large as 350 and 300 W m^{-2} are computed over Antarctica in January, and over Greenland and Arctic Ocean areas in July, respectively, i.e. over areas with surface albedo values larger than 70%. Also, large OSR values are found over regions having large cloud amounts; thus, OSR values as high as 220 W m^{-2} are found over the

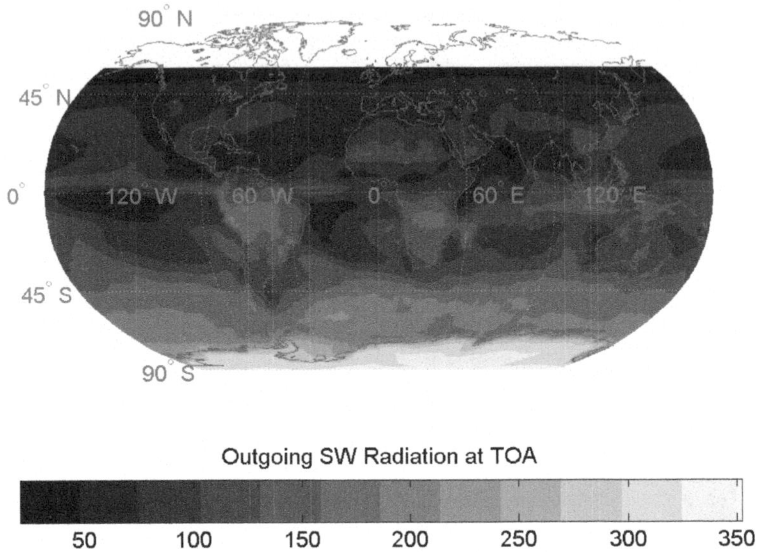

FIG. 8.9. Long–term (1984–1997) average global distribution of outgoing shortwave radiation (in W m^{-2}) at the top of the atmosphere for January. (Hatzianastassiou et al. 2004)

storm track zone of the Southern Hemisphere, around 60S, in January, whereas values between 130–200 W m^{-2} are computed in October (Fig. 8.9). A corresponding feature, with OSR up to 200 W m^{-2} in April and July, appears over the northern Pacific oceanic areas with large cloud amounts.

8.4.3 *Zonal–seasonal variation*

Monthly mean 10–degree latitude zonal fluxes were computed by averaging along latitude, by considering the fraction of surface area contained in each 2.5–degree zone. To a very good approximation the surface area can be taken to vary as $\cos\theta$, where θ is the latitude and so the surface area fraction is given by $E_j = \sin 10j - \sin 10(j-1)$ and equal to 0.1736 for $j = 1$ corresponding to zone 10-0°, and sequentially equal to 0.1684, 0.1580, 0.1428, 0.1232, 0.10, 0.0737, 0.0451, 0.0152, through to zone 90–80°. The lowest seasonal variation of OSR occurs in the tropics, where values remain near 100 W m^{-2} throughout the year, as can be seen in Fig. 8.10. The seasonal variability increases towards the poles, with the largest values of about 300 W m^{-2} occurring during polar summer, and is primarily associated with the solar zenith angle. The latitudinal variation of OSR depends also on scattering and absorption processes and on atmospheric conditions such as the variation in amounts and types of clouds with latitude.

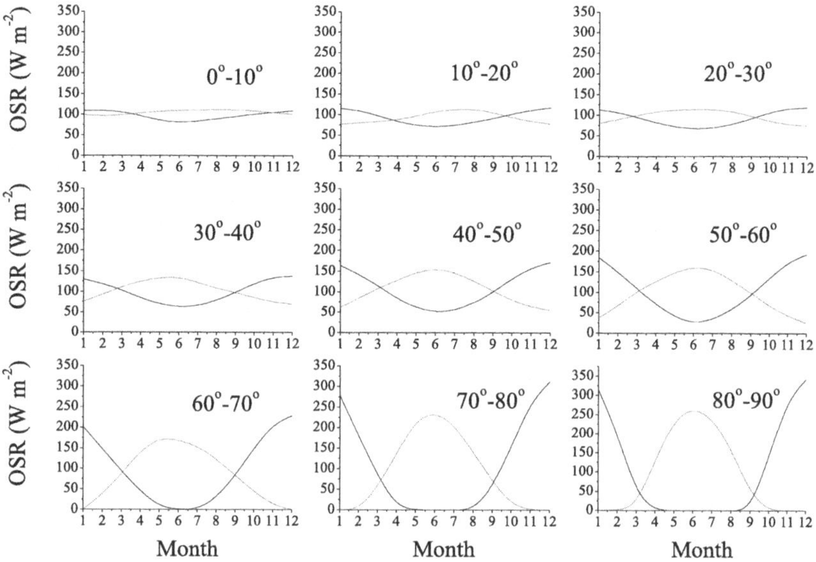

FIG. 8.10. Long–term (1984–1997) zonal–seasonal variation of outgoing shortwave radiation (W m^{-2}) at TOA, for both hemispheres, NH (dotted lines) and SH (solid lines).

8.4.4 *Mean annual latitudinal variation*

Monthly mean 10–degree latitude zonal fluxes were computed by averaging along latitude, by considering the fraction of surface area contained in each 2.5–degree zone. Subsequently, annual mean quantities were computed by summing the corresponding monthly means for each 10–degree latitudinal zone over the 12 months of the year. The reflected OSR flux (Fig. 8.11) is maximum near the equator, has a minimum of about 93 W m^{-2} in the sub–tropics (10–20° N and S), and increases towards the poles, with secondary minima of 100 and 90 W m^{-2} near 60–70° S and N, respectively. In the tropics, about 25% of the incoming mean annual solar radiation is scattered back to space ($R_p \approx 0.25$, Fig. 8.12). The planetary albedo increases from the equator to the poles, with a small minimum between 10 and 20° N and S, reaching values of 65% and 50% at the south and the north pole, respectively. The general equator–to–pole increase in R_p is caused by increasing surface albedo values, associated with larger land–to–sea fractions and snow and ice–covered areas, increasing solar zenith angle, and large cloud amounts. The secondary R_p minimum near the equator is due to large cloud amounts along the intertropical convergence zone (ITCZ).

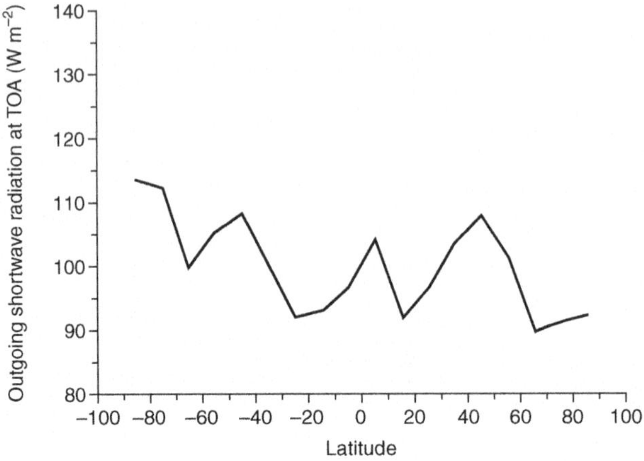

FIG. 8.11. Long-term (1984–1997) annual average latitudinal variation of outgoing shortwave radiation (W m^{-2}) at the top of the atmosphere. (Hatzianastassiou *et al.* 2004b)

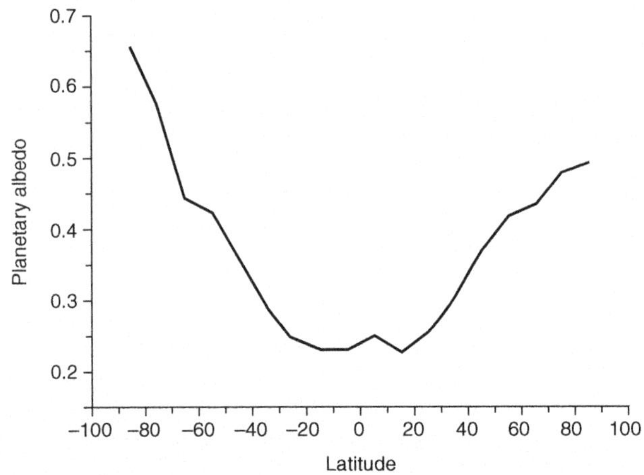

FIG. 8.12. Long-term (1984–1997) annual average latitudinal variation of planetary albedo at the top of the atmosphere. (Hatzianastassiou *et al.* 2004b)

8.4.5 *Seasonal variation*

The seasonal range of OSR for the Southern Hemisphere is slightly larger than for the Northern Hemisphere. It ranges from about 60 to about 140 W m^{-2} for the Northern Hemisphere, and between 55 and 155 W m^{-2} for the Southern Hemisphere. The mean global OSR (Fig. 8.13) has a very small seasonal variability of about 15 W m^{-2}, i.e. between 95 and 110 W m^{-2}, with a minimum in late summer–early autumn. The mean hemispherical planetary albedo (Fig. 8.14) does not show great seasonality; it varies between 27.5% (minimum value in October) and 30.2% (maximum value in May) for the Northern Hemisphere, and between 26.6% (minimum in June) and 31.5% (maximum in December) for the Southern Hemisphere. The global mean planetary albedo ranges from about 28% in September and 30.5% in December.

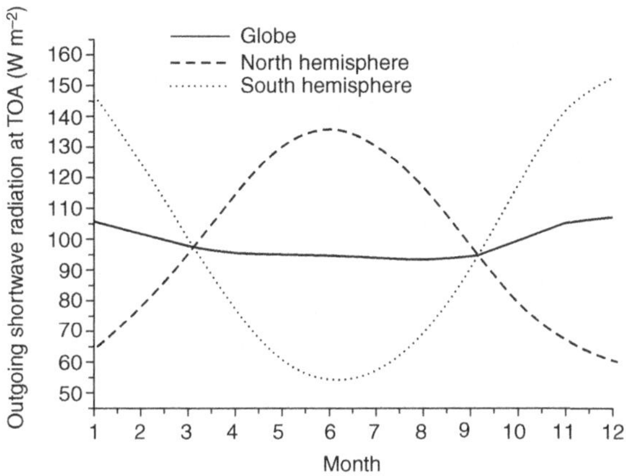

FIG. 8.13. Long–term (1984–1997) latitudinal average seasonal variation of outgoing shortwave radiation (W m^{-2}) at the top of the atmosphere. (Hatzianastassiou *et al.* 2004b)

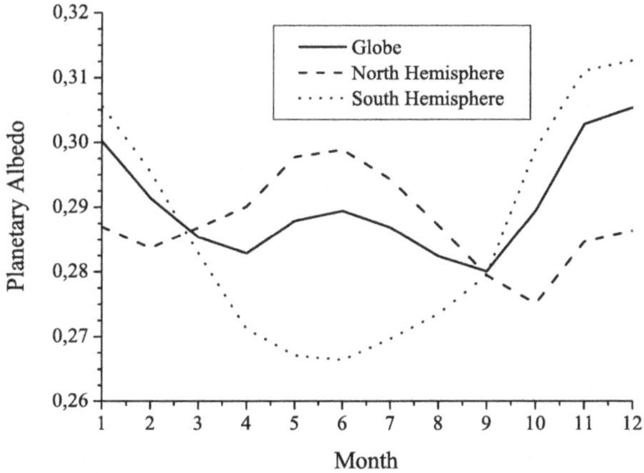

FIG. 8.14. Long–term (1984–1997) latitudinal average seasonal variation of planetary albedo at the top of the atmosphere. (Hatzianastassiou *et al.* 2004b)

8.4.6 *Mean annual hemispherical variation*

Hemispherical means are computed by weighting the latitudinal variations of the fluxes F with the fraction of surface area, E_j (§8.4.3), contained in each latitudinal zone j

$$F = \frac{\sum_{j=1}^{9} F_j E_j}{\sum_{j=1}^{9} E_j}. \tag{8.6}$$

The mean annual hemispherical and global average values over the 14–year period from 1984 to 1997, for the incoming, outgoing, net incoming (i.e. absorbed) SW radiation at TOA, and planetary albedo, are given in Table 8.8 and compared against ERBE data and other published values. In Table 8.9 are given the incoming and outgoing SW radiation, and the net incoming and planetary albedo for the hemispheres and the globe. On a 14–year basis (1984–1997), the Earth is found to reflect back–to–space 101.2 W m^{-2} from the global mean received 341.5 W m^{-2}, resulting in a long–term planetary albedo equal to 29.6%. The interhemispherical differences are equal to 0.34 W m^{-2} for ISR, 0.1 W m^{-2} for OSR, 1.43 W m^{-2} for the net incoming SW radiation at TOA, and 0.35% for R_p, implying slightly larger Northern than Southern Hemispherical values.

Table 8.8 *Model mean annual hemispherical and global averages of outgoing shortwave radiation at the top–of–atmosphere (OSR) and planetary albedo. NH is Northern Hemisphere, SH is Southern Hemisphere. The radiative fluxes are expressed in W m^{-2}, while the planetary albedo is in per cent. NCEP/NCAR, ECMWF and NASA results are from reanalysis, model results are from Hatzianastassiou et. al. (2004b).*

Study	OSR			Planetary Albedo		
	NH	SH	Globe	NH	SH	Globe
Model (1984–1997)	100.63	101.72	101.17	29.45	29.80	29.62
Model (1985–1989)	101.30	102.36	101.83	29.65	29.99	29.82
ERBE (1985–1989)	101.04	100.76	100.90	30.10	29.10	29.6
NCEP/NCAR (1982–1994)			115.75			33.87
ECMWF (1985–1993)			103.75			30.36
NASA (1981–1992)			94.75			27.72
Yang *et al.* (1999)			115.3			
Wild *et al.* (1998 – ECHAM4)			105.0			30.7
Chen and Roeckner (1996)			106.4			31.0
Rossow and Zhang (1995)			111.55			32.6

8.4.7 *Time series of planetary albedo*

The time series of monthly average planetary albedo (in %), for the 14–year (1984–1997) period for the globe is shown in Fig. 8.15. The mean global planetary

Table 8.9 *Model mean annual hemispherical (NH is Northern Hemisphere, SH is Southern Hemisphere) and global average: incoming shortwave radiation ($F_{\downarrow}$), outgoing shortwave radiation ($F_{\uparrow}$) and net incoming shortwave radiation ($F_{net,\downarrow}$) at the top-of-atmosphere, and planetary albedo (R_p), for the 14-year (1984–1997) period. The radiative fluxes are expressed in W m^{-2}, while the planetary albedo in %.*

	$F_{\downarrow}$	$F_{\uparrow}$	$F_{\mathrm{net},\downarrow}$	R_{p}
NH	341.67	100.63	241.04	29.45
SH	341.33	101.72	239.61	29.80
Globe	341.50	101.17	240.32	29.62

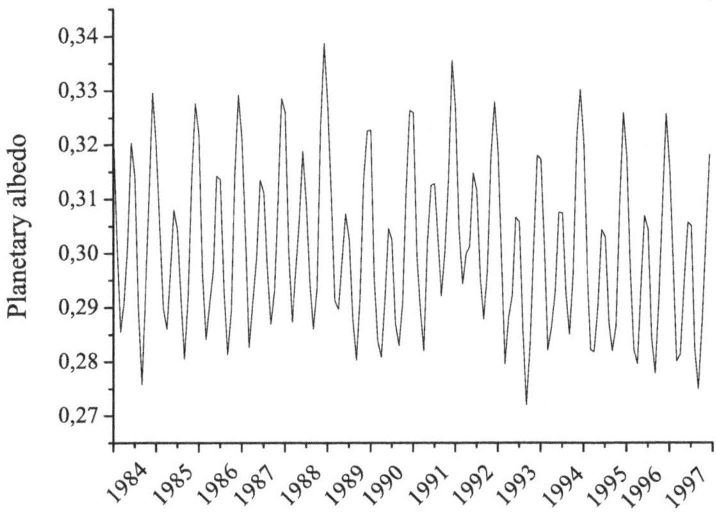

FIG. 8.15. Long-term (1984–1997) time series of monthly average planetary albedo for the globe.

albedo, R_p, exhibits a decreasing trend of about 0.6 % from January 1984 through December 1997. The mean global R_p varies between about 0.285 and 0.330, over this period. The decrease of mean global values of R_p arises mainly in the tropical and subtropical areas, related to decreasing cloudiness.

8.5 The shortwave radiation budget at surface

8.5.1 *Global distribution*

Figure 8.16 shows the geographical distribution, on a 2.5×2.5 degree resolution, of the 17-year (1984–2000) average downwelling shortwave radiation (DSR) at

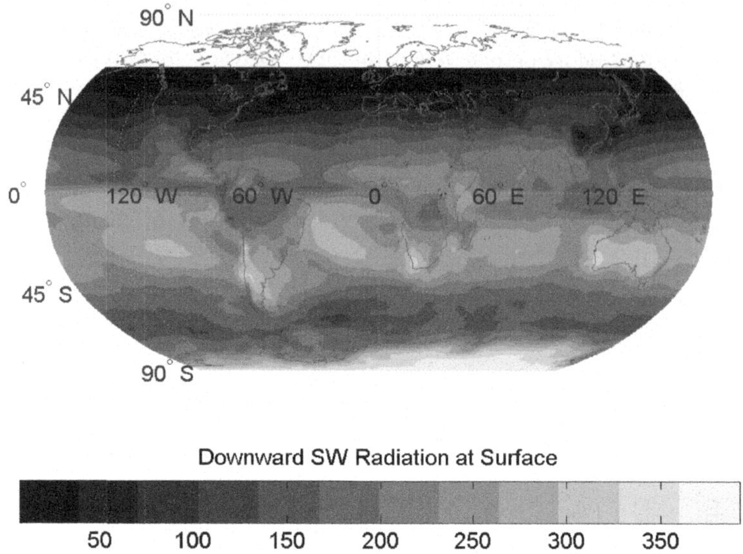

FIG. 8.16. Long–term (1984–2000) average global distribution of downward shortwave radiation (W m^{-2}) at the Earth's surface for the month of January. (Hatzianastassiou *et al.* 2005)

the Earth's surface for the month of January. The latitudinal gradient of DSR is primarily determined by the incoming solar flux at TOA, and secondarily by clouds, while the patterns of longitudinal variation are mostly determined by cloud and surface properties. There is a gradual DSR decrease from the summer to the winter pole in January and July. Overall, the DSR has maximum values over the *subsidence regions* associated with anticyclonic conditions and small cloud amounts, such as oceanic areas in low latitudes of the summer hemisphere, as well as over the polar areas of the summer hemisphere. In contrast, small DSR values are found over regions with large cloud amounts, such as the middle latitudes of the summer hemisphere. Note that there are small DSR values off the western coasts of South America and South Africa in July, being smaller than corresponding DSR fluxes in adjacent regions of the same latitude, which are attributed to large cloud amounts (of about 80%). Also, there are relatively small DSR values over south–eastern Asia in July (150 W m^{-2}), where large cloud cover (80%) occurs, associated with the monsoons. Apart from oceanic areas, there are some extended continental regions with large DSR values, such as the United States, Southern Europe, North Africa and Middle–East in July, or correspondingly South America, South Africa and Australia in January.

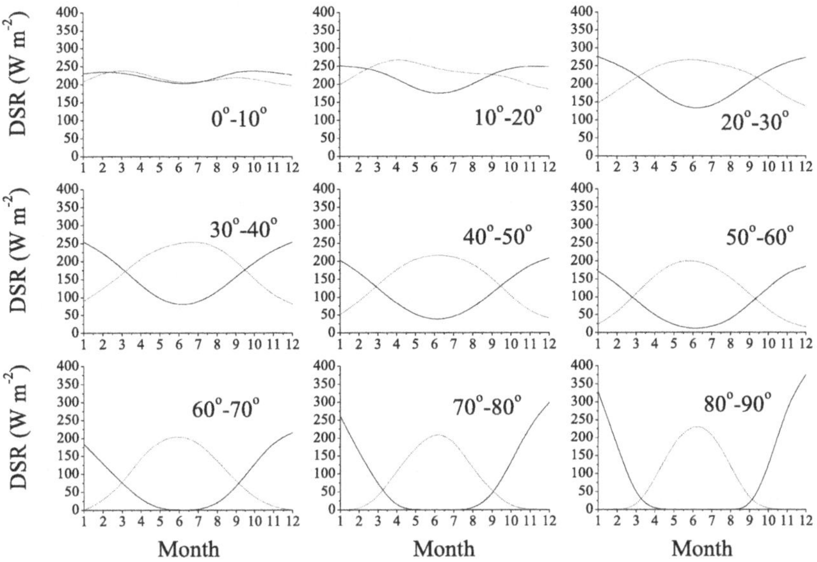

FIG. 8.17. Long–term (1984–1997) zonal–seasonal variation of downward shortwave radiation (DSR in W m^{-2}) at the Earth's surface, NH (dotted lines) and SH (solid lines).

8.5.2 *Zonal–seasonal variation*

The seasonal variation of DSR is similar to that of OSR, except that there is a larger variation in the subtropics and that the Arctic variation is smaller than the Antarctic, as shown in Fig. 8.17. Large DSR fluxes are found in polar regions during local summer, equal to about 300–350 W m^{-2} over Antarctica in January, and 200–250 W m^{-2} over the Arctic in July. The larger values over Antarctica, compared with the Arctic, are due to the larger incoming solar radiation at TOA during perihelion ($>$ 500 W m^{-2}), but also to the smaller ISCCP-D2 summer cloudiness in Antarctica than in the Arctic (20% against 60%, respectively) and drier atmosphere which affects the near-infra–red. In both hemispheres, the maximum of DSR values outside the polar regions occur in subtropical areas (between 10° and 35° latitude) rather than in the tropics. This is due to the fact that total cloudiness has minimum values (about 50%) over the subtropical areas, and not along the equator, where the ITCZ involves total cloud–cover values of about 60–70%. In the equatorial region (10°S–10°N), the double peaks in DSR occur during spring and autumn when the Sun is overhead the equator.

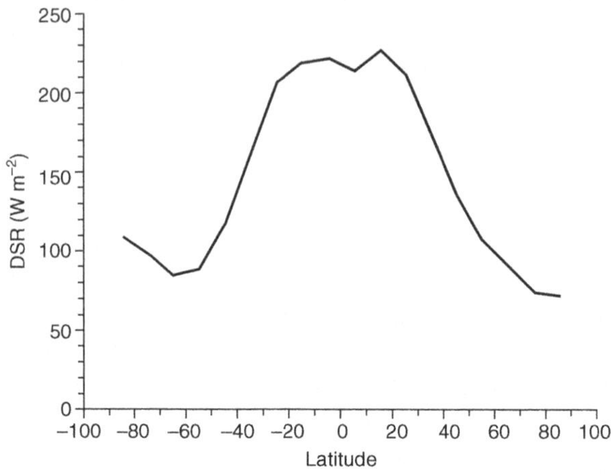

FIG. 8.18. Mean annual downward shortwave radiation (DSR in W m^{-2}) at the Earth's surface, for the period January 1984 through to December 2000. (Hatzianastassiou *et al.* 2005)

8.5.3 *Latitudinal and seasonal variation*

Monthly mean 10–degree latitude zonal fluxes were computed by averaging first along 2.5°–width longitudinal circles, then along latitude, by considering the fraction of surface area contained in each 2.5–degree zone. Subsequently, annual mean quantities were computed by summing the corresponding monthly means for each 10–degree latitudinal zone over the 12 months of the year. The latitudinal variation of mean annual DSR flux has maximum values of about 230 W m^{-2} in subtropical areas decreasing rapidly to about 80 W m^{-2} towards the poles (Fig. 8.18). There is a small local minimum in mean annual DSR at the equator (equal to about 220 W m^{-2}), due to clouds associated with the ITCZ. There is also a local maximum of 230 W m^{-2} in DSR around 15°N. Note that in the Southern Hemisphere there is a local minimum due to the persistent and extended cloudiness occurring over the oceanic zone (storm track zone) surrounding Antarctica. The mean hemispherical DSR fluxes have opposite seasonalities, varying within the range 110–230 W m^{-2}, thus resulting in a mean global DSR flux ranging from 165 to 176 W m^{-2} throughout the year (Fig. 8.19). The seasonal variability is slightly larger in the Southern than in the Northern hemisphere.

8.5.4 *Validation with observations*

Model computed DSR fluxes need to be validated through comparison with corresponding extensive measurements from ground–based stations. A comparison of observed versus calculated monthly mean DSR fluxes within each 2.5×2.5

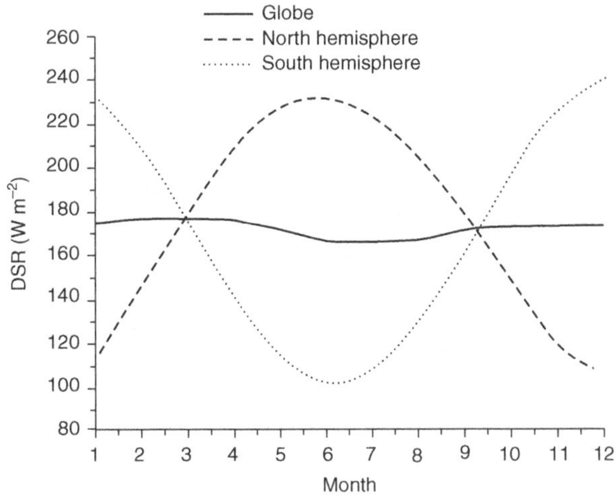

FIG. 8.19. Long–term (1984–2000) average seasonal distributions of downward shortwave radiation (W m^{-2}) at the Earth's surface for the Northern Hemisphere, Southern Hemisphere and the Globe. (Hatzianastassiou *et al.* 2005)

degree cell containing a GEBA station is shown in the scatterplot of Fig. 8.20. The scatterplot shows a comparison of all measured monthly values against the model values within each cell containing a GEBA station. The overall bias is equal to –6.5 W m^{-2}, the root–mean–square error (RMS) is 23.4 W m^{-2}, the slope of the one–to–one line is 0.974, the correlation coefficient is 0.99 and the number of measurements is N = 27858. The model underestimates DSR in the polar areas, but we note that the detection of clouds is problematic over highly reflecting surfaces.

8.5.5 *Mean annual hemispherical variation*

Downward and net downward (absorbed) SW radiation at the Earth's surface, for the four midseason months of the year, are given in Table 8.10, for the 17–year period from 1984 to 2000.

The long–term model results indicate that the Earth's surface receives an annual average of 171.6 W m^{-2} and absorbs 149.4 W m^{-2} (Table 8.11), resulting in a long–term surface albedo equal to 12.9%. The interhemispherical differences are equal to 4.6 W m^{-2} for DSR and only 0.3 W m^{-2} for net DSR, implying slightly larger Northern than Southern hemisphere surface solar radiative fluxes. Overall, according to model results using ISCCP–D2 data, on a mean annual basis and at the global scale, the Earth's surface receives 50.2%, while it absorbs 43.7% of the incoming SW radiation entering the Earth–atmosphere system. Model estimates have varied over the years, with the more recent estimates being closer to those

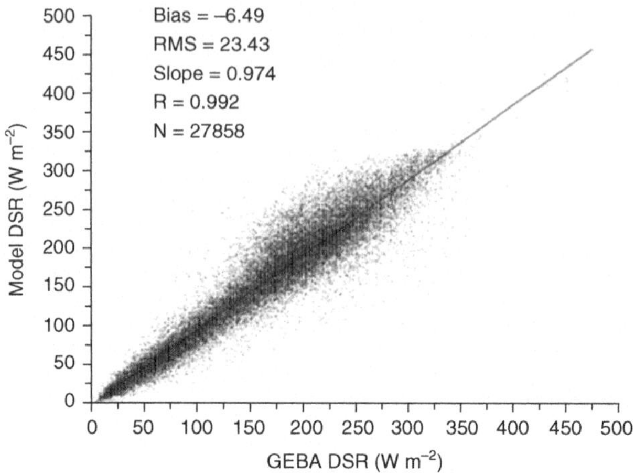

FIG. 8.20. Scatterplot comparison between model monthly DSR, against GEBA surface measurements. (Hatzianastassiou *et al.* 2005)

Table 8.10 *Model mean annual hemispherical (NH is Northern Hemisphere, SH is Southern Hemisphere) and global averages of downward shortwave radiation (DSR) and net downward shortwave radiation (net DSR) at the Earth's surface for January, April, July, October, and the whole year, for the period 1984–2000. The radiative fluxes are expressed in W m*$^{-2}$. *Numbers in parentheses are standard deviations and represent interannual variabilities of the means. (Hatzianastassiou et al. 2005)*

	NH	SH	Global
Downward shortwave radiation			
January	113.8(9.8)	234.8(20.7)	174.3(15.3)
April	209.8(16.5)	140.4(11.5)	175.1(14.0)
July	224.8(17.2)	105.6(7.7)	165.2(12.5)
October	148.6(12.0)	197.0(16.6)	172.8(14.3)
Annual	173.9(14.0)	169.3(14.3)	171.6(14.1)
Net downward shortwave radiation			
January	98.6(2.7)	204.5(6.5)	151.6(4.6)
April	177.0(4.6)	127.5(3.4)	152.2(4.0)
July	193.6(5.1)	94.9(2.1)	144.2(3.6)
October	130.9(3.5)	171.6(4.3)	151.2(3.9)
Annual	149.6(3.9)	149.3(4.0)	149.4(4.0)

given by the GEBA surface observations, as can be seen in Table 8.11.

8.5.6 *Long-term anomaly*

The computed *anomaly* (defined as the DSR per cent differences from the long-term mean value) in mean hemispherical and global DSR flux over the 17-year

Table 8.11 *Model long-term mean annual hemispherical and global averages of downward shortwave radiation (W m^{-2}) at the Earth's surface (DSR), and net downward shortwave radiation (absorbed) at the Earth's surface (Net DSR) for the Northern Hemisphere, Southern Hemisphere and the Globe. (Hatzianastassiou et al. 2005)*

Study	DSR			Net DSR		
	NH	SH	Globe	NH	SH	Globe
Hatzianastassiou *et al.* (2005)	173.9	169.3	171.6	149.6	149.3	149.4
Wild *et al.* (1998 – ECHAM4)			170.0			147.0
Liou (2002)			189.0			161.0
Global Energy Balance Archive			169.0			151.0
NCAR/GCM3 (Gupta *et al.* 1999)	194.4	194.4	194.4	169.0	173.4	171.2
Garratt *et al.* (1998)			195.5			167.0
Kiehl and Trenberth (1997)						168.0
Fowler and Randall (1996)						172.0
Li *et al.* (1995)						157.0
Rossow and Zhang (1995)			193.4			165.1
Hartmann (1994)						171.0
Li and Leighton (1993)				155.0	159.0	157.0
Darnell *et al.* (1992)			173.0			151.0

period, is given in Fig. 8.21. The hemispherical mean DSR anomalies vary from –9 to 9 W m^{-2}, while the mean global DSR anomaly has variations of up to 6 W m^{-2}, about the 17–year mean value. Large negative anomalies, i.e. reduced SW radiation reaching the Earth's surface, were found for the period 1991–1993, which can be attributed to the Mount Pinatubo eruption in June 1991, and to the 1991/1992 El Niño event. During the same period, the outgoing SW radiation increased by about 4 W m^{-2}. The induced rapid decrease in DSR is followed by a recovery period, with positive DSR anomalies through 1994. In the figure, there are other interesting features corresponding to climatic events, such as El Niño and La Niña, associated with negative and positive DSR anomalies, respectively. For example, negative DSR anomalies as large as 6 W m^{-2} follow the 1986/1987 El Niño event. The large negative DSR anomalies of about –5 W m^{-2} during 1984 can be attributed to the influence of the El Chichon eruption that took place in 1983. Also, note that in 1998 the DSR anomalies pass from large positive values (of about 4 W m^{-2}) to negative ones (equal to –2 W m^{-2}), attributed to the 1997/1998 El Niño event. In general, the variation of DSR anomalies is similar but opposite in magnitude to that of outgoing shortwave radiation anomalies.

Also given is a 4th-order polynomial fit to 17–year time–series of global averages of DSR anomalies, which shows significant positive DSR anomalies (increasing DSR) in the period starting from year 1992. In contrast, in the period 1984–1992, negative DSR anomalies mostly occur, indicating a decreasing DSR flux at the Earth's surface. The situation was reversed after the early 1990s. This, according to an analysis based on ISCCP–D2 data (we note that the use of such data for trend analysis has been questioned, but here we are more concerned with the use

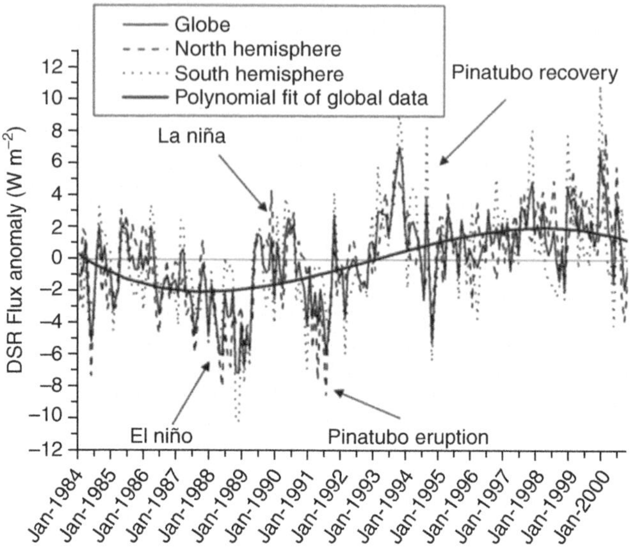

FIG. 8.21. Time–series of global and hemispherical averages of downward shortwave radiation (DSR) flux anomalies (W m^{-2}) for the period 1984–2000. (Hatzianastassiou *et al.* 2005)

of anomalies), is due to a decrease in cloudiness, especially low–level, primarily in tropical and subtropical regions.

8.5.7 *Sensitivity analysis*

The assessment and quantification of the role of various key climatic parameters that determine the downwelling shortwave radiation (DSR) at the surface is very important for SW radiation budget studies. A series of sensitivity tests are given in Table 8.12. In each test, the relevant parameter, X, was modified by a certain amount, ΔX in relative percentage terms, and the resultant radiation change Δ DSR, with respect to defined reference cases, was computed in absolute terms (W m^{-2}). The year 1988 was chosen to represent the reference case. The tests were performed on a monthly basis and at the grid-cell (2.5–degree latitude-longitude) level, and results are given in terms of the maximum Δ(DSR) values encountered. Clearly, cloud cover plays an important role in determining changes to DSR, and also the cloud asymmetry factor. Aerosol single scattering albedo and surface reflection also have a significant effect on DSR.

Table 8.12 *Sensitivity analysis for maximum change in DSR observed globally within a grid–cell. Climatic parameters X are: A_c, cloud cover; τ_c^s, cloud scattering optical depth; τ_c^a, cloud absorption optical depth; g_c, cloud asymmetry parameter; W_{H2O}, total precipitable water; W_{O3}, total ozone column abundance; W_{CO2}, total column carbon dioxide; R_g, surface albedo; ISR, incoming SW radiation at TOA; AOT, aerosol extinction optical depth; ω_{aer}, aerosol single scattering albedo; g_{aer}, aerosol asymmetry parameter. (Hatzianastassiou et al. 2005)*

X	ΔX (%)	Δ(DSR) (W m^{-2})
Low-A_c	10	–22.05
Middle-A_c	10	–17.51
High-A_c	10	–15.30
Low-τ_c^s	10	–5.92
Middle-τ_c^s	10	–5.21
High-τ_c^s	10	–5.84
Low-τ_c^a	10	–1.64
Middle-τ_c^a	10	–1.52
High-τ_c^a	10	–1.37
g_c	5	20.64
W_{H2O}	10	–6.65
W_{O3}	10	–3.37
W_{CO2}	10	–0.51
R_g	10	13.41
ISR	1	4.63
AOT	10	–2.62
ω_{aer}	10	11.81
g_{aer}	10	4.79

8.6 Shortwave aerosol radiative forcing

The SW *aerosol radiative forcing*, ΔF, or more precisely the 'aerosol flux change', is the effect of aerosols on the SW radiation budget at TOA, at the Earth's surface, or within the atmosphere, and it is given by

$$\Delta F_i = F_i - F_{\text{clear},i} \tag{8.7}$$

where F_i and $F_{\text{clear},i}$ are the SW radiative fluxes with and without the presence of aerosols, respectively. The index i involves various aerosol forcings defined in terms of the corresponding SW fluxes. The forcings ΔF_{TOA}, ΔF_{atmo}, ΔF_{surf}, and ΔF_{nsurf}, represent the effect of aerosols on the outgoing SW radiation at TOA, the SW radiation absorbed within the atmosphere, the downward SW radiation at the Earth's surface, and the net downward (or absorbed) SW radiation at the surface.

Aerosols cool the Earth's surface by scattering to space and through atmospheric absorption, thus reducing the solar radiation that reaches the surface. Therefore, the presence of aerosols has a significant effect on the thermal dynamics of the

Earth–atmosphere system. By cooling the surface and warming the atmosphere, aerosols act also to produce more stable atmospheric conditions by decreasing convective activity. They also reduce evaporation from the surface, and so can have a significant effect on the hydrological cycle by suppressing cloud formation and associated precipitation. The aerosol redistribution of radiative energy between the Earth's surface and the atmosphere highlights the role of increasing loads of particulate matter (amount and composition) in the atmosphere on climate by enhancing desertification processes, especially in semiarid regions such as the Mediterranean basin.

8.6.1 *Aerosol forcing at TOA*

In general, aerosols increase through reflection the outgoing solar radiation at TOA, by up to 6 W m^{-2} on a mean monthly basis, producing thus a planetary cooling (Fig. 8.22). However, aerosols can produce a planetary warming by decreasing, on a mean monthly basis, the outgoing solar radiation at TOA, by up to 4 W m^{-2}. This is due to important absorption of atmospheric solar radiation by aerosol particles, such as those characterized by mineral components, over desert areas (e.g. Sahara, Arabia) or soot particles over continental areas (e.g. Europe, North Asia, North America). In general, the sign of the effect of aerosols on the outgoing SW radiation at TOA (aerosol forcing ΔF_{TOA}) under a clear sky is determined by the values of the single scattering albedo, ω, and the surface albedo, R_{g}. Thus, whereas planetary warming is found over Siberia in winter, there is planetary cooling in summer. The role of R_{g} is also shown by the contrast between the planetary cooling produced by mineral aerosols over oceanic areas of low R_{g} or over continental areas of R_{g} lower than 0.3 (sub-Sahel), and the planetary warming produced by similar mineral aerosols over highly reflecting deserts (R_g >0.3). The aerosol–induced planetary warming over highly reflecting surfaces (e.g. polar regions) is due to particle absorption caused by pollution aerosols (haze), which is enhanced through multiple reflections between ice– or snow–covered surfaces and the aerosols above. Field measurements have verified this effect, for example during the Aerosol Characterization Experiment (ACE–2). An example is the Arctic haze containing a considerable amount of anthropogenic impurities, including soot, (originating from the industrialized centres in northern Europe, North America, and Asia), and undergoing long–range transport under winter anticyclonic conditions (high–pressure regions with sinking air and low pressure gradients favouring gentle winds) that decrease substantially the albedo and act against global cooling due to anthropogenic sulphates. The Arctic aerosol type, including soot particles transported from midlatitude continental areas, can be found north of 70°N. The anticyclonic situation, favourable for this transport, persists mainly during winter through springtime, when an intense high–pressure system pushes the Arctic front to the south. Large polluted agricultural and industrialized areas of Eurasia, Japan, and North America are then within the Arctic air mass, which can move aerosols across the North

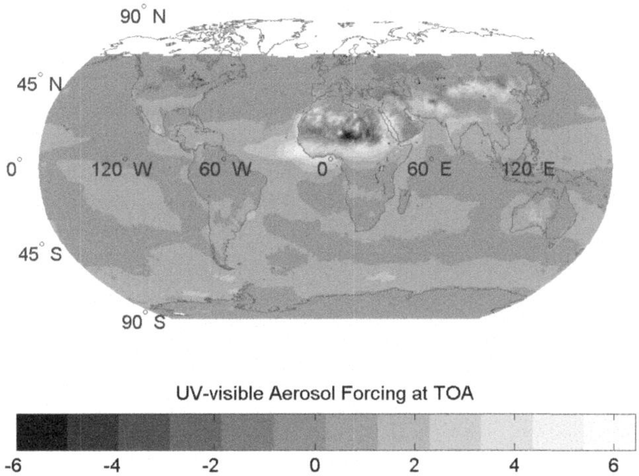

FIG. 8.22. Global distribution of the aerosol effect on the outgoing UV–visible radiation at top–of–atmosphere (aerosol forcing ΔF_{TOA}, W m^{-2}) for January, using GADS aerosol climatological data. (Hatzianastassiou *et al.* 2004a)

Pole. This atmospheric activity is further amplified by the lack of clouds and precipitation. The magnitude of ΔF_{TOA} is determined by the incoming solar radiation, the reflectance of the aerosol layer, directly associated with AOT, and the clear–sky fraction. The maximum summer tropical and subtropical ΔF_{TOA} values are associated with relatively small cloud cover (e.g. over India in January and over the Mediterranean in July), and large incoming solar radiation values, along with relatively large AOT values. The ΔF_{TOA} values in midlatitudes of the Northern Hemisphere (Europe, USA) are larger in summer than in winter, mainly because of larger incoming solar fluxes at TOA and less cloudiness during this season.

8.6.2 *Aerosol forcing of atmospheric absorption*

Significant aerosol–induced increase in solar atmospheric absorption can be seen in Figs. 8.23 and 8.24. Aerosols are found to cause increases, on a mean monthly basis, up to 25 W m^{-2} in the UV-visible and 7 W m^{-2} in the near–infra–red in areas (e.g. Sahara) characterized by significant amounts of absorbing aerosols (such as mineral dust) over surfaces with large surface albedo (>0.3). The solar atmospheric absorption is also found to be increased by less than 10 W m^{-2}, due to aerosols, over Europe, North America, South, and South-East Asia, but also over South Africa, the Amazon basin and Australia. Note that model computations indicate a significant aerosol-induced solar atmospheric absorption of 5–12 W m^{-2} in the sub–Sahelian region (5–10°N) in January, which is the local dry season. According to the Global Fire Atlas, fires are largely concentrated in this

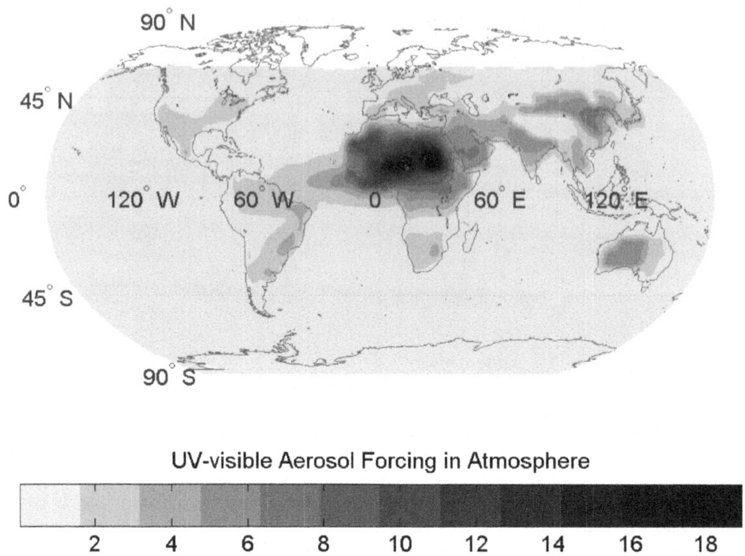

FIG. 8.23. Global distribution of the aerosol effect on atmospheric absorption of UV–visible radiation (aerosol forcing ΔF_{atmo}, W m^{-2}) for January, using GADS aerosol climatological data. (Hatzianastassiou *et al.* 2004a)

area in winter (January). Of course, the effect of aerosols on the SW atmospheric absorption (ΔF_{atmo}) in the sub–Sahelian region can also be attributed to desert dust advected from Sahara southward toward the equator, by the Harmattan trade winds occurring in the dry season. In contrast, during Northern Hemisphere summer (July), with the dry season shifted to the Southern Hemisphere, the biomass burning is dominant to the south of the equator, with maximum concentration between 5 and 15° S. Model computations indicate an aerosol–induced increase in solar atmospheric absorption during summer, of about 5–8 W m^{-2} in South Africa and within the Amazon basin.

8.6.3 *Aerosol forcing at the Earth's surface*

Model results indicate that aerosols can decrease regionally the monthly mean downward UV–visible radiation at the Earth's surface by up to 28 W m^{-2}, and the absorbed solar radiation by the surface by as much as 23 W m^{-2}, thus producing surface radiative cooling. The largest decreases in surface solar radiation are found over areas with significant aerosol amounts (AOT); thus, AOT, ΔF_{surf}, and $\Delta F_{\mathrm{nsurf}}$ values are correlated. Therefore, a large effect of aerosols on the downward solar radiation at the Earth's surface is encountered over land areas (especially over deserts) while smaller forcing values are found over oceans, due to optically thin aerosol layers above. According to model computations,

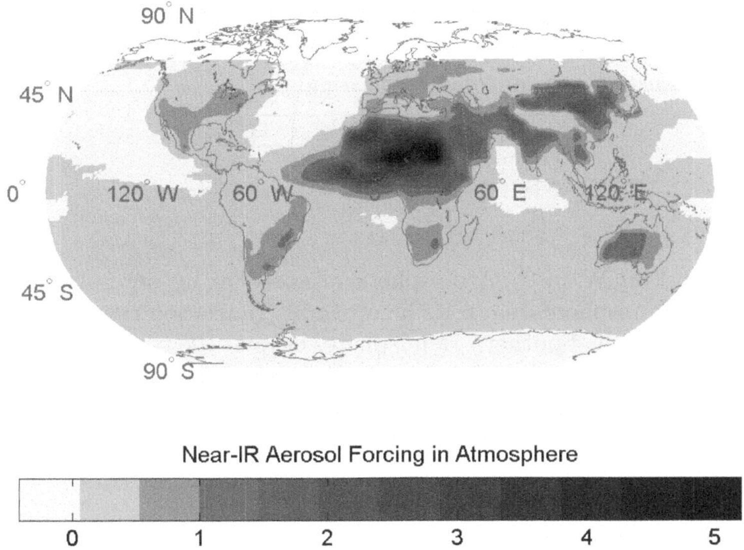

FIG. 8.24. Global distribution of the aerosol effect on atmospheric absorption of near–infra–red radiation (aerosol forcing $\Delta F_{\rm atmo}$, W m^{-2}) for January, using GADS aerosol data. (Hatzianastassiou *et al.* 2006)

aerosols can produce locally an effect on solar surface radiation that is up to three times larger than their effect at TOA.

The radiative effect of aerosols on the absorbed solar radiation at the Earth's surface is determined by the forcing $\Delta F_{\rm surf}$ and the surface albedo $R_{\rm g}$. Therefore, the features of $\Delta F_{\rm nsurf}$ are similar to those of $\Delta F_{\rm surf}$, except in the high–albedo polar regions. Note that in cases of small $\Delta F_{\rm atmo}$ values, the forcings $\Delta F_{\rm TOA}$ and $\Delta F_{\rm nsurf}$ are almost equal in magnitude. Under such circumstances, the aerosol forcing at the surface could be estimated from the corresponding forcing at TOA, as suggested by Ramanathan *et al.* (2001), for purely scattering aerosols.

8.6.4 *Mean annual hemispherical aerosol forcings*

In Table 8.13 are given mean annual model estimates of the three types, $\Delta F_{\rm TOA}$, $\Delta F_{\rm atmo}$ and $\Delta F_{\rm nsurf}$ of direct aerosol forcings (combined natural and anthropogenic), under all-sky conditions, for the Northern Hemisphere (NH), Southern Hemisphere (SH), and the globe. We note that for energy conservation we have

$$\Delta F_{\rm TOA} + \Delta F_{\rm atmo} + \Delta F_{\rm nsurf} = 0. \tag{8.8}$$

The global cooling to space, due to all aerosols, is estimated at 1.62 W m^{-2}, while the surface cooling is estimated to be –3.22 Wm^{-2}, thus giving an atmospheric

Table 8.13 *Model mean annual global and hemispherical (NH, SH) all–sky shortwave direct aerosol radiative forcing (W m^{-2}), for combined natural and anthropogenic aerosols. Forcings are given in terms of: outgoing SW radiation at TOA, atmospheric absorption, and net at surface. (Hatzianastassiou et al. 2007)*

	ΔF_{TOA}	ΔF_{atmo}	$\Delta F_{\mathrm{nsurf}}$
Globe	1.62	1.60	–3.22
NH	1.72	2.63	–4.35
SH	1.51	0.58	–2.09

absorption of 1.60 W m^{-2}. Atmospheric absorption by aerosols is greater in the Northern Hemisphere due to its heavy industrialization (more carbonaceous aerosols), whereas the Southern Hemisphere, beyond the tropical biomass burning, has natural marine sulphate aerosols that are primarily scattering. This can be seen in the ΔF_{atmo} values. Similar radiative forcing values are also reported in the review by Yu *et al.* (2006), based on a measurement-based assessment.

The changes to global gradients in OSR will affect general circulation patterns in the atmosphere, and hence climate. Many studies of the direct aerosol SW forcing have concentrated upon the distributions and radiative effects of anthropogenic scattering sulphate aerosols that are believed to produce a substantial cooling effect. Natural aerosols are equally important to climate. On the other hand, attention has been directed to mineral-dust aerosols, especially those originating from changes in land use, because of their large contribution to atmospheric aerosol loading and absorption of solar radiation. Many global studies of direct forcing due to anthropogenic sulphate, black (elemental) carbon, carbonaceous and soot aerosols from biomass burning and fossil fuels, nitrate aerosols, soil dust, and organic matter have been carried out. We note that mineral dust (from deserts) and black carbon (industrial pollution sources) actually reduce the outgoing solar radiation and hence contribute to global warming. The global direct aerosol forcing might be comparable to the anthropogenic greenhouse forcing of longwave radiation, i.e. global warming, which is estimated to reduce the outgoing thermal infra–red radiation by about 2.4 W m^{-2} (IPCC 2001).

8.7 Longwave radiation budget at TOA

Over the past three decades, considerable effort has been expended in the global measurement of the TOA longwave radiation budget. The Earth Radiation Budget Experiment (ERBE) provides the most comprehensive set of data for the TOA energy budget. The calculated global long–term (1984–2004) average of the outgoing longwave radiation (OLR) at the top of the atmosphere is found to be 239 W m^{-2}, based on model runs with ISCCP climatological data.

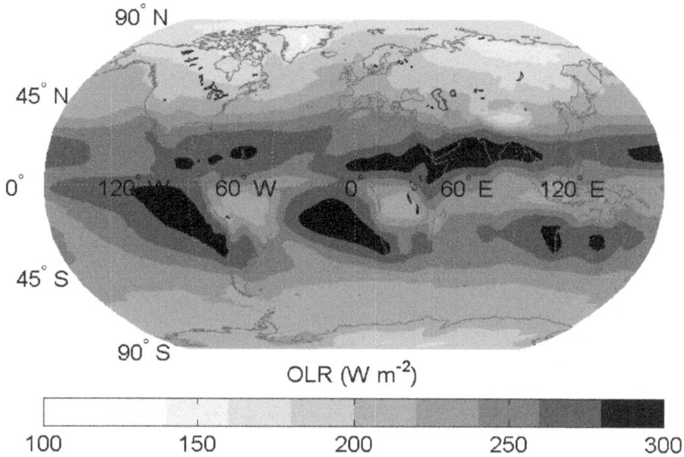

FIG. 8.25. Model long–term (1984–2004) outgoing longwave radiation (OLR) in W m^{-2} at TOA for January.

8.7.1 *Global distribution*

An example of the global distribution of the OLR at TOA for the month of January (long–term average 1984–2004) is shown in Fig. 8.25. Tropical regions covered by clouds in January, such as the Amazon, central Africa and Indonesia during the southern wet season exhibit lower OLR (lower emission temperature) than the desert regions of the subtropics and oceanic areas west of Africa, Australia and South America (Eastern Pacific). The lowest OLR is in Greenland during the polar night because of its altitude.

8.7.2 *Zonal–seasonal variation*

The seasonal variability of the outgoing longwave flux at TOA is relatively small in the tropical and subtropical zones, especially in the zone 20–30°, and increases at higher latitudes poleward of 30°, with the largest variability in the polar zones (Fig. 8.26). Antarctica has the coldest and driest climate resulting in the lowest OLR, especially during the polar night where in July it drops to about 120 W m^{-2}. Similar values are found for Greenland during January.

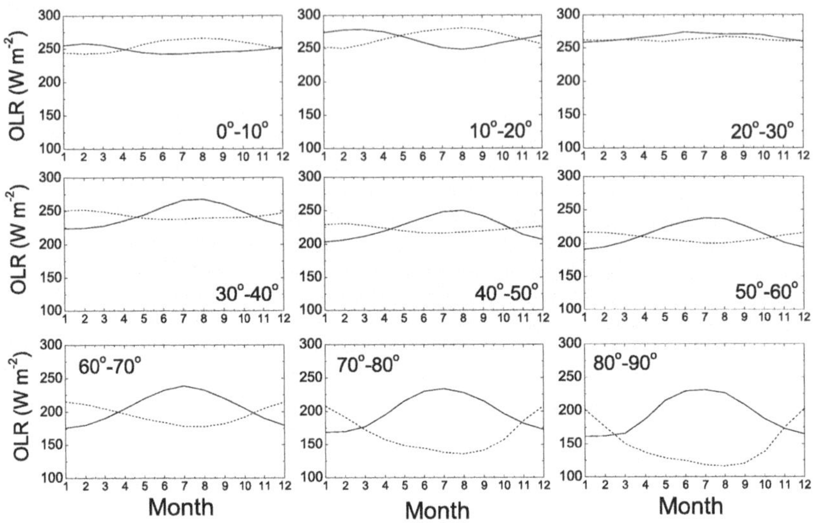

FIG. 8.26. Model long–term (1984–2004) seasonal–zonal variation of outgoing long-wave radiation (W m^{-2}) for the Northern Hemisphere (solid line), and Southern Hemisphere (dotted line).

8.7.3 *Latitudinal and seasonal variation*

The mean annual OLR increases slightly from the equator up to 25° latitude (where it is equal to about 250 W m^{-2}) and then decreases gradually towards the poles to values between 150–180 W m^{-2} (Fig. 8.27). We also note that the decreasing OLR towards the polar regions is negatively correlated with the increasing planetary albedo, principally due to the presence of clouds. The small increase in OLR from 0° to 30° is due to the presence of deserts in these latitudes, where cloud cover is at a minimum. The hemispherical and global mean seasonal variation of OLR is shown in Fig. 8.28. The global mean OLR is 239.1 W m^{-2} with a peak in summer because of the larger OLR in summer of the Northern Hemisphere. The Southern Hemisphere generally exhibits a weaker seasonal variation due to the larger surface fraction of oceans.

8.7.4 *Long–term hemispherical and global means*

In Table 8.14 is shown the global radiation balance at TOA . The net incoming solar radiation appears to be larger then the outgoing longwave radiation by 4.9 W m^{-2}, according to ERBE measurements and only 1.2 W m^{-2} based on model calculations using ISCCP climatological data. This imbalance is within the

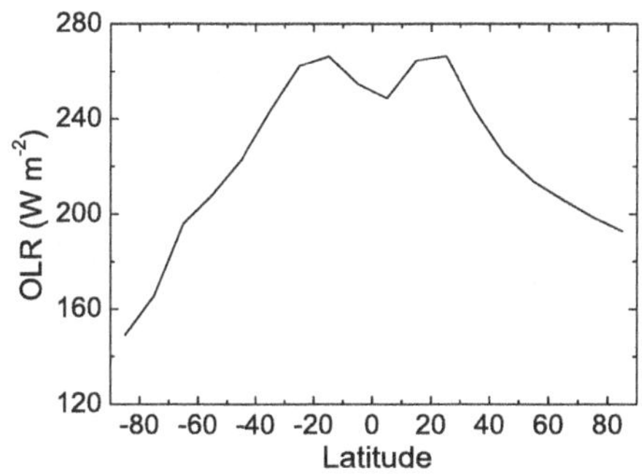

FIG. 8.27. Model long-term (1984–2004) average annual mean latitudinal variation of OLR (W m^{-2}).

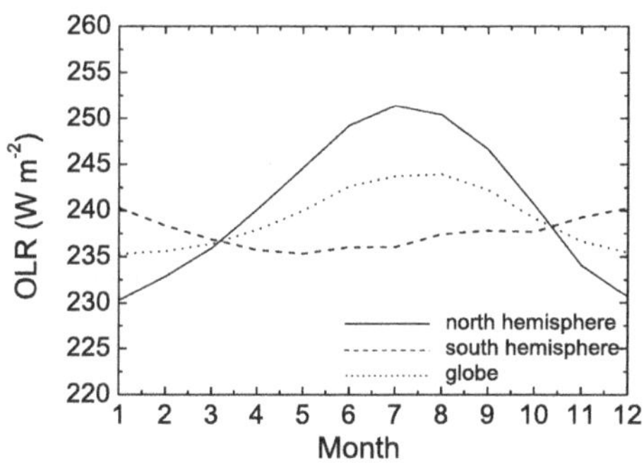

FIG. 8.28. Model long-term (1984–2004) average global mean seasonal variation of OLR (W m^{-2}).

errors of the model estimates of each component. However, there appears to be an imbalance between planetary solar heating and planetary thermal cooling to space in the Southern Hemisphere of 2 W m^{-2}, while the Northern Hemisphere appears to be in radiative balance.

8.8 Longwave radiation budget at surface

The estimation of the Earth's longwave radiation budget represents a major objective of the World Climate Research Programme as demonstrated by its Global Energy and Water Cycle Experiment (GEWEX), and in particular the

Table 8.14 *Model computed mean annual hemispherical (NH Northern Hemisphere, SH Southern Hemisphere) and global average: outgoing longwave radiation, OLR (1984–2004) and net incoming shortwave radiation NISR (1984–1997) at the top-of-atmosphere, and net incoming all-wave radiation, NET. The corresponding values from ERBE (Feb 1985 - May 1989) are also shown. The radiative fluxes are expressed in W* m^{-2}.

	ERBE OLR	model OLR	ERBE NISR	model NISR	ERBE NET	model NET
NH	235.5	240.6	234.7	241.0	–0.8	0.4
SH	234.7	237.6	245.4	239.6	10.7	2.0
Globe	235.1	239.1	240.0	240.3	4.9	1.2

GEWEX Surface Radiation Budget Project. The amount of downwelling longwave radiation reaching the surface of the Earth and the outgoing flux at TOA are indicators of the strength of the atmospheric greenhouse effect and hence they are key parameters in climate modelling.

The only reliable direct measurements of downwelling longwave flux at the surface, are those provided by well-calibrated surface instruments. Current archives of such measurements have a very limited temporal and geographical coverage. For example, downwelling longwave fluxes (DLR) reaching the surface, exist for less than 20 BSRN stations around the world and in most cases data exist only since the mid-1990s. Therefore, in order to monitor the surface downwelling longwave radiation on a global scale and over a long enough period to identify climate change impacts, one needs to rely on satellite data, in conjunction with radiative–transfer models, with validation against surface measurements. The reliability of the model DLR at the surface is primarily affected by its sensitivity to cloud cover and cloud properties, and to the vertical profiles of temperature and humidity, especially in the lower troposphere. Clouds, which are the most important determinants of the surface radiation budget, represent a major uncertainty in climate modelling (Intergovernmental Panel on Climate Change, IPCC 2001). The International Satellite Cloud Climatology Project (ISCCP) provides one of the most extensive and comprehensive global cloud climatologies currently available and also provides surface and atmospheric parameters for longwave radiation transfer.

8.8.1 *Global distribution*

Model long–term (1984–2004) results with spatial resolution of 2.5 degrees are given in Fig. 8.29 for January. The input data were cloud properties from ISCCP-D2, and temperature and humidity profiles from the NCEP/NCAR Reanalysis project. As expected, in January, the maxima of the DLR occur over a broad swath along the equator, mainly over tropical and subtropical oceans along the intertropical convection zone. In these regions cloud amounts, water vapour and air temperatures are high, while cloud bases are low. These maxima are shifted

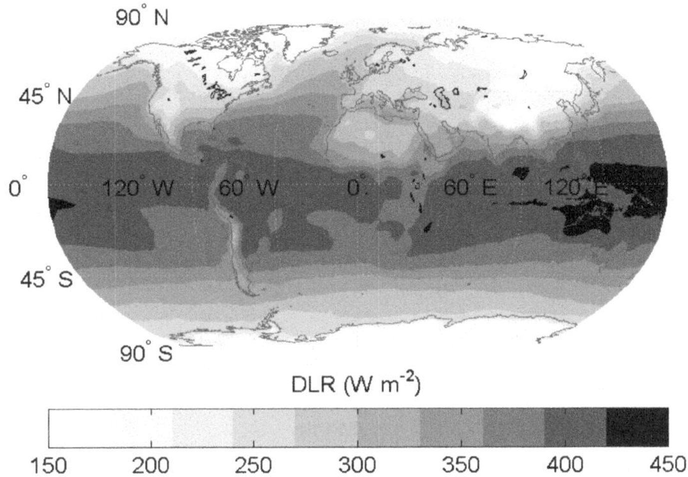

FIG. 8.29. Model long–term (1984–2004) downwelling longwave flux (DLR) in W m^{-2} at the Earth's surface for January.

slightly northwards in July. Minima occur, as expected, in the polar regions, with Antarctica and Greenland having the lowest values, and over an extended area in the Northern Hemisphere in winter. There are regional minima of the DLR in winter over dry desert areas (Sahara, Atacama, Kalahari, Central Australia), characterized by clear–sky conditions, as well as over high–altitude areas (Tibetan Plateau, Rocky Mountains, Andes, Greenland, Antarctica), with low cloud cover, low air temperatures, and low moisture content. Very high values of DLR are found in the West Pacific, a region strongly influenced by the El Niño effect.

8.8.2 *Zonal, latitudinal and seasonal variations*

8.8.2.1 *Zonal–seasonal variation* A lower seasonal variability in DLR is seen in the Southern Hemisphere, owing to the larger fraction of its surface covered by oceans, compared with that in the Northern Hemisphere. The seasonal variation is small in the tropics, owing to the smaller variation in atmospheric water vapour and temperature, for both hemispheres, with increasing variability at higher latitudes in the Northern Hemisphere (Fig. 8.30).

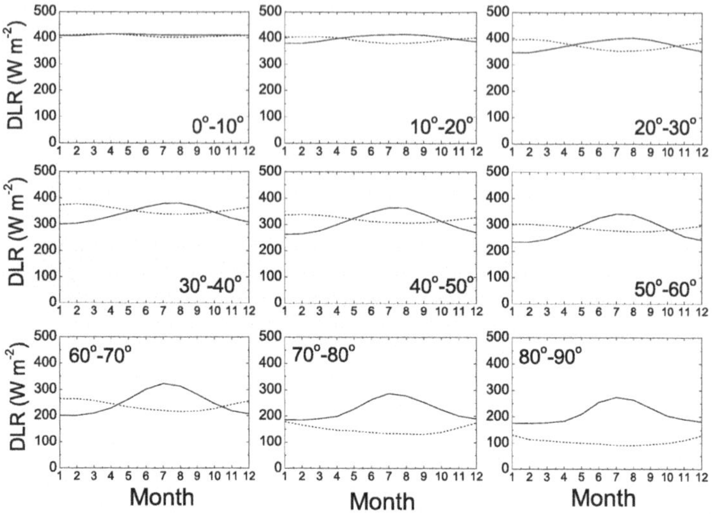

FIG. 8.30. Model zonal mean long–term (1984–2004) seasonal variation of DLR in W m^{-2} at the Earth's surface for the Northern Hemisphere (solid line), and Southern Hemisphere (dotted line).

8.8.2.2 *Seasonal variation* In Fig. 8.31 is shown the seasonal variation of the DLR for both hemispheres and the globe. The variation for the hemispheres is approximately between 320 and 380 W m^{-2}, or about 60 W m^{-2} over the year compared to about double that for the DSR. The DLR for the northern hemisphere exhibits greater variation owing to its greater land fraction that results in greater heterogeneity in atmospheric conditions, compared to the largely ocean covered southern hemisphere.

8.8.2.3 *Latitudinal variation* In Fig. 8.32 is shown the model global mean long–term (1984–2004) latitudinal variation of DLR in W m^{-2} at the Earth's surface. The DLR peaks in the tropics at about 400 W m^{-2} owing to higher tropospheric temperatures, more water vapour and the effects of warmer cloud–base temperatures. Antarctica is colder and drier then the Arctic so receives substantially lower, by about 100 W m^{-2}, DLR at the surface.

8.8.3 *Long–term anomaly*

The long–term trend (1984–2004) in the monthly downwelling longwave radiation anomaly, based on ISCCP–D2 2.5–degree cloud climatologies and NCEP/NCAR specific humidity and atmospheric temperature profiles, is shown in Fig. 8.33.

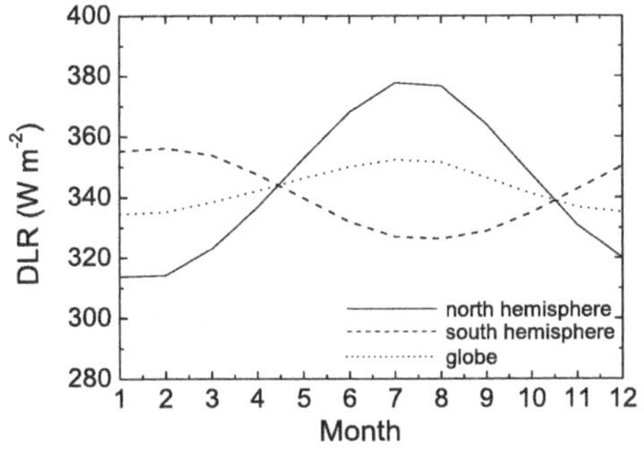

FIG. 8.31. Model global mean long–term (1984–2004) seasonal variation of DLR in W m^{-2} at the Earth's surface.

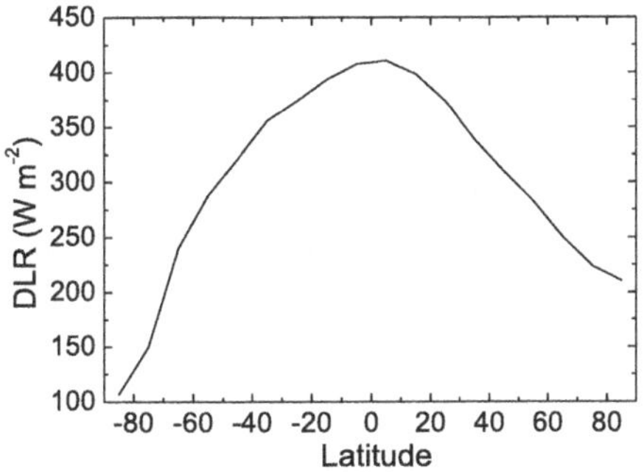

FIG. 8.32. Model global mean long–term (1984–2004) latitudinal variation of DLR in W m^{-2} at the Earth's surface.

There appears to be no trend in the DLR in comparison with the trend in the DSR. Also shown are the results of a more detailed spectral model using NASA–Langley 1–degree resolution cloud climatologies and GEOS–1 Reanalysis humidity and temperature profiles for the available years of data (1986-1994). Both sets of model results indicate no trend in the DLR.

8.8.4 *Long–term hemispherical and global means*

In Table 8.15 are shown the long–term mean Northern (NH), Southern (SH) Hemisphere and global downwelling flux (W m^{-2}), estimated by different methods. The model results of Pavlakis *et al.* (2004) were based on ISCCP–D2 data for the 10-year period 1984–1993. The climate models give values close to the es-

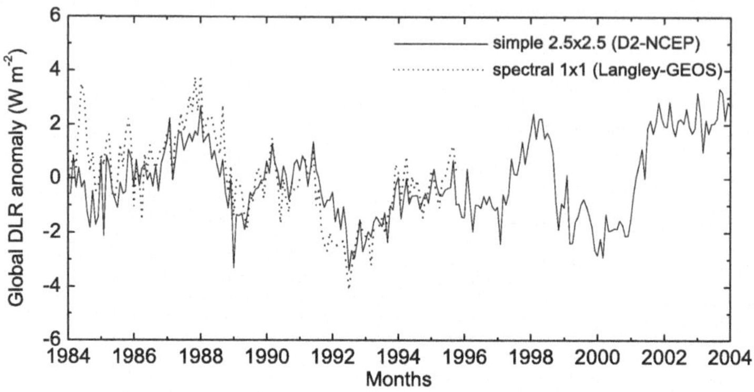

FIG. 8.33. Model global mean long–term (1984–2004) trend in downwelling longwave flux anomaly in W m^{-2} at the Earth's surface, using ISCCP–D2 2.5×2.5 degree cloud data and NASA–Langley 1.0×1.0 degree cloud data.

timate based on the GEBA ground–based data. Radiation models give a slightly higher DLR for the Northern Hemisphere compared with the Southern Hemisphere. The corresponding DLR values for the 21–year period 1984–2004 are 343.8, 341.2 and 342.5 W m^{-2}, for NH, SH and the globe.

Table 8.15 *Mean Northern (NH), Southern (SH) Hemisphere and global downwelling flux (W m^{-2}), estimated by a selection of different methods.*

Source	NH	SH	Global
Radiation-transfer models			
Pavlakis *et al.* (2004)	343.3	341.1	342.2
Gupta *et al.* (1999)	351.0	344.6	347.8
Rossow and Zang(1995)			348.0
General circulation models			
Wild *et al.* (2001)			344.0
Garratt *et al.* (1998)			339.0
ECMWF	339.3	339.9	339.6
GEBA&BSRN stations			345

8.8.5 *Sensitivity analysis*

8.8.5.1 *Atmospheric temperature profile* A sensitivity test examining the effect on the DLR to an increase (or decrease) in the air temperature by 2 K (the entire profile is moved to higher/lower temperature) shows that such an increase causes an increase of the global average DLR by 9 W m^{-2} (or decrease of 8.9 W m^{-2}), as shown in Table 8.16. This effect depends on the climatic conditions, i.e. on temperature, cloud cover and water–vapour content. The weakest effect (about 2 W m^{-2}) is observed in very cold climates with very low cloud cover (practically clear sky), while the largest effect (about 10 W m^{-2}) occurs in hot and cloudy (mainly tropical) regions.

8.8.5.2 *Specific humidity* When the specific humidity is increased (or decreased) by 25%, in each atmospheric layer for each latitudinal–longitudinal grid–cell, the result is a global increase in the DLR of 6.2 W m^{-2} (or decrease by 8.3 W m^{-2}) on average, with differences ranging from 1 to 14 W m^{-2}. Regions with originally low water–vapour content are the least affected. The response of the DLR to water–vapour changes clearly depends also on temperature, with the colder regions the least affected, and on low and middle cloud cover, since, when cloud is present, water vapour thermal emission that reaches the surface comes from a much smaller atmospheric column. The largest sensitivity of the DLR on water vapour is encountered in the dry zones of the subtropics and the Eastern Pacific ocean. For tropical regions with large water–vapour content, larger than approximately 3 g cm^{-2}, the DLR sensitivity becomes saturated, as the infra–red emissivity of the water–vapour layer increases towards unity.

8.8.5.3 *Cloud cover* Satellite cloud–cover uncertainties can be quite large as thresholds in temperature and reflectivity are used to identify clouds from satellite measurements. As a result, the fractional cloud cover can differ greatly among various satellite data sets depending upon the threshold values used. The sensitivity of the DLR to possible errors in the cloud data is given in Table 8.16. When the total cloud amount is changed by ±30% of its value for each pixel the global mean value of the DLR at the surface changes by about ±10 W m^{-2}. If just the low cloud amount is changed by ±30%, the effect on the global average DLR is about ±6 W m^{-2}, while if the middle cloud amount is changed by ±30%, the effect on the DLR is less but comparable to the low cloud case (about ±4 W m^{-2}). For the high cloud cover the effect is, expectedly, much less significant (±0.8 W m^{-2}). Close examination of the effect on a regional basis shows that the difference in the DLR ranges from 0 to 25 W m^{-2}, depending on the original (low and middle) cloud cover, water–vapour content and temperature. For the same amount of cloud cover, the smaller the amount of water vapour in the lower atmosphere, the higher the increase in the DLR caused by the increase in cloud cover. The warmest regions, in which the mean temperature of the lower part of the atmosphere is larger than 290 K, show the lowest sensitivity to the cloud cover increase. This is due to the fact that in these same regions generally the water vapour–content of the atmosphere is very high, which in conjunction with high atmospheric temperatures leads to a significant contribution to the DLR under clear–sky conditions. On the other hand, cold regions (mean T <270 K) with low water–vapour content are the most affected by an increase in cloud cover. This is an important result, given that the most severe discrepancies in cloud cover between different data sets occur in cold climates (Pavlakis *et al.* 2004).

8.8.5.4 *Cloud overlap scheme* One known limitation of the ISCCP-D2 data set is the assumption that the clouds are classified into non–overlapping layers. From the satellite point of view, if there are low–level clouds under optically thick middle–level clouds, they will not be observed. This fact leads to a systematic underestimation in low–level cloud amount, and consequently in DLR at the surface. To examine the possible effect that this assumption has on the DLR, two cloud–overlap schemes, based on random and maximum overlap, can be employed. The random–overlap scheme assigns lower cloud amounts under higher clouds, based on cloud cover in areas where the satellite's view to the lower cloud is not obscured. The maximum–overlap scheme assumes that a lower cloud always exists under the higher cloud. Overall, the random overlap assumption seems to be the preferable procedure. The maximum effect on the DLR reaches about 10 W m^{-2}, for the random–overlap scheme and is observed in areas where originally both low and middle cloud cover is significantly high.

Table 8.16 *Changes in mean global downwelling flux DLR, for January, for corresponding ± changes in key climatic parameters, based on model runs with ISCCP climatological data. (Pavlakis et al. 2004)*

Parameter	Change (±)	DLR difference (W m^{-2})
Precipitable water	25%	+6.2/–8.3
Air temperature	2 K	+9.0/–8.9
Surface temperature	2 K	±2.0
Total cloud cover	30%	+9.8/–11.0
Low–cloud cover	30%	+5.9/–6.0
Middle–cloud cover	30%	+4.1/–4.2
High–cloud cover	30%	+0.8/–0.8
Cloud–base height	20%	+1.8/–1.7
Clouds	Random cloud–overlap scheme	+1.2
Clouds	Maximum cloud–overlap scheme	+3.2

8.9 Mediterranean Sea heat budget

The Mediterranean Sea, being a semienclosed system, is an ideal test basin for the climate–research community, because it provides a good opportunity to investigate the closure of its hydrological and heat budgets. The closing of such budgets engenders more confidence that our understanding in many geophysical fields, such as sea–air interactions, modelling and measuring solar and terrestrial fluxes and hydrological quantities, is adequate.

Evaporation losses from the Mediterranean Sea are larger than the freshwater gains, creating large salinity values. The effect of the above is: a) an inflow of warmer, less salty Atlantic water from Gibraltar at the surface; and b) an outflow of cooler, saltier Mediterranean water at a larger depth. The warm inflow adds to the Mediterranean heat budget between 5 and 7 W m^{-2}. The outflow spreads across the Atlantic, with some of it moving northwards and affecting the thermohaline circulation of the Northern Atlantic. It has been argued that the greenhouse effect coupled with the damming of the Nile, will increase the Mediterranean outflow, displacing the North Atlantic Drift and creating conditions similar to the ones that triggered the last ice age. Therefore, a better understanding of the Mediterranean Sea heat and hydrological budgets would help us to better understand climate change on a regional scale.

8.9.1 *The heat budget equation*

The heat budget, Q_t is determined from the radiative and turbulent energy components. The radiative ones are the net solar (shortwave) flux Q_s absorbed by the sea and the net terrestrial (longwave) flux Q_l emitted by the sea. The turbulent components are the latent Q_e and the sensible heat flux, Q_h corresponding to evaporative and convective energy losses of the sea, respectively,

$$Q_t = Q_s - Q_l - Q_e - Q_h. \tag{8.9}$$

Q_t should balance the heat flux advected into the sea from Gibraltar, A_h, the heat flux entering the sea from the bottom surface G, and the change ΔH in the heat content H of the sea, so that

$$Q_t + A_h + G - \Delta H = 0. \tag{8.10}$$

G can be assumed to be zero. Although on an annual basis ΔH can be neglected, on a monthly scale it is very important, as will be shown later. The advected heat fluxes from the Black Sea and freshwater influx via rivers are estimated to be less than 1 W m^{-2} each, and so are considered negligible.

The heat content H of the Mediterranean Sea has a strong seasonality and affects significantly the seasonal variation of the heat budget. To derive it, we need to divide the Mediterranean into $J \times L \times K$ three-dimensional water cells, with J, L and K corresponding to the partition with respect to latitude, longitude and depth, respectively. H can then be estimated from (e.g. Krahmann *et al.* 2000)

$$H = \sum_{j=1}^{J}\sum_{l=1}^{L}\sum_{k=1}^{K} c_p(p, s, \theta) \cdot \rho(p, s, \theta) \cdot V \cdot (\theta - \theta_0), \tag{8.11}$$

where p is the average pressure, s the salinity, θ the potential temperature, ρ the density, V the volume, c_p the specific heat content of the water cell in location

(j, l, k), and θ_0 is a reference potential temperature. Quantities p, θ, ρ, and c_p can be calculated using the MATLAB Seawater Library of the Commonwealth Scientific & Industrial Research Organisation (CSIRO) Marine Research Division given the salinity, temperature and depth for a specific water cell. Knowing the seasonal distribution of H we can easily derive the amount ΔH of thermal energy that the Mediterranean absorbs from or provides to the environment each month. These amounts are used later in the indirect derivation of the evaporation E.

8.9.2 *Latent and sensible heat flux*

We divide the Mediterranean Sea into two–dimensional latitude–longitude water surface cells. In each cell the sensible heat flux Q_{h} and aerodynamic latent heat flux Q_{ea} are estimated from

$$Q_{\mathrm{h}} = \rho_{\mathrm{a}} c_{\mathrm{pa}} U C_{\mathrm{H}}.(\theta_{\mathrm{a}} - T_{\mathrm{w}}), \tag{8.12}$$

$$Q_{\mathrm{ea}} = \rho_{\mathrm{a}} L U C_{\mathrm{E}}.(q_{\mathrm{s}} - q), \tag{8.13}$$

where ρ_{a} is the air density, c_{pa} the moist air specific heat capacity, U the scalar wind, θ_a the potential air temperature. The potential temperature is defined by

$$\theta = T \left(\frac{p_o}{p} \right)^{R/c_p}, \tag{8.14}$$

where R is the gas constant and p_{o} is a reference pressure, taken as 1 bar. Thus the potential temperature of a gas is the temperature that it would have if the gas is compressed or expanded adiabatically to the pressure p_{o}. T_{w} is the water surface temperature, L the latent heat of evaporation, q_{s} the saturation humidity at the water surface temperature, and q the observed specific humidity. The saturation humidity q_{s} over salt water (assuming a salinity of 34 psu) is 98% of q_{s} over pure water. L can be estimated from

$$L = 2.501 \times 10^6 - 2350 T_{\mathrm{a}} \qquad (\mathrm{J/kg}), \tag{8.15}$$

with the air temperature T_{a} in C. C_{H} and C_{E} are turbulent exchange coefficients, estimated by a variety of methods. The derivation of C_{H} and C_{E} is relatively simple for neutral atmospheric stability conditions and take values of the order of 10^{-3} (see Brutsaert 1984), but once atmospheric instability is included in the analysis, it becomes considerably more complex. However, over periods of a day or longer the neutral forms of C_{H} and C_{E} can be used. We verify the validity of the above assumption in §8.9.8.

The aerodynamically derived latent heat flux Q_{ea} is one way to estimate the evaporation, quantifying the drying power of the air. Another method to derive evaporation is the heat–balance method, which assumes that the evaporation process uses all energy that is available to it. Thus, the second estimate for the evaporation is:

$$Q_{\mathrm{er}} = Q_{\mathrm{s}} - Q_{\mathrm{l}} - \Delta H + G + A_{\mathrm{h}} - Q_{\mathrm{h}}. \tag{8.16}$$

The heat–balance method closes the heat budget by definition.

The evaporation rate E can be derived from the latent heat flux Q_{e} from

$$E = \frac{Q_{\mathrm{e}}}{\rho L}, \tag{8.17}$$

with ρ and L being the water density and the latent heat of evaporation, as defined before. The equation above is used to calculate the bulk aerodynamic evaporation rate E_{a} from Q_{ea} and the heat–balance evaporation rate E_{r} from Q_{er}.

8.9.3 *Sea data for computing heat storage*

For the calculation of the heat storage, ΔH we need the seawater temperature, salinity and pressure in a three–dimensional grid. We can use such data from two sources. The first is the Mediterranean Data Archaeology and Rescue (MEDAR) MEDATLAS 2002 database, a comprehensive data product of multidisciplinary *in–situ* data and information on the Mediterranean and Black Seas, through a wide co-operation of the Mediterranean countries. The MEDAR data are climatological values at a monthly 0.2×0.2 degree resolution, down to a depth of 4000 m. The second is the World Ocean Atlas 2001 from the National Oceanographic Data Center (NODC) of the National Oceanic and Atmospheric Administration (NOAA), which gives climatological temperature and salinity data down to a 1500 m depth, at a monthly 1×1 degree resolution.

The temperature structure at the surface from MEDAR is shown in Fig. 8.34 for March. Similarly, we present the temperature structure at a depth of 75 m in Fig. 8.35. The warmest water is in the southeast basin, while the coldest is generally in the north. There are large seasonal differences in the surface water temperature, which oscillates between 10–25 °C. It is apparent that the temperature seasonality becomes attenuated already at 75 m.

We include also an image of the July salinity at the surface (Fig. 8.36). The salinity is not characterized by a seasonality as strong as that of the water temperature. An area with little salinity is to the east of Gibraltar, showing the less salty Atlantic water flowing into the Mediterranean. Over a smaller spatial extent, the North Aegean and North Adriatic also have low salinity. In the first case we see less salty water flowing from the Black Sea and in the second, freshwater from the river Po diluting the shallow seawater.

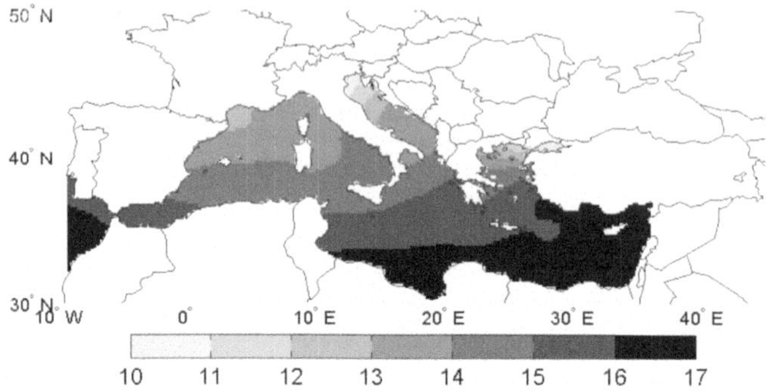

FIG. 8.34. MEDAR water temperature (°C) at 0 m depth for March. (Matsoukas *et al.* 2005)

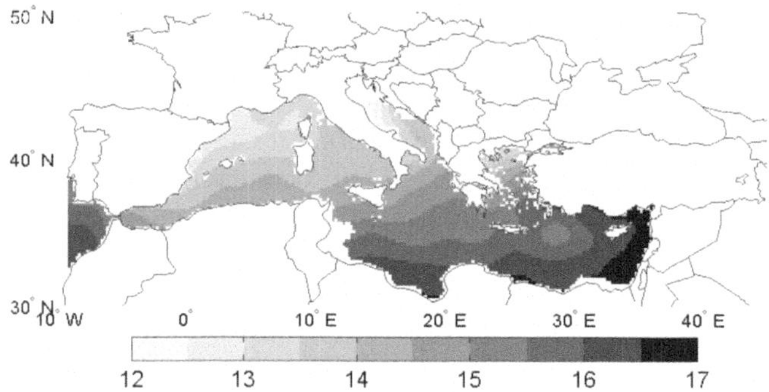

FIG. 8.35. MEDAR water temperature (°C) at 75 m depth for March. (Matsoukas *et al.* 2005)

8.9.4 *Data for computing turbulent fluxes*

The heat–balance latent heat Q_{er} can be estimated from data already mentioned above. For the bulk aerodynamic latent heat Q_{ea} we can use eqn (8.13), with two data sources for the necessary climatic quantities. The first is the 40–year European Centre for Medium–range Weather Forecasts (ECMWF) Re–Analysis (ERA-40) and the second is the International Comprehensive Ocean–Atmosphere Data Set (ICOADS), which is a collection of global marine surface observations and their monthly summary statistics. ICOADS observations are taken primarily

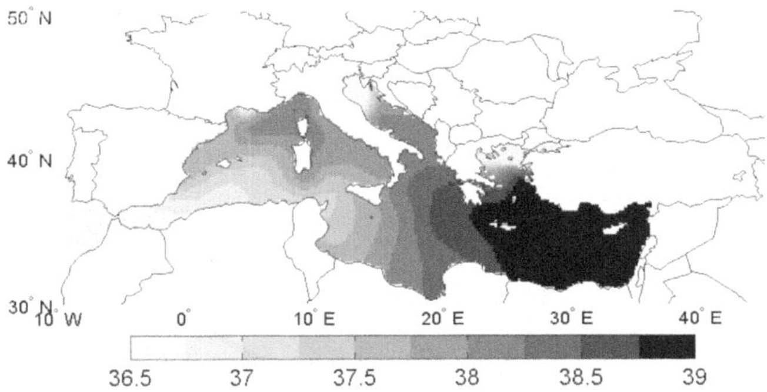

FIG. 8.36. MEDAR salinity (psu) at the surface for July. (Matsoukas *et al.* 2005)

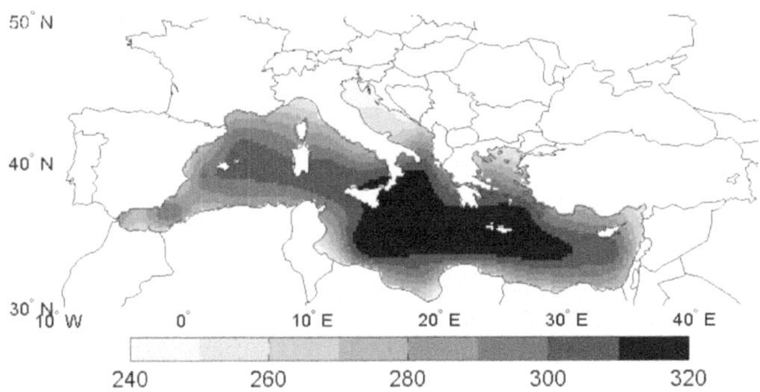

FIG. 8.37. Model Q_s (W m^{-2}) for July 1990. (Matsoukas *et al.* 2005)

from ships (merchant, ocean research, fishing, navy, etc.) and from moored and drifting buoys. Data from an earlier version of ICOADS in the period 1980–1993 were used to generate a global air–sea heat flux climatology by the Southampton Oceanography Centre (SOC). The aerodynamic latent heat flux Q_{ea} is also available directly from ERA–40, SOC, and NESDIS (NOAA Environmental Satellite Data Information Service) databases. The sensible heat flux Q_h is available directly from ERA–40.

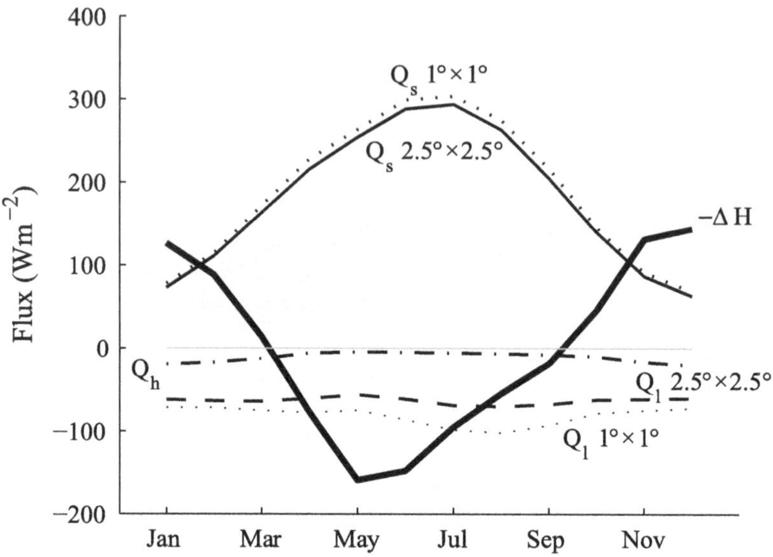

FIG. 8.38. Monthly climatology of the heat–budget components. Here, positive values correspond to available energy for evaporation, while negative values to energy losses to the atmosphere or heat storage. (Matsoukas *et al.* 2005)

8.9.5 *Radiation fluxes*

As an example we show a 2.5×2.5 model net solar flux Q_s, for July 1990 in Fig. 8.37. The values shown are averaged over 24 h. As expected, Q_s generally decreases from south to north and the summer values are much larger than those in winter. The Q_s values over sea are larger than over land in July, because the sea albedo is much smaller than the land albedo. Thus, the largest part of the downwelling component is absorbed by the sea. On the other hand, in January the solar zenith angle is low and the sea reflectivity increases due to the Fresnel effect. Also, the downwelling component is not very large and therefore Q_s differences between sea and land are small. Model monthly components of the heat budget using ISCCP-D2 monthly 2.5×2.5 degree and daily 1×1 degree NASA–Langley climatological data are shown in Fig. 8.38. The seasonality of Q_s is quite pronounced, with winter values around 70 W m^{-2} and summer values just under 300 W m^{-2}. The coarser data span the period 1984–2000, while the finer–resolution data are for 1984–1995. Nevertheless, the two climatologies are very close and differ by only a few W m^{-2} for any month.

The net longwave flux Q_l for July 1990 is shown in Fig. 8.39. In this figure and in Fig. 8.38 one can see that the seasonality of Q_l is considerably less pronounced compared to Q_s.

8.9.6 *Heat storage*

We can derive the monthly climatological heat–storage change using the MEDAR Atlas temperature and salinity climatological data and eqn (8.11). Positive ΔH values denote increase in the heat content, i.e. water temperature increase, and correspond to spring and summer, when the solar radiation is at its maximum. Negative values are seen during winter and autumn, when the solar radiation decreases. We plot the change in the negative of heat storage, $-\Delta H$ in Fig. 8.38 to allow a visual check that the sum of all terms of eqn (8.10) are close to zero. The heat content itself H is at its maximum in October and at its minimum in March. The H maximum and minimum, coincide with the points of intersection between ΔH and the zero line in Fig. 8.38. The World Ocean Atlas 2001 (WOA1) dataset produces similar monthly ΔH values to MEDAR, although the amplitude of the WOA01 seasonal cycle is stronger by 20 W m^{-2}.

During spring and summer the Mediterranean stores solar energy at a rate of 100–150 W m^{-2}, a value of the same order of magnitude as $Q_{\rm s}$. On the other hand, during autumn and winter, it gives out this energy back to the environment. On an annual basis, ΔH has a value of zero. We note here that the gridded MEDATLAS 2000 data are given as long–term monthly climatologies. Therefore, we can only estimate the ΔH maps as 1×1 degree monthly climatologies and not as monthly time series for the study period 1984–2000.

8.9.7 *Turbulent fluxes*

8.9.7.1 *Sensible heat flux* The sensible heat flux $Q_{\rm h}$ is taken directly from ERA-40 and its monthly climatological behaviour can be observed in Fig. 8.38 (dot-dash line). It has its maximum (absolute) values during winter and its min-

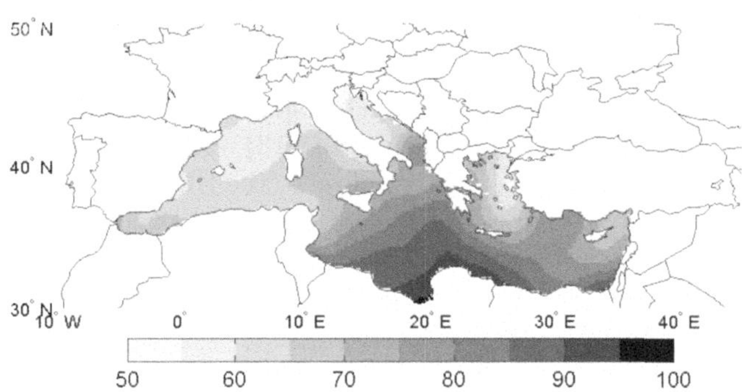

FIG. 8.39. Model $Q_{\rm l}$ (W m^{-2}) for July 1990. (Matsoukas *et al.* 2005)

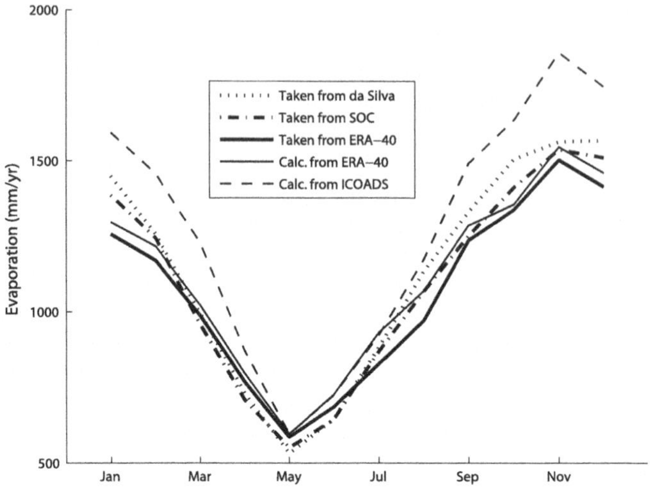

FIG. 8.40. Monthly estimates of E_a. (Matsoukas *et al.* 2005)

imum during summer, while its magnitude is the smallest of all components.

8.9.7.2 *Latent heat flux* We can use two methodologies to estimate the latent heat flux Q_e and evaporation E. One is the bulk aerodynamic and the other the heat–balance method. With Q_{ea} and E_a we denote the bulk aerodynamic latent heat and evaporation estimates, while with Q_{er} and E_r, we denote the heat–balance values. The latent–heat flux is presented in Figs. 8.40 and 8.41. In these plots the latent–heat estimates are translated into evaporation rates, using eqn (8.17). Figure 8.40 shows the evaporation estimates that are given directly by data sources, namely the monthly climatological evaporation rates from NOAA/NESDIS, SOC, and ERA–40. All three are estimates of bulk aerodynamic evaporation E_a, derived using variations of eqn (8.13). Also, the E_a calculated by the model using standard climatological quantities and eqn (8.13) are shown in the same figure. We present the aerodynamic evaporation E_a calculated from ERA-40 and from the ICOADS data. The ICOADS derived E_a is larger than the ERA–40 due to its higher wind speeds. Note that the evaporation that we derive from ERA–40 data using eqn (8.13) and assuming neutral atmospheric conditions (thin solid line) is very close to the ERA-40 evaporation (thick solid line) although ERA–40 does not make the neutral atmosphere assumption. Thus, Brutsaert (1984) seems justified in saying that for monthly evaporation the neutral atmospheric stability assumption is valid. A further check on this assumption was performed, including atmospheric instability in our derivation of E_a. The derived values (not shown) were similar to the ones calculated with the neutral stability assumption, although larger by about 5% during winter.

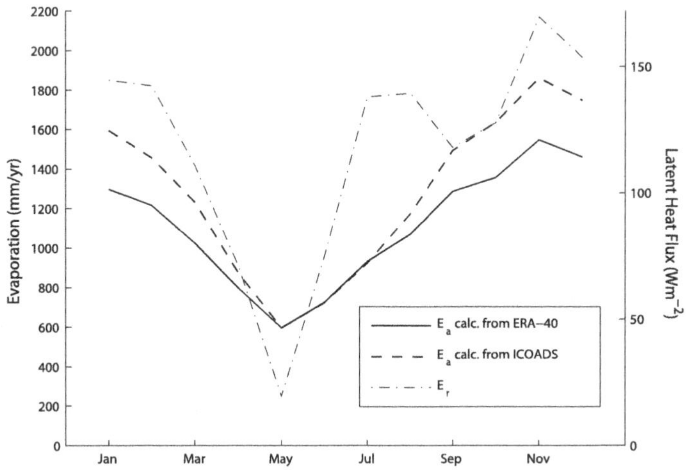

FIG. 8.41. Monthly estimates of E_a and E_r. The E_a are calculated from ICOADS and ERA–40 data. (Matsoukas *et al.* 2005)

8.9.8 *Seasonal evaporation rate*

We now examine the evaporation derived from the heat–balance method E_r, using eqn (8.16) and eqn (8.17). In Fig. 8.41 we show E_r and for comparison purposes, we also include calculations of E_a estimates using ERA-40 and ICOADS data. It is interesting to note that the minimum evaporation in May given by the bulk aerodynamic formula, is also reproduced by the heat–balance method. Interpreting this behaviour from the energy standpoint, we see in Fig. 8.38 that most of the solar energy that month is used to increase the sea heat content. After subtracting the longwave sea–emitted radiation, the energy available to drive the evaporation process is very small. This results in the May evaporation minimum. On the other hand, the maximum evaporation in winter can be linked to the very large heat content release from the sea during the same time. The released thermal energy is larger even than the absorbed solar energy then, and is the primary energy source for the strong evaporation process.

8.9.9 *Annual evaporation rate*

Estimates of the mean annual evaporation are shown in Table 8.17. In the first row E_r is the evaporation determined by the heat balance method. Rows 2, 3, and 4 correspond to evaporation estimates from the NOAA/NESDIS, SOC and ERA–40 databases. Finally, rows 5 and 6 give the E_a estimated from ERA–40 and ICOADS climatological data.

The differences between E_r and E_a are rather large, and can rise to more than 350 mm year^{-1}. There are also qualitative differences as seen in Fig. 8.41, where

Table 8.17 *Annual Mediterranean evaporation rates. (Matsoukas et al. 2005)*

Index	Source	Value (mm $year^{-1}$)
1	E_r	1500
2	E_a from NOAA/NESDIS	1130
3	E_a from SOC	1090
4	E_a from ERA-40	1060
5	E_a from eqn (8.13) and ERA-40	1110
6	E_a from eqn (8.13) and ICOADS	1280

E_a has one maximum (during winter), while E_r has two. The largest maximum is in winter, when the surface winds achieve their largest speeds and also there is a peak in energy released from the heat content of the sea. In this case both E_r and E_a take their largest values. The second E_r maximum is in July and August, when more energy is provided to the sea by the Sun than is stored as heat content. This leads to a secondary maximum in the available energy for evaporation Q_{er}.

8.9.10 *Comparison with Red and Black Seas*

We compare the annual heat budget components of the Mediterranean, Red and Black Seas in Table 8.18. The Red Sea is located in the warmest region of the three seas and so its heat budget components are generally the highest, followed by the Mediterranean, and Black Seas. In Table 8.19 we compare the annual evaporation from each sea. The energy balance method gives higher values than the bulk aerodynamic method, and again the evaporation rate is highest for the Red Sea which has the highest solar heating of the three seas.

Table 8.18 *Annual heat budget components for the Red, Black and Mediterranean Seas, in W m^{-2} (Matsoukas et al. 2005, 2007).*

Component	Red	Medi.	Black
Q_s	238	186	139
Q_l	70	63	50
Q_h	12	11	14
Q_{ea}	130	90	60
Q_{er}	164	117	75

Table 8.19 *Annual evaporation rates (mm year^{-1}) for the Red, Black and Mediterranean Seas. Comparison of energy balance estimates (E_r) from Matsoukas et al. (2005, 2007) and bulk aerodynamic estimates (E_a) from other databases.*

Source	Red	Medi.	Black
Matsoukas *et al.*	1960	1500	830
da Silva *et al.*	1680	1130	–
SOC	1750	1090	–
ERA-40	1280	1060	700

8.10 Bibliography

8.10.1 *Notes*

For the early work on the Earth's radiation budget see Abbot and Fowle; Dines; Houghton; London; and Budyko's work.

For WCRP GEWEX projects see Rossow and Schiffer, Rossow *et al.* (ISCCP), and Stackhouse *et al.* (SRB). For ISCCP cloud optical properties see Rossow *et al.*, and for cloud parameterizations see Slingo; Ebert and Curry; Del Genio *et al.*

For correlations between TOA and surface radiative fluxes to derive surface fluxes directly from the satellite measurements of TOA fluxes see for example; Ramanathan; Schmetz; Weare; Cess *et al.*; Darnell *et al.*; Pinker and Laszlo; and Li and Leighton.

Early work using models and climatological data from satellites, following the early work of London, include the studies of Manabe and Strickler; Vonder Haar and Suomi; Sasamori *et al.*; Raschke *et al.*; Ellis and Vonder Haar; Jacobowitz *et al.*; Gautier *et al.*; Stephens *et al.*

For a review of early estimates of planetary albedo see Hunt et al. For cloud physical thickness see early work of Peng *et al.*

Studies related to ERBE data include those of Barkstrom *et al.*; for ScaRaB see Kandel *et al.*; for CERES see Wielicki *et al.*; for MSG see Sandford *et al.*

References for the reanalyses projects include; Kistler *et al.* for NCEP/NCAR; Uppala *et al.* for ECMWF ERA–40; Schubert *et al.* for GEOS-1.

For the GADS aerosol climatology database see Koepke *et al.* For model results using GADS and actual global relative humidity data see Hatzianastassiou *et al.* For changes to atmospheric general circulation patterns by aerosols based on GCM studies see Kristjansson *et al.* and Lau *et al.*

See Vitale et al. for the ACE–2 aerosol field experiment; for the Global Fire Atlas see Arino and Melinotte. For other field studies on aerosols see Kaufmann *et al.*; Ramanathan *et al.*; Lelieveld *et al.* Also, see the Indian Ocean Experiment, INDOEX.

For the GEBA database see Gilgen and Ohmura, and for the BSRN database see Ohmura *et al.*

For trends in radiation fluxes see: Gilgen *et al.*; Stanhill and Cohen; Liepert; Allan and Slingo; Wild *et al.*; Pinker *et al.*, Hatzianastassiou *et al.*; Wielicki *et al.*; Fotiadi *et al.*

For cloud overlap schemes see Chen *et al.*; and Zhou and Cess.

Climatic consequences of the greenhouse effect on Mediterranean outflow and NAO are discussed in Johnson.

For the estimation of turbulent–exchange coefficients required for the aerodynamic approach for the computation of sensible and latent heat flux see Brutsaert; Miller *et al.*; Gilman and Garrett; Angelucci *et al.* For the estimation of the sea heat content see Krahmann *et al.*

For oceanic databases see Levitus. For the ICOADS database see Woodruff *et al.* and for the SOC air–sea flux climatology see Josey. Aerodynamic latent heat flux data are available from ERA–40, SOC and NOAA/NESDIS (da Silva *et al.*) databases. Sensible heat flux data are available from ERA–40.

For the estimation of lake evaporation from standard meteorological measurements using the Penman method see Vardavas and Fountoulakis.

8.10.2 *References and further reading*

Abbot, C. G. and Fowle Jr, F. E. (1908). Income and Outgo of Heat from the Earth, and the Dependence of its Temperature Thereon. *Annals of the Astrophysical Observatory (Smithsonian Institution, Washington DC)*, **2**, 159–176.

Allan, R. P. and Slingo, A. (2002). Can current climate model forcings explain the spatial and temporal signatures of decadal OLR variations? *Geophys. Res. Lett.*, **29**, 1–4.

Angelucci, M. G., Pinardi, N. and Castellari, S. (1998). Air-sea fluxes from operational analyses fields: Intercomparison between ECMWF and NCEP analyses over the Mediterranean area. *Phys. Chem. Earth*, **23**, 569—574.

Arino, O. and Molinette, J.-M. (1998). The 1993 Africa Fire Map. *Int. J. Remote Sens.*, **19**, 2019–2023.

Barkstrom, B., Harrison, E., Smith, G., Green, R., Kibler, J., Cess, R. and the ERBE Science Team (1989). Earth Radiation Budget Experiment (ERBE) archival and April 1985 results. *Bull. Am. Meteorol. Soc.*, **70**, 1254–1262.

Brutsaert, W. (1984). *Evaporation into the atmosphere.* Reidel, Dordrecht.

Budyko, M. I. (1956). *Heat balance of the Earth's surface.* U.S. Weather Bureau Memo. PB131692.

Budyko, M. I. (1982). *The Earth's climate: Past and future*, Academic Press, New York.

Cess, R. D., Dutton, E. G., DeLuisi, J. J. and Jiang, F. (1991). Determining surface solar absorption from broadband satellite measurements for clear skies: Comparison with surface measurements. *J. Climate*, **4**, 236–247.

Cess, R. D. and coauthors (1995). Absorption of solar radiation by clouds: Observations versus models. *Science*, **267**, 496–499.

Chen, C.-T., and Roeckner, E. (1996). Validation of the Earth radiation budget as simulated by the Max Planck Institute for Meteorology general circulation model ECHAM4 using satellite observations of the Earth Radiation Budget Experiment. *J. Geophys. Res.*, **101**, 4269–4287.

Chen, T., Zhang, Y. and Rossow, W. B. (2000). Sensitivity of atmospheric radiative heating rate profiles to variations of cloud layer overlap. *J. Climate*, **13**, 2941–2959.

Darnell, W. L., Staylor, W. F., Gupta, S. K. and Denn, F. M. (1988). Estimation of surface insolation using Sun-synchronous satellite data. *J. Climate*, **1**, 820–835.

Darnell, W. L., Staylor, W. F., Gupta, S. K., Ritchey, N. A. and Wilber, A. C. (1992). Seasonal variation of surface radiation budget derived from International Satellite Cloud Climatology Project C1 data. *J. Geophys. Res.*, **97**, 15741–15760.

da Silva, A., Young, A. C. and Levitus, S. (1994). *Atlas of surface marine data 1994. Volume 1: Algorithms and procedures.* NOAA Atlas NESDIS 6, U.S. Department of Commerce, Washington, D.C.

Del Genio, A. D., Yao, M.-S., Kovari, W. and Lo, K.-W. (1996). A prognostic cloud water parameterization for global climate models. *J. Climate*, **9**, 270–304.

Dines, W. H. (1917). The heat balance of the atmosphere. *Q. J. R. Meteorol. Soc.*, **43**, 151–158.

Ebert, E. E. and Curry, J. A. (1992). A parameterization of ice cloud optical properties for climate models. *J. Geophys. Res.*, **97**, 3831–3836.

Ellis, J. S. and Vonder Haar, T. H. (1976). Zonal average earth radiation budget measurements from satellites for climate studies. *Atmos. Sci. Rep.*, **240**, Colorado State University, Boulder.

Fotiadi, A., Hatzianastassiou, N., Matsoukas, C., Pavlakis, K. G., Drakakis, E.,

Hatzidimitriou, D. and Vardavas, I. M. (2005). Analysis of the decrease in the tropical mean outgoing shortwave radiation at the top of atmosphere for the period 1984–2000. *Atmos. Chem. Phys.*, **5**, 1721–1730.

Fowler, L. D. and Randall, D. A. (1996). Liquid and ice cloud microphysics in the CSU general circulation model. Part II: Impact of cloudiness, the Earth's radiation budget, and the general circulation of the atmosphere. *J. Climate*, **9**, 530–560.

Garratt, J. R., Prata, A. J., Rotstayn, L. D., McAvaney, B. J. and Cusack, S. (1998). The surface radiation budget over oceans and continents. *J. Climate*, **11**, 1951–1968.

Gautier, C., Diak, G. and Masse, S. (1980). A simple physical model to estimate incident solar radiation at the surface from GOES satellite data. *J. Appl. Meteorol.*, **19**, 1005–1012.

Gilgen, H., Wild, M. and Ohmura, A. (1998). Means and trends of shortwave irradiance at the surface estimated from Global Energy Balance Archive. *J. Climate*, **11**, 2042–2061.

Gilgen, H. and Ohmura, A. (1999). The Global Energy Balance Archive. *Bull. Am. Met. Soc.*, **80**, 831–850.

Gilman, C. and Garrett, C. (1994). Heat flux parameterizations for the Mediterranean Sea: The role of atmospheric aerosols and constraints from the water budget. *J. Geophys. Res.*, **99**, 5119–5134.

Gupta, S. K., Ritchey, N. A. Wilber, A. C. and Whitlock, C. H. (1999). A climatology of surface radiation budget derived from satellite data. *J. Climate*, **12**, 2691–2710.

Hartmann, D. L. (1994). *Global physical climatology.* Academic Press, New York.

Hatzianastassiou, N., Katsoulis, B. and Vardavas, I. M. (2004a). Global distribution of aerosol direct radiative forcing in the ultraviolet and visible arising under clear skies. *Tellus*, **56B**, 51–71.

Hatzianastassiou, N., Fotiadi, A., Matsoukas, C., Drakakis, E., Pavlakis, K. G., Hatzidimitriou, D. and Vardavas, I. M. (2004b). Long–term global distribution of Earth's shortwave radiation budget at the top of atmosphere. *Atmos. Chem. Phys.*, **4**, 1217–1235.

Hatzianastassiou, N., Matsoukas C., Fotiadi, A., Pavlakis, K., Drakakis, E., Hatzidimitriou, D. and Vardavas I. M. (2005). Global distribution of Earth's surface shortwave radiation budget. *Atmos. Chem. Phys.*, **5**, 2847–2867.

Hatzianastassiou, N., Matsoukas C., Fotiadi, A., Drakakis, E., Stackhouse Jr, P. W., Koepke, P., Pavlakis, K., Hatzidimitriou, D. and Vardavas I. M. (2006).

Modelling the direct effect of aerosols in the near-infrared on a planetary scale. *Atmos. Chem. Phys. Disc.*, **6**, 9151–9185.

Hatzianastassiou, N., Matsoukas C., Drakakis, E., Stackhouse Jr, P. W., Koepke, P., Fotiadi, A., Pavlakis, K., and Vardavas I. M. (2007). The direct effect of aerosols on solar radiation based on satellite observations, reanalysis datasets and spectral aerosol optical properties from the Global Aerosol Data Set (GADS). *Atmos. Chem. Phys. Disc.*, **7**, 1–31.

Houghton, D. D. (1985). *Handbook of applied meteorology.* John Wiley & Sons, New York.

Houghton, H. G. (1954). On the annual mean balance of the Northern Hemisphere. *J. Meteor.*, **11**, 1–9.

Hunt, G. E., Kandel, R. and Mecherikunnel, A. T. (1986). A history of pre-satellite investigations of the Earth's radiation budget. *Rev. Geophys.*, **24**, 351–356.

IPCC, 2001: *Climate change 2001: The scientific basis. Contribution of working group I to the Third Assessment Report of the Intergovernmental Panel on Climate Change.* Houghton, J. T., Ding, Y., Griggs, D. J., Noguer, M., van der Linden, P. J., Dai, X., Maskell, K. and Johnson, C. A. (ed.), Cambridge University Press, Cambridge.

Jacobowitz, H., Smith, N. L., Howell, H. B., Nagle, F. W. and Hickey, J. R. (1979). The first 18 months of planetary radiation budget measurements from the Nimbus 6 ERB experiment. *J. Atmos. Sci.*, **36**, 502–507.

Johnson, R. G. (1997). Ice age initiation by an ocean-atmospheric circulation change in the Labrador Sea. *Earth Plan.*, **148**, 367–379.

Josey, S. A., Kent, E. C. and Taylor, P. K. (1999). New insights into the ocean heat budget closure problem from analysis of the SOC air-sea flux climatology. *J. Climate*, **12**, 2856–2880.

Kandel, R., Viollier, M., Raberanto, P., and coauthors (1998). The ScaRab earth radiation budget dataset. *Bull. Am. Met. Soc.*, **79**, 765–783.

Kaufman, Y. J., Tanré, D. and Boucher, O. (2002). A satellite view of aerosols in the climate system. *Nature*, **419**, 215–223.

Kiehl, J. T. and Trenberth, K. E. (1997). Earth's annual global mean energy budget. *Bull. Am. Met. Soc.*, **78**, 197–208.

Kistler, R. and coauthors (2001). The NCEP-NCAR 50-year reanalysis: Monthly means CD-ROM and documentation. *Bull. Am. Met. Soc.*, **82**, 247–268.

Koepke, P., Hess, M., Schult, I. and Shettle, E. P. (1997). *Global aerosol data set.* Report No 243, Max-Planck Institut für Meteorologie, Hamburg.

Krahmann, G. F., Schott, F. and Send, U. (2000). Seasonal heat content changes in the western Mediterranean Sea as a means for evaluating surface heat flux formulations. *J. Geophys. Res.*, **105**, 16941–16950.

Kristjansson, J. E. Iversen, T., Kirkevåg, A., Seland, Ø, and Debernard, J. (2005). Response of the climate system to aerosol direct and indirect forcing: Role of cloud feedbacks. *J. Geophys. Res.*, **110**, D24206, doi: 10.1029/2005JD006299.

Lau, K. M., Kim, M. K. and Kim, K. M. (2006). Asian summer monsoon anomalies induced by aerosol direct forcing. The role of the Tibetan plateau. *Climate Dyn.*, **26**, 855–864.

Lelieveld, L. and coauthors (2002). Global air pollution crossroads over the Mediterranean. *Science*, **298**, 794–799.

Levitus, S. (2002). *World ocean database 2001.* Volumes 1-4, NOAA Atlas NESDIS, U.S. Government Printing Office, Washington D.C.

Liepert, B. G. (2002). Observed reductions of surface solar radiation at sites in the United States and worldwide from 1961 to 1990. *Geophys. Res. Lett.*, **29**(12), doi:10.1029/2002GL014910.

Li, Z. and Leighton, H. G. (1993). Global climatologies of solar radiation budgets at the surface and in the atmosphere from 5 years of ERBE data. *J. Geophys. Res.*, **98**, 4919–4930.

Li, Z., Barker, H. W. and Moreau, L. (1995). The variable effect of clouds on atmospheric absorption of solar radiation. *Nature*, **376**, 486–490.

Liou, K. N. (2002). *An introduction to atmospheric radiation.* International Geophysics Series, Vol. 84, Academic Press, New York.

London, J. (1957) *A study of the atmospheric heat balance.* Final Report, Contract AF 19(122)-165, Coll. of Eng., N. Y. Univ., New York.

Manabe, S. and Strickler, R. F. (1964). Thermal equilibrium of the atmosphere with a convective adjustment. *J. Atmos. Sci.*, **21**, 361–385.

Matsoukas, C., Banks, A. C., Hatzianastassiou, N., Pavlakis, K. G., Hatzidimitriou, D. , Drakakis, E., Stackhouse, P. W. and Vardavas, I. M. (2005). The seasonal heat budget of the Mediterranean Sea. *J. Geophys. Res. - Oceans*, **110**, C12008, 1–15.

Matsoukas, C., Banks, A. C., Hatzianastassiou, N., Pavlakis, K. G. , Drakakis, E., Stackhouse Jr, P. W. and Vardavas, I. M. (2007). The seasonal heat budget of the Red and Black Seas. *J. Geophys. Res. - Oceans*, in press.

Miller, M. J., Beljaars, A. C. M. and Palmer, T. N. (1992). The sensitivity of the ECMWF model to the parameterization of evaporation from the tropical oceans. *J. Climate*, **5**, 418–434.

Ohmura, A. and coauthors (1998). Baseline Surface Radiation Network: New precision radiometry for climate research. *Bull. Am. Met. Soc.*, **79**, 2115–2136.

Pavlakis, K. G., Hatzidimitriou, D., Matsoukas, C., Drakakis, E., Hatzianastassiou, N. and Vardavas, I. M. (2004). Ten-year global distribution of downwelling longwave radiation. *Atmos. Chem. Phys.*, **4**, 127–142.

Peng, L. I., Chou, M.-D. and Arking, A. (1982). Climate studies with a multilayer energy balance model I. Model description and sensitivity to the solar constant. *J. Atmos. Sci.*, **39**, 2639–2656.

Pinker, R. T. and Laszlo, I. (1992). Modelling surface solar irradiance for satellite applications on a global scale. *J. Appl. Meteorol.*, **31**, 194–211.

Pinker R. T. , Zhang B., and Dutton, E. G. (2005). Do satellites detect trends in surface solar radiation? *Science*, **308**, 850–854.

Ramanathan, V. (1987). The role of Earth Radiation Budget studies in climate and general circulation research. *J. Geophys. Res.*, **92**, 4075–4094.

Ramanathan, V., Crutzen, P. J., Kiehl, J. T. and Rosenfeld, D. (2001). Aerosols, climate, and the hydrological cycle. *Science*, **294**, 2119–2124.

Ramanathan, V. and coauthors (2001). The Indian Ocean Experiment. An integrated assessment of the climate forcing and effects of the great Indo-Asian haze. *J. Geophys. Res.*, **106**, 28371–28399.

Raschke, E., Gratzki, A. and Rieland, M. (1987). Estimates of global radiation at the ground from the deduced data sets of the International Cloud Climatology Project. *J. Climatol.*, **7**, 205–213.

Rossow, W. B. and Zhang, Y.-C. (1995). Calculation of surface and top of atmosphere radiative fluxes from physical quantities based on ISCCP data sets 2. Validation and first results. *J. Geophys. Res.*, **100**, 1167–1197.

Rossow, W. B., Walker, A. W., Beuschel, D. E. and Roiter, M. D. (1996). *International Satellite Cloud Climatology Project (ISCCP). Documentation of new cloud datasets.* Wold Meteorol. Org., Geneva.

Rossow, W. B. and Schiffer, R. A. (1999). Advances in understanding clouds from ISCCP. *Bull. Am. Met. Soc.*, **80**, 2261–2287.

Sandford, M. C. W. and coauthors (2003). The geostationary Earth radiation budget (GERB) instrument on EUMETSATs MSG satellite. *Acta. Astron.*, **53**, 909–915.

Sasamori, T., London, J. and Hoyt, D. V. (1972). Radiation budget of the Southern Hemisphere. *Meteor. Monogr.*, **35**, 9–23.

Schubert, S., Park, C.K., Wu, C., Higgins, W., Kondratyeva, Y., Molod, A.,

Takacs, L., Seablom, M. and Rood, R. (1995). A multi-year assimilation with the GEOS-1 system: Overview and results. *The NASA technical report series on global modeling and data assimilation.* **6**, M. Suarez, ed., NASA TM 104606.

Schmetz, J. (1989). Towards a surface radiation climatology: Retrieval of downward irradiance from satellites. *Atmos. Res.*, **23**, 287–321.

Slingo, A. (1989). A GCM parameterization for the shortwave radiative properties of water clouds. *J. Atmos. Sci.*, **46**, 1419–1427.

Stackhouse Jr, P. W., Cox, S. T., Gupta, S. K., Dipasquale, R. C. and Brown, D. R. (1999). *The NCRP/GEWEX Surface Radiation Budget Project Release 2: First results at 1 degree resolution.* Paper presented at 10th Conference on Atmospheric Radiation: A Symposium with Tributes to the Works of Verner Suomi. Am. Met. Soc., Madison, Wisc.

Stanhill, G. and Cohen, S. (2001). Global dimming: A review of the evidence for a widespread and significant reduction in global radiation with discussion of its probable causes and possible agricultural consequences. *Agricult. Forest. Meteorol.*, **107**, 255–278.

Stephens, G. L., Campbell, G. G. and Vonder Haar, T. H. (1981). Earth radiation budgets. *J. Geophys. Res.*, **86**, 9739–9760.

Uppala, S. and coauthors (2005). The ERA-40 re-analysis. *Quart. J. Roy. Meteor. Soc.*, **131**, 2961–3012.

Vardavas, I. M. and Koutoulaki, K. (1995). A model for the solar radiation budget of the Northern Hemisphere: Comparison with Earth Radiation Budget Experiment data. *J. Geophys. Res.*, **100**, 7303–7314.

Vardavas, I. M. and Fountoulakis, A. (1996). Estimation of lake evaporation from standard meteorological measurements: Application to four Australian lakes in different climatic regions. *Ecol. Model.*, **84**, 139–150.

Vitale, V., Tomasi, C., Lupi, A., Cacciari, A. and Marani, S. (2000). Retrieval of columnar aerosol size distributions and radiative-forcing evaluations from sunphotometric measurements taken during the CLEARCOLUMN (ACE 2) experiment. *Atmos. Env.*, **34**, 5095–5105.

Vonder Haar, T. H. and Suomi, V. E. (1971). Measurements of the Earth's radiation budget from satellites during a five-year period. *J. Atmos. Sci.*, **28**, 305–314.

Weare, B. C. (1989). Relationships between net radiation at the surface and the top of atmosphere derived from a general circulation model. *J. Climate*, **2**, 193–197.

Whitlock, C. H. and coauthors (1995). First global WCRP shortwave surface radiation budget dataset. *Bull. Am. Met. Soc.*, **76**, 905–922.

Wielicki, B. A., Barkstrom, B. R., Harrison, E. F., Lee III, R. B., Smith, G. L. and Cooper, J. E. (1996). Clouds and the Earth's Radiant Energy System (CERES): An Earth Observing System Experiment. *Bull. Am. Met. Soc.*, **77**, 853–868.

Wielicki, B. A. and coauthors (2002). Evidence for large decadal variability in the tropical mean radiative energy budget. *Science*, **295**, 841–844.

Wild, M., Ohmura A., Gilgen H., Roeckner E., Giorgetta M. and Morcrette J.-J (1998). The disposition of radiative energy in the global climate system: GCM-calculated versus observational estimates. *Clim. Dyn.*, **14**, 853–869.

Wild, M., Ohmura, A., Gilgen, H., Morcrette, J.-J. and Slingo, A. (2001). Evaluation of downwelling longwave radiation in general circulation models. *J. Climate*, **14**, 3227–3239.

Wild, M., Gilgen, H., Roesch, A. and coauthors (2005). From dimming to brightening: decadal changes in solar radiation at Earth's surface. *Science*, **308**, 847–850.

Woodruff, S. D., Diaz, H. F. Elms, J. D. and Worley, S. J. (1998). COADS Release 2 data and metadata enhancements for improvements of marine surface flux fields. *Phys. Chem. Earth*, **23**, 517–527.

Yang, S.-K., Hou, Y.-T., Miller, A. J. and Campana, K. A. (1999). Evaluation of the earth radiation budget in NCEP-NCAR Reanalysis with ERBE. *J. Climate*, **12**, 477–493.

Yu, H. and coauthors (2006). A review of measurement-based assessment of aerosol direct radiative effect and forcing. *Atmos. Chem. Phys.*, **6**, 613–666.

Zhou, Y. P. and Cess, R. D. (2000). Validation of longwave atmospheric radiation models using Atmospheric Radiation Measurement data. *J. Geophys. Res.*, **105**, 29703–29716.

9

THEORY OF RADIATION MEASUREMENTS

9.1 Introduction

Radiation measurements are crucial for understanding the climate system and monitoring its behaviour. Increasingly, the most important measurements are made from instruments in space, orbiting the Earth on platforms that offer global coverage of the atmosphere and surface beneath. These *remote sensing* instruments are the subject of the following chapter. Here, we consider the basic design and operating principles of the instruments that produce both remote and *in–situ* climate measurements.

We are interested mainly in devices that determine the intensity of the radiation as a function of its wavelength or wave number (λ^{-1}). The instruments of interest for climate research operate mostly, although not exclusively, in the infra–red part of the spectrum, which is conveniently broken down into two regimes, the *near* and the *thermal* infra–red. In the climate system, virtually all of the solar energy flux is contained within the range from about 0.2 to about 5.0 μm, which includes the *visible* and the *ultraviolet* range as well as the near–infra–red, while the planetary or terrestrial emission occurs between 5.0 and 100 μm, which is therefore defined to be the thermal infra–red region.

Most common infra–red optical systems can be categorized as either radiometers or spectrometers, although the terms are used fairly loosely and the following definitions are not always adopted in the literature. In most contemporary usage, *radiometers* are characterized by absolute radiometric performance and relatively low spectral resolving power. They are calibrated before and during use, so that they can measure flux or radiance directly. They do not scan the spectrum, but employ a fixed wavelength band or bands. In radiometers, typically the ratio of bandwidth to wavelength is not less than a few per cent. *Spectrometers* differ from radiometers in that they scan in wavelength, have high spectral resolution and are not necessarily absolute radiometric instruments. They may operate either in absorption or emission and measure the position and intensity of spectral lines in absorption on a scale of transmitted or absorbed energy versus wavelength. Calibration in wavelength is usually accomplished with reference scans of a reference gas or material, such as polyethylene, which has a number of distinctive absorption features at well-defined wavelengths. A spectrometer that is absolutely calibrated in energy units is sometimes called a *spectroradiometer*. The term is also used for radiometers with unusually high spectral resolution, and perhaps with some limited spectral scanning capability.

The fundamental principles by which instruments operate are best appreciated by considering simple, idealized systems first, then the practical factors that must be taken into account in reality, and finally (in Chapter 10) some examples of real instruments and their applications. Infra–red systems making geophysical measurements often have signal–to–noise ratios that are marginal for their particular application. This arises from the low flux of infra–red photons from relatively cool targets, the effects of background fluxes, especially from warmer or more extended objects internal or external to the system, as well as fundamental limitations including statistical effects in the detector and in the photon flux itself. It is necessary, therefore, to characterize the performance of a system in terms of measurable or calculable quantities in order that a device can be matched to a desired application. Failure to do so would risk ending up with an experiment in which the signal was drowned in noise, or a waste of resources through the use of unnecessarily sophisticated detectors, perhaps employing cooling where cheaper room–temperature devices would have done. Instead, we apply the known physics of the radiation field, of materials and of devices such as detectors and multilayer filters, to the design of a device with a specific application in mind.

9.2 Detectors

Detectors are the heart of any measurement system: any device that converts a stream of photons into an electrical signal can be used as an infra–red detector. For example, the nerves in the human skin act as a detector, albeit a relatively insensitive one; a blind person can tell that there is a heat source, such as a fire, in a room, and roughly how hot it is, by holding a hand towards it, with no other source of information. There are many different types of solid-state detector, operating on a variety of principles, which have been used at one time or another. We will concern ourselves here only with the types that are found in modern infra–red systems. All of these fall into one of two main categories, *thermal detectors* (where the heating effect of the radiation is measured) and *photon detectors* (where the quanta excite electronic transitions in the detector material – usually a semiconductor – directly). Before discussing how they work, we will look at some useful formulae and definitions that apply to all types.

A conceptual detector consists of a detector element of a material that responds to incident radiation by producing a voltage or current. In the scheme shown in Fig. 9.1, the element of area A cm^2 produces a signal voltage V_s volts at the output on exposure to a flux F W cm^{-2}. There will also in general be an unwanted voltage V_n at the output, consisting of random noise. The *voltage responsivity*, R_v (V W^{-1}) of the detector is defined by

$$R_v = \frac{V_s}{FA}, \tag{9.1}$$

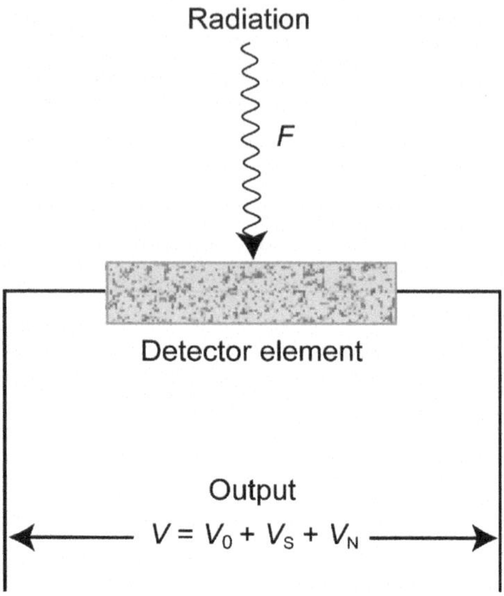

FIG. 9.1. A conceptual infra–red detector. When the element is exposed to a flux of radiation F, a voltage V_s is induced across the leads to the element. In general, a constant offset voltage V_0 and a random noise voltage V_n (which may or may not be functions of F) will also be present.

so that R_v is the number of volts that appear at the output of the detector for every watt of power incident upon it, and the signal–to–noise ratio of the measurement of F is SNR$= V_s/V_n$. There may also be a constant offset voltage, V_0, in which case the total voltage measured is

$$V = V_0 + (V_s + V_n). \tag{9.2}$$

V_0 can be determined by noting the voltage across the detector when the incoming flux is zero, i.e. by calibration of the detector or instrument using a shutter or, on a satellite, a view of cold space.

Most detectors take a finite time to reach a steady output after being exposed to a flux of radiation. The time constant for this process τ, which determines how rapidly measurements can be repeated, is defined by the expression

$$V = V_{ss}\left[1 - \exp(-t/\tau)\right], \tag{9.3}$$

so $t = \tau$ is the time taken to reach 63% (= 1–1/e) of the steady-state output V_{ss} after the detector is first exposed to the radiation stream.

The importance of the time constant, apart from reminding us that detectors cannot respond instantaneously (especially thermal detectors, which need time to warm up to their equilibrium temperature), is because the incoming flux is usually *modulated* or *chopped*, making V an a.c., not a d.c., signal. If τ is significantly greater than the reciprocal of the chopping frequency f, then the reponsivity of the detector will be less than its value for $\tau = 0$ because the output will not have time to respond to the fluctuating input. We discuss below why chopping is a good idea, even though it can cause problems for 'slow' detectors. Assuming that the rate of chopping is not so fast that the detector cannot follow the fluctuations, the responsivity will normally be independent of the chopping frequency over a wide range, and is defined in terms of the peak (or r.m.s) value of the output voltage.

Generally, the signal that emerges from a detector is not the main concern – what matters is the ratio of the information–carrying signal to the meaningless fluctuations or *noise* that always accompany it. The sources of noise are different for different types of detector, and will be discussed in the relevant context below. For now, the crucial thing to recognize is that, in many practical infra–red systems of interest, the signal is often not much greater than the noise, even though great trouble may have been taken to eliminate as many sources of noise as possible. Noise originating in the detector, which gets amplified along with the signal, is the most fundamental of all. To characterize this, we use a parameter called the *noise equivalent power* or NEP, which is simply the amount of power that has to fall on the detector before the signal output voltage is exactly equal to the noise voltage V_n. Then

$$\mathrm{NEP} = \frac{V_n}{R_v\sqrt{\Delta f}}, \tag{9.4}$$

where Δf(Hz) in eqn (9.4) is the *bandwidth* in which the signal, and the noise, are measured. The bandwidth is usually defined inside the amplifier, which boosts the small signal from the detector. Intuitively, it makes sense to have a small bandwidth because it excludes some of the noise (which is often present at all frequencies) while the signal (at a single frequency f at the centre of Δf) need hardly be attenuated at all. In fact, it is shown later in this chapter that Δf is determined by the *integration time* t_o, such that $\Delta f = 1/2t_o$.

The physical meaning of t_o is simply the time the amplifier spends summing the signal (and averaging out the random component of the noise) from the detector output. For this reason it is sometimes called the *dwell time*. It is the amount of time a radiometer spends dwelling on one scene before it is available to switch to another, or in the case of a scanning spectrometer, before it steps to the next wavelength. In practice, t_o will be set by an R–C circuit in the amplifier. The larger the value of the time constant of the amplifier, the longer it sums the signal voltage and the smaller is Δf. Now, the reason why Δf appears to the half power in eqn (9.4) can be appreciated; the noise, if random, is reduced by the square root of the number n of independent samples taken, and clearly n is proportional to t_o and inversely proportional to Δf.

Generally, NEP is used to describe the performance of a given, individual detector. The user buying a detector for a specific application will decide what NEP is needed, and expects the value to be included in the specification of the device. Nominally identical detectors often have quite different values of NEP, since small flaws introduced in the manufacturing process introduce additional noise in ways that are often obscure to manufacturer and user alike. The *detectivity* (W^{-1} $Hz^{1/2}$), defined by

$$D = 1/\mathrm{NEP}, \tag{9.5}$$

is preferred by some manufacturers because better detectors have higher values of D. However, a more useful and better-known parameter is the *specific detectivity* D^* (W^{-1} cm $Hz^{1/2}$) defined by

$$D^* = \sqrt{A}/\mathrm{NEP}. \tag{9.6}$$

D^* is approximately independent of detector area, since most (but not all) sources of V_n are proportional to $A^{1/2}$, while V_s is proportional to A. Hence D^* is used to compare different *types* of detectors.

9.3 Thermal detectors

Thermal detectors work by measuring the heating effect of a beam of radiation. The three main types available commercially are thermocouples, normally with multiple junctions (hence thermopiles), thermistor bolometers, and pyroelectric bolometers.

Thermocouples use the *thermoelectric effect* in the junction between dissimilar metals, traditionally copper and constantan, which have a very large value of the thermoelectric coefficient that determines the responsivity. Thermopiles have had something of a renaissance recently because modern micromachining techniques allow large numbers of small junctions in a given detector size, leading to a fairly rapid response and high D^* in a robust and reliable device that does not require cooling. A recent design (Foote *et al.* 1998) uses 2–μm widths and 2–μm spacings between conductors to place about 50 thermocouples in a 100 μm wide detector element. This has a resistance of ~250 kΩ and achieves a responsivity of 1.5 kV/W, response time of 1.7 ms, and detectivity D^* of 2.4×10^8 W^{-1} cm $Hz^{1/2}$.

A *thermistor* is a resistor with a high temperature coefficient of resistance, α. The word *bolometer* strictly refers to anything that measures radiation: here, it does this by measuring the change in the resistance of the detector element due to the heating effect of the radiation. Platinum has a high coefficient of resistance ($a \sim 0.003$ Ω K^{-1}) and used to be the commonest thermistor material, but in recent years this has been deposed by a sintered semiconductor mixture of manganese, nickel and cobalt oxides that achieves $a \sim 0.056$ Ω K^{-1}. A steady bias voltage must be applied to maintain a current through the detector element, which is normally in series with a load resistance, as shown in Fig. 9.2. The latter can be a second detector element, identical to the first, but shielded

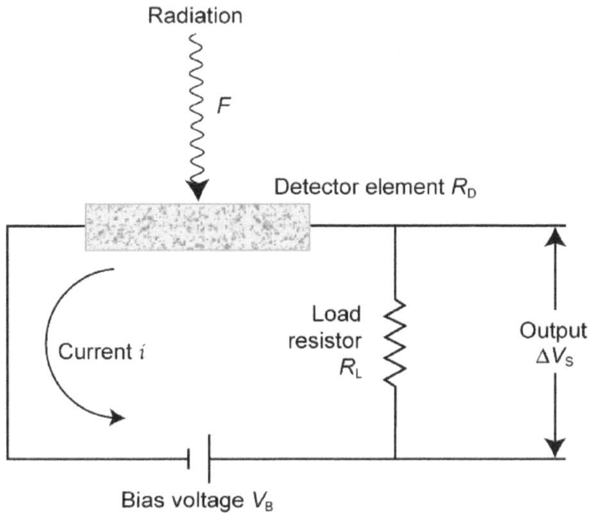

FIG. 9.2. A bolometer detector consists of an element with a high temperature coefficient of resistance, a fixed load resistor, and a bias voltage to provide a current through both. Changes in the current due to the heating effect of the radiation are detected as a change in the voltage across the load resistor.

from the radiation, an arrangement that provides additional stability should the background temperature vary.

The change in the signal voltage, ΔV_s across the load resistor produced by a change in detector resistance, ΔR_D, induced by the incoming radiation from the target is given by

$$\Delta V_s = \frac{V_B R_L}{R_D + R_L} - \frac{V_B R_L}{R_D + \Delta R_D + R_L} \approx V_B R_L \frac{\alpha \Delta T}{R_D}, \tag{9.7}$$

assuming a log–linear dependence, with coefficient α, between the temperature of the detector element, T, and its resistance R_D, and $R_D \gg R_L$. In general, V_S is accompanied by V_N and V_0, the noise and offset components of the output voltage as before. The advantages of thermistor bolometer detectors include sensitivity to a wide range of wavelengths, and reasonable performance ($D^* \approx$ few $\times 10^8$ W^{-1} cm Hz$^{1/2}$) without cooling. The main drawback is the need to provide a high (noise–free) bias voltage.

Superconducting versions have been developed that use a detector element held at its normal–to–superconducting transition temperature. Enormous coefficients of resistivity are thereby obtained, but maintaining a very precise operating temperature without noise-inducing fluctuations presents practical difficulties, especially for deployment on satellites.

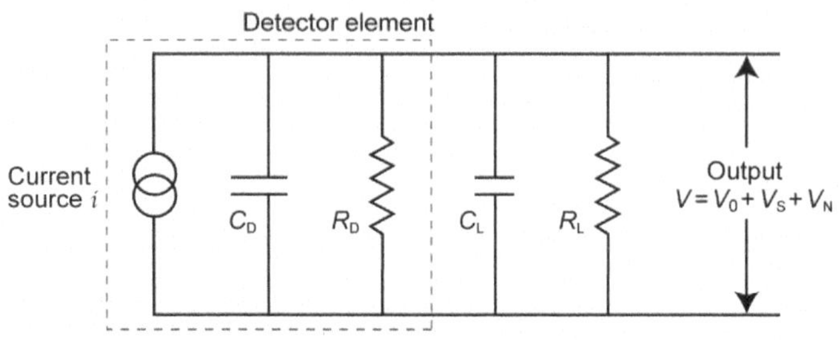

FIG. 9.3. Pyroelectric detector equivalent circuit. The components inside the box represent the detector element; R_L is the load resistor connected across the element, and C_L the stray capacitance of the external circuit. The output voltages V are r.m.s. values of an a.c. signal.

Pyroelectric detectors offer good performance requiring neither bias voltage nor cooling by exploiting the properties of ferroelectric materials, which exhibit a permanent electric dipole moment. Pyroelectric crystals are a subclass of these characterized by a high coefficient of the temperature dependence of polarizability, p. A common example, used for instance as motion detectors in domestic burglar alarms, is lithium tantalate $LiTO_3$, but the material most used in practice is deuterated triglycine sulphate, known as DTGS, which has both a large value of p (about 10^{-8} C cm^{-2} K^{-1}) and a small dielectric constant, which corresponds to small internal losses.

In pyroelectric detectors used in the a.c. mode the chopped incoming radiation induces fluctuations in the polarization of the crystal that appears as a varying charge separation across the detector element. The detector behaves like a capacitor with a charge that varies at the modulation frequency, and can be modelled as a capacitance of a few pF in parallel with a current source (Fig. 9.3). Externally, the element is connected across a large load resistor R_L through which the induced charge can flow, and the signal is then the voltage across this load. A practical model of the detector will include the internal resistance of the element R_o and the stray capacitance C_L of the external circuit. If the modulation (chopping) frequency is $\omega = 2\pi f$ so that the incident power is $P(t) = P_o \exp(i\omega t)$, the corresponding detector temperature T as a function of time t will be the solution of

$$C^* \frac{dT(t)}{dt} + G^* T = P(t), \tag{9.8}$$

which is

$$T(t) = \frac{P_o \exp(i\omega t)}{G^* + i\omega C^*}, \tag{9.9}$$

where G^* and C^* are the thermal conductance and capacitance, respectively, of the element. The amplitude of the temperature cycle is

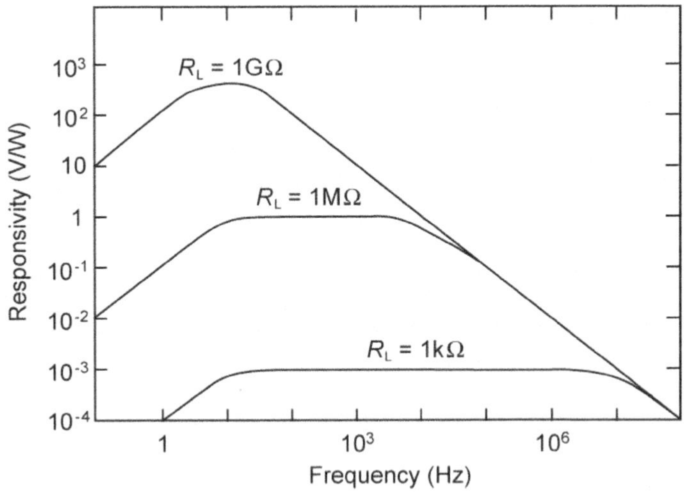

FIG. 9.4. The response of a pyroelectric detector as a function of frequency for several values of the load resistance R_{L}. The responsivity can be improved by increasing R_{L}, but at the expense of frequency response. The device will normally be operated in the plateau region, so that output is less sensitive to frequency fluctuations. (After Boyd 1983).

$$\left|\frac{\mathrm{d}T}{\mathrm{d}t}\right| = \frac{\omega P_{\mathrm{o}}}{G^* \sqrt{1+(\omega\tau)^2}}, \tag{9.10}$$

which gives rise to an output current

$$i = pA\frac{\mathrm{d}T}{\mathrm{d}t}, \tag{9.11}$$

by the definition of p, so the current response (amps W^{-1}) is

$$R_i = \frac{|i|}{P_o} = \frac{\omega pA}{G^* \sqrt{1+(\omega\tau)^2}}, \tag{9.12}$$

where τ is the thermal time constant C^*/G^* (s).

The load resistor R_{L} is in parallel with the internal resistance of the detector R_0 so the output voltage depends on the combination $R = R_0R_{\mathrm{L}}/(R_0 + R_{\mathrm{L}})$. Similarly the capacitance is $C = C_0 + C_{\mathrm{L}}$, if there is any external capacitance C_{L}, stray or otherwise. The electrical time constant is RC, so the (voltage) responsivity is

$$|R_{\mathrm{v}}| = \frac{|R_{\mathrm{i}}|R}{\sqrt{1+(\omega RC)^2}}. \tag{9.13}$$

The responsivity is plotted as a function of the load resistance R_{L} in Fig. 9.4. It is desirable to operate at frequencies in the 'plateau' regime where $\omega RC \ll 1$

and $\omega\tau \gg 1$, because then the voltage responsivity is independent of chopping frequency

$$|R_{\rm v}| = \frac{pAR}{G^*\tau}. \tag{9.14}$$

From Fig. 9.4 it can be seen that a device operated at around 100 Hz with a load resistance of 10 MΩ will operate on the plateau and maximize the responsivity. Typical values of the thermal and electronic time constants are $\tau = 16$ ms and $RC = 100\ \mu$s ($R = 10^7$ ohms, $C = 10$ pF).

The main source of noise in pyroelectric bolometers, as in most thermal detectors, is *Johnson noise* in the equivalent resistance. This is due to thermally excited random motions of the free electrons in a conductor, produced by collisions with the atoms of the lattice. Johnson (1928) showed empirically that the noise power, expressed as the square of the noise voltage $V_{\rm n}$, in a resistance R is proportional to R and to the temperature, T. If the noise is independent of frequency, the power also depends directly on the bandwidth Δf(Hz). A statistical analysis, due originally to Nyquist (1928), shows that the constant of proportionality is 4 times Boltzmann's constant k. Hence *Nyquist's formula*

$$V_{\rm n} = \sqrt{4kRT\Delta f}. \tag{9.15}$$

We can now write down an expression for the NEP (W Hz$^{-1/2}$) of this, the best-performing and most commonly used room–temperature thermal detector, in terms of its physical properties and the operating conditions

$$\text{NEP} = \frac{\sqrt{2kT/t_0R}}{pA/G^*\tau}. \tag{9.16}$$

G^* and τ are minimized by making the flake very thin (a few $\times 10\ \mu$m). Typically then $pA/G^*\tau \approx 1\ \mu A\ \text{W}^{-1}$; R is maximized (10^8 ohm is about the largest practical value), $\Delta f = 1$ Hz by definition of NEP (signal power to match noise per unit bandwidth) and NEP is typically $\approx 10^{-8}$ to 10^{-9} W Hz$^{-1/2}$.

The *signal–to–noise ratio* (SNR) is often used to describe detector performance in a situation where it is receiving r.m.s. power P

$$\text{SNR} = \frac{P}{\text{NEP}\sqrt{\Delta f}} = \frac{P\sqrt{2t_{\rm o}}}{\text{NEP}}. \tag{9.17}$$

As we have already noted in our definitions of NEP and SNR, the dependence on the bandwidth Δf arises because the signal is at a single frequency (the chopping frequency $f_{\rm c}$) while the noise is present at all frequencies (Fig. 9.5). Many common noise sources, including Johnson noise, are 'white', that is, they have the same mean amplitude at all frequencies. Electronic filtering thus enhances SNR, and we can calculate its expected value for a given set of measurement parameters. In particular, it is useful to replace the obscure quantity Δf with

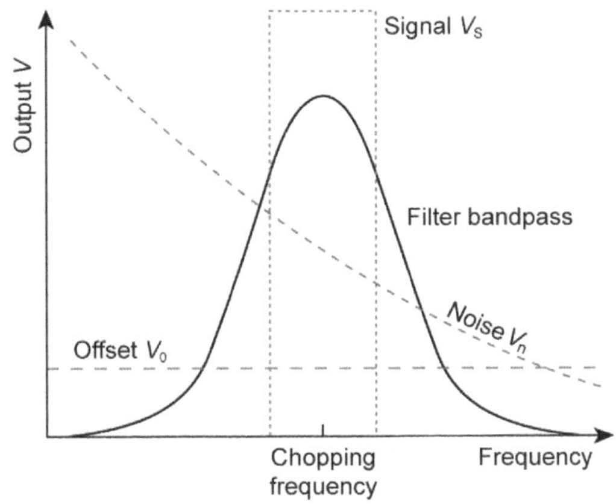

FIG. 9.5. Measurements made in a limited bandwidth centred on the signal frequency have better signal–to–noise ratio than unfiltered measurements, since white noise is present at all frequencies.

the convenient t_o, so that the signal–to–noise ratio is expressed in quantities that are likely to be well known. To derive the relevant expression, the bandwidth is defined in terms of the power of the signal

$$\Delta f = \int_{\infty}^{0} \left| \frac{R_\text{v}(f)}{R_\text{v}(f_\text{c})} \right|^2 \mathrm{d}f, \tag{9.18}$$

where $R_\text{v}(f)$ is the response of the system at frequency f, and is given by

$$R_\text{v}(f) = \frac{R(f_\text{c})}{t_\text{o}} \int_{t_\text{o}/2}^{-t_\text{o}/2} \exp\left(-\text{i}2\pi f t\right)\mathrm{d}t = \frac{R(f_\text{c}) \sin \pi f t_\text{o}}{\pi f t_\text{o}}. \tag{9.19}$$

Thus,

$$\Delta f = \int_{\infty}^{0} \left| \frac{R_\text{v}(\pi f t)}{\pi f t} \right|^2 \mathrm{d}f = \frac{1}{2t_\text{o}}, \tag{9.20}$$

where t_o is the integration time (s) of the measurement. Figure 9.6 illustrates graphically the effect of integrating the signal (i.e. reducing the bandwidth) in observations of the signal returned from a lidar (optical radar) system viewing the atmosphere vertically upwards from the ground. This is an example of an *active system*, that is, one that includes its own source. A real lidar system, used to observe the backscatter from thin cloud layers in the atmosphere, uses two lasers, one to generate the transmitted beam and the other as a local oscillator in a heterodyne detection system. The random noise is reduced by the square root of the number of pulses averaged, and the signal is clearly seen emerging from the noise as a result. In a single measurement, the information is nearly lost in the noise; after 100 pulses (an improvement of a factor 10 in SNR) it stands out clearly.

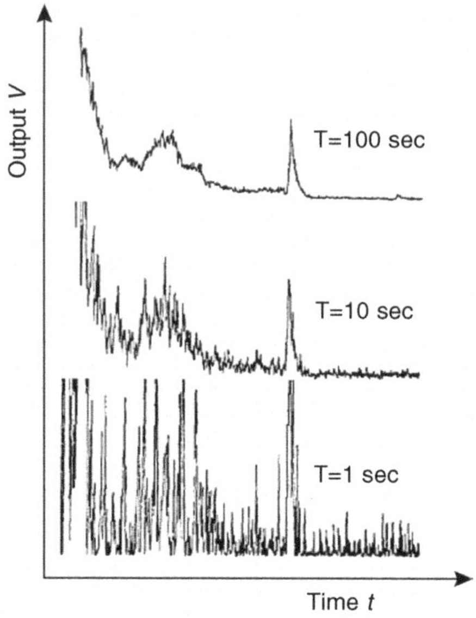

FIG. 9.6. Cloud backscatter measurements with an upward–viewing infra–red lidar, showing the reduction of noise by integrating signals for various times. The effect on the signal–to–noise ratio of varying the bandwidth can be seen clearly. With a dwell time of 100 s (top), the backscattered signal from layers of low and high cloud can be discerned; in a 1–s measurement, these are largely lost in the noise.

9.4 Photon detectors

There are two main categories of photon detector, photoconductive and photovoltaic. *Photoconductive detectors* are chips of semiconductor that rely for their operation on the excitation by incident photons of current carriers that flow in the external circuit under the influence of an applied bias voltage. *Photovoltaic devices* are usually semiconductor diodes of a design that allows carriers to cross the junction against the potential barrier when they are excited by absorbed photons. When this happens, a voltage appears across the terminals of the device, which can be measured directly and no bias is required.

Photovoltaic devices are generally superior to photoconductive ones for most applications because the bias voltage is an inconvenience, as well as a source of extra noise if it is not perfectly stable. Photoconductors also dissipate more power, obviously; this is a further disadvantage because the devices usually have to be cooled and the dissipation makes extra demands on the cryogen supply or refrigerator used for this. Furthermore, we shall see below that an ideal photovoltaic detector has a lower noise voltage than an ideal photoconductive device, by a factor of $\sqrt{2}$. On the other hand, photovoltaic detectors (often called photodiodes) are in practice more prone to noise sources having their origin in imperfections introduced during manufacture, and, while the state–of–the–art of making them advances still further, photoconductors are still used extensively, especially at the longer infra-red wavelengths.

Both kinds of photodetector offer generally higher performance, in terms of responsivity, NEP, and time constant, in comparison to even the best thermal detectors. However, they will not replace the latter entirely, because thermal detectors have the great advantage that they do not need to be cooled. They also have a nearly uniform performance over a very wide spectral range – essentially the whole of the infra–red in most cases – whereas photon detectors peak sharply at a particular wavelength and cut off altogether at the long wavelength side of this peak. There is also the practical consideration that thermal detectors are cheap and robust, and very high performance is not always essential anyway. As we continue with our analysis of the performance of different detector types in terms of their physical properties, we will build up the criteria that allow us to decide what detector type and characteristics are best for a given application.

9.4.1 *Photoconductive detectors*

These are semiconductors with bandgaps chosen to be smaller than, but close to, the energy of the photons to be detected. Incident radiation then excites transitions into the conduction band and a current flows when a bias voltage is applied. At infra–red wavelengths and room temperatures, $h\nu$ is comparable to kT so thermal excitation into the conduction band also occurs, resulting in an additional noise source. The detectors are usually cooled to suppress this, often to around 77 K using a liquid–nitrogen reservoir, and the band–gap adjusted to the energy of the photons to be measured by using different types of material, intrinsic and with n–type and p–type doping.

Many photoconductive detectors are based on the properties of the 'solid solutions' *lead tin telluride* and *mercury cadmium telluride* (MCT). The wavelength at which the performance is maximized can be varied by adjusting the mixture of components. For $HgCd_xTe_{(1-x)}$, for example, the gap energy (eV) is

$$E_{\rm g} = -0.25 + 1.59x + 0.327x^3 + 5.233 \times 10^{-4} T(1 - 2.08x), \tag{9.21}$$

where x is the fraction of cadmium in the mixture, and T is temperature. Doped germanium is another commonly used material.

The basic setup for a photoconductive detector is similar to the generalized bolometer shown in Fig. 9.2, although a modest (but still accurately constant) bias voltage of a few volts now suffices. Let ΔS (cm^{-2}) be the increase in conduction–band electron surface density produced by the photon flux of power P falling on a detector of active surface area A cm^2 so that

$$\Delta S = \frac{\eta P t_{\rm l}}{h\nu A}, \tag{9.22}$$

where η is the efficiency of a photon for excitation and t_l the lifetime of a carrier in the conduction band. The current i that flows is

$$i = -evl\Delta S \tag{9.23}$$

amps, where e is the charge on the electron, v its drift velocity, and l the distance across the active element ($l^2 = A$, usually). Hence

$$i_\mathrm{s} = \frac{\eta PeG}{h\nu}. \tag{9.24}$$

$G = \nu t/i$ is called the *photoconductive gain*, defined as the number of carriers that flow in the external circuit per successful photon. G can be greater than one, because once an excited *state* has been created many carriers can flow through the detector element in the conduction band until a transition in the reverse direction takes place.

There are three important sources of noise current in photoconductors, and each dominates at a different frequency. The noise at low frequencies is very high, due to a phenomenon called $1/f$ noise (or *'flicker' noise*). The cause of this effect is not sufficiently understood to allow it to be calculated from any well-founded physical theory. From measurements, however, it has been found that the noise power is inversely proportional to the chopping frequency; and so for practical purposes it can be overcome simply by chopping fast enough. A few times 10 Hz is generally sufficient.

GR (generation–recombination) noise is due to statistical fluctuations in the number of carriers available to conduct at any given instant. A statistical analysis, assuming that generation and recombination are stochastic and independent gives

$$[i_\mathrm{n}]^2 = 4Gei\Delta f \tag{9.25}$$

for an intrinsic detector at low temperatures. The current i has two components: that produced by the radiation of interest, and that produced by background radiation from the rest of the system and its surroundings. If the total power falling on the detector is P then

$$[i_\mathrm{n}]^2 = \frac{4\eta Pe^2G^2\Delta f}{h\nu}, \tag{9.26}$$

and the noise equivalent power (W $\mathrm{Hz}^{-1/2}$) is

$$\mathrm{NEP} = \sqrt{\frac{4Ph\nu\Delta f}{\eta}}, \tag{9.27}$$

i.e. this is the signal power from the target required to produce an output equal to the noise induced by the total power falling on the detector. The condition

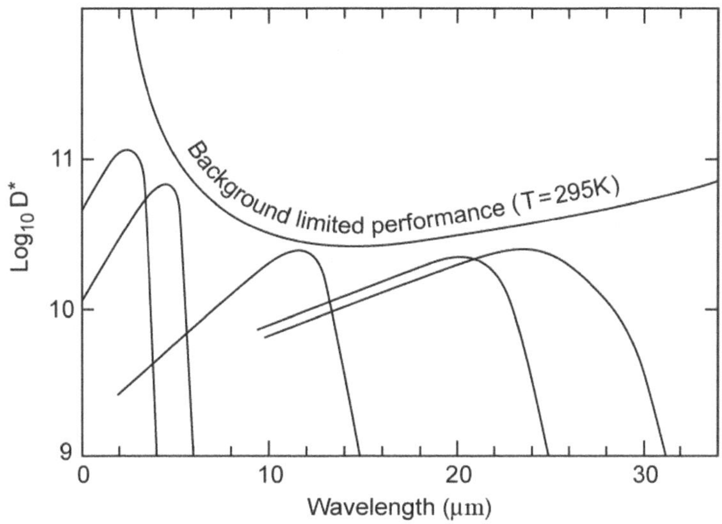

FIG. 9.7. Detectivity D^* as a function of wavelength for several common detector materials used in photoconductive mode. These are (left to right) lead sulphide (PbS) at a temperature of 295 K; indium antimonide (InSb) at 77 K; mercury cadmium telluride ($Hg_{0.8}Cd_{0.2}Te$) at 77 K; and germanium doped with cadmium, copper and zinc, respectively, all at liquid helium temperature (~4 K). The ideal (background limited) performance for a room-temperature target is also shown.

for the contribution from Johnson noise due to the resistance in the circuit to be negligible is

$$\frac{4kT\Delta f}{R} \ll \frac{4\eta PG^2e^2\Delta f}{h\nu}, \tag{9.28}$$

or

$$\frac{(kT/e)(h\nu/e)}{\eta G^2 R} \ll P, \tag{9.29}$$

which, in addition to choosing material with large η and G, implies a low temperature T and a large equivalent resistance R. The detector is then said to be *photon–noise limited.*

If the power P originates mainly in the background rather than the source then the detector is said to be operating in the 'BLIP' (background limited intrinsic photoconductor) condition. Figure 9.7 shows how most detectors approach this condition at their wavelength of peak performance.

9.4.2 *Photovoltaic detectors*

Photovoltaic mode detectors are fast and sensitive, and operate without the need for any bias voltage. The commonest examples are semiconductor p-n diodes made of silicon, indium antimonide, or indium arsenide. The principle of operation is that pair production by photons induces a measurable voltage across the junction, which is related to the incoming photon flux. For wavelengths longer than about 3 to 4 μm, they have to be cooled to prevent the output from being swamped by spontaneous electron–hole pairs produced as a result of thermal excitation.

For a diode in the photoconductive mode, with an applied bias voltage V_a, the current i is given by the normal expression for a diode, plus a photocurrent in the opposite direction, i.e.

$$i = i_{\mathrm{sat}} \left(\exp\left(\frac{eV_{\mathrm{a}}}{kT}\right) - 1\right) - \frac{\eta ePG}{h\nu}. \tag{9.30}$$

In photovoltaic mode the device is open circuit, with a voltage

$$V = \left(\frac{kT}{e}\right) \log\left(1 + \frac{\eta eP}{h\nu i_{\mathrm{sat}}}\right) \tag{9.31}$$

appearing across the terminals. Note that this type of detector is inherently nonlinear; for this reason, it is generally arranged in a suitable circuit where the diode acts as a current source, where there is no voltage across the detector.

Photon noise in these devices is known as *shot noise*, which has its origin in the finite nature of the charge on the electron. Suppose that the detection system is characterized by a time t_0 during which time N photon events occur. The mean current is

$$\bar{i} = \frac{e\overline{N}}{t_0}, \tag{9.32}$$

while the noise current i_{n} is given by

$$(i_{\mathrm{n}})^2 = (i - \bar{i})^2 = \frac{e^2}{t^2}(N - \overline{N})^2. \tag{9.33}$$

If N is assumed to have a Poisson distribution about the mean, a standard identity gives

$$(i_{\mathrm{n}})^2 = \frac{e^2}{t_0^2}\bar{N} = \frac{e\bar{i}}{t_0} = 2e\bar{i}\,\Delta f, \tag{9.34}$$

which is the expression known as *Schottky's formula.*

In photovoltaic mode, the resistive (Johnson) noise contribution is generally negligible compared to that due to the photon flux P. If the background contribution to the power at the detector is also negligible compared to that originating in the source, i.e. $P \approx P_{\mathrm{s}}$, then the noise current is

$$(i_{\mathrm{n}})^2 = \frac{2\eta e^2 P_{\mathrm{s}}}{h\nu}\,\Delta f, \tag{9.35}$$

while the signal current is

$$i_s = \frac{\eta e^2 P_s}{h\nu}, \tag{9.36}$$

so the signal–to–noise ratio, in terms of dwell time t_0, is

$$\mathrm{SNR} = \frac{i_s}{i_n} = \sqrt{\frac{\eta t_0 P_s}{h\nu}}. \tag{9.37}$$

The noise equivalent power, defined as the signal power that results in a signal–to–noise ratio of unity, is then

$$\mathrm{NEP} = \frac{2h\nu}{\eta}\Delta f = \frac{h\nu}{\eta t_0}. \tag{9.38}$$

This device, used in a mode where the only important source of noise is that intrinsic to the photon stream being measured, is described as an *ideal photon detector*.

9.5 Detector arrays and charge coupled devices

Detectors are increasingly being supplied as *arrays*, i.e. assemblies of more than one sensitive element on the same substrate. Area arrays with $m \times n$ elements are mostly used for infrared imaging, and may contain tens of thousands of elements. Linear arrays, with $n\times 1$ elements, can be scanned across a target ('push–broom' imaging), as a cheaper and more easily manufactured alternative, if the extra time is available. An alternative use of linear arrays is to measure (or 'multiplex') different wavelengths simultaneously by locating them in the plane of dispersion of a spectrometer. A problem with arrays was the large number of pre–amplifiers required, if each detector element has its own. The pre–amplifiers are normally an integral part of the detector, mounted near the sensitive element to reduce the possibility of noise generation at a point of small signal and high impedance. Clearly installing 10000 of them on a 100×100 element array measuring less than 1 cm^2 represents a challenge, which has led to the rapid adoption of the charge coupled device or CCD. These can read the signal from many detectors through a single pre–amplifier.

9.6 Properties of IR systems

The three 'dimensions' of an infra–red device are represented by its spectral, geometrical and radiometric properties. It is useful to extend the concept of the responsivity of a detector to complete systems. The responsivity R_v is then the ratio between the change in output voltage, ΔV, and the change in target radiance, ΔI, hence

$$R_v = \frac{\Delta V}{\Delta I}, \tag{9.39}$$

and

$$V = R_{\mathrm{v}}(I - I_0) + V_{\mathrm{n}}, \tag{9.40}$$

where I_0 is the radiance from a reference source and V_{n} is the output noise voltage. The relationship between input radiance and output voltage is assumed to be linear (the system design will be such as to make sure this is the case) and the contributions to the net responsivity of the system from spectral, geometrical, and radiometric factors are independent of each other, so that

$$R_{\mathrm{v}} = R_0 \times R_1(\lambda) \times R_2(f) \times R_3(x,y) \times R_4(\theta,\phi)\cdots. \tag{9.41}$$

Thus, the overall response R_{v} of the system depends on all three domains, so that, for example, increasing spectral resolution reduces the energy throughput and hence the signal. In general however the variables are not coupled, so that this change should not directly affect the field of view, for example.

System definition begins with a consideration of which properties are fixed by the application and which are free parameters in the optimization of the design. It is also necessary, of course, to take account of purely practical factors such as weight and cost: a design which has been optimized only on the basis of physics is not necessarily the best overall.

9.6.1 *Spectral properties*

Spectral properties can be placed into three categories: wavelength definition or selection, spectral resolution, and wavelength calibration. The first two are intrinsic properties of the wavelength–selection technique and the optical materials and detectors employed; the last involves the separate provision of one or more reference wavelengths.

Usually, radiation consisting of a wide range of wavelengths will be incident on the entrance aperture. The degree to which each is attenuated in the output from the detector depends on: (i) the wavelength–dependent throughput (transmissivity or reflectivity) of each lens, mirror, and window in the optical chain; (ii) the wavelength-dependent response of the detector itself, and (iii) the performance of the wavelength selective components specifically included for the purpose. The last may include gratings, etalons, and/or blocking filters as appropriate. The wavelength response of the device is the product of the response of all of these.

The resulting spectral bandwidth is generally dominated by a single component such as a filter or grating. The key variable is not just the resolution ($\Delta\lambda$ or $\Delta\nu$), defined and discussed below, but also the precise shape of the wavelength–response curve of the system as a whole, which is generally nearly the same as that of the principal spectrally selective component. In the simplest case, the spectral profile of the system is a rectangle or 'boxcar' of width $\Delta\lambda$. Some filter profiles approximate this behaviour, while grating instruments employing slits give rise in general to a triangular transmission function. For interferometers the

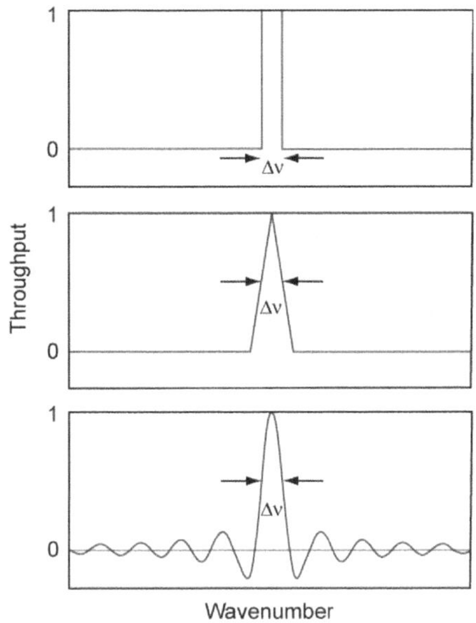

FIG. 9.8. Generalized wavelength responses of; (a) a filter instrument, (b) a grating spectrometer, and (c) a Fourier spectrometer, all with the same spectral resolution $\Delta\lambda$.

corresponding profile is the Fourier transform of a boxcar profile, which can be shown to be the sinc function $\sin(x)/x$ (Fig. 9.8).

The boxcar profile is obviously the ideal, in that it maximizes spectral resolution and throughput simultaneously. The spectral response function $R_1(\lambda)$ is simply

$$R_1(\lambda) = 1 \quad \text{if} \quad \lambda_1 \leq \lambda \leq \lambda_2 \tag{9.42}$$

$$= 0 \quad \text{if} \quad \lambda \leq \lambda_1 \quad \text{or} \quad \lambda \geq \lambda_2, \tag{9.43}$$

and the spectral resolution is defined as $\Delta\lambda = \lambda_2 - \lambda_1$. For other profile shapes, it is useful to introduce the concept of equivalent width ΔW, where $\Delta W = \int R_\lambda \mathrm{d}\lambda$ so that ΔW is the width of the boxcar profile with the same area under a curve of response versus wavelength. For the boxcar ΔW and $\Delta\lambda$ are obviously identical; for other shapes the two are interchangeable in the expression for throughput whenever the source radiance is slowly varying across the spectral passband, which is often the case if the source is a continuum emitter such as a blackbody, or when the spectral resolution is high, or both.

In spectroscopy, the key factor in defining spectral resolution is the ability of a scanning instrument to separate the signals from two line sources separated by $\Delta\lambda$. The features can be considered to be resolved if a spectral scan shows a minimum or dip between the maximum intensity corresponding to each. For spectral profiles of more-or-less regular shape, the resolution is then the width

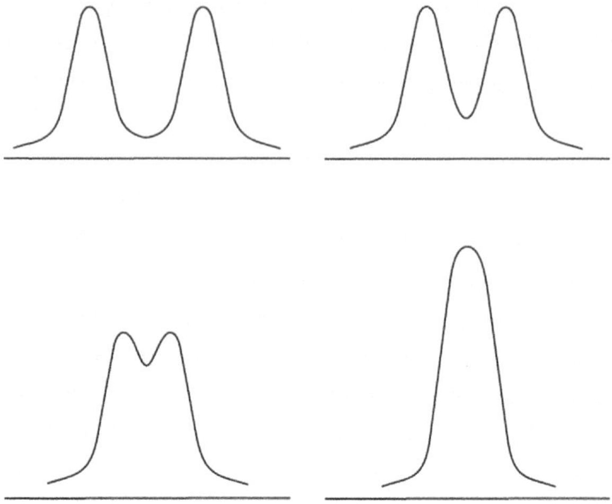

FIG. 9.9. Features in an intensity scan; (a) fully, (b) well, (c) barely, and (d) not resolved.

of the profile between the half-maximum points. The convention for Fourier transform spectrometers is to take the full width at half–height of the central peak, which can be shown from the properties of the sinc function to have a value of $0.6043/L$ in wave number units, where L is the maximum path difference in cm. In the convention introduced by Rayleigh, the requirement for resolution of two features is for a drop of 17% in the signal midway between the two lines, and this can be adopted for passbands of complex or irregular shape (Fig. 9.9).

9.6.2 *Wavelength calibration*

Wavelength calibration of a spectrometer involves inserting an emission or absorption source of known properties in place of the target. For best results the spectral features of the reference source should be sharp, strong, stable, numerous and distributed across the operating range of the instrument. Laser diodes, discharge lamps, cells containing gases, and thin plastic films are among the candidates available.

9.6.3 *Geometrical optical properties*

9.6.3.1 *Aperture and field of view* The most fundamental geometric optical properties of a system are the optical axis – the direction in which the detector is viewing the target – and the aperture, defined as the area of the entrance pupil measured perpendicular to the axis. The functions $R_3(x, y)$ and $R_4(\theta, \phi)$ specifying the relative response for displacements around the axis, spatially or angularly, which define the *field of view*, describe the part of the target that is sampled. This can be complex and unevenly weighted, but for the common case of a system with cylindrical symmetry, entrance aperture of area A, and a well–defined field of view of solid angle Ω, we expect the number of photons entering the system to be proportional to the product $A\Omega$. Many instruments that view external targets have a scanning ability, i.e. can alter the direction of the optical axis by moving internal mirrors, etc. In such cases, an additional term in $\cos\mu$, where μ is the angle between the optical axis and the normal to the surface of the source, enters as a consequence of the definition of radiance. Then we have

$$R_3(x,y)R_4(\theta,\phi) = \int\int \cos\theta \mathrm{d}\Omega \mathrm{d}A. \tag{9.44}$$

The product $A\Omega$ is sometimes called the *etendue* or energy grasp of the system. In an ideal system, etendue is conserved throughout; in real systems, the value corresponding to the whole is its lowest value in any stage of the optical train.

9.6.3.2 *Beam modulation* The geometric properties of an optical system include not only the direction and field of view, but also factors such as modulation of the beam, which can be deliberate (chopping) or undesired (vibration due to external disturbances). In the most common form of beneficial modulation, mechanical chopping, the beam is interrupted by a vibrating or rotating blade in order to convert the output from the detector from a d.c. to an a.c. signal, for ease of amplification. If the modulator has a period very much shorter than the thermal time constants of the sources of the background, chopping enables the output due to the background including its low–frequency fluctuations to be rejected. In a real instrument there will be sources of additional low–frequency fluctuations in the detector and the analogue part of the signal–processing subsystem. The use of modulation at a sufficiently high frequency also overcomes these effects as far as their influence on measurement of the mean signal flux is concerned.

We can express the same idea in frequency–domain language as follows. The noise power spectrum of the background is flat at high frequencies and rises at low frequencies. In d.c. systems (i.e. without modulators) the signal–amplitude information captured occupies a bandwidth (its baseband) extending from zero Hz (d.c.) up to a value of the order of the reciprocal of the response time of the signal processing electronics. A modulator shifts the signal information (but not the background) from its baseband to a higher frequency where the excess low–frequency noise has fallen below the white level, i.e. the unavoidable fluctuations. The wanted information (the amplitude of the signal flux) becomes the modulation on a carrier at the modulator frequency. The amplitude information is recovered by demodulation in the signal–processing electronics. By this means the majority of the drifts in the optical system, the detector and the analogue part of the signal–processing subsystem (except gain drifts) can be avoided.

Usually, the modulator has no spectral dependence of significance to the system but in the special case where this is a feature of the design, it is called a selective modulator. Selective modulation can be used to distinguish not only the scene radiation from the background but also different components of the scene emission, for example those from a particular atmospheric gas. Physically, a non-selective modulator is usually either a metal disc with sectors cut out rotating about its axis or an oscillating metal shutter, hence the name 'chopper' for these mechanisms. The surfaces that actually interrupt the beam may be either black on the side facing the optical system or reflecting and angled so that the field of view of the system is directed intermittently to a stationary black reference source. The stability of the temperature of the moving parts is less important if the second choice is made. Selective modulation can be achieved using interrupting surfaces that have spectrally dependent transmissions or reflectivities. Selective modulation restricted to the minute spectral intervals occupied by the absorption lines of a gas can be obtained by using a modulator consisting of a cell containing the gas whose pressure is caused to oscillate. By this means the emission from a particular gas in the scene can alone be detected in the presence of other emissions.

Chopping is used to 'label' the incoming beam and reduce the sensitivity of the system to fluctuations in the background flux. Even when an instrument is cooled to the point where the background photon noise is negligible, there may still be a large background signal. In this situation, slow drifts in the background signal level, due for example to small changes in the temperature of the instrument housing, can upset the radiometric calibration of the device. The solution in this case is, obviously, simply to cool the instrument further; however, extensive cooling is awkward to provide in practice and is employed only when absolutely essential. An alternative approach is to chop the incoming signal beam against a cold reference target, to modulate the signal but not the background. The a.c. component of the detector output is amplified by a phase–sensitive amplifier with a band–pass filter tuned to the frequency of the mechanical chopper.

A disadvantage of chopping is that it reduces the throughput of the system by at least a factor of 2, since half of the incoming energy is rejected. This can be important in systems where the source is weak and signal is at a premium. In fact, for the most common kind of chopper, a rotating toothed wheel, the blades take a finite time to cover and uncover the beam and the net effect is a reduction of a factor between 2 and 4. The latter value applies when the chopper segment is the same size as the beam, so the total obscuration (and total passage) of the flux is only momentary, resulting in a triangular waveform at the detector. An intermediate value of $2\sqrt{2}$ applies if the modulation is sinusoidal.

9.6.3.3 *Ray tracing* The geometrical properties of a system may be determined by ray tracing, graphically for simple systems or using one of the many advanced computer packages for complex instruments. For infra–red systems, where visible alignment is usually not possible and where unwanted sources and reflections may be dispersed through the instrument, ray tracing is an essential part of the design process. The procedure is to launch a number of rays, some parallel to the optical axis and some skewed to it, each defined by their direction cosines X, Y, and Z and the co-ordinates x, y, and z of the point where they intercept the entrance aperture. At each optical surface, the direction of reflection is calculated geometrically and of transmission using Snell's law for each ray. They are then traced to their point of impact with the next surface, and the process repeated, until they reach the detector (or not, as the case may be). Many thousands of rays may be traced in this way until all of the aperture has been mapped out in space and in incident angle to the required detail. It is often most instructive to launch the rays backwards through the system, starting at the detector; a uniform package of rays ends up forming a pattern in object space, the distortions in which show the departure of the field of view from the specification. Minute adjustments can then be made in the positions or properties of each component, often automatically by the program, until the result is acceptable within some criterion specified by the experimenter.

9.6.4 *Radiometric properties*

9.6.4.1 *Signal–to–noise ratio and throughput* In addition to the other factors discussed above, the responsivity of the system depends on an efficiency factor ε, usually called the *throughput*. Throughput is defined simply as the probability that a photon entering the system within the spectral pass–band and field of view will reach the detector. It can be estimated by calculating the product of the transmission and reflection losses of each component in the system, and determined experimentally by comparing the measured to the calculated signal from a standard source.

Given knowledge of the throughput and the geometrical and spectral factors determining the energy flux onto the detector, the signal at the output of the system can then be calculated. In practice, the important factor is not the size of the signal but the ratio of this to the noise on the output from the detector. The dominant source of this noise is usually the detector itself or fluctuations in the radiation field, signal and background, falling on the detector.

9.6.4.2 *Response time* Along with signal–to–noise ratio, the radiometric performance is defined by the response time of the system. Where signal levels are high relative to the noise, the important factor determining the repeat time between successive measurements is the time constant of the detector. A system with time constant τ requires an integration time of 5τ to achieve an output equal to 99.3% of the input. More often, the controlling factor is the dwell time required to achieve the required signal–to–noise ratio by integrating the signal electronically. This can be achieved with a network of time constant RC, which smoothes the signal and the noise so that, if the latter is random, an improvement in signal–to–noise ratio proportional to $\sqrt{RC}$ is achieved. Further improvements

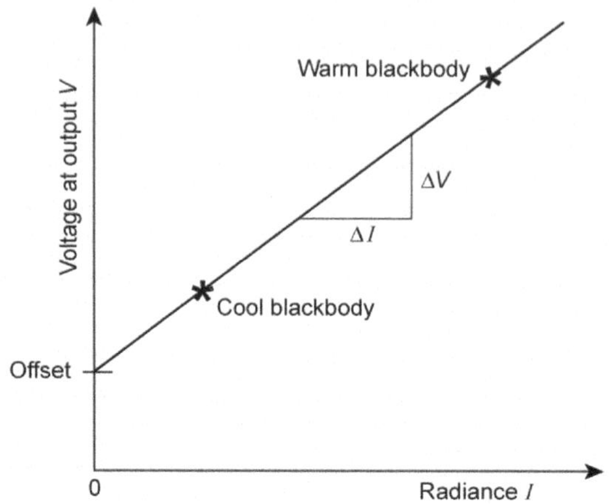

FIG. 9.10. Response of a linear system to input radiance.

can be achieved, and the time required for the measurement of a weak source usefully reduced, by using phase sensitive detection and other active techniques.

9.6.4.3 *Signal processing and linearity* Depending on the type of detector used, each component of the background and signal photon currents will give rise to a mean output from the detector that is proportional to either the mean number of photons incident in the measuring time or to their energy. The signal–processing subsystems that follow the detector consist of an analogue part followed by a digital part, and these are designed to provide an output count that is proportional to the current or voltage output from the detector, i.e. to provide linear signal processing. Then the entire system is linear from end to end. Considerable care must be applied in practice to the selection of detectors and their operating ranges, and in the design of signal–processing electronics, to ensure that the output from the system is in fact a linear function of the incoming intensity. If this is not the case, the relationship between signal and output can never be determined accurately since, even if the transfer function is measured, it cannot be assumed not to change, particularly if the source of the non-linearity is a detector being operated at a point in its range that is approaching saturation.

9.6.4.4 *Intensity calibration* The usual principle is to ensure linearity of the device and then to use observations of two standard sources of known intensity to fix the constants of the straight–line response function (Fig. 9.10). Absolute radiometry is impossible without radiometric calibration, since, while it is possible to build a radiometer in which there is a high degree of confidence that at any instant the output will be linearly related to the scene radiance, calculating the position of this straight–line transfer characteristic to the accuracy desired for the intended measurements will almost always be impossible given the amount of detailed knowledge required about the components and their physical state. The usefulness of a radiometric model of the system is limited to detailed verification of the radiometric performance of an instrument during its final tests. In operation, the transfer function (i.e. the straight line that describes output voltage in terms of target radiance) can be established by allowing the instrument to view two sources of known brightness, ideally at each end of the range of target brightnesses, which provide standard radiance levels. The only accurate and reliable reference sources in the infra–red are blackbodies of various designs. Radiance calibration of the system can be performed by pointing its optical system at blackbodies at different temperatures before or after viewing the scene. This can be achieved by moving the entire optical system or steering its field of view with a mirror. The latter method avoids the need to move the (possibly massive) optical system, but requires that changes in the effective emissivity of the mirror, its temperature, and the amount of scattered radiation entering the optical system from the surroundings remain sufficiently small as the three views are selected. For instruments in orbit, which can view cold space, an ideal zero-radiance background is available and this is generally used for the lower calibration point. How long the calibration established in this way remains useful depends on how rapidly changes in the optics transmission, detector response, and the gain and offset of the analogue part of the signal–processing subsystem take place due to temperature changes, aging, power–supply variations, etc.

9.6.4.5 *Reducing the background* Ideally, the background is made negligible by cooling the whole optical system sufficiently. Placing a modulator in front of the optical system is a useful alternative method of reducing the effects of the background where cooling the system is impractical, although of course this does not reduce the fundamental fluctuations in the background. In many practical designs the fore optics are warm and the detector is cooled to enhance its performance. It is often worth increasing the amount of the system that is cooled to include the narrowest band–limiting filter and the system stops (which are usually placed just in front of the detector) as this can reduce the background and its fundamental fluctuations considerably. The ideal is to cool the entire instrument, when this is practical; see the example of the tropospheric emission spectrometer in Chapter 10. An additional advantage is gained by doing this, in that drifts in the background signal no longer require calibrating out.

9.7 Radiometric performance

9.7.1 *Signal–to–noise ratio*

The performance of a measurement system is characterized by the signal–to–noise ratio at its output. In a well-designed infra–red system, this is usually controlled by the noise in the detector, other sources of noise such as vibration, thermal or amplifier noise having been rendered negligible by careful design and

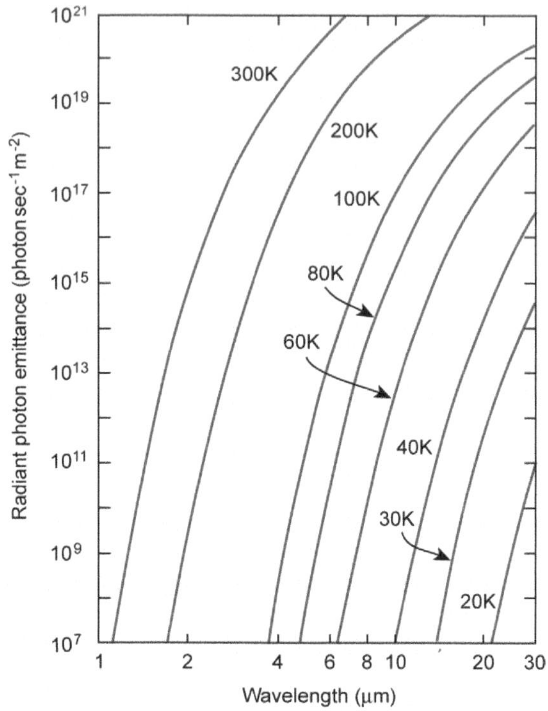

FIG. 9.11. Photon emittance versus wavelength for various temperatures.

implementation. We distinguish between intrinsic detector noise, such as that due to lattice vibrations in the detector element, or the statistics of current flow through the detector, and extrinsic noise, produced by fluctuations in the photon flux from the background, and ultimately in that from the source itself. The latter generally dominates in the most sensitive observing systems, where thermal effects and the background flux of radiation have both been reduced by cooling parts, or all, of the instrument.

Thus, in the case where the limit on the signal–to–noise ratio attainable in a measurement of the spectral mean intensity of a photon stream is set by the fluctuations in the stream itself, we distinguish two cases, the ideal one in which the entire stream being detected originates in the scene being viewed and the *background limited* case in which the dominant stream is generated by the measuring instrument rather than the scene. In less–favourable circumstances still the limit may be set by excess noise generated in the detector, the *detector* or *technology* limit. While many different configurations are possible, as a general rule moderately cryogenic systems (typically those cooled with liquid nitrogen or with mechanical refrigerators to around 80 K) can reach background limited performance at wavelengths up to about 10 μm or so, while at longer wavelengths they employ cooling with liquid helium or advanced refrigerators to much lower temperatures of a few K. Uncooled devices are generally technology limited at all wavelengths.

Several types of photon detectors, for example extrinsic germanium and silicon, are background limited down to quite low levels of background photon flux (photons $cm^{-2}\ s^{-1}$). The amount by which the background is reduced by cooling depends not only on the temperature but is also a function of wavelength, as shown in Fig. 9.11. At some point, reducing the temperature produces no further improvement as photon noise due to the source itself takes over.

For example, the current noise i_n for a photovoltaic detector can be written

$$i_n = \sqrt{\eta N_b\, 2\Delta f\, e^2 + \eta N_s\, 2\Delta f\, e^2 + i_0^2}\,, \tag{9.45}$$

where N_b and N_s are the total numbers of photons falling on the detector from the background and source, respectively, i_0 represents the sum of all sources of noise other than photon noise ($1/f$ noise, Johnson noise, etc.) and the other symbols have the same meaning as before. The signal–to–noise ratio is

$$\text{SNR} = \frac{\eta N_s e}{\sqrt{\eta N_b\, 2\Delta f\, e^2 + \eta N_s\, 2\Delta f\, e^2 + i_0^2}}. \tag{9.46}$$

For a sufficiently bright source, the second term in the denominator will dominate the third and for some sufficiently low background this expression reduces to

$$\text{SNR} = \sqrt{\frac{\eta N_s}{2\,\Delta f}} \tag{9.47}$$

and the signal–to–noise ratio is proportional to the square root of the signal strength.

On the other hand, for a very weak signal and a low background situation, like that in most astronomical observations, or an uncooled system limited by technology noise, the third term in eqn (9.45) dominates and

$$\text{SNR} = \frac{\eta N_s e}{i_0}. \tag{9.48}$$

Now, the signal–to–noise ratio is directly proportional to the signal. This is also the case for an inadequately cooled system where $N_s \ll N_b$

$$\text{SNR} = \frac{\eta^{1/2} N_s}{2r N_b \Delta f}. \tag{9.49}$$

In designing a system for a particular application, it clearly is necessary to evaluate each term first for a particular configuration, and to decide what cooling is necessary. Providing the cooling is often difficult or expensive and can introduce other practical problems, hence the designer wants to keep it to a minimum.

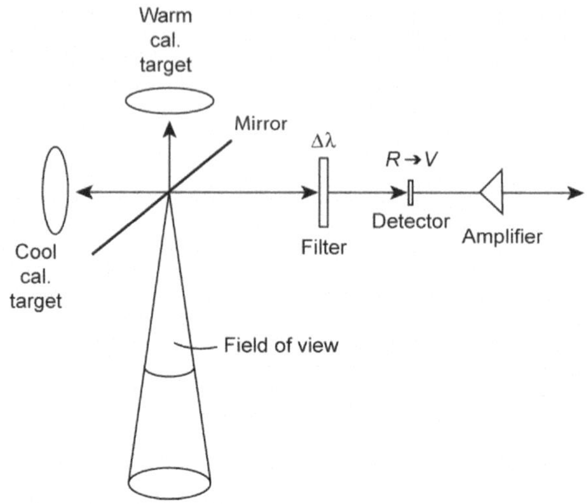

FIG. 9.12. Schematic diagram of a simple radiometer.

9.7.2 *A generalized radiometer*

Consider the simplest kind of radiometer in which the dominant fluctuation in the electrical signal from the detector is produced by processes occurring within the detector, not by fluctuations in the incoming photon stream. This is often the case in real systems where the source is bright and achieving the ultimate sensitivity is unnecessary, for instance net–flux radiometers. The diagram in Fig. 9.12 represents the simplest concept of a radiometer – some light-gathering optics, a detector and its amplifier, and a readout device, in this case a voltmeter. There is also some kind of wavelength filter – if this is not provided explicitly, then the spectral bandpass will be dominated by the wavelength dependence of the components in the optical chain, including the detector, and the complex wavelength-dependent transmission of the intervening atmosphere if the device is in air.

The system is assumed to observe an extended source, that is, one that fills the field of view of the instrument, which is determined by its optical design, and specified by the solid angle Ω within which radiation is accepted at the aperture of area A. If G is the gain of the amplifier and V the reading in volts at the output then the energy falling on the detector of responsivity R_v is V/GR_v W. If the target is approximately a blackbody of temperature T, using the Stefan-Boltzmann law with constant σ, the signal power S reaching the detector is

$$S = \frac{\sigma T^4 \, \Omega A}{\pi} = \frac{V}{GR_\mathrm{v}}. \tag{9.50}$$

From which T can be calculated in terms of the measured voltage, the properties of the detector and the geometry. If the detector has a noise equivalent power with the value NEP [W Hz$^{-1/2}$] then the signal–to–noise ratio (SNR) is given by

$$\mathrm{SNR} = \frac{S}{\mathrm{NEP}\sqrt{\Delta f}} = \frac{V}{GR_\mathrm{v}\,\mathrm{NEP}\sqrt{\Delta f}}. \tag{9.51}$$

The lowest temperature that can be measured is that which produces SNR = 1; this is known as the *noise equivalent temperature* or NET and is given by

$$\mathrm{NET} = \left(\frac{\pi\,\mathrm{NEP}\sqrt{\Delta f}}{\sigma\Omega A}\right)^{0.25}. \tag{9.52}$$

A more useful quantity is the *noise equivalent temperature change* (NEΔT), given by

$$\mathrm{NE\Delta T} = \mathrm{NEP}\left(\frac{\mathrm{d}S}{\mathrm{d}T}\right)^{-1}\sqrt{\Delta f} = \frac{T\sqrt{\Delta f}}{4\,\mathrm{SNR}}. \tag{9.53}$$

The NEΔT is the amount by which T has to change to produce the smallest detectable change in the detector output, which is by definition equal to the NEP. It is therefore a measure of the sensitivity of the radiometer, and of the uncertainty in any measurement. For a given instrument and target, it can be improved only by integrating the measurement for a longer time, which reduces Δf proportionately.

9.7.3 *A realistic radiometer*

A more practical optical system is one where only the detector and band–limiting filters are cooled and beam modulation has been introduced. The background is often much more intense than the stream from the scene and its fluctuations often limit the sensitivity of the radiometer to the latter. Given that the objective is to measure the strength of the photon stream from the scene a problem this immediately presents is how to distinguish this stream from the background. The technique usually employed is to label the scene radiation by mechanical modulation or chopping of the radiation beam to convert the signal from d.c. to a.c., making it easier to amplify and filter, and reducing some sources of detector noise, particularly those that are most important at low frequencies (§9.6.3.2).

In a chopped system, the photon flux entering the optical system then alternates between that originating in the scene and the reference flux originating from or reflected by the chopper. The resulting flux at the detector can be regarded as the superposition of a steady flux equal to the mean of the scene flux and the reference flux and a square wave flux of peak–to–peak amplitude equal to their difference. Corresponding steady and square wave currents will be present in the detector output. The steady component is rejected. The square waveform may or may not be filtered in the amplifiers so that only the fundamental component is retained. In either case the signal is then synchronously rectified at the chopping frequency and then integrated to produce the output. The output of such a system will be zero when the reference flux is the same as the scene flux and maximum when these fluxes are most different.

The fluctuations in the background flux that give rise to noise at the output are those that lie in a frequency band centred on the chopping frequency that has a width of the order of the reciprocal of the integration time. The similar fluctuations in the reference flux also contribute to the noise level. In critical applications the optical system and its surroundings will be cooled, in extreme cases to helium temperatures. If a black chopper is used it will be cooled also, as will the source of reference flux used with a mirror chopper. A good example of this latter technique is the 'chopping to cold space' often employed in satellite–borne infra–red radiometers. As noted above, the necessary use of choppers to label the scene radiation flux leads to at least half the photons emitted from the scene being wasted.

Chopping alone will not guarantee good performance if the background flux is very large and/or variable compared to the signal. This leads to a requirement to cool, not only the detector itself, but its surroundings and key elements in the optical chain. The detector has a view to its surroundings of solid angle 2π around the beam cone. For infra–red wavelengths, the Planck function is very much smaller at 77 K and at 220 K than it is at 290 K and the contribution to the background falling on the detector from its cold surround and the fluctuation due to the signal flux are both negligible. It is therefore the fluctuation in the background lying in the beam cone (due to emission from the optical components and their surroundings the latter entering the beam via reflections and scattering at the surfaces of the components), which is the dominant source of noise.

Of the parameters of the system, Ω may well be fixed by the spatial resolution required, t_0 by the required response time and the linear dimensions of aperture A and the ability to cool the whole optical system will be limited by cost. Obtaining a good optical transmission is obviously always desirable.

In practice, the realistic system described above would also be modified to include at least some of the following refinements:

(i) A band–limiting filter that passes a spectral range $\Delta\nu$ centred on ν cm^{-1}. Measurements that attempt to integrate over too wide a range of wavelengths may maximise the signal–to–noise ratio, but are generally more difficult to interpret, except in situations where the integrated radiance from the target is the essential measurement, for example in energy–balance studies. Narrow-band measurements are essential for radiometric temperature sounding, or for spectroscopic composition measurements, for example.

(ii) A method of radiometric calibration, to account for the uncertainties and changes in the values of key instrument parameters such as background flux, optical system throughput, and detector response.

(iii) Thermal control. The detector performance is generally temperature dependent; if the temperature of the radiometer housing is not held constant the gain, offset and noise will drift.

(iv) Electromagnetic shielding and vibrational isolation. Sensitive radiometers need protection against extraneous noise sources, such as electromagnetic interference and microphonic noise induced by vibration.

The *noise equivalent radiance* (NER) of a realistic system can be written

$$\mathrm{NER} = \frac{2\,\mathrm{NEP}}{A\Omega\,\Delta\nu\,\varepsilon\,\sqrt{t_0}}, \tag{9.54}$$

where A is the aperture or collecting area, Ω the solid angle of the field of view, $\Delta\nu$ the spectral resolution (or, more generally, the spectral bandwidth) in cm^{-1}, ε the efficiency or throughput of the optical components, and t_0 the measurement time in seconds. A modulation factor of 1/2 appears (in the denominator) because the chopper, usually a rotating toothed wheel with equally spaced teeth and gaps, passes this fraction of the total energy if the beam has a circular cross-section. The signal–to–noise ratio for a one-second measurement is then equal to the power P_s at the detector due to the source radiance I_s divided by the noise equivalent power of the detector, or

$$\mathrm{SNR} = \frac{P_s}{\mathrm{NEP}} = \frac{I_s\,A\Omega\,\Delta\nu\,\varepsilon}{2\,\mathrm{NEP}}. \tag{9.55}$$

If the signal is integrated for a dwell time t_0 then the output is increased in proportion while the noise, assumed random, accumulates with a $\sqrt{t_0}$ dependency, so we can write

$$\mathrm{SNR} = \frac{P_s\,t_0}{\mathrm{NEP}\sqrt{t_0}} = \frac{I_s\,A\Omega\,\Delta\nu\,\varepsilon\,\sqrt{t_0}}{2\,\mathrm{NEP}}. \tag{9.56}$$

As before, the NET is the value of T for which SNR = 1 when viewing a blackbody, for which $I_s(\nu, T)$ is given by the Planck function $B_s(\nu, T)$, hence NET is found by solving

$$B(\nu,\,\mathrm{NET}) = \frac{2\,\mathrm{NEP}}{A\Omega\,\Delta\nu\,\varepsilon\,\sqrt{t_0}}. \tag{9.57}$$

Unlike NET, NEΔT has significance only in terms of viewing a specific source, and is a function of source temperature, T_s. If the target has an emissivity of e_ν at wave number ν cm^{-1} then

$$e_\nu\,\mathrm{NE}\Delta T(T_s)\,\frac{\partial B(\nu,\,T_s)}{\partial T} = \frac{2\,\mathrm{NEP}}{A\Omega\,\Delta\nu\,\varepsilon\,\sqrt{t_0}}. \tag{9.58}$$

It follows that, in principle, the NET and the NEΔT have the same value at T_s = 0 K, for a blackbody source with $e_\nu = 1$.

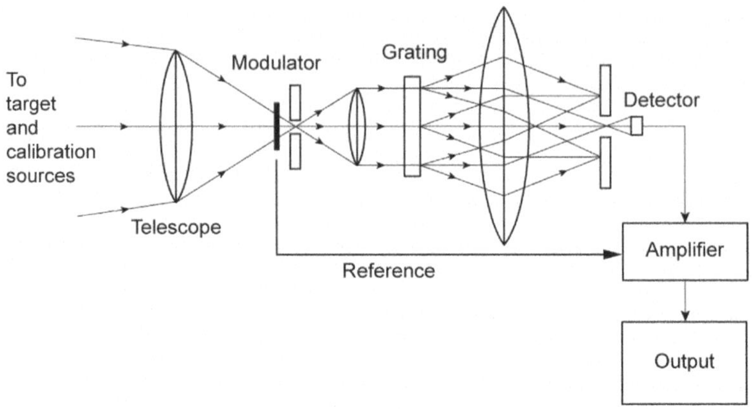

FIG. 9.13. Conceptual grating spectrometer layout, showing entrance and exit optics and associated optics.

In summary, in much the same way that NEP describes the performance of a particular detector and D^* describes a detector type or family (§9.2), NET describes the performance of an instrument or system, while SNR and NEΔT both characterize the quality of a specific measurement made using that system.

9.7.4 *Spectrometers and interferometers*

Many of the instruments used for climate measurements are essentially infra–red radiometers with the properties described above. Most of them have refinements of one kind or another, which will be described when we discuss specific sensors and measurement programmes in the next chapter. For high spectral resolution work, however, spectrometers using gratings or prisms and interferometers, usually of the Fabry–Perot or Michelson type, are required. These are more complex devices in general than radiometers, and, although the same principles apply, performance limitations are more acute, especially in geophysical applications, including climate studies, where distant observations of relatively cool sources are often involved.

Spectrometers are categorized by the wavelength–selective element and the means by which it is tuned or scanned over a range of wavelengths. The most familiar examples are those employing slits in conjunction with gratings or prisms, which disperse the radiation and then select the desired region geometrically. These are generally simpler and more robust than slitless interferometric devices such as the Michelson interferometer and its derivatives, although the latter offer a significantly higher throughput of energy and are capable of higher spectral resolution. The main elements of a simple slit–plus–grating spectrometer are shown in Fig. 9.13. The dispersive element is represented as working in transmission, as some gratings and all prisms do, but in reality it is more likely to be a reflecting grating as these have fewer intrinsic losses and a wider wavelength response. For the same reason, the lenses used to collimate the beam before incidence on the grating, and to focus it onto the output slit in front of the detector, are often in reality conic–section mirrors. The arrangement most commonly used in laboratory infra–red spectrometers has a single detector, and the grating tilts to scan in wavelength. These systems may have uncooled detectors for convenience and cheapness.

A commercial device of this type, developed for field research and applications by the oil and mineral exploration industry, uses a dual–beam approach in which the target is continuously compared to a reference source, using the same optical system, with a 200–Hz tuning–fork chopper and a sophisticated control and data–logging system. Three gratings and two thermoelectrically cooled detectors, silicon for the short wavelengths and lead sulphide for the longer, give a spectral range from 0.35 to 3.0 μm. The gratings and order–sorting filters are switched automatically by stepper motors under microprocessor control, to give a smooth scan over the range, while the operator trains the field of view onto the target of interest using a telescopic gunsight type of arrangement. The whole device weighs less than 16 kg including batteries for one day of operation.

Optimum performance of a dispersive spectrometer, in terms of signal–to–noise ratio for a given resolving power $\Delta\lambda/\lambda$, is obtained when the entrance and exit slit widths are equal. If the design is free of effects due to diffraction or aberrations, then the instrumental transmission function $\varepsilon(\lambda)$ is a triangle of width (meaning full width at the half-maximum power points) $\Delta\lambda$.

The origin of the triangular shape can be understood by imagining the passage of quasi–monochromatic radiation through an infinitely narrow entrance slit. This produces a uniform response across the finite exit slit, corresponding to a rectangular instrument function. If the actual entrance slit is now seen as being made up of the sum of a large number of narrow slits, the sum of the transmission functions is triangular.

The flux passing through the exit slit is given by

$$F(\lambda) \;=\; \varepsilon(\lambda)B(\lambda,T)\;A_{\mathrm{s}}\Omega\;\Delta\lambda\;, \tag{9.59}$$

where B is the radiance from the source, A_{s} is the area of either slit, and $\varepsilon(\lambda)$ is the throughput. The solid angle Ω is equal to the area of the grating, A_{g}, divided by the square of the focal length f of the collimator and condenser. The flux can be rewritten in terms of the dispersion at the exit slit, $\partial\alpha/\partial\lambda$, thus

$$F(\lambda) \;=\; \varepsilon(\lambda)B(\lambda,T)\frac{h}{f}\;A_{\mathrm{g}}\frac{\partial\alpha}{\partial\lambda}\;(\Delta\lambda)^2\;. \tag{9.60}$$

So, the flux obtainable at a given resolution is improved by minimizing the ratio of focal length f to slit height h (f-number) and the area and dispersive power of the grating. Optical–design considerations generally limit the f-number to about unity, and the dispersion to ≤ 1 mr nm^{-1}. Then, the only means remaining to improve the signal–to–noise ratio for a given source is to use larger cross–section optics.

The general problem of obtaining adequate signal–to–noise ratio with instruments viewing the Earth from space can be illustrated with some rough order of magnitude numbers. Suppose that the grating in this example has an area of 10×10 cm^2 and is 10 cm from an entrance slit of area 0.1 cm^2, illuminated by a 1000 K source. These are typical order–of–magnitude numbers for real systems. Suppose further that the device has a spectral resolution of 0.01 cm^{-1} and that we wish to obtain the spectrum from 1000 to 10000 cm^{-1} with a maximum SNR of 100 using an uncooled thermistor detector with NEP $= 10^{-10}$ W Hz$^{-1/2}$. A straightforward calculation shows that such a scan would take on the order of one day to complete, illustrating the problem of making measurements with simultaneous high spatial and spectral resolution from a satellite that orbits the Earth more than 10 times per day. Cooling the system, and thereby gaining an improvement of 2 or more orders of magnitude in signal–to–noise ratio, becomes essential. Limiting the spectral range to those wavelengths that contain most of the desired information, for example the 667–cm^{-1} carbon dioxide band for atmospheric temperature sounding, further improves the practicality of the measurements. Finally, large–aperture optics are often used. These essential improvements raise the mass of space instruments making climate observations, like those discussed in the next chapter, to more than ten times that of the device described above that is typical of those used for field geology on the Earth's surface.

The principle of the diffraction grating involves optical interference, of course, but the term interferometer is generally reserved for slitless devices like the moving–mirror Michelson, where the full circular aperture can be used and a gain in throughput of up to two orders of magnitude results. The basic principle of Fabry–Perot and Michelson–type devices is to divide the incoming beam into two using a partially reflecting beamsplitter and then modulate the path difference traversed by the two components before they recombine. This is achieved in the Michelson device by moving one or both of the mirrors that return the two components of the divided beam to the beamsplitter for recombination. The use of corner cubes instead of plane mirrors to return the two beams means that any tilting of the reflector affects only the translation of the beam and has no first-order spectral effects, making the device less sensitive to alignment errors. Various schemes are employed to fold and multipass the beam in order to maximize the optical–path difference, while keeping the overall dimensions of the instrument reasonably compact, which is particularly important for use on satellites and if cooling is employed to boost the sensitivity. The maximum path difference achievable in practice is about 1 m, enabling spectra to be obtained with a resolution of 0.005 cm^{-1}. In a practical system, provision may be made for the selection of different filters and detectors for different wavelength regions, each of limited extent to exclude source photons from parts of the spectrum not being scanned. This improves the signal–to–noise ratio by reducing the photon noise in the cooled photon detectors.

Interferometers have a number of general advantages over other types of spectrometric instrument, including higher spectral resolution, higher throughput, and wider wavelength coverage. They are, however, prone to additional sources of noise that can annul the advantages in a realistic, as opposed to an ideal, system. The first of these is phase errors. An interferogram may not be the simple sum of sine waves that it is in theory, if the path difference inside the instrument is a function of frequency due to optical dispersion in the beam splitter or other components. Phase shifts may also be introduced in the electronics. A second problem area in interferometers is channelling, a periodic modulation of the spectral baseline caused by optical interference inside individual components, and by stray light inside the system.

The effects of detector non-linearity on an interferogram are difficult to untangle, and this type of instrument is more likely to suffer from the problem in the first place because of the large dynamic range required of the detector. Fourier spectrometers are also extremely susceptible to vibration, electromagnetic interference, and to sampling errors. In summary, the design of an instrument of this type is an extremely complex and specialized affair, which explains the enduring popularity of grating and filter instruments in spite of their inferior theoretical performance.

9.8 Bibliography

9.8.1 *Notes*

An introduction to the topics in this chapter may be found in the books by Taylor (Chapter 9), Harries, Stevens and Houghton *et al.* Beer provides an excellent overview of the interferometric technique and Chantry of infra–red techniques in general; Boyd focuses on detector theory. There is, of course, a large technical literature on the subject of satellite remote sensing instrumentation; see for instance the website of the Society of Photo-optical Instrument Engineers (SPIE).

9.8.2 *References and further reading*

Atkins, P. W. and Friedman, R. S. (1996). *Molecular quantum mechanics.* Oxford University Press, Oxford.

Beer, R. (1992). *Remote sensing by Fourier transform spectrometry.* John Wiley & Sons Inc., New York.

Boyd, R. W. (1983). *Radiometry and the detection of optical radiation.* Wiley Interscience, New York.

Chantry, G. W. (1984). *Long wave optics.* Academic Press, New York.

Foote, M.C. and Jones, E. W. (1998). High performance micromachined thermopile linear arrays. Proc. SPIE, 3379, 192-197. *Infrared Detectors and Focal Plane Arrays V.* E. L. Dereniak, R. E. Sampson (ed.) Jet Propulsion Laboratory, Pasedena.

Harries, J. E. (1994). *Earthwatch: Climate from space.* Wiley-Praxis Series in Remote Sensing, New York.

Houghton, J. T., Taylor, F. W. and Rodgers, C. D. (1984). *Remote sounding of atmospheres.* Cambridge University Press, Cambridge.

Johnson, J. B. (1928). Thermal agitation of electricity in conductors. *Phys. Rev.*, **32**, 97–109.

Nyquist, H. (1928). Thermal agitation of charges in conductors. *Phys. Rev.*, **32**, 110–113.

Schottky, W. (1918). On spontaneous current fluctuations in different electrical conductors. *Ann. Phys. (Leipzig)*, **57**, 541–567.

Steinfeld, J. I. (1985). *An introduction to modern molecular spectroscopy.* MIT Press, Massachusetts.

Stephens, G. L. (1994). *Remote sensing of the lower atmosphere.* Oxford University Press, Oxford.

Taylor, F. W. (2005). *Elementary climate physics.* Oxford University Press, Oxford.

10

CLIMATE OBSERVATIONS BY RADIOMETRY–SPECTROMETRY

10.1 Introduction

Quantitative and spectroscopic measurements of radiation are at the heart of observations made to understand and monitor the climate system (Fig. 10.1). The time-dependent complexity of the system under observation requires that the measurements span the globe, including the vertical dimension. Many of the important processes, for example those involved in ozone depletion, are associated with transient dynamical phenomena and occur on local and diurnal scales, calling for high–resolution measurements in space and time. Energy–budget studies of the Earth require measurements at the surface and external to the planet, ideally with global coverage that is comprehensive in angular as well as geographical space, and time. The Earth radiates about the same total energy to space as a 255–K blackbody, with a spectrum rich in molecular vibration-rotation lines that contain information about the physical state of the atmosphere, and continuum fluctuations that represent weak and/or complex atmospheric bands plus surface and cloud emissivity variations. The most pronounced of these spectral features are those caused by absorption and emission in the bands of the principal atmospheric minor constituents, especially water vapour and carbon dioxide. Infra–red spectroscopy and radiometry from satellites observing the reflected and thermally emitted radiation as a function of latitude, longitude, time of day, and season, can yield global fields of vertical temperature and minor constituent distributions (e.g. water and ozone), cloud cover, thickness and height, and ocean temperature.

Individual measurements or profiles typically take a second or less to acquire, and can be repeated as the scene changes below, allowing the parameters to be mapped in three dimensions. Satellites in high-inclination orbits cover the whole planet approximately every day, make it possible to study processes at work in large-scale weather systems, the stratospheric ozone layer, pollution in the lower layers of the atmosphere, and the 'greenhouse' gases that drive global warming. Mineralogy, vegetation (including agricultural products and some of the species in the sea), the icy cryosphere, and volcanism and its products are also accessible to sensors in space. Radar and lidar are rapidly being adopted to investigate land and ice topography, sea state, and to infer wind fields.

In this chapter we discuss the use of radiometry and spectroscopy to measure

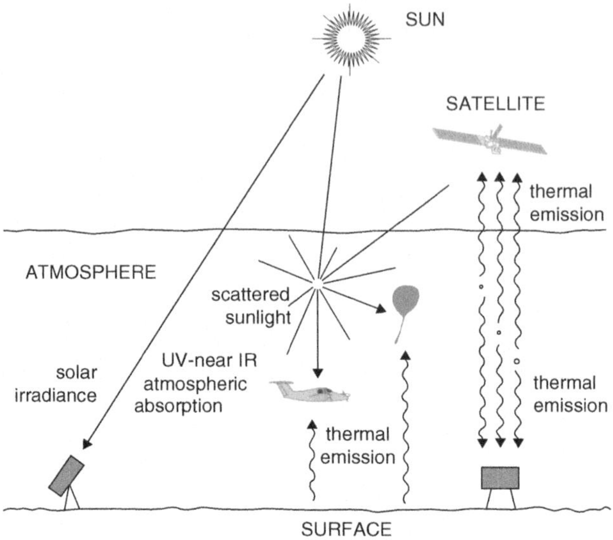

FIG. 10.1. An overview of some of the main ways in which climate data are collected using radiation measurements. Broadly speaking, the measurements are of two types: radiometric, to investigate the flux of energy in the climate system, and spectroscopic, to map temperature and composition using remote–sensing techniques for dynamical and process studies. The source of the radiation may be the Sun, viewed directly in the UV, visible or near–infra–red spectrum through the atmosphere or by scattering from atmospheric molecules, cloud particles or the surface, or it may be longwave thermal emission from these components of the climate system. The observing platform may be on the surface, airborne, or orbiting in space.

quantities useful for climate research, focusing on specific real instruments as much as possible. The goal is to define the state–of–the–art in climate measurements, and show how the fundamentals developed in earlier chapters are applied in practice.

10.2 Surface radiation budget: the pyranometer

The Earth–atmosphere system and global climate depend crucially on the surface radiation budget (SRB), i.e. the balance between shortwave and longwave irradiances and the outgoing fluxes at the Earth's surface. The SRB depends on the amount of incident solar radiation at the top of the atmosphere (TOA), the amount of energy absorbed, scattered or reflected towards the surface or back into space within the atmosphere and the reflection, absorption and longwave radiation from the ground at the surface itself. Radiative–transfer processes within the atmosphere depend critically on the concentrations of water vapour, trace gases (greenhouse gases) and aerosols, while the surface reflectance/emittance properties largely determine radiative processes at the Earth's surface.

The role of Earth's SRB for the global freshwater budget and the impacts of climate change and aerosol loads on a global scale are essential for a number of important issues. A reliable estimate of energies absorbed by the atmosphere and the surface layers of the Earth through an assessment of the SRB in combination with the radiation budget at the TOA will enable the calculation/modelling of heat transports in the different parts of the Earth–atmosphere system under current conditions and will ultimately improve our understanding of climate processes on the global to the regional scale. Such an understanding is an essential prerequisite to the modelling of Earth–atmosphere heat transports in atmospheric general circulation models (AGCMs, §11.4). At present, these models provide the only means of enhancing our understanding of climate processes and of possible future climate developments. The latter is particularly relevant with regard to anthropogenic climate change as a consequence of increased concentrations of atmospheric greenhouse gases.

A long-term SRB is also important because of the many processes within the climate system and the biosphere that are affected by the radiative fluxes. The net surface solar and terrestrial radiations affect the heating/cooling of the surface and act to control sensible and latent heat fluxes. The amount of solar energy and moisture on the Earth's land surface determine its ability to sustain life processes and these life processes affect in turn the net fluxes.

The SRB is also important from an applied point of view, since it governs the use of clean renewable solar energy, e.g. related to photovoltaic arrays and industrial applications such as solar–energy cookers and refrigeration technologies that are important to improving the quality of the environment.

Thus, knowledge of the SRB is required by many fields of science; these include: regional surface climatology, ocean modelling and energy budgets, polar climatology, validation of global climate models (GCMs), plant–growth models, understanding of forest ecosystems and forest fires and regional water resources, to just name a few.

However, the radiation budget not only exerts an influence on the climate system and the Earth's surface. There are important feedbacks that have to be taken into account, when considering the influence of surface and climate processes on the Earth's radiation budget. In particular, rapid changes in surface reflectance/emittance properties, arising, for example, from a retreating ice/snow layer due to global warming, have profound effects on net fluxes at the surface and may cause regional and global climatic feedbacks. Equally important are changes in cloudiness, aerosol concentrations, and radiatively active trace gas concentrations that regulate the surface fluxes in profound ways.

Thus, quantifying the changes in the SRB arising from changes in surface properties and atmospheric chemical composition is an important pre–requisite in understanding the processes of global change and its impact on ecosystems and societies.

When considering these problems, two issues become particularly important. Clouds, their dynamics and properties remain the major uncertainty in determining the radiation budget (IPCC 2001). The International Satellite Cloud Climatology Project (ISCCP), addresses this problem and provides the most extensive and comprehensive global cloud climatologies.

However, under clear–sky conditions, aerosols become a dominant determinant of the SRB and thus of climate development. Unlike greenhouse gases that drive global climate change though, aerosols are fundamentally different. A major difference is that aerosols have much shorter atmospheric lifetime (from 10^{-4} days for natural, to tens of days for anthropogenic aerosols) compared with the important greenhouse gases (decades to centuries). This, together with microphysical and mixing processes, results in larger spatial and temporal aerosol variability.

Hatzianastassiou *et al.* (2004a,b, 2006) have conducted detailed modelling studies to examine globally clear–sky shortwave aerosol forcing over mainly absorbing surfaces (such as oceans and vegetation-covered surfaces) versus low–absorbing/high–reflecting surfaces typical for the snow– and ice–covered polar regions. Their results demonstrate that an increase in relative humidity enhances the planetary cooling effect of aerosols over oceans and low–albedo land areas, whilst over polar regions and highly reflecting land surfaces, the expected warming effect of aerosols changes to a cooling effect. Thus, global warming and an associated increase in relative humidity, would lead to enhanced aerosol cooling worldwide. The sensitivity results also demonstrate that an increase in surface albedo due to, for example, a reduction in land vegetation cover as a result of enhanced desertification would lead to increased atmospheric warming by aerosols. This, in turn, will result in a reduction of cloud formation and a further enhancement of the desertification process.

With regard to the radiative properties of the Earth's surface, the high albedo of snow and ice surfaces in the polar regions are particularly important. However, the Arctic sea–ice cover, an important component in this regard, is apparently shrinking in area and decreasing in mean thickness. This can be considered the beginning of a positive feedback process where decreasing sea–ice surfaces lead to increasing heat uptake by the ocean, which results in further atmospheric warming. Thus, there are possible interlinkages between the SRB, climate change and a deteriorating Arctic sea-ice cover.

The Baseline Surface Radiation Network (BSRN) and the Global Energy Balance Archive (GEBA) are important examples of global data acquisition and archiving systems for data on observed climatologies, radiation and other fluxes

FIG. 10.2. The Eppley pyranometer is typical of the instruments used for the measurement of global, reflected, and diffuse shortwave radiation. The detector is a differential thermopile with the hot–junction receivers blackened and the cold–junction receivers whitened. Temperature–compensation circuitry frees the instrument from the effects of ambient temperature changes. A precision ground optical glass hemisphere of Schott glass (WG295) uniformly transmits energy from 0.285 to 2.800 μm over zenith angles from 0–70°. The sensitivity is 11 microvolts per W m^{-2} with a response time of 5 s. The size is 5 3/4 in. diameter, 3 3/4 in. high and the weight is 2 lbs.

at the Earth's surface. BSRN has measuring sites covering major climatic zones between the Arctic and the Antarctic. At these sites the main radiative fluxes, direct solar, diffuse sky, global radiation (Fig. 10.2), reflected shortwave radiation, longwave in– and outgoing radiation are measured at 1 Hz frequency with 1–min statistics. Many stations have spectral measurements of direct solar radiation for the estimation of the aerosol optical depth, total ozone, and precipitable water vapour. Most stations have synoptic and radiosonde observations. This is the only network of radiation measurements with a defined accuracy with collocated atmospheric observations.

GEBA is a database with monthly means of energy fluxes at the Earth's surface including direct solar, diffuse sky, global radiation, reflected shortwave radiation, longwave in– and outgoing radiation, net radiation, sensible heat flux, latent heat flux, subsurface heat flux, and heat of fusion. At present, GEBA possesses about 250,000 monthly means for about 1600 sites, including monthly mean irradiances transferred from the World Radiation Data Center in St. Petersburg. The oldest data in the archive is the August 1922 global radiation thermoelectrically measured at University of Stockholm by A. Ångström, followed by that from the IGY in the late 1950s.

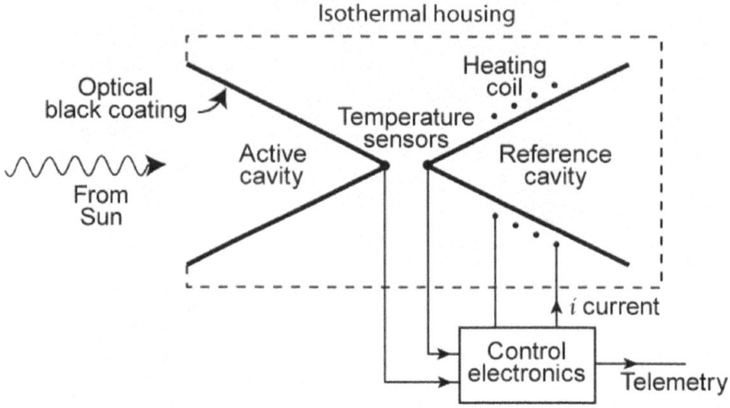

FIG. 10.3. The ACRIM II instrument contains two nominally identical conical cavities back–to–back, as shown here. The one on the left faces the Sun and is heated by radiation; the one on the right is blanked off and is heated electrically until the temperatures of the two cones, monitored by sets of sensitive thermistors, are identical. The power supplied to the electrically heated cone, divided by the area of the aperture of the radiation-heated cone, is an accurate measure of the solar irradiance.

10.3 Solar irradiance: ACRIM

One of the first applications of radiometry from space was the measurement of the solar constant, that is to say the total amount of energy from the Sun falling upon the Earth at all wavelengths. Obviously, this is a fundamental quantity that is difficult to measure from the surface because of atmospheric absorption and scattering from clouds and aerosols. The most successful technique is the 'active cavity method', in which the heating effect of the radiation is compared very precisely to the electrical energy required to maintain an identical cavity at the same temperature.

Figure 10.3 shows the details of the ACRIM II (Active Cavity Radiometer Irradiance Monitor) device that operates on this principle and that has been used to obtain a long series of measurements of solar constant since the late 1960s. An absolute accuracy of 99.9% and a long-term precision of a few parts per million is claimed for this sensor. It operates by controlling both the active and the reference cavities at the same relative temperature using an electronic servo. A shutter then repeatedly admits and excludes solar radiation from the active cavity for periods of about one minute each, and derives a value for the energy in the solar flux by comparing the difference between the asymptotic values of the electrical power drawn during the two periods. While conceptually simple, it will be clear that obtaining the high precision required to track the small changes in the solar output (typically around 0.1%) that occur under normal conditions requires very precise manufacturing tolerances in the blackness of the cavities, the temperature sensors, and the size of the aperture. Note that there is no transmitting component in the optical path, a key factor in permitting the desired precision to be achieved.

10.4 TOA radiation budget: ERB, CERES and GERB

Instruments on satellites in high-inclination orbits can measure the net reflected and thermally emitted radiation as a function of latitude, longitude, time of day, and season, covering the whole planet. The 'top of atmosphere' (TOA) radiative energy budget of the Earth, on both a global and regional basis, can then be built up from the balance between the measured ingoing and outgoing fluxes of radiation. Since the heating by the Sun and the cooling by the Earth are separated in wavelength, falling to the short– and long–wave sides, respectively, of about 4 μm, in principle the sensor need have only minimal spectral resolution, provided, however, that a high degree of radiometric accuracy is achieved.

Apart from the ubiquitous problems of absolute calibration in the hostile environment of space where the gradual deterioration of reference targets cannot be checked, real instruments cannot obtain a uniform spectral response over the wide wavelength ranges (roughly 0.4 to 4.0 μm for solar and 4.0 to 100 μm for thermal fluxes) required to measure the energy balance. If the radiation from the Earth or calibration target changes its distribution with wavelength, as well as its overall intensity, these are difficult to separate when analysing the data. Usually it is found necessary to break the wavelength range into segments that are measured and calibrated separately.

While blackened cavities, cones or grooved plates make reasonable reference sources for thermal–IR measurements, it is much more difficult to come up with a reliable and stable source of near–IR radiation of known intensity. Standard lamps and plates that diffuse the reflected light of the Sun into the instrument are commonly used, but are seldom completely reliable, requiring redundant sources to allow for failure or changes due to aging. In addition, integration of the measured radiances over a 2π solid angle is required, and where reflected sunlight is involved these may have strong directional components that are only imperfectly sampled due to the limitations of the observing geometry achievable from a single orbit. This limitation can be partially overcome by using angular scans that later can be integrated into hemispherical fluxes, and by fitting the data to empirical models of the reflectance properties of different regions of the Earth and integrating the model, with a corresponding addition to the error budget.

The first successful meteorological instrument on an orbiting satellite made radiation–budget measurements. Explorer 7, launched on October 13, 1959 carried a simple radiometer with five hemispherical thermistor detectors and crude wavelength selection by the use of black, white and gold coatings with differing sensitivities to solar and terrestrial radiation. With continuous data as the satellite circled the Earth at an inclination of 50° to the equator, it was possible to see for the first time how the energy from the Sun that is absorbed by the Earth is redistributed within the climate system, and finally re-emerges as longwave infra–red radiation.

NASA's Earth Radiation Budget Experiment (ERBE) scanners have three broad spectral channels, covering the wavelength ranges from 0.2 to 5.0 μm (i.e. effectively the entire solar spectrum), 5 to 50 μm (most of the Earth's thermal infra–red spectrum) and 0.2 to 50 μm (both). ERBE sensors flew on two Sun-synchronous polar orbiting satellites (NOAA-9 and NOAA-10) and on the Earth Radiation Budget Satellite (ERBS). The first of an improved version known as CERES (Clouds and the Earth's Radiant Energy System) was launched into a 35° inclination orbit in 1998 and the second in 2000 on Terra, part of the EOS series of satellites discussed further below. In Europe, radiation–budget sensors have been included on the geostationary platform Meteosat, using a field of view that covers the same part of the Earth at all times of day.

The error in the individual measurements of each component of the radiation budget in a single direction, place and time has been estimated at about ± 5 W m^{-2}, or about 2%, but the uncertainty in the overall balance of the planet is considerably larger than this, perhaps $\pm 10\%$ at best. By comparison, the so–called cloud–radiative forcing, a useful quantity in climate studies that is defined as the flux averaged over all conditions minus the clear-sky flux, is typically about 10 to 20% of the total. Thus, useful measurements of the local energy balance are achievable but global measurements with adequate precision are still beyond the current state–of–the–art, especially since our most important goal is to detect and understand changes in the global energy balance.

Measurements of the surface–radiation budget plays a key role in determining the surface temperature and evaporation and hence the hydrological cycle, with practical applications in weather forecasting, hydrology and agriculture over a wide range of time–scales. However, for climate monitoring and for evaluating climate models, high quality measurements of the radiation balance at the top of the atmosphere as well as at the surface must be available. In fact, algorithms exist for deriving the surface–radiation budget from the top-of-atmosphere radiation–budget measurements from satellites in both the solar and thermal regimes. Within the context of climate impact research, the distinct contributions from clear and cloudy skies can be investigated and identified.

Not only are the ERB at TOA and at the surface important, but also the difference between them, the atmospheric radiation budget. The concept of budgeting the radiative components of the Earth–atmosphere system is generally used to investigate the forcing exerted by changes in aerosol contents or variable or cloud cover and the vertical profile of the radiation divergence that determines the available potential energy for the dynamic processes of the atmosphere, allowing intercomparison with climate–model results and a better understanding of dynamical atmospheric processes. The main contribution to the profile comes from clouds and aerosols; for example, large plumes of Saharan dust and of smoke from biomass burning are frequently seen in satellite imagery. These have a significant impact on the radiation budget and on the retrieval of surface temperatures that need to be quantified and understood.

The available observations include detailed chemical composition measurements, aerosol physical characterization (size–distribution measurements), aerosol optical properties (scattering and absorption) and results from lidar profiles. Satellite data from the new Meteosat Second Generation satellites includes improved images that give a global climatology of cloud parameters, allowing studies of the diurnal cycle of convective clouds at an unprecedented temporal resolution, and, over the decadal lifetime of the MSG programme, an exciting insight into the interannual variability of this cycle and its influence on precipitation.

Clouds, their dynamics and properties remain the major uncertainty in determining the radiation budget. The precise effects of cloud feedback in climate trends are uncertain enough to completely overwhelm the forcing. The International Satellite Cloud Climatology Project (ISCCP), addresses this problem and provides the most extensive and comprehensive global cloud climatologies to provide information on how the cloud field is changing with time, either due to natural variability or to changed greenhouse forcing. Our understanding of the occurrence of cirrus clouds as well as of their microphysical and optical properties is incomplete. Furthermore, there are indications that the steadily growing air traffic may significantly increase high–level cloudiness. The availability of MSG observations will allow the existing uncertainties on the radiative forcing of natural as well as on aircraft–induced cirrus to be reduced. The combination of the instruments SEVIRI and GERB on MSG will be ideally suited to investigate the radiative forcing of cirrus clouds, because it provides the necessary information to infer cirrus optical and microphysical parameters from SEVIRI and to determine accurately and independently the radiative fluxes from GERB.

Aerosols affect the Earth's radiation budget both directly and indirectly, inducing an average global negative radiative forcing, a cooling effect, which may counteract global warming, positive forcing due to increases of the well–mixed greenhouse gases, estimated to be 2.43 W m^{-2} (IPCC 2001). Aerosols exhibit strong temporal and spatial variability, producing globally uneven patterns of radiative forcing (compared to greenhouse gases). Besides, there is a large uncertainty in the crucial aerosol radiative parameters (e.g. optical thickness and single–scattering albedo), which along with the small residence time of atmospheric aerosols (few days to weeks), and their variable concentrations, makes difficult the characterization of aerosols as climate-forcing agents, and the quantification of their radiative forcing difficult.

Although significant progress has been achieved in better characterizing the radiative forcing by different types of aerosols (e.g. sulphate but also carbonaceous or organics), the level of scientific physical understanding of climate forcing due to aerosols is still very low and remains one of the largest uncertainties in climate variability and climate–change studies. Relative–humidity changes affect aerosol liquid–water content, size and number distribution, total concentration, and composition, and hence change the extinction coefficient and refractive indices and thus the aerosol microphysical, optical, and radiative properties. The forcing is highly variable in consequence. The effect of reduced cloud cover due to the indirect effect of aerosols is to increase the magnitude of aerosol forcing, since this implies an increased clear–sky fraction, and hence, clear–sky direct aerosol forcing at TOA, within the atmosphere, and at the surface. Thus, there is a larger clear–sky effect of aerosols on the outgoing solar radiation at TOA (higher planetary cooling), more atmospheric warming and surface cooling, i.e. reduced downward solar radiation at the surface.

Finally, satellite data allow an assessment of current sea–ice variability and change in the polar regions and its possible modification by, and feedbacks to, climate change, as well as ramifications for the radiation budget.

The ERB instrument flown on Nimbus 7 is shown in Fig. 10.4. A special feature of this type of instrument is the response function for the thermal infra–red passband, which has to be wide enough to include effectively all of the energy from a 240 K blackbody, and also spectrally flat over the whole range (approximately 0.2 μm to 50 μm). This is achieved in the ERB design by using reflecting optics, except for a filter made of diamond. The thermal and reflected solar components are separated by measuring the latter separately with channels that include a filter made of Suprasil-W fused silica, which cuts off at about 3.8 μm, and taking the difference to get the longwave flux. The two bands are separated using a reflecting chopper, arranged so that a reference blackbody fills the beam during the part of the cycle when the target is not in view. Note the extensive baffling of the primary optics to cut out stray light.

The ERB instrument was calibrated before flight against active cavity radiometers similar to ACRIM, for the solar channels, and a specially fabricated cavity blackbody with an emissivity in vacuum of 0.995 and an operating range of 180 to 390 K, for the thermal component. The solar calibration is particularly difficult because the Earth's atmosphere modifies the wavelength distribution of the energy reaching the detector, so that if the instrument response function is not perfectly flat, as in practice must be the case, the calibration against the cavity instrument can be in error. This error is small in a well–designed instrument, and can be further reduced by calibrating with several targets (in this case solar simulators and tungsten bulbs) with different effective temperatures, or by using filters to check the calibration in different segments of the wavelength range.

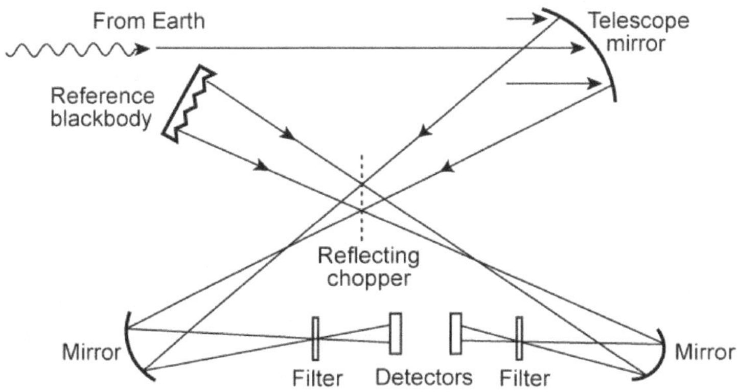

FIG. 10.4. Schematic of the ERBE (Earth Radiation Budget Experiment) instrument. The modulator or 'chopper' is a rotating blade, polished to be highly reflecting on both sides. The two sets of infra–red detectors, arranged symmetrically as shown, alternate between looking out at the Earth or at an internal reference blackbody every 20 ms as the blade rotates. The different spectral channels are selected by filters in front of the detectors; here, only one detector–filter pair is shown on each side for clarity.

10.5 Sea surface temperature: ATSR

The surface temperature of the Earth's oceans is a key climate parameter and also a tracer for warm and cold currents near the surface. Again, rather precise measurements are required, with precisions of <0.1 K and absolute accuracies of better than 0.5 K. Achieving these fairly demanding values from space is made particularly difficult by the presence of the intervening atmosphere, which has absorption features even in the most transparent spectral windows, plus very variable attenuation due to thin clouds and aerosols.

A recent instrument designed to surmount these difficulties by using two views of the surface at an angle of 60° to each other, is the Along–Track Scanning Radiometer (ATSR). The radiance measured from the same footprint from different points in the orbit is composed of essentially the same component from the surface (assuming the surface emissivity is independent of direction, which is generally true to a good approximation), but with the atmospheric absorption and emission weighted by the cosine of the zenith angle as a result of the longer path length. ATSR has channels in the windows at 3.7, 10.8 and 12.0 μm (Fig. 10.5). Within the instrument, the direction of view is actually swept by a rotating mirror through a cone (Fig. 10.6), skewed so that one side is in the local vertical to give the nadir view. The functions of modulation and calibration are combined by employing a sufficiently rapid scan that encompasses the scene and the two standard sources, producing a video–type output. This approach eliminates the effect of background drift, as well as the need for a separate

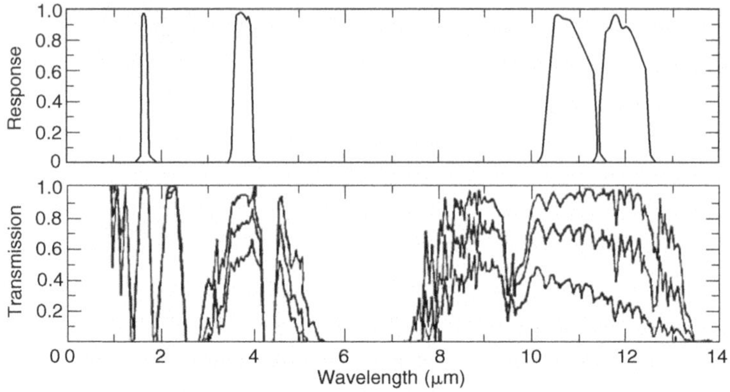

FIG. 10.5. The filter bandpasses that define the ATSR spectral channels are shown at the top; a response of 1.0 means the filter is perfectly transmitting at that wavelength while 0 means it is opaque. Perfectly square filters are difficult to manufacture so real instrument profiles have complex shapes as shown here. The bottom panel shows the transmission of the atmosphere, from the sea surface to the satellite, as a function of wavelength. Comparing the upper and lower panels shows how the filters select radiation from the surface in spectral 'windows', i.e. regions of high transparency, so the obscuring effect of the atmosphere and the contribution emission from the atmosphere makes to the signal are minimized.

chopper and calibration mirror. It also has the multiple advantages of scanning from side to side of the subspacecraft track, of allowing views of two calibration blackbodies positioned between the downward– and forward–viewing apertures every time the Earth–viewing direction is switched, and of accomplishing the switching quickly and smoothly. Otherwise, ATSR is conceptually a fairly conventional modern infra–red radiometer, using interference filters to define the spectral bands and closed–cycle refrigerators to cool the HgCdTe detectors.

The key to success in measuring the subtle effects of ocean temperature on the climate system, and in particular to have hope of measuring secular changes, is careful calibration of the instrument before flight. Figure 10.7 shows the arrangement whereby ATSR was geometrically and radiometrically calibrated under vacuum. The field of view measurement used a hot point source in the form of a tungsten ribbon filament, capable of temperatures up to 3200 K, illuminating a pinhole of diameter 70 μm in the focal plane of a fixed offaxis paraboloid, with a two–axis gimbal–mounted mirror in the configuration shown in Fig. 10.7 to provide a scanning, collimated input beam. The size of the mirrors was such as to completely fill the aperture of the space instrument during testing, which is essential since the optical train is not symmetric about its axis.

For radiometric calibration, two conical blackbodies were used, arranged so that

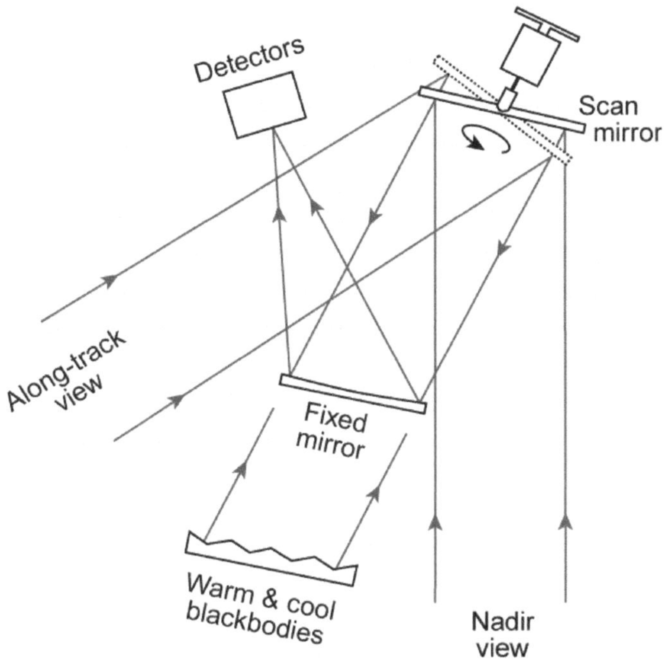

FIG. 10.6. Schematic of the Along–Track Scanning Radiometer. The filters described in Fig. 10.5 are in the housing with the detectors, which is cooled to 77 K by a miniature Stirling–cycle refrigerator. The scan mirror is spun by a motor, and is tilted, as shown, with respect to the rotation axis so that the view seen by the detectors describes a circle. Within that circle it scans across (i) a nadir (vertically downwards) view of the Earth, (ii) an along–track view at 53° to the vertical along the orbit track, so that the same area is seen twice about 2.5 minutes apart as the spacecraft moves in its orbit; (iii) a cool blackbody (inside the page in this diagram) and (iv) a warm blackbody (above the page). The blackbodies are kept at temperatures representative of the upper and the lower values of the sea–surface temperature (265 and 305 K).

both were viewed simultaneously through the two Earth–viewing ports, and mounted on a rotating plate that allowed the two calibration targets, at different temperatures representing the range expected from the Earth, to be interchanged during the course of a test run. Calculations established the emissivity of the reference blackbodies to be 0.9985 at 3.7 μm and 0.9995 in the longer-wavelength channels. Other practical details of importance are the complex system of blackened, temperature-controlled baffles that represent the operating environment of the instrument when in space by simulating the radiative background due to the spacecraft, the Earth, and cold space.

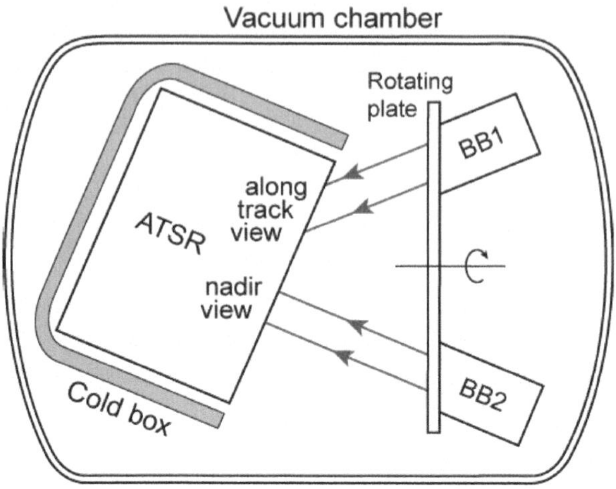

FIG. 10.7. For pre–flight laboratory calibration, the Along-Track Scanning Radiometer (ATSR) was installed in a vacuum chamber at an angle so it could view two standard radiation sources (blackbodies) BB1 and BB2 at different temperatures that span the range of expected target temperatures on the Earth. The internal scan mirror of the instrument views the onboard calibration blackbodies and the standard sources in the chamber in rapid succession. The sources in the chamber are of superior construction to those inside the instrument, but are themselves too bulky to be carried into space. Hence they are used to cross–calibrate the simpler inflight blackbodies and determine their emissivity. BB1 and BB2 are mounted on a rotating plate that allows them to be interchanged, so that any radiometric differences between the downward and along–track views can be assessed. The inside of the chamber has liquid–nitrogen–cooled walls and other thermally controlled surfaces to simulate conditions on the satellite in space.

10.6 Surface properties: Thematic Mapper and MODIS

For observations from space of the land, the key variables are the reflectivity (in the near–IR) and the emissivity (in the thermal–IR). Figure 10.8 shows how these 'fingerprints' vary for some common materials. Careful choice of wavelength filters (for example, isolating the chlorophyll bands) enhances sensitivity to vegetation types and different stages of growth, or can address the mineralogy and physical state of the surface, with a view to studying desertification, for example, or even prospecting for oil or other valuable deposits. Problems that have to be addressed include the contaminating and obscuring effects of the atmosphere, the relatively bland spectra of most minerals, and the fact that material of different kinds is always mixed and layered in any scene. Spectroscopic data are also sensitive to parameters such as fractional cover, shadowing and dampness, and analysis often requires the application of advanced image processing and empirical models to separate the wide range of variables.

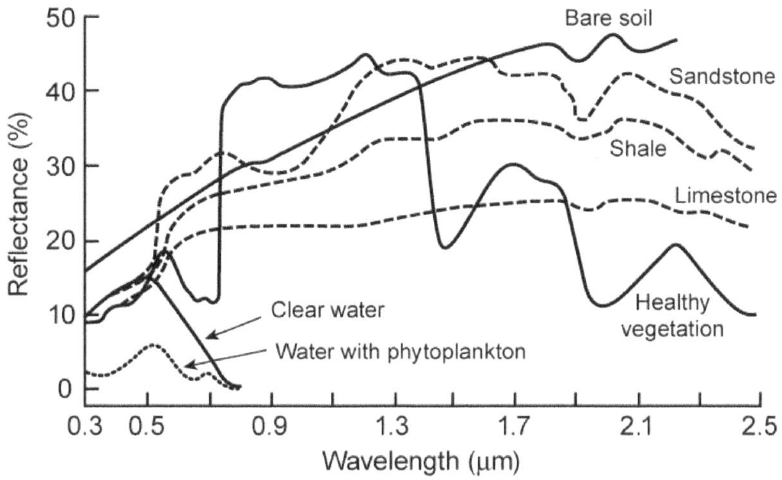

FIG. 10.8. Spectral signatures of some common materials in the near–IR. Similar differences apply in the thermal–IR and microwave spectral regions: for instance, the dielectric constants of snow, water, and ice are so different that the effect on emission is easily detected. At a wavelength of 1.55 cm, for example, the emissivity of sea water is 0.44, whereas that of ice is approximately twice as great.

One of the earliest instruments to work on this principle was the Thematic Mapper on the *Landsat* series of Earth satellites. This had six spectral channels in the visible and near–IR from 0.45 to 2.35 μm and one in the thermal–IR 'window' at 10.4 to 12.5 μm. The instantaneous field of view was 30×30 m^2 at the surface from the orbital altitude of 705 km, with a swath width of 185 km comprising twenty simultaneous measurements using a linear array of detectors in the focal plane aligned perpendicular to the scan direction. The large aperture telescope – 50 cm across – makes for an instrument measuring $0.9\times0.9\times1.8$ m^3, weighing 325 kg, and with a power consumption of 250 W. In analyzing the data, the experimenters normally worked with the ratio of signals from pairs of channels, one inside the selected feature, the other just outside, to reduce unwanted effects such as those due to variations in surface illumination or the intervention of thin clouds.

Remote sensing from Landsat enabled the rapid acquisition of land–surface data that otherwise would have been costly, if not impossible, to collect. Recently, the Terra and Aqua satellites (the first two components of the Earth Observing System, EOS, launched in 1999 and 2002), have carried two new instruments, ASTER (Advanced Spaceborne Thermal Emission and Reflection Radiometer) and MODIS (Moderate Resolution Imaging Spectrometer) for multispectral thermal imaging. MODIS has 490 detectors in 36 spectral channels, 19 covering the visible and near–IR and the rest in the thermal–IR. As with ATSR, a contin-

Table 10.1 *Summary of MODIS spectral bands and uses, from Ward (1997). In the RH column, the small numbers (<1) are noise equivalent temperature change (K) and the large numbers (>10) are signal–to–noise ratios, for typical targets in each case.*

Primary use	Band	Bandwidth (μm)	SNR or NEΔT
Land/cloud	1	.620 – .670	128
boundaries	2	.841 – .876	201
	3	.459 – .479	243
	4	.545 – .565	228
	5	1.230 – 1.250	74
	6	1.628 – 1.652	275
	7	2.105 – 2.155	110
Ocean colour	8	.405 – .420	880
Phytoplankton	9	.438 – .448	838
Biogeochemistry	10	.483 – .493	802
	11	.526 – .536	754
	12	.546 – .556	750
	13	.662 – .672	910
	14	.673 – .683	1087
	15	.743 – .753	586
	16	.862 – .877	516
Atmospheric	17	.890 – .920	167
water vapour	18	.931 – .941	57
	19	.915 – .965	250
Surface/cloud	20	3.660 – 3.840	0.05
temperature	21	3.929 – 3.989	2.00
	22	3.929 – 3.989	0.07
	23	4.020 – 4.080	0.07
Atmospheric	24	4.433 – 4.498	0.25
temperature	25	4.482 – 4.549	0.25
Cirrus clouds	26	1.360 – 1.390	150
Water vapour	27	6.535 – 6.895	0.25
	28	7.175 – 7.475	0.25
	29	8.400 – 8.700	0.05
Ozone	30	9.580 – 9.880	0.25
Surface/cloud	31	10.780 – 11.280	0.05
temperature	32	11.770 – 12.270	0.05
Cloud top	33	13.185 – 13.485	0.25
altitude	34	13.485 – 13.785	0.25
	35	13.785 – 14.085	0.25
	36	14.085 – 14.385	0.35

uously rotating mirror views the Earth and internal calibration targets in one rotation, but in this case also views cold space for a zero–radiance reference.

The radiation from the scan mirror is reflected via a telescope onto a series of beam splitters that divide the photons spectrally into four focal planes, as shown in Fig. 10.9. The energy is directed through focusing optics onto the arrays of detectors, each array coated with an individual bandpass filter. The 36 bands are summarized in Table 10.1.

The calibration philosophy for MODIS differs from that of ATSR because the former observes the sea surface, where the emissivity is quite well known in advance. For land observations, it is still useful to have rigorous pre–launch calibration against traceable sources and an onboard calibration system using an aluminium v–groove plate with a calibrated emissivity and accurately monitored temperature as one point and the space view as the other. In addition to knowing absolute radiance, a ground truth method is used in which an airborne imaging system, with similar characteristics as MODIS on an aircraft, images the same scene as MODIS at the same time. Since the error in the airborne system is much less than that of the satellite instrument, the airborne measurements are taken as truth and the calibration routine can be modified to correct for the difference between calculated and measured thermal imaging surface emissivity. At the same time, the surface campaign can identify visually and otherwise what the material is in the field–of–view of both instruments and allow a database of spectral fingerprints to be built up using real data.

MODIS data is finding wide application in many areas of natural and sustainable resource management, monitoring land–cover dynamics and land use, carbon–budget modelling, coastal and ocean–water quality monitoring, aerosol budgets, production in agriculture, forestry and fisheries including crop–yield monitoring and forecasting. It also has value in the emergency–response area, for the management of emergencies such as fires and floods and following the evolution of volcanic ash plumes.

10.7 Atmospheric temperature: HIRS and AIRS

Atmospheric vertical temperature profiles obtained by infra–red remote sensing from satellites are now routinely used by the meteorological services to augment traditional sources such as balloon–borne radiosondes. The High-resolution Infrared Radiation Sounder or HIRS, which has been in use for many years for this purpose, featured a large number of channels (20) with the highest spectral resolution that can be obtained with interference filters (about 10 cm^{-1}), in order to separate the emission intensities from different parts of the vibration-rotation bands of CO_2 at 15 and 4.3 μm. These originate from different height ranges in the atmosphere, and permit the reconstruction of a vertical profile of temperature. Crude profiles of water vapour are achieved by the inclusion of three channels in the 6.7 μm band of H_2O.

The HIRS optical system, like MODIS, uses dichroic beamsplitters to provide three channels in the visible, near–IR and mid–IR, further multiplexing being by means of a single filter wheel carrying two rings of filters. The latest, advanced temperature sounders use fixed-grating spectrometry or interferometry to have large numbers of narrow channels sampled simultaneously. The AIRS (Atmospheric Infra-Red Sounder) on the Aqua spacecraft alongside MODIS measures simultaneously in 2378 spectral channels from 0.4 to 1.7 μm and 3.4 to 15.4 μm, with a spectral resolution of $\lambda/\Delta\lambda \approx$1200, which corresponds to about 0.01 μm in the temperature–sounding channels. A diffraction grating disperses the

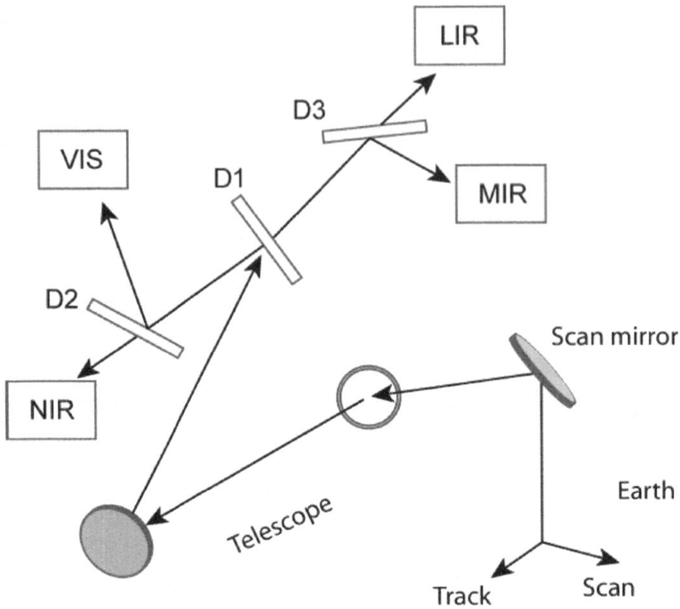

FIG. 10.9. The MODIS optical layout. The scan mirror reflects the Earth view into the telescope assembly and a system of dichroic beamsplitters separates the incoming radiation by wavelength into four channels; visible (VIS) (0.412 to 0.551 μm), near–infra–red (NIR) (0.650 to 0.940 μm), short/mid–infra–red (MIR) (1.240 to 4.565 μm), and longwave infra–red (LIR) (6.715 to 14.235 μm).

radiation from the scene onto 17 linear arrays of HgCdTe detectors on the focal plane (Fig. 10.10), giving 90 simultaneous ground footprints across the flight track of the satellite. Each detector element has a field of view of 1.1°, which corresponds to a spatial resolution at the surface of about 10 km. The detectors are cooled to 60 K by a Stirling cycle pulse tube cryocooler, while the rest of the spectrometer is cooled to 155 K by a two-stage passive radiative cooler to reduce the background flux and achieve the required instrument sensitivity. A scan mirror that makes a full cycle every 2.67 s provides spatial coverage and views of cold space, hot calibration targets and a visible radiometric source. The scan mirror is radiatively coupled to the spectrometer so that it operates at a temperature of 250 K. Since the mirror by definition has a low emissivity, this is cool enough to reduce the flux emitted from the mirror into the instrument to an acceptable level.

The AIRS instrument also includes four visible/near–IR channels between 0.40 and 0.94 μm, primarily for the diagnostics of cloud in the IR field–of–view. Cloud contamination is a major problem for tropospheric sounders: even with the relatively high spatial resolution of AIRS, it has been estimated that only about

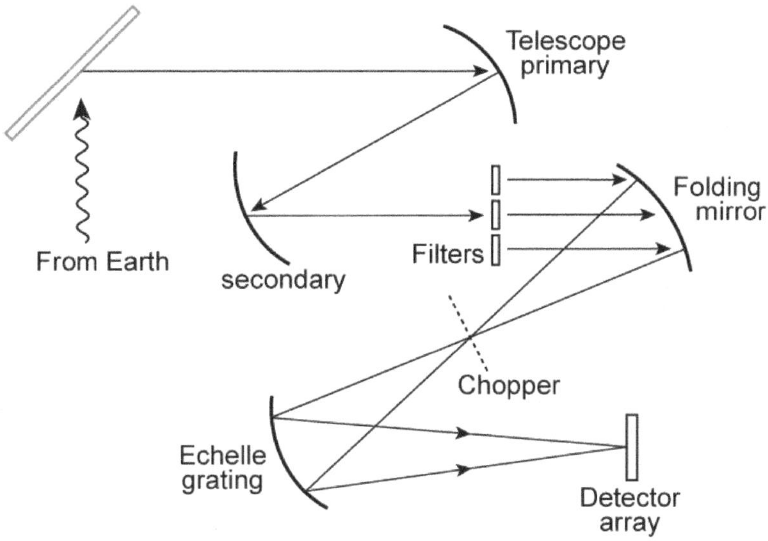

FIG. 10.10. Diagram of the AIRS instrument. The principle is similar to that of ERBE and MODIS (Figs. 10.4 and 10.9), except that the spectral channels are defined not only by filters, as shown, but then by an eschelle grating (a diffraction grating on a curved substrate, so that it focuses as well as disperses the radiation). The result is high spectral resolution and the capability to have a large number of spectral channels (see Fig. 10.11)

5% of measurements made over the ocean have less than 0.6% change in radiance due to the presence of cloud in the field–of–view. For this reason, AIRS flies in tandem with a pair of microwave sounders, AMSU (Advanced Microwave Sounding Unit) and HSB (Humidity Sounder for Brazil) that sense millimetre-wavelength radiation that is insensitive to the presence of most types of cloud. Together, the three sounders can achieve temperature–profile retrievals with an accuracy of better than 1 K in 1–km vertical layers in the troposphere under clear and partly cloudy conditions, the goal set in the 1990s by the World Meteorological Organisation for accurate numerical weather prediction. In addition to atmospheric temperature profiles, AIRS routinely produces data sets on each of the following:

sea-surface temperature;

land-surface temperature and emissivity;

relative humidity profiles and total precipitable water vapour;

fractional cloud cover;

cloud spectral IR emissivity;

cloud-top pressure and temperature;

total ozone burden of the atmosphere;

column abundances of minor atmospheric gases such as CO_2, CH_4, CO, and N_2O;

outgoing longwave radiation and longwave cloud–radiative forcing;

precipitation rate.

In nadir temperature sounding, as practiced by HIRS and AIRS, the upwelling radiation is measured by an instrument that is looking vertically downwards, and so viewing the complete atmospheric column. Vertically resolved information about the temperature, including the surface temperature at the lower boundary, is obtained by selecting several spectral intervals of different opacity in which to measure the radiance. Inside the strong absorption bands of infra-red-active atmospheric gases such as carbon dioxide the atmosphere is highly opaque, and the radiation measured originates near the top. In a more transparent region, the radiation originates in the lower atmosphere and from the surface. At wavelengths where the absorption has an intermediate value, the radiance originates somewhere in the middle of the atmosphere. The weighting of the emission that contributes to the measured radiance thus varies with the wavelength at which the measurement is made, and the contribution as a function of height is described by a wavelength–dependent weighting function.

The radiance at the top of the atmosphere in a narrow interval $\Delta\lambda$ centred on wavelength λ is

$$R_\infty(\lambda, \Delta\lambda) = \int_0^\infty B(\lambda, T)\mathrm{d}t(z, \lambda, \Delta\lambda) \tag{10.1}$$

$$= \int_0^\infty B(\lambda, T)\frac{\mathrm{d}t(z, \lambda, \Delta\lambda)}{\mathrm{d}z}\mathrm{d}z + B(\lambda, T)\, t(0, \lambda, \Delta\lambda), \tag{10.2}$$

where the Planck function, B and the transmittance from level z, t depend strongly on λ but more weakly on $\Delta\lambda$, provided $\Delta\lambda$ is narrow. If $\Delta\lambda$ is in a strongly absorbing region, $t(0, \lambda, \Delta\lambda)$ tends to zero, the second term that represents the contribution from the surface to space vanishes and the radiance leaving the top of the atmosphere is just the integral over height of the Planck function weighted by the derivative with respect to height of the atmospheric transmittance along the viewed path. By choosing a series of adjacent spectral intervals along the edge of an atmospheric absorption band, where the absorption coefficient changes rapidly with wavelength, it is possible in principle to select a set of weighting functions that span the atmosphere from top to bottom.

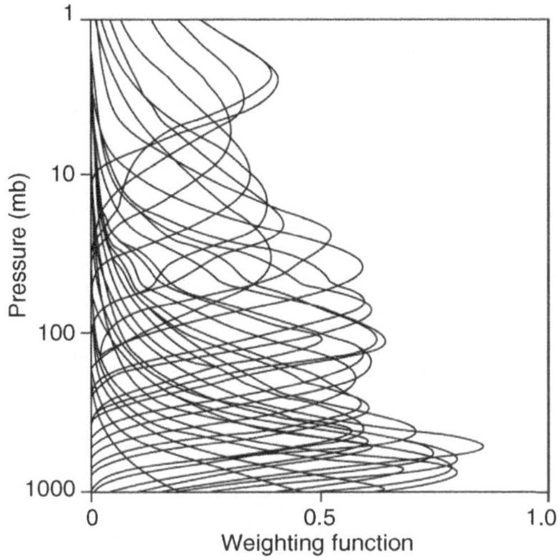

FIG. 10.11. Weighting functions corresponding to a few of the 2378 spectral channels in the AIRS instrument. Those shown here are in the 1100 cm^{-1} to 600 cm^{-1} (about 9 to 16 μm) spectral region. The log pressure scale is approximately proportional to height: 100, 10 and 1 mb correspond to about 15, 30 and 50 km above the surface, respectively.

Figure 10.11 shows what is actually achieved by AIRS, which has no less than 2378 infra–red channels sampled every 2.7 s.

The calculation of the weighting functions requires knowledge of the composition of the atmosphere, at least the abundances of the principal absorbers in the spectral intervals chosen for the sounding instrument. This is one of the reasons why wavelengths dominated by CO_2, or in the microwave O_2, are chosen in general, since these gases are nearly uniformly mixed. However, if a simultaneous set of observations is made at wavelengths dominated by an absorber of unknown concentration, for example water vapour, then a second retrieval can be performed in which the profile of the absorber amount is the unknown to be retrieved. Since most of the radiation originates near the peak of the relevant weighting function, a set of radiance measurements contains information about the variation with temperature with height. A radiometric instrument on a satellite can make such a set of measurements, which can then be used to reconstruct the temperature profile with a vertical resolution that is related to the number and width of the weighting functions. An optimized set of nadir (downward–viewing) radiances can yield profiles with a temperature accuracy of one or two K with a vertical resolution of around 10 km. Limb–viewing instruments have much sharper weighting functions that can resolve layers only 1 to 2 km thick, although of course the horizontal resolution is poorer, in the direction along the line–of–sight (but not perpendicular to that), generally around 200 km. Often, this is an acceptable sacrifice since the most interesting features are those that vary over relatively short distances in the vertical.

The conversion of radiometric measurements into geophysical quantities uses retrieval methods based on radiative–transfer theory to find the best solution for the values of the variables of interest that are consistent with the observed radiation field leaving the top of the atmosphere, and quantify the errors and uncertainties. Several different approaches are possible (see Rodgers 2000) but all involve fitting calculations of the spectrum of the outgoing radiation to the atmospheric parameters that produce a match to the measurements. Most of the retrieval methods used in practice are based on least-squares fitting of measured to calculated radiances, and often include a number of refinements. For example, instead of finding the profile from the measurements alone, statistical 'a priori' information about the expected value of the profile, based on a history of previously measured values ('climatology'), can be folded in. One way to do this is to use the climatological value of the temperature profile as the starting point for a retrieval by iteration, or to use independent data from another instrument (as AIRS does with AMSU and HSB, as discussed above) to tie down the effects of clouds, aerosols and humidity on the observed radiances. These quantities can also be products of the retrieval; the realistic representation of the time-dependent behaviour of clouds, their formation processes and their radiative effects, is a key problem in climate modelling and prediction, and a major source of uncertainty that only more and better data will resolve.

10.8 Atmospheric composition: IRIS, ATMOS and TES

Infra-red and microwave instruments also measure atmospheric composition. The intensity of the emitted thermal flux depends on the temperature and the abundance of the emitter (i.e. on the probability per molecule of emission of a photon of given wavelength, and on the number of emitting molecules present). So, once the temperature has been determined from the emission by a molecule of known abundance, such as CO_2, measurements of the same atmospheric path at a different wavelength in an emission band of a different species can yield information on the concentration of the second species.

Among the first targets for global measurement campaigns were the oxides of nitrogen, N_2O, NO, NO_2, and N_2O_5, following the discovery that they are involved in the catalytic destruction of ozone, posing a threat to the Earth's shield against solar ultraviolet radiation. Most members of the NO_y family are very reactive and the amounts available to interact with ozone vary dramatically with space and time and cannot be followed effectively by traditional measurements from aircraft or balloons. Global measurements of nitrogen dioxide reveal large gradients vertically, diurnally, and with latitude, and nitric oxide in the upper stratosphere shows a strong correlation with solar activity. NO_2 is removed by conversion to 'reservoir' species such as dinitrogen pentoxide and nitric acid in the long polar night. Reactions involving chlorine and other halogen–bearing species also have a major role and measurements of such reactive molecules as chlorine oxide (ClO) have also become a priority. An understanding of the chemistry going on in all regions of the atmosphere is crucial to predicting the fate of the ozone layer at all latitudes.

The lower atmosphere is experiencing a gradual change due to the accumulation of minor constituents such as carbon dioxide, carbon monoxide, chlorofluorocarbons, nitrous oxide, and methane. These are radiatively active gases that affect the energy balance at the surface, in part by radiating back absorbed radiative energy (the greenhouse effect), leading to a general warming. Changes in reactive gases such as ozone have a profound effect on the chemical mix that affects not only the 'greenhouse' but also leads to phenomena such as acid rain. The first remote–sensing experiments to measure tropospheric carbon monoxide globally for the first time in 1976 found high concentrations, not over the urban industrial centres of N America and Europe, as expected, but downwind of the forests of S America and Africa. Since then, the importance of biomass burning has been in the headlines. Today, remote–sounding instruments on satellites are monitoring a wide range of greenhouse gases, ozone–destruction precursors, and other pollutants in a global check on the health of the planet.

The Infrared Interferometer Spectrometer (IRIS) was a very early example of an instrument designed to conduct spectral 'surveys' of the Earth, that is, high-resolution measurements spanning a wide spectral range, which could be used for a variety of investigations, not all of which need to be planned in advance. A number of versions have flown, not only in Earth orbit but also to Mars, Jupiter, Uranus and Neptune, where they made excellent progress in understanding atmospheric structure and composition.

IRIS was launched on Nimbus 3 in 1969, becoming the first Michelson interferometer in space. A similar version with higher spectral resolution (2.8 versus 5 cm^{-1}) and a narrower field–of–view (5° versus 8°) flew a year later on Nimbus 4. Both had potassium bromide beamsplitters, giving a spectral range of 400 to 1600 cm^{-1} (6.3 to 25 μm); the relatively modest spectral resolution corresponds to a total mirror travel of only a few mm. This travel occurs at constant speed, with sampling of the signal from the detector at regular small intervals. The mirror position at the sample points is required to very high accuracy, and is achieved by using a second interferometer with a monochromatic source at a frequency much higher than the greatest infra–red frequency being measured. In IRIS, this was a neon discharge lamp, or a He-Ne laser in some versions. The sampling point was then tracked by counting the zero crossings of the signal from the reference source when it passed through the same interference path as the infra–red beam.

Background radiation from the instrument components was reduced by thermally isolating the baseplate from the warm spacecraft, and allowing the housing to cool radiatively to space, down to an electrically thermostatted value of 250 K. The thermistor bolometer detector was not cooled below this temperature, and had a sensitivity that dictated an integration time of 10.9 s to scan the spectrum. In this time an orbiting spacecraft moves nearly 100 km, tending to change the scene below during the course of an integration, and leading to aliasing of any spatial variations into the spectrum. This was prevented by image-motion compensation, using a scan mirror in front of the interferometer. The same mirror scanned to cold space and to a warm target for two–point radiometric calibration.

ATMOS (atmospheric trace molecular spectroscopy) differed from IRIS principally by using the Sun as a source, and by employing cooled detectors, so that the signal–to–noise ratio obtained was very high. It took advantage of this by having an enormous physical mirror displacement, which provided the very high spectral resolution needed to detected pollution molecules present in the atmosphere in quantities of less than one part per billion by volume, and the short scan time required to obtain spectra before the tangent path through the atmosphere changed significantly due to the orbital motion of the space shuttle to which it was mounted. By moving both mirrors through a maximum physical displacement of 25 cm in 2.2 s, spectra covering from 600 to 4800 cm^{-1} with a resolution of 0.013 cm^{-1} were achieved, while the line of sight between the instrument and the Sun moved only about 2 km relative to the surface of the Earth at the tangent point. Figure 10.12 shows a schematic of the optical layout. Most of the complexity results from the need to fold and focus the beams, and to drive both mirrors. Note also the inclusion of a camera that images the Sun continuously in the visible, in order to verify *post facto* the positioning of the field–of–view in the centre of the disc, and to check for sunspots, or refractive effects in the Earth's atmosphere, both of which affect the spectrum and its baseline. The detector is cadmium mercury telluride, cooled by a mechanical refrigerator.

The Tropospheric Emission Spectrometer (TES) is an infra-red imaging Fourier transform spectrometer with mechanically cooled optics and detectors, operating in the spectral range 650–3050 cm^{-1}(3.3–15.4 μm). TES is designed to measure thermal emission from tropospheric minor constituents with high spectral and high spatial resolution from the Earth Observing System satellite AURA, launched in June 2004.

The main focus is on tropospheric ozone, a key constituent in pollution studies for which no good global data exist, but the wide spectral range means dozens of species (including NO_y, CO, SO_2 and other sulphur compounds) are observed to address the atmospheric chemistry part of the greenhouse puzzle. An advanced feature of TES is that it can scan continuously between the nadir and the limb. The interesting tropopause region is observed with high vertical resolution, high spectral resolution and continuous spectral coverage, allowing problems of stratosphere–troposphere exchange to be addressed. Convective transport across the tropopause is inhibited by the temperature structure, and the ways in which

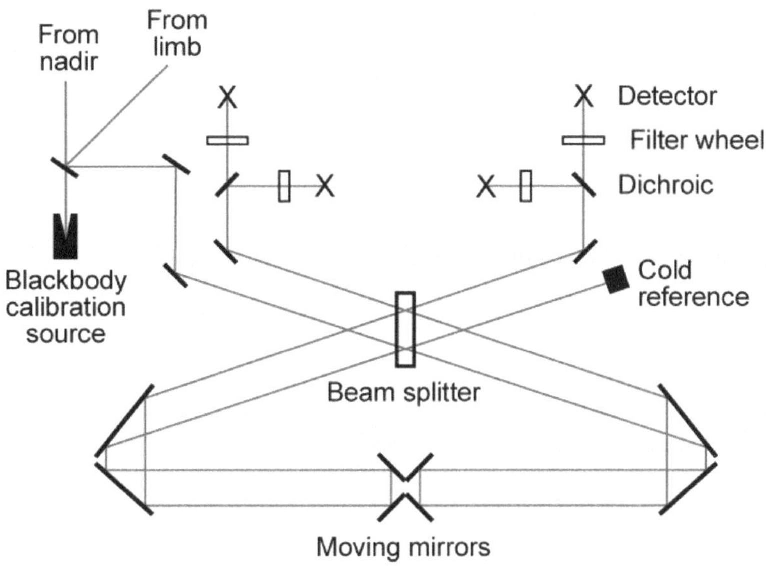

FIG. 10.12. A schematic diagram of the Tropospheric Emission Spectrometer, showing the main subsystems. TES is basically a Michelson interferometer with two pairs of moving mirrors to give a long path difference in a compact instrument. The purpose of the filters, which, along with the detectors, are cooled to around 85 K, is to limit the total amount of radiation falling on the detectors and hence reduce the photon–induced noise. Several filters are mounted on wheels, as shown, which can be rotated to define different spectral regions. A high–resolution spectrum is then obtained within that region by interferometry. Four spectra, in different spectral regions, are obtained simultaneously by the use of dichroics (wavelength–sensitive beam splitters) and multiple detector/filter assemblies, as shown.

pollutants from the surface reach the stratosphere, and stratospheric ozone enters the troposphere, are still very poorly understood.

In many ways, TES is the ultimate remote–sensing instrument, having nadir and limb viewing, cryogenic detectors and optics, wide spectral coverage and high spectral resolution capable of resolving individual spectral lines with good signal–to–noise ratio. In principle, TES measurements should be capable of making accessible to the experimenter virtually all of the information the Earth's passively emitted thermal radiation field contains.

The principal components and general layout of TES are shown in Fig. 10.12. These gather, focus, modulate, filter, detect, amplify and calibrate the signal from a well–defined field of view in the atmosphere, either downwards or sideways to the limb. The detectors feature 41×16 element optically conjugated focal plane arrays each optimized for a different spectral region and operating at a temperature of 63 K using mechanical refrigerators. In addition, each focal plane is equipped with interchangeable cooled filters that limit the instantaneous spectral bandwidths to about 250 cm^{-1}. This provides essential control over the instrument thermal background and reduces the data rate. Except for two external mirrors (part of the pointing system), the entire optical path is radiatively–cooled to about 165 K, to reduce further the instrument background.

10.9 Detection of climate change

The application of the instruments described in the previous sections to studying the *processes* involved in climate change is fairly obvious. For example, absolute radiometers (ACRIM) measure the output of the Sun; broadband radiometers (ERBE, GERB, CERES) monitor the Earth's radiation budget; spectroradiometers (ATSR, MODIS, HIRS, AIRS) measure surface and atmospheric temperatures; interferometers (IRIS, ATMOS, TES) measure composition, including greenhouse-gas amounts. Collectively and in various combinations, these instruments and others can monitor and study the general circulation of the atmosphere and its cloud and aerosol regimes, both crucial but elusive components of the climate system.

However, the measurement and detection of climate *change* is much more difficult. The traditional method has been to synthesize a global mean surface temperature from thousands of measurements made by different thermometric instruments around the world, over an extended period of time. This is obviously an arduous procedure, and one that is subject to many different types of error. Given the urgency of detecting quite small changes and trends before they become large enough to be detrimental to human affairs, there is a clear need to bring modern technology to bear. Radiometric instruments in space offer convenient global coverage by a single instrument, or at least a set of nominally identical and intercalibrated instruments, and therefore a potential solution to many of the problems inherent in the use of fixed, ground-based networks.

What, however, are the best parameters to measure, and how should the data be utilized, in order to produce the most sensitive and least error–prone estimates of global change, if indeed change is occurring? Since we are interested in future projections of climate change, we have no choice but to use forecasting models. The role of measurement then is to test and validate the models using relevant global data that are accurate and calibrated absolutely. The more obvious approaches, like using an ATSR–type instrument to map global temperatures and then taking global and annual averages, work in principle but in practice are affected by surface emissivity variations and cloud contamination of the radiances. Also, this approach does not use all of the available information in radiance measurements from space. Attempts to use radiometric measurements of the tropospheric temperature profile have focused on microwave sounders, which are less prone to interference by cloud effects, since most clouds are transparent at these long wavelengths. However, these also have problems with variable surface emissivity, and in particular with the vertical response function, which covers the whole troposphere and part of the lower stratosphere. The effect of this is to confuse and mask the climate signal since the temperature changes at various heights, which can oppose each other, are convolved together in an inconvenient way.

Goody *et al.* (1998) considered this whole question carefully and published a summary with recommendations. In this they stress the difference between observing *forcings* (changes in the key climate processes, such as those described in the opening paragraph of this section) and observing a *response* (such as a change in global or regional mean surface temperature or precipitation patterns). The latter requires the climate–change signal to be specified and benchmarks for this signal to be recorded at different times. The existence of anthropogenic forcing is not in serious doubt, and therefore we expect that the signal should, in time, emerge from the natural variability. The natural variability in the signal is itself of interest, both intrinsic and because knowledge of the statistics of natural variability is required to determine when the signal will emerge. Natural variability on timescales longer than the measurement period cannot be distinguished from an emerging signal; we must have independent data on the natural variability before we can interpret the data. This may come from the record to be found in archives of climate data, but because of the many limitations of old data it is more likely to be obtained from the chaotic behaviour of a general circulation model.

The chances of detecting a climate–change signal tend to improve if the signal selected is as complex as possible, in order to use as much measured information as possible and to maximize the chance that the signal will be distinct from natural variability. Goody *et al.* consider two examples of such 'fingerprints': the atmospheric refractivity, measured by transmitting radio waves between satellites in orbit along a path that intersects the atmosphere as one enters occultation behind the Earth as seen from the other, and the spectral radiance emerging from the top of the atmosphere, as measured by an interferometer equipped with an absolute calibration system. In order to maximize its complexity they manufacture the fingerprint from refractivities at 20 different heights below 25 km, 30 different locations around the globe, and all four seasons. For a fingerprint based on radiances the 20 heights are replaced by 1000 different quasimonochromatic wavelengths between 400 and 1400 cm^{-1}, a range that accounts for 98.3% of the radiance from a blackbody at the mean radiating temperature of the Earth.

Figure 10.13 illustrates the power of these complex fingerprints. The four frames show the change in the radiance leaving the top of the atmosphere, at the spectral resolution of the IRIS instrument (§10.8) calculated for four different forcing factors. These factors are (a) a doubling of CO_2 from 330 to 660 ppm; (b) an increase in the water–vapour amount, through the effect of temperature on the

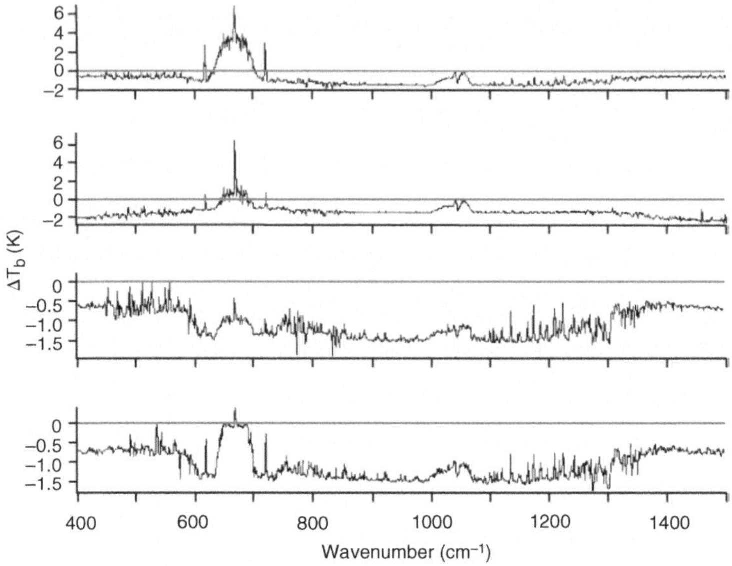

FIG. 10.13. The calculated effect on the outgoing radiance spectrum of the Earth of (a) doubling the CO_2 concentration, (b) increasing the relative humidity of the entire troposphere, (c) increasing the solar flux, and (d) changing the cloud cover. In each case the change in radiance over the unperturbed value is shown. The perturbations were chosen in cases (b), (c) and (d) to give the same change in surface temperature as in (a), namely 1.63 K. (Goody *et al.* 1998)

relative humidity; (c) an increase in the solar flux from 300 to 305.5 W m^{-2}; (d) the introduction of a model cirrus cloud. These all produce a readily measurable change in the climate signal, i.e. one that is much larger than the instrumental noise or calibration uncertainty, and are chosen so they all change the surface temperature by the same amount, namely 1.63 K. However, even before numerical analysis it is clear that they produce different and distinguishable fingerprints. With some assumptions, probably optimistic, Goody *et al.* found that the signature of global warming due to increasing CO_2 should be detectable when the concentration of the gas has risen by just 6.2% and, in the model they used, the surface temperature has risen by 0.2 K.

Attempts to apply this promising approach to real measurements have been made by Harries *et al.* (2001), Anderson *et al.* (2003), Brindley and Harries (2003), and Brindley and Allan (2003). The data used were from IRIS and IMG, which offered a time separation of 27 years, during which time the theoretical expectation would be that change should be detected, since CO_2 increases of nearly 10% have been measured in that time. The task is harder with real, as opposed to calculated, data, however. Corrections have to be applied for the different

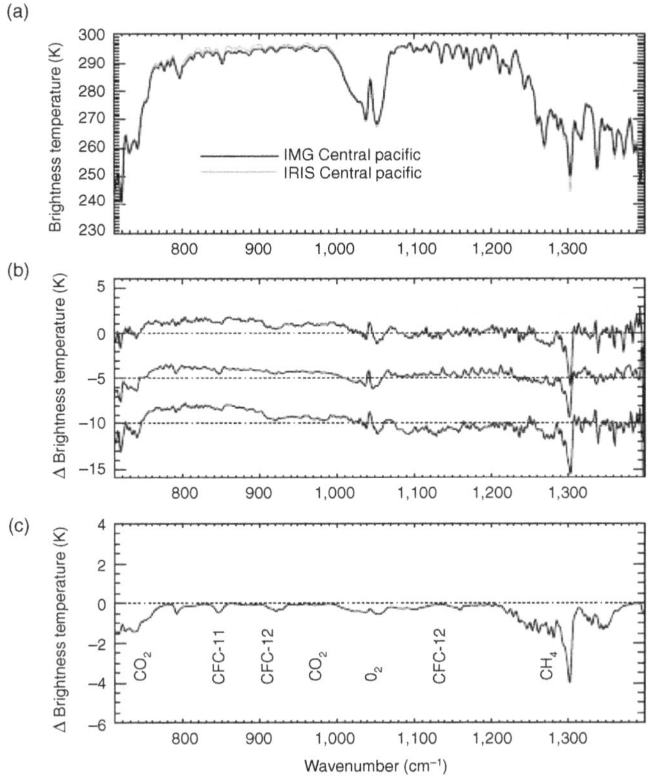

FIG. 10.14. The results of a comparison between IRIS data from 1970 and IMG data from 1997, by Harries *et al.* (2001). The top frame compares a 3–month average of the two over selected areas; the central frame is the difference spectrum for three areas of the planet; the bottom frame shows the calculated difference due only to the known change in greenhouse gases, with no response by the climate.

spectral resolutions of the instruments and their footprints on the planet, as well as for the limitations in space and time of the sampling in each case. The difficulties of doing this are shown to be formidable, and lead to large random errors.

The results of Harries *et al.* (2001) are summarized in Fig. 10.14. They show that the largest differences between 1970 and 1997 are seen in the strong bands of CO_2 and (especially) CH_4, and are attributable to the increases in the concentrations of those gases that have been measured chemically on the ground. These, of course, are examples of increases in forcing; any indications of a response to these forcings are most likely to appear as an increase in radiance seen in the 1100 to 1200 cm^{-1} window region, but very little is apparent. At the same time, an unexplained cooling is seen in the 800 to 900 cm^{-1} region, which could be contamination by thin cloud, and that may also be offsetting the global-warming signal in the shorter–wavelength window.

The problem with clouds is a crucial one for this approach. Not only does it make the result very sensitive to instrument field–of–view and sampling characteristics, it also raises the possibility that secular changes in cloud characteristics, for example in the abundance of an otherwise invisible thin layer of high ice clouds, or in the general prevalence of background aerosols, might mask other changes and remain undetected if their spectral signature is not sufficiently specific. A separate means of monitoring these may need to be contemplated, while simultaneously the measurements might be made at a large distance from the Earth, where most of a hemisphere remains in the field–of–view while measurements are made over an extended period by a single instrument with a reliable long-term calibration system. Alternatively, a less cloud-dependent approach may be better, for instance the refractivity measurements that Goody *et al.* originally postulated as an alternative to the spectral radiance approach. No instrumentation is in place at present that could be used to test and assess this.

Overall, the message is that the spectral radiance method for monitoring climate change (as opposed to forcing) is more complicated and difficult than it may at first seem, and so far has not proven conclusive. However, an approach to the reliable and early detection of climate change is still needed urgently, and this remains one of the most promising. Much better results could be expected from an observing system that was designed specifically for the purpose, rather than trying to reconcile disparate data sets originally generated for other purposes.

10.10 Bibliography

10.10.1 *Notes*

A comprehensive description of information retrieval from satellite observations can be found in the book by Rodgers; see also Chahine. Observations by some recent state–of–the–art instruments are discussed by Aumann *et al.* (AIRS) and Ward (MODIS); see also Kondratyev *et al.*

The detection of climate change using satellite remote sensing is covered by Anderson *et al.*, Goody *et al.*, and two papers by Harries *et al.* Cloud–climate uncertainties are discussed in Rossow and Schiffer, and IPCC 2001. For the shrinking of the Arctic sea–ice cover see, for example, Johannessen *et al.*

10.10.2 *References and further reading*

Anderson, J. G., Dykema, J. A., Goody, R. M., Hu, H. and Kirk-Davidoff, D. B. (2002). Absolute spectrally resolved radiance: A benchmark for climate monitoring from space. *J. Quant. Spect. Rad. Trans.*, **85**, 367–383.

Aumann, H. H., Chahine, M. T., Gautier, C., Goldberg, M. D., Kalnay, E., McMillin, L., Revercomb, H., Rosenkranz, P. W., Smith, W. L., Staelin, D. H., Strow, L. L. and Susskind, J. (2003). AIRS/AMSU/HSB on the Aqua mission: Design, science objectives, data products, and processing systems. *IEEE Trans. Geosci.*, Special Issue on EOS Aqua, **41**, 2, 253–264.

Beer, R. (1992). *Remote sensing by Fourier transform spectrometry.* John Wiley & Sons Inc., New York.

Brindley, H. E. and Allan, R. P. (2003). Simulations of the effects of interannual and decadal variability on the clear-sky outgoing long-wave radiation spectrum. *Quart. J. Roy. Met. Soc.*, **129**, 2971–2988.

Brindley, H.E. and Harries, J. E. (2003). Observations of the infrared outgoing spectrum of the Earth from space: The effects of temporal and spatial sampling. *J. Clim.*, **16**, 3820–3833.

Chahine, M. T. (1972). A general relaxation method for inverse solution of the full radiative transfer equation. *J. Atmos. Sc.*, **29**, 741–747.

Goody, R. M. and Yung, Y. L. (1996). *Atmospheric radiation: Theoretical basis.* Oxford University Press, Oxford.

Goody, R., Anderson, J. and North, G. (1998). Testing climate models: an approach. *Bull. Am. Met. Soc.*, **79**, 2541–2549.

Harries, J., Brindley, H. and Geer, A. (1998). Climate variability and trends from operational satellite spectral data. *Geophys. Res. Lett.*, **25**, 3975–3978.

Harries, J., Brindley, H., Sagoo, P. and Bantges, R. (2001). Increases in greenhouse forcing inferred from the outgoing longwave spectra of the Earth in 1970 and 1997. *Nature*, **410**, 355–357.

Hatzianastassiou, N., Katsoulis, B. and Vardavas, I. M. (2004a). Global distribution of aerosol direct radiative forcing in the ultraviolet and visible arising under clear skies. *Tellus B*, **56B**, 51–71.

Hatzianastassiou, N., Katsoulis, V. and Vardavas, I. M. (2004b). Sensitivity analysis of aerosol direct radiative forcing in the ultraviolet–visible wavelengths and consequences for the heat budget. *Tellus B*, **56B**, 368–381.

Hatzianastassiou, N., Matsoukas C., Fotiadi, A., Drakakis, E., Stackhouse Jr, P. W., Koepke, P., Pavlakis, K., Hatzidimitriou, D. and Vardavas I. M. (2006). Modelling the direct effect of aerosols in the near–infra–red on a planetary scale. *Atmos. Chem. Phys. Disc.*, **6**, 9151–9185.

Houghton, J. T., Taylor, F. W. and Rodgers, C. D. (1984). *Remote sounding of atmospheres.* Cambridge University Press, Cambridge.

IPCC, 2001: *Climate change 2001: The scientific basis. Contribution of working group I to the Third Assessment Report of the Intergovernmental Panel on Climate Change.* Houghton, J.T., Ding, Y., Griggs, D.J., Noguer, M., van der Linden, P.J., Dai, X., Maskell, K. and Johnson, C.A. (ed.), Cambridge University Press, Cambridge, United Kingdom and New York, NY, USA.

Johannessen, O. M., Shalina, E. V. and Miles, M. W. (1999). Satellite evidence for an Arctic Sea ice cover in transformation. *Science*, **286**, 1937–1939.

Koepke, P., Hess, M., Schult, I. and Shettle, E. P. (1997). *Global aerosol data set.* Report No 243, Max-Planck Institut für Meteorologie. Hamburg.

Kondratyev, K., Cracknell, A. P. and Kondratyev, K. Ya. (1997). *Observing global climate change.* Taylor & Francis, London.

Ramanathan, V., Crutzen, P. J., Kiehl, J. T. and Rosenfeld, D. (2001). Aerosols, climate and the hydrological cycle. *Science*, **294**, (5549), 2119–2124.

Rodgers, C. D. (2000). *Inverse methods for atmospheric sounding, theory and practice.* Series on Atmospheric, Oceanic and Planetary Physics, **2**, World Scientific, New York.

Rossow, W. B. and Schiffer, R. A. (1999). Advances in understanding clouds from ISCCP. *Bull. Am. Met. Soc.*, **80**, 2261–2287.

Stackhouse Jr, P. W., Cox, S. T., Gupta, S. K., Dipasquale, R. C. and Brown, D. R. (1999). *The NCRP/GEWEX Surface Radiation Budget Project Release 2: First results at 1 degree resolution*, paper presented at 10th Conference on Atmospheric Radiation: A Symposium with Tributes to the Works of Verner Suomi. Am. Met. Soc., Madison, Wisc.

Stephens, G. L. (1994). *Remote sensing of the lower atmosphere.* Oxford University Press, Oxford.

Taylor, F. W. (2005). *Elementary climate physics.* Oxford University Press, Oxford.

Ward, K. (1997). *MODIS Technical Specifications.* [http://modarch.gsfc.nasa.gov/ MODIS/ INSTRUMENT/MODIS_specs.html].

11

CLIMATE MODELLING

11.1 Introduction

The complexity of the climate system means computer models are essential tools for analysing observations and for predicting the future as a function of the various changes that can be anticipated in the input parameters or boundary conditions. Simple models have a role to play, particularly for education, but also to develop parameterizations of specific features, such as radiative fluxes in atmospheres with high concentrations of aerosols or clouds of different types. These can then be incorporated into more complete climate models, including coupled atmospheric and oceanic GCMs with physical processes and parameters such as concentrations of greenhouse gases, the salinity of the deep layers of the oceans, the biosphere, glaciers and ice sheets, rates of volcanic emission and the topography of continents and mountains.

Even the most complex models must make assumptions and simplifications, including compromises in the resolution in space and time represented by the grid on which the equations are solved. Subgrid–scale phenomena, such as radiative transfer in irregular cloud formations or turbulent heat transfer, cannot be represented purely by the relevant laws of physics and have to be parameterized. A wide range of schemes is in use in state–of–the–art climate forecasting, and the Intergovernmental Panel on Climate Change uses the spread between forecasts from control runs by research laboratories in several countries as a way of estimating the uncertainty they introduce. The uncertainty in the input data itself, such as the expected rate at which greenhouse gases will build up in the atmosphere, is treated by IPPC 2001 using 'scenarios' ranging from strict controls over emissions and new clean technology through 'business as usual' to pessimistic assumptions about population growth and per capita power consumption increases.

Some processes that are likely to be important are not properly included even in state of the art models. The melt rate of large ice masses like the Greenland ice sheet, which potentially can raise by around 7 m the global sea level in the relatively near future, and the evolution of cloud cover in the atmosphere in response to other changes such as a general warming, are examples currently receiving much attention. So are changes in the important ocean currents like the Gulf Stream and the snow and sea-ice cover, with their important contributions to the Earth's albedo, and chemical and biological processes in response to factors such as air pollution and deforestation.

To date, the combined efforts of experimentalists, data analysts, and modellers, have revealed much of the basic physics that produce the present-day climate. We are still a long way from the point where past, let alone future, climatic variations can be reliably explained, nor can we use the models we have developed for the Earth to explain many of the major climate phenomena on the neighbouring Earth-like planets. These failures clearly indicate that the physics in even apparently successful terrestrial climate and weather–forecasting models may be fundamentally deficient.

11.2 Simple climate models

11.2.1 *Global energy balance models*

The simplest relevant climate model follows from a consideration of the energy balance of the Earth as a whole. The solar constant $S_\odot$ (the radiant energy falling on unit area at the mean distance of the Earth from the Sun) is given by (see §5.2)

$$S_\odot = \left(\frac{R_\odot}{r_o}\right)^2 \sigma T_\odot^4, \tag{11.1}$$

where $R_\odot$ is the radius of the Sun, of effective emitting temperature $T_\odot$ and r_o its mean distance from the Earth. The total energy radiated from the Earth, of radius $R_\oplus$ and albedo α, is

$$4\pi R_\oplus^2 \sigma T_\oplus^4 = (1-\alpha) S_\odot \pi R_\oplus^2, \tag{11.2}$$

from which we find that the effective emitting temperature of the Earth, $T_\oplus$, is

$$T_\oplus = \left(\frac{(1-\alpha)S_\odot}{4\sigma}\right)^{\frac{1}{4}}. \tag{11.3}$$

From this it is found that $T_\oplus$ is about 255 K, and that a 1% change in solar constant corresponds to a change in the radiometric temperature of the Earth of about 0.65 K. Since the emission from the Earth to space is predominantly from the troposphere, with a lapse rate that is at least approximately constant with height, this will be close to the change in the surface temperature to be expected for a 1% change in solar luminosity.

In fact, observations of the total output of the Sun, made with accurate radiometric instruments orbiting above the atmosphere (§10.4), show that the solar constant has varied by less than 0.1% during the last three decades. Most of even that small fluctuation is caused by periodic variations in sunspot number that are so rapid as to have no likely effect on the climate. The remaining, longer–term fluctuations, including those associated with the Sun's 11–year cycle, are predicted by this model to affect the Earth's surface temperature by < 0.01 K.

The first generation of advanced climate models emphasized *gradual* change, for example the steady increase in global mean temperature that results from the buildup of greenhouse gases, ameliorated by a simultaneous buildup in sulphate and other kinds of aerosol, and possible changes in the amount and type of cloud cover. A more recent goal is the modelling and prediction of *abrupt* climate change, in which large changes may occur in only a few decades. The challenge is to understand what the stable states of the climate are, how stable the current state is, what transitions are possible, and how and when they are likely to occur.

This new focus has come about because; (a) definite evidence has emerged that rapid climate change, such as the Younger Dryas event, actually occurred in the relatively recent past (around 12000 years ago), and (b) modelling of complex systems in general suggests that the climate system may have multiple equilibria or 'eigenstates' that are stable, separated by unstable or metastable versions that prevail, if at all, relatively briefly. The implication is that climate change may involve gradual evolution for a while, but, once the limits of a stable state are reached, it will then make a rapid transition to the next stable state, which may be quite different. Examples of this kind of climate 'flip' might be from a mainly ice–free planet to one that is heavily glaciated, or a gradual warming of a couple of degrees over a century, say, followed by a much faster increase to a Venus–like state (the 'runaway greenhouse' scenario).

Two simple models, one radiative and the other dynamical, both of which represent some aspects of the real climate system, will be considered to show how multiple equilibria can occur, even in uncomplicated systems. They arise basically because of non–linearities in the dependences between variables, analogous to the behaviour of quadratic and higher–order equations, which have multiple solutions. This will become clear from the following examples.

A simple energy–balance model can also illustrate the concept of multiple climate equilibria, with the implication that rapid climate change can occur between these states. If we now consider the situation where the mean temperature T of the Earth is changing with time t, and the albedo α is a function of T, the balance equation becomes

$$C\frac{\mathrm{d}T}{\mathrm{d}t} = \frac{S_\odot}{4}\left(1 - \alpha(T)\right) - e\sigma T^4, \tag{11.4}$$

where C is a representative heat capacity. The dependence of albedo on temperature is realistic, since the amount of ice on the surface of the Earth will increase or decrease as the temperature changes. At equilibrium $\mathrm{d}T/\mathrm{d}t = 0$, so

$$(1 - \alpha(T))S_\odot = 4e\sigma T^4. \tag{11.5}$$

For almost any form of the function $\alpha(T)$ between values that are constant when the planet is completely ice covered or completely ice free, this expression has at least three solutions for T.

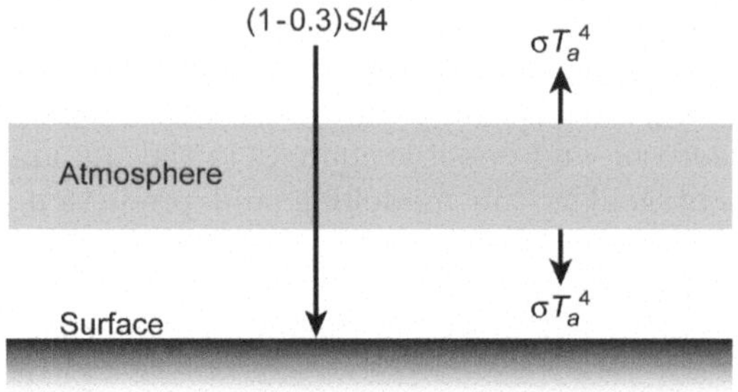

FIG. 11.1. A 'single-slab atmosphere' model of the greenhouse effect, in which the atmosphere is treated as a homogeneous layer of temperature T_a that is perfectly transparent to solar radiation and perfectly opaque in the thermal infra–red. The surface receives the equivalent of two solar constants, raising its mean temperature from 255 to 303 K.

While such a simple model has no great significance for the real Earth, it shows how multiple equilibria, some stable and some unstable, can arise in a system that has some attributes of the climate. In a more realistic, i.e. complicated, representation, there could be quasistable equilibria that are stable against small or short-lived perturbations but respond catastrophically to larger or more sustained forcing. There is some evidence from the geological record that Earth's climate has behaved in this way in the not–too–distant past, for instance, during the 'Younger Dryas' event around 12000 years ago.

11.2.2 *Simple greenhouse models*

The energy–balance model can be extended so that the surface temperature and the effective emitting temperature are calculated separately. The actual behaviour of the cloud-free atmosphere, which is largely transparent at wavelengths corresponding to most of the incoming solar energy, and largely opaque at those wavelengths at which the Earth emits most of its thermal energy, is approximated by a bimodal transmission function with $t = 1$ at wavelengths shorter than about 4 μm, and $t = 0$ at longer wavelengths. The flux of energy F from the Sun that is absorbed by the planet is the same as before, and all of it heats the surface. The outgoing energy to space, which in this model must be entirely from the atmosphere, since its opacity prevents the surface from radiating to space, must also be equal to F in order to achieve energy balance. However, the flux from the atmosphere occurs in both the upward and downward directions (Fig. 11.1), and so in equilibrium the surface receives a second contribution equal to that from the Sun, raising its temperature to $2^{1/4}\times$ 255 or 303 K. This calculated greenhouse enhancement of 48 K is larger than the observed 33 K, not surprisingly in view of the simplicity of the model.

If the atmosphere were homogeneous in reality, this enhancement would be the

Table 11.1 *Constants used to calculate model temperature profiles, and the radiative equilibrium, stratospheric and surface temperatures that result for present-day CO_2 abundances.*

Distance, Earth to Sun	r_o	150000000	km
Radius of Sun	$R_\odot$	696000	km
Radius of Earth	$R_\oplus$	6380	km
Effective temperature of Sun	$T_\odot$	5777	K
Solar constant	$S_\odot$	1366	W m^{-2}
Albedo of Earth	α	0.30	
Effective temperature of Earth	$T_\oplus$	255	K
Temperature of stratosphere	T_s	215	K
Temperature of surface	T_o	288	K
Pressure at tropopause	p_*	227	mb
Tropospheric lapse rate	Γ	6.6	K km^{-1}
Scale height	H	8.45 (surface) 6.35 (tropopause)	km

upper limit on what could be achieved. In fact, it is clear (from studying Venus, for example) that the greenhouse enhancement can be not one but many times the direct solar input. The effective emitting temperature of Venus is about 240 K, while the surface temperature is 730 K. The downward flux at the surface must therefore be of the order of $(730/240)^4$ or about 85 times that at the top of the atmosphere. This becomes possible in more realistic models in which the lower atmosphere, still opaque in the infra-red, is warmer in its lower regions (which warm the surface) than its upper (which radiate to space). This means representing the atmosphere by many layers, rather than just one. If a large number of isothermal layers is specified, the temperature profile becomes quasi-continuous and a realistic representation can be achieved.

Alternatively, a model with a more accurate representation of the vertical structure of the atmosphere can be constructed very simply by assuming an adiabatic (convective) lapse rate in the troposphere (§11.3.2), and a radiative equilibrium temperature in the stratosphere (§11.3.3). Such a model is completely characterized by a fixed tropospheric lapse rate $\Gamma = g/c_p$, a stratospheric temperature, related to the effective radiative temperature of the Earth $T_\oplus$ by $T_s = T_\oplus/2^{1/4} = 255/2^{1/4} \approx 215$ K, and either a tropopause height z_* or a surface temperature T_o (Table 11.1). The independent variables are the amount of infra-red absorber present in the atmosphere, and the mean albedo A of the Earth.

If the absorber amount in the atmosphere increases, the temperature at every level in the troposphere must move to a lower pressure to keep the same overlying optical depth and hence the same net emission to space. For the case where the amount of absorber is doubled, assuming the strong limit of absorption, see

Table 11.2 *Model results for surface temperature (K) corresponding to different values of Earth's albedo α, and different atmospheric CO_2 mixing ratios (pre–industrial 290 ppmv, present 367 ppmv, possible future 580 ppmv), assuming that the abundances of all greenhouse gases (including water vapour) vary in the same proportion.*

Albedo	290 ppmv	367 ppmv	580 ppmv
0.25	288	294	306
0.30	284	290	302
0.35	280	287	299

§4.2.4, eqn (4.52), the temperature $T(\mathrm{p})$ will now be found at pressure p', where $p' = p/\sqrt{2}$. Since the tropospheric lapse rate g/c_p must remain the same, there has to be an increase ΔT in the surface temperature in the model. Physically what is happening is that the depth of the convective layer increases to offset the greater opacity of the atmosphere, so that heat is still brought up to the level where it can radiate to space. The stratospheric temperature remains unchanged because, in the optically thin approximation, this depends only on the equilibrium temperature of the planet, $T_\oplus$.

If, however, the composition of the atmosphere is constant and the albedo of the planet changes, the temperatures at every pressure level must adjust to maintain an overall energy balance between the planet and the Sun. If the albedo increases, $T_\oplus$ will decrease, according to eqn (11.3), and this requires the same decrease in temperature at every level (including the surface) since the lapse rate must remain the same, at g/c_p in the troposphere and zero in the stratosphere.

Table 11.2 and Fig. 11.2 show some calculated values for this scheme, which show that changes in the planetary albedo from 0.25 to 0.35 alter the surface temperature by about 7 K; while doubling the greenhouse–gas abundance has an effect that is more than twice as large, at 18 K. The implicit assumption has been made that doubling of CO_2 is accompanied by doubling of all other greenhouse gases, including water vapour, since the simple model cannot separate the individual contributions (and indeed we have no way of knowing what change has taken place in mean concentration of the most important greenhouse gas, tropospheric water vapour, between pre–industrial times (left–hand column in Table 11.2) and the present (central column). Nor do we have much idea how the albedo of the Earth has changed during that period.

Suppose, however, purely as an exercise, that the water–vapour and other greenhouse-gas abundances do roughly track that of CO_2 as the atmosphere and surface warm up. Suppose further that the increased temperature and humidity give rise to extra cloud cover that has increased the mean albedo of the planet, from 0.25 in 1760 to 0.3 today, and that the increase will continue until a value of 0.35 is reached when CO_2 and the other gases reach twice their pre–industrial value later in the present century. Finally, account needs to be taken of the lag that the surface temperature experienced following greenhouse–induced changes, due to the thermal inertia of the climate system, primarily in the oceans. On a timescale of the order of a century, only about half of any change in forcing is reflected in the surface air temperature.

With all of these assumptions, the change in mean surface temperature of the Earth has increased by about 1 K since 1760, and will increase by a further 4.5 K in the twenty–first century. These figures are close to observations and IPCC 2001 predictions, respectively. Of course, we must recognize that this model is too crude to tell us with certainty what is really happening in the extremely complicated system that it represents. It can only give us a very general idea of how the climate *might* respond to known changes, and a feeling for how those responses might reinforce each other, or alternatively tend to cancel each other out.

The next step is to recognize that possible changes are linked to each other, not only in the sense that an increase in aerosol is expected along with an increase in CO_2, because of the nature of the industrial processes involved, but because certain changes will *force* other quantities to change, with further consequences for the climate. An example already considered is where increased air pollution induces higher surface temperatures, resulting in increased evaporation of water from land and sea, which in turn results in higher atmospheric humidities and more rapid cloud formation. Assuming that more cloud means a higher planetary albedo, this is an example of *negative feedback* in the climate system.

Of course, water vapour is itself a powerful greenhouse gas. So, in the above scenario, the increased humidity of the atmosphere is tending to amplify any warming through adding water vapour to the atmosphere, at the same time as it might be reducing it through increased albedo due to clouds. The water–vapour enhancement is an example of *positive feedback*.

Another example of positive feedback would be the reduction in the solubility of carbon dioxide in water with temperature. It is known that a large fraction (about one third) of the CO_2 released into the atmosphere is taken up by the ocean and does not contribute directly to global warming. As the ocean gets warmer, however, and as it gets saturated with CO_2, it will tend to respond by adding, rather than removing, CO_2 from the atmospheric greenhouse, thus amplifying the warming. To be useful as predictors of future climate change, models have to include these and other feedback processes in their computational schemes. What is difficult is to understand the physics of the feedback process in enough detail to produce a formulation that can be incorporated into the model. In the simple model just described, a parameterization was implicitly assumed where the formulae

$$\frac{\Delta[\mathrm{H_2O}]}{[\mathrm{H_2O}]} = \frac{\Delta[\mathrm{CO_2}]}{[\mathrm{CO_2}]} \quad \text{and} \quad \alpha = 0.25 + \left(\frac{\Delta[\mathrm{CO_2}]}{[\mathrm{CO_2}]}\right)^2 \qquad (11.6)$$

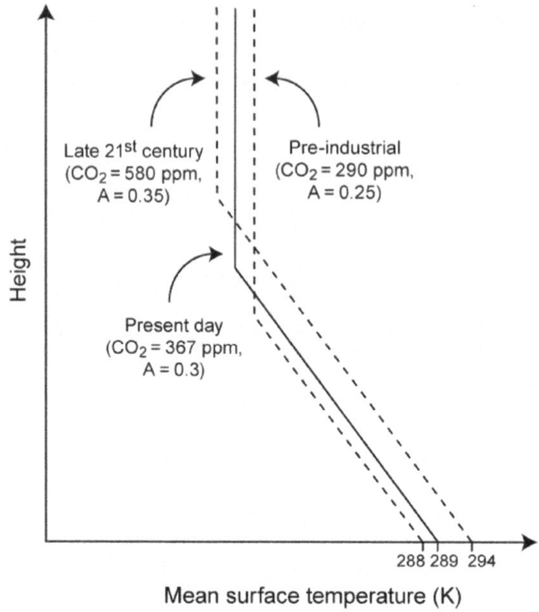

FIG. 11.2. Simple greenhouse models for the present (solid line) and pre–industrial and late–twenty–first century Earth, as discussed in the text.

were found to fit the past data and hence deemed appropriate for extrapolation into the future. Such crude methods may be used for some of the processes, even in a full GCM, while research goes on to develop a better parameterization, based on an improved understanding of the detailed physics. The need for three–dimensional and time–dependent representations, especially when tackling the forecasting problem, is perhaps the largest single reason why advanced climate models are so complicated and expensive to design and run.

Simple models give some insight into the principal processes at work, and allow us to begin to account for recently observed small changes and of predicting major changes in the future. They also show, quite convincingly, that the basic physics implies non–trivial changes in mean surface temperature and/or global cloud cover in response to a doubling of the atmospheric CO_2 concentration or other significant changes in greenhouse gas concentration. Finally, they illustrate the principles and main features of the feedback processes that make the final outcome of any global change scenario so uncertain.

What simple models cannot do is to give us a reliable quantitative prediction of the likely changes the Earth will experience in response to the expected ongoing increases in greenhouse forcing. For that we require much more sophisticated models, with dynamical schemes based on coupled ocean–atmosphere and a realistic treatment of radiative transfer.

11.3 Radiative–convective climate models

One–dimensional radiative–convective (RC) climate models are very powerful diagnostic tools for understanding climatic sensitivity to changes in the physical and chemical structure of the Earth's atmosphere and surface. The usefulness of RC models stems from their relative simplicity and their ability to provide a quantitative estimate of climate change produced by variations in, for example, the atmospheric content of carbon dioxide or water vapour, changes in the solar flux and in the Earth's surface albedo. Most RC models incorporate the concept of a *convective adjustment* and parameterizations for solar and terrestrial radiation transfer, as discussed in Chapters 4 and 6. In RC models, the atmosphere is divided into two sections. The lower atmosphere or troposphere is taken to be in convective equilibrium and the upper atmosphere (stratosphere and mesosphere) is in radiative equilibrium. The upper boundary of the convective zone is determined by concepts of atmospheric stability.

11.3.1 *Convective versus radiative equilibrium*

We have seen in §2.3 that the atmospheric temperature gradient determines the stability of the atmosphere against convection. An atmosphere in radiative equilibrium (see Fig. 2.11) produces essentially a discontinuity (of about 20 K) between the Earth's surface temperature and the near–surface atmospheric temperature. This is a consequence of having a non–ideal blackbody atmospheric layer overlaying a blackbody surface. We can make an estimate of this discontinuity by considering the simple case of a grey atmosphere in radiative equilibrium (see §3.5.10). In the two–stream approximation the mean intensity is given by $J = (I^+ + I^-)/2$ and the net upwelling flux, f, is constant in radiative equilibrium and given by $f = \pi(I^+ - I^-)$. Thus the mean intensity can be expressed as $J = I^+ - f/2\pi$. At the Earth's surface, with temperature T_s, $I^+ = B(T_s)$, where $B(T_s) = \sigma T_s^4/\pi$, and we have seen that within the atmosphere, eqn (3.138),

$$J(\tau_s) = B(T(\tau_s)) = B(T_o)\left(1 + \frac{3}{2}\tau_s\right), \tag{11.7}$$

where $T(\tau_s)$ is the near–surface atmospheric temperature. Now $f/2\pi = B(T_o)$ where T_o is the atmospheric skin temperature, equal to $(1/2)^{\frac{1}{4}}\, T_\oplus$, where $T_\oplus$ is the effective temperature of the Earth, so that the Earth's surface temperature can be calculated from

$$B(T_s) = B(T_o)\left(2 + \frac{3}{2}\tau_s\right). \tag{11.8}$$

If we set $\tau_s = 0.85$, for the total grey optical depth of the present atmosphere, and $T_\oplus = 255$ K then we obtain $T_o = 214$ K, $T_s = 288$ K and the near–surface air temperature is then $T(\tau_s) = 263$ K. We see that there is a discontinuity of 25 K between the surface temperature and the near–surface air temperature. We also see that for $\tau_s \gg 1$, the atmospheric layer above the blackbody surface is also a blackbody, eqn (3.83), and so from the above equations we have $T(\tau_s) \approx T_s$.

Thus, in radiative equilibrium, the temperature rapidly decreases with altitude so that the atmospheric density increases with altitude. We thus have colder more dense layers over warmer less dense layers. This initiates convective transport of heat as the colder layers sink towards the surface and the warmer layers rise up in the atmosphere. If we now define the adiabatic atmospheric lapse rate by $\Gamma = -\mathrm{d}T/\mathrm{d}z$ and the corresponding radiative equilibrium value by Γ_{rad}, then when $\Gamma_{\mathrm{rad}} > \Gamma$ the atmosphere is unstable against convection and hence its temperature is determined by convective equilibrium as given by the adiabatic lapse rate. We can then use this criterion to determine the altitude when the atmosphere is stable against convection, that is, it is in radiative equilibrium.

11.3.2 *Convective equilibrium*

The lower atmosphere or troposphere is considered to be in convective equilibrium. The temperature structure is then determined by the surface temperature, T_s, and the atmospheric lapse rate, Γ. The adiabatic lapse rate in saturated air was derived in Chapter 2 and is given by

$$\Gamma_{\mathrm{moist}} = -\left(\frac{\mathrm{d}T}{\mathrm{d}z}\right)_{\mathrm{sat}} = f_c \Gamma_{\mathrm{dry}}, \tag{11.9}$$

with $\Gamma_{\mathrm{dry}} = g/c_{\mathrm{pd}} = 9.76$ K/km, where g is the gravitational acceleration and c_{pd} is the specific heat of dry air at constant pressure for the gas mixture that constitutes the atmosphere. In the presence of large amounts of water vapour c_{p} will increase above the dry air value. In §2.6, we obtained an expression for the tropospheric lapser rate when water vapour is a minor constituent, as it is in the present atmosphere. For atmospheres at higher surface temperatures, as would be for elevated CO_2 levels (possibly in the past during the Precambrian period 4.5–2.5 billion years ago) or larger solar incoming flux (expected in the future on timescales of billions of years as the Sun ages), the atmospheric water–vapour content can increase substantially with the Earth's surface temperature. For the general case, the moist lapse rate factor for a reversible saturated adiabatic expansion is given by (Iribarne and Godson 1981)

$$f_c = \frac{(p/p_{\mathrm{d}})\,(1 + r_w u_1)}{1 + u_2 + u_3}, \tag{11.10}$$

$$u_1 = \frac{L_{\mathrm{v}}}{RT} + 1 - 1/\epsilon, \tag{11.11}$$

$$u_2 = \frac{c_{\mathrm{pv}} r_{\mathrm{w}} + c_{\mathrm{w}}(r_{\mathrm{tw}} - r_{\mathrm{w}})}{c_{\mathrm{pd}}}, \tag{11.12}$$

$$u_3 = \frac{L_{\mathrm{v}}^2 r_{\mathrm{w}}(\epsilon + r_{\mathrm{w}})}{c_{\mathrm{pd}} R_{\mathrm{d}} T^2}. \tag{11.13}$$

The saturated lapse rate is always less than the dry value, and the ratio f_c can fall as low as 0.21 in a hot atmosphere of 400 K (Fig. 11.3). The total atmospheric pressure is $p = p_d + p_v$, $r_w = \epsilon p_v/p_d$ is the water-vapour mixing ratio by mass, where $\epsilon = M_{H_2O}/M_{air} = 0.62$, is the ratio of the molecular weight of water vapour (18.02 g mole^{-1}) to that of dry air (28.96 g mole^{-1}). The total water mixing ratio r_{tw} includes liquid water that for pseudo-adiabatic conditions can be set equal to r_w. The specific heat of dry air, c_{pd}, is equal to 0.24, $R = 1.9865/M$ and $R_d = 1.9865/M_{air}$ are the gas constants, for the mixture and dry air, respectively, all in cal g^{-1}K^{-1}. For water vapour, the gas constant is $R_v = 0.1103$ and the specific heat capacity is $c_{pv} = 0.443$. The specific heat capacity of water is $c_w = 1.008$, while for ice there is a very weak temperature dependence

$$c_i = 0.503 + 1.767 \times 10^{-3}(T - T_o). \tag{11.14}$$

L_v is the latent heat of evaporation (cal g^{-1}) over a water body ($T \geq T_o$), or that of sublimation over a frozen water body ($T \leq T_o$) and is given by Kirchoff's law

$$\frac{dL_{12}}{dT} = c_2 - c_1, \tag{11.15}$$

where c_1 and c_2 correspond to phases 1 and 2 of the water substance, and we note that a change of phase occurs at constant pressure. On integrating, we get

$$L_{wv} = 597.3 - 0.565(T - T_o) \quad T \geq T_o, \tag{11.16}$$

$$L_{iv} = 667.0 - 0.06(T - T_o) - 8.835 \times 10^{-4}(T - T_o)^2 \quad T \leq T_o, \tag{11.17}$$

where $T_o = 273.15$ K $= 0.0$ °C, is the freezing point of fresh water and the latent heat of melting is 79.7 cal g^{-1} at $T = T_o$. We note that sea water freezes at -1.9 °C. Assuming a linear dependence of the latent heat on temperature, we can integrate the Clasius-Clapeyron equation (eqn (2.14)) to obtain the saturation water vapour pressure, at T, in the form

$$\ln p_v = a_1 - a_2/T + a_3 \ln T. \tag{11.18}$$

To a very good accuracy we can set $a_1 = 55.49369$, $a_2 = 6814.413$, and $a_3 = -5.12239$, for $T \geq T_o$, and $a_1 = 27.54365$, $a_2 = 6195.730$, and $a_3 = -0.543971$, for $T \leq T_o$. At $T = T_o$ we have $p_{vo} = 6.107$ mbar. In Fig. 11.3 are shown; the growth of the water vapour pressure with temperature for saturated conditions (relative humidity, $rh = p_{H_2O}/p_v$, equal to unity), the rise in the atmospheric specific heat capacity, the decrease in the mean molecular weight and the decrease in the moist lapse rate for a dry air atmosphere component at 1013.25 mbar (1 atm). As expected, the water–vapour pressure rises rapidly to 1 atm at the standard boiling point of 100 °C (373.15 K), and by 400 K it reaches 2342 mbar, while the moist lapse rate remains fairly constant above 100 °C at 2.1 K km^{-1}. Thus, by 400 K, under saturated conditions, the atmospheric pressure would rise to 3.3 atm. The discontinuity in the moist lapse rate at 273.15 K is due to the phase change from ice to liquid water.

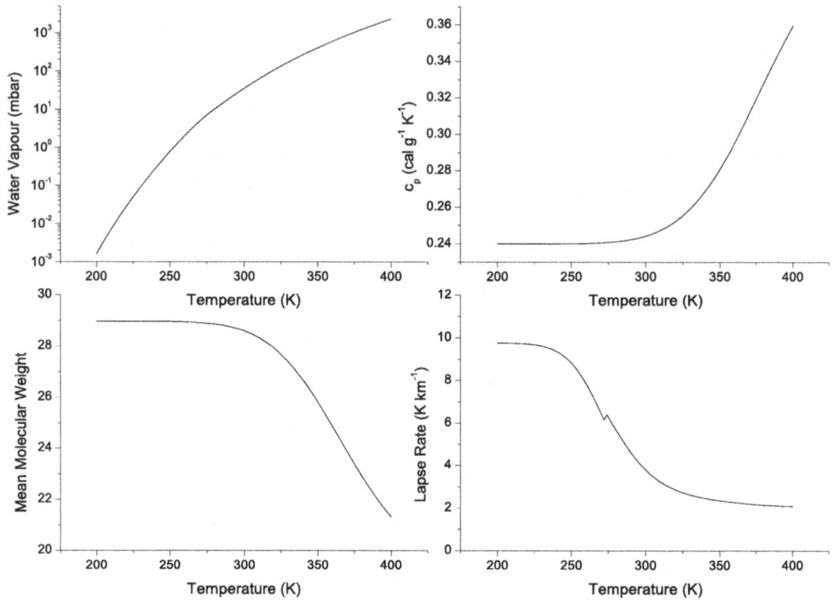

FIG. 11.3. Variation of saturated water–vapour pressure, specific heat capacity, mean molecular weight, and moist lapse rate for a dry atmosphere component at 1013.25 mbar.

11.3.3 *Radiative equilibrium*

The upper atmosphere is in radiative equilibrium. This is achieved by a balance between solar radiation heating, mainly due to atmospheric ozone absorption (see Chapter 6) and radiative cooling to space by carbon–dioxide emission in the optically thin upper atmosphere (see Chapter 4). At low pressures the coupling between the infra–red rotational excited states of a molecule and the thermal energy of the atmosphere is weakened and radiative transitions between discrete energy states are faster than collisional excitation and de–excitation. The populations of the molecular states are then determined primarily by radiation processes rather than collisional ones and the atmosphere is not in local thermodynamic equilibrium (Non–LTE). The problem of computing the atmospheric temperature then becomes more complex (see Lopez–Puertas and Taylor 2001). Effects of Non–LTE for the present atmosphere become important above about 70 km altitude. Parameterizations for computing the radiative equilibrium temperature in LTE are given in Chapters 4 and 6 (see also Vardavas and Carver 1984a).

The mean global vertical heating rate, $H(z)$, at altitude z, can be calculated from

$$H(z) = \frac{1}{\rho c_p} \frac{dF(z)^{\downarrow}_{SW}}{dz} \tag{11.19}$$

where $F(z)^{\downarrow}_{SW}$ is the net downwelling solar radiation flux at level z, ρ is the atmospheric density and c_p the atmospheric specific heat capacity at constant pressure. In Fig. 11.4 is given the mean global vertical heating rate due to solar radiation absorption for an RC model with mean cloud cover of 0.5, surface albedo of 0.1, with the relative humidity falling with altitude according to (Manabe and Whetherald 1967)

$$\mathrm{rh} = \mathrm{rh_s}\frac{p/p_s - 0.02}{1 - 0.02} \qquad p/p_s > 0.02 \tag{11.20}$$

with a surface relative humidity of 0.8, planetary albedo of 0.33, a tropospheric moist lapse rate determined according to eqn (11.10) with a resulting water column of 2.09 g cm^{-2} and ozone vertical distribution as given in Fig. 7.10, and surface CO_2 of 365 ppmv. The heating rate follows the ozone profile in the upper atmosphere, while in the troposphere the heating rate shown is due to the single cloud layer of the model and absorption of water vapour in the near–infra–red. Note, however, that within the convective equilibrium zone the radiative heating is not used for the determination of the atmospheric temperature, which is determined by the lapse rate. The surface temperature T_s and the temperature structure, $T(z)$, within the radiative equilibrium zone are determined by balancing the net outgoing terrestrial longwave flux and the net incoming solar shortwave flux at all levels in the radiative equilibrium zone, which includes the top of the atmosphere,

$$F^{\uparrow}_{LW}(T(z)) = F^{\downarrow}_{SW}(z). \tag{11.21}$$

Thus, the radiative equilibrium temperature structure at any altitude depends on the radiative processes at all levels from which radiation can reach the specific level. The vertical temperature structure, $T(z)$, within the convective equilibrium zone is determined by the surface temperature and the moist lapse rate at each altitude (Vardavas and Carver 1984b).

11.3.4 *Climatic effects of increasing CO_2 levels*

In Fig. 11.5 are shown mean global vertical temperature profiles obtained by an RC–photochemical model (described in Vardavas and Carver 1984a, and Lavvas *et al.* 2007), without climatic feedbacks associated with cloud cover and relative humidity, for increasing CO_2 levels from 1 PAL = 365 ppmv, the present atmospheric level, to 8 PAL. As the CO_2 level rises the lapse rate decreases and the convection zone extends to higher altitudes. The decrease in the lapse rate acts as a negative feedback that limits the rise in the surface temperature due

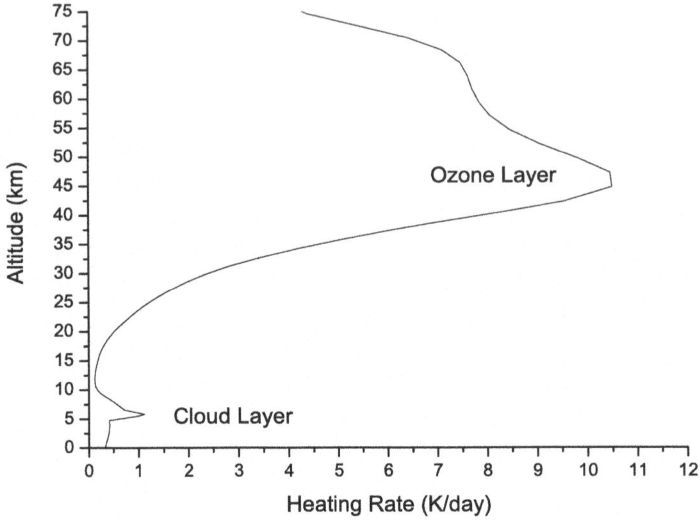

FIG. 11.4. Variation of the global mean vertical heating rate due to solar radiation absorption primarily by ozone in the upper atmosphere. The heating rate follows the variation in the ozone mixing ratio with altitude. Heating due to water vapour and heating within the cloud layer occurs in the lower troposphere.

to the greenhouse effect. As the lapse rate decreases the troposphere is overall warmer and so contributes more towards the outgoing longwave flux that is needed to balance the net incoming solar flux. In the other extreme, a surface temperature that is determined by radiative equilibrium would need to be significantly higher to compensate the greenhouse reduction in outgoing longwave flux, as the troposphere would be relatively cooler.

In the stratosphere, the atmosphere is optically thin with regard to the thermal infrared absorption by CO_2 and so any photons emitted upwards in this spectral range will escape to space. The cooling to space by the CO_2 increases as the mixing ratio of CO_2 increases with the result that the stratosphere becomes colder.

In Table 11.3 are shown the effects of increasing CO_2, without cloud and relative humidity feedbacks, from its present atmospheric level on various climatic parameters of the atmosphere. The model gives a surface temperature increase of 1.3 K for a doubling of the present level of CO_2. As the CO_2 content of the atmosphere increases the water–vapour content increases significantly, and this introduces an important feedback in the climate model. The

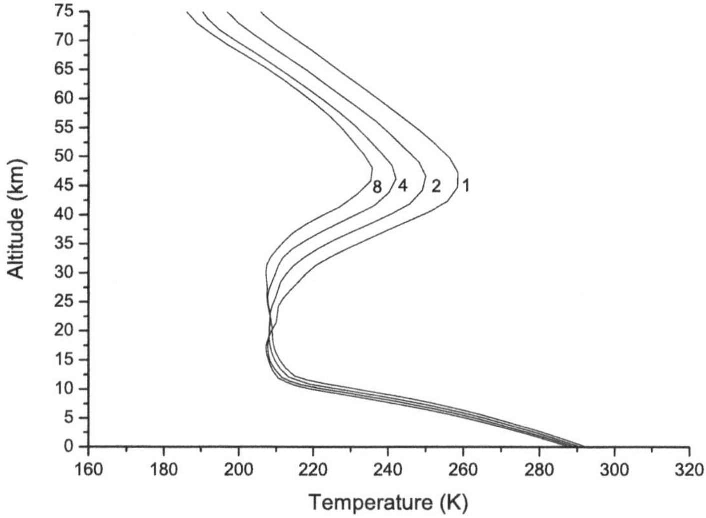

FIG. 11.5. Variation of the global mean vertical atmospheric temperature with increasing atmospheric carbon dioxide levels (1 PAL = 365 ppmv). The RC model has fixed cloud cover of 0.5, surface relative humidity of 0.8, and surface albedo of 0.1. Water vapour and moist lapse rate feedbacks are included, as are changes to atmospheric composition through the inclusion of the photochemistry.

Table 11.3 *Climatic effects of increasing atmospheric CO_2 levels (1, 2, 4 and 8 PAL), without feedbacks. Water–vapour column W_{H_2O} in g cm^{-2}. Stratospheric peak water–vapour mixing ratio H_2O St in ppmv.*

Parameter	1	2	4	8
T_s	288.15	289.43	290.90	292.50
W_{H_2O}	2.09	2.35	2.65	3.02
H_2O St	5.75	5.43	4.79	3.19

troposphere–stratosphere cold trap (temperature minimum) rises in altitude as the stratosphere cools, and so the stratosphere becomes drier. Cloud and relative humidity will also play a significant role in controlling the Earth's mean global surface temperature.

11.3.5 *Climatic effects of increasing CH_4 levels*

In §7.9.2 we saw that the methane mixing ratio increases non-linearly with the surface emission rate of methane. Results from an RC–photochemical model,

Table 11.4 *Climatic effects of increasing atmospheric CH_4 emission (1, 1.2, 1.4, 1.6, 1.8 and 2 PR), where 1 PR = 1080 Tg year^{-1}. Water–vapour column W_{H_2O} is in g cm^{-2}. Stratospheric peak water–vapour mixing ratio H_2O St is in ppmv.*

Parameter	1	1.2	1.4	1.6	1.8	2.0
CH_4	1.71	2.49	3.48	4.74	6.27	8.10
CO	0.101	0.123	0.149	0.179	0.212	0.249
H_2	0.539	0.653	0.771	0.892	1.011	1.128
T_s	288.15	288.32	288.51	288.70	288.90	289.11
W_{H_2O}	2.06	2.09	2.13	2.16	2.20	2.24
H_2O St	5.75	6.88	8.45	10.45	12.75	15.50

Table 11.5 *Climatic effects of increasing atmospheric CO emission (1.0, 1.5, 2.0, 3.0 PR), where 1 PR = 3295 Tg year^{-1}. Water vapour column W_{H_2O} is in g cm^{-2}. Stratospheric peak water–vapour mixing ratio H_2O St is in ppmv.*

Parameter	1.0	1.5	2.0	3.0
CO	0.101	0.173	0.273	0.620
CH_4	1.71	2.25	2.95	5.16
H_2	0.539	0.612	0.673	0.765
T_s	288.15	288.29	288.44	288.82
W_{H_2O}	2.06	2.13	2.15	2.23
H_2O St	5.75	6.80	8.13	12.21

without feedbacks, are given in Table 11.4 for increasing values of methane emission up to twice a present rate 1 PR = 1080 Tg year^{-1}. The global mean surface temperature rise is about 1 K for a doubling in the methane emission rate which results in a methane mixing ratio that is almost 5 times the present mixing ratio. We note that a doubling of the methane mixing ratio results in about 0.4 K rise in the mean global surface temperature. As tropospheric methane levels increase, there is an increase in tropospheric CO and H_2, due to a reduction in OH by methane. On the other hand, increased oxidation of methane in the stratosphere results in higher water–vapour mixing ratios there.

11.3.6 *Climatic effects of increasing CO levels*

Results from an RC–photochemical model, without feedbacks, are given in Table 11.5 for increasing values of CO emission up to three times a present rate 1 PR = 3295 Tg year^{-1}. The global mean surface temperature rise is about 0.29 K for a doubling in the CO emission rate that results in a CO mixing ratio that is 2.7 times the present mixing ratio. As tropospheric CO levels increase, reaction with OH increases and so there is a reduction in methane oxidation by OH in the troposphere, thus methane levels rise and so does stratospheric water vapour.

Table 11.6 *Climatic effects of increasing atmospheric H_2 emission (1.0, 1.5, 2.0, 3.0 PR), where 1 PR = 47 Tg year^{-1}. Water–vapour column W_{H_2O} is in g cm^{-2}. Stratospheric peak water vapour–mixing ratio H_2O St is in ppmv.*

Parameter	1.0	1.5	2.0	3.0
H_2	0.539	0.682	0.827	1.123
CH_4	1.71	1.75	1.80	1.88
CO	0.101	0.103	0.105	0.109
T_s	288.15	288.16	288.17	288.20
W_{H_2O}	2.06	2.10	2.10	2.11
H_2O St	5.75	5.93	6.13	6.51

11.3.7 *Climatic effects of increasing H_2 levels*

Results from an RC–photochemical model, without feedbacks, are given in Table 11.6 for increasing values of H_2 emission up to three times a present rate 1 PR = 47 Tg year^{-1}. The global mean surface temperature rise is about 0.02 K for a doubling in the H_2 emission rate that results in a H_2 mixing ratio that is 1.5 times the present mixing ratio.

11.3.8 *Climatic effects of cloud–cover feedbacks*

In Table 11.7 are shown the climatic effects of increasing the global cloud cover, for CO_2 at 1 PAL with the surface albedo fixed at 0.1. The global mean surface temperature decreases as the cloud cover increases and the mean global planetary albedo rises. We see that about 10% increase in cloud cover results in about 2 K global cooling. Methane, CO and H_2 increase as the water vapour in the atmosphere, and hence oxidation by OH, decreases. The rise in methane partly offsets the global cooling. If cloud cover increases with global warming due to a rise in greenhouse gas concentrations, any rise in temperature would be limited in part by the cloud–cover increase. Cloud feedbacks are complex not only because of the uncertainty in the magnitude change in cloud cover brought on by rising levels of greenhouse gases but also uncertainties in the changes in cloud optical properties and cloud types (see Chapter 8).

11.4 General circulation models

While radiative–convective models are very useful tools for investigating how the atmospheric composition affects the radiation field, and in turn the role of the radiation field on atmospheric photochemistry, the fact that they have only one spatial dimensional (height) poses some fundamental limitations. The real Earth is a three–dimensional distribution of atmospheric and surface properties that are linked by atmospheric dynamics and ocean circulation, whose accurate

Table 11.7 *Climatic effects of changing global cloud cover, from a present value set at 0.50. Methane, carbon monoxide and hydrogen levels in ppmv. Water–vapour column W_{H_2O} in g cm^{-2}. Stratospheric peak water–vapour mixing ratio H_2O St in ppmv. The planetary albedo is α.*

Parameter	0.44	0.46	0.48	0.50	0.52	0.54	0.56
CH_4	1.24	1.38	1.54	1.71	1.92	2.15	2.42
CO	0.083	0.089	0.095	0.101	0.109	0.117	0.126
H_2	0.450	0.481	0.510	0.539	0.571	0.601	0.630
T_s	290.00	289.32	288.72	288.15	287.45	286.88	286.31
W_{H_2O}	2.46	2.32	2.21	2.09	1.99	1.89	1.80
H_2O St	5.33	5.44	5.57	5.75	6.00	6.30	6.67
α	0.302	0.312	0.322	0.332	0.342	0.352	0.363

representation requires the inclusion of all three dimensions, in what has become known as a general circulation model (GCM). Because the physics of the atmosphere and of the ocean are quite different, we usually produce separate atmospheric and oceanic GCMs and link them together to form a coupled general circulation model (CGCM).

In order for 3D general circulation models to adequately treat the dynamics of the atmosphere, they have to have quite high resolution in space and time, which imposes an enormous computational burden. To compensate, they can include only a limited number of chemical species and reactions, while the radiation budget is treated through broadband parameterizations, that is they use only a limited number of spectral intervals to perform the radiative–transfer calculation. There is still scope, therefore, for radiative-convective models, since they allow a rapid sensitivity analysis of the effects of radiation transfer and chemistry on the mean global vertical temperature structure. The RC models can be linked with a photochemical model to examine in detail the spectral radiation effects on the vertical photochemistry and temperature structure, in a way that is beyond the scope of present–day GCMs. There is a need for both kinds of model, therefore, although we may expect that in time the two will merge. For the present, RC models lead when studying radiative or chemical processes, while GCMs are essential for most dynamical studies and, crucially, for long–term climate predictions.

Richardson (1922) reported one of the earliest attempts to simulate atmospheric motions mathematically knowing, however, that the complexity of modelling the global atmosphere required significant computational power. An early computer pioneer, von Neumann (see Aspray 1990, Charney *et al.* 1950), played a key role in the development of numerical weather prediction. The Royal Swedish Air Force Service was the first to begin real–time numerical weather forecasting in December 1954, using a model developed by the Institute of Meteorology at the University of Stockholm. By the 1960s, the computing power available was

Table 11.8 *3D global and regional climate models used by the Hadley Centre, UK.*

Acronym	Type	Details
AGCM	Atmosphere GCM	Coupled to land-surface and cryosphere, used for atmospheric process studies
OGCM	Ocean	Ocean circulation and sea ice
AOGCM	Atmosphere–ocean	Coupled atmospheric–oceanic circulation
RCM	Atmosphere–land	Regional climate with boundaries from AOGCMs
CCYM	Carbon cycle	Land–surface cycle from AOGCM, marine cycle from OGCM
STOCHEM	Atmospheric chemistry	Atmospheric composition changes

sufficient for a first attempt at modelling the global general circulation. By the mid– to late 1960s, the Geophysical Fluid Dynamics Laboratory at Princeton was established, and Smagorinsky, Manabe, and their collaborators (Manabe *et al.* 1965) had implemented a nine–level, hemispheric primitive–equation GCM and used it to perform carbon–dioxide doubling experiments (e.g. Manabe and Bryan, 1969). The mid-1980s saw the first serious attempts at climate modelling that involved coupled atmospheric and oceanic circulation models.

As models and research groups have proliferated around the world, liaison between groups and comparisons of their products has become more important, and also more difficult. In 1990, the World Climate Research Programme established the Atmospheric Model Intercomparison Project, which lists details of the existing AGCMs and the centres worldwide that perform climate–simulation experiments with them, and establishes a common protocol for intercomparisons, to assist in this endeavour. The Intergovernmental Panel on Climate Change (IPCC) was established in 1988 by two United Nations organizations, the World Meteorological Organization (WMO) and the United Nations Environment Programme (UNEP) to assess the 'risk of human–induced climate change'. Its reports are widely cited in almost any debate related to climate change and generally regarded as authoritative, although obviously considerable disagreement remains, particularly about the evolution of such a complex system as the global climate.

11.4.1 *Types of climate models*

An example of the various models used by one leading climate–modelling institute is given in Table 11.8 that lists those used at the UK Meteorological Office Hadley Centre. The report by Roeckner *et al.* (2003) on the ECHAM5 model of the Max Planck Institute for Meteorology also gives full details of the complexity of GCM modelling and the relevant equations.

Due to the complexity of radiative transfer, numerous assumptions and simplifications are made within all GCMs. Generally, these include

1. The plane–parallel assumption, i.e. that columns of atmosphere and ocean, within a resolution element, have no curvature.
2. Gas constituents are homogeneously mixed, i.e. that there is no compositional variation (including cloud properties) or temperature contrast within each cell.
3. Gas constituents are in local thermodynamic equilibrium (LTE), which is generally true in the troposphere and lower stratosphere but not at higher levels.
4. The spectral resolution for both shortwave and longwave radiation transfer is broadband rather than detailed monochromatic.
5. For efficiency reasons the radiative–transfer computation in GCMs is performed less frequently than the dynamics and other parameterizations.
6. Shortwave scattering/absorption is usually parameterized by implementing either the delta–Eddington approximation (§3.5.6 and §6.7), or the two–stream approximation (§3.5.7), while longwave absorption/emission is parameterized by absorptance formulations (§4.5). Also, for infra–red and near–infra–red bands the correlated–k approximation (§6.2.2) is used to convert the non–linear dependence of optical depth on absorber amount to a linear dependence.

Below, we consider some specific examples of the treatment adopted by some of the world's leading climate modelling groups. These illustrate the commonality and also the diversity of methods and assumptions within models.

11.4.2 *Solar radiation transfer in GCMs*

11.4.2.1 *HadAM3 Scheme* This is the atmospheric component of the HadCM3 coupled GCM that was used extensively by the IPCC in its evaluation of global climate predictions, as we discuss below. In HadAM3, the shortwave fluxes are calculated by a two–stream method, where the spectrum is divided into 6 bands with the wavelength limits: 0.20–0.32; 0.32–0.69 (2 bands); 0.69–1.19; 1.19–2.38; and 2.38–5.00 μm. Within each band, data are treated using the correlated–k approximation method using gaseous absorption data from the HITRAN and LOWTRAN databases.

Rayleigh scattering by gases is explicitly represented but scattering by clouds and aerosols are much more difficult to incorporate. The amount of background aerosol at any given time or location is unknown, so a calculation of a fixed amount of climatological background aerosol is incorporated as an interim approach. Cloud–radiative interactions are parameterized separately for water and ice as functions of water content and effective radius. Ice particles are treated as spheres with a fixed radius of 30 μm. Stratified clouds are treated as overlapping layers and convective clouds as vertical towers; mixed–phase clouds are represented by horizontally adjacent regions of ice and water cloud.

11.4.2.2 *NCAR CAM 3.0 Scheme* This is a state-of-the-art model from the National Center for Atmospheric Research in the USA. In this, the shortwave scattering/absorption is parameterized by the delta–Eddington approximation applied in 19 discrete spectral intervals; 7 for O_3, 1 for the visible, 7 for H_2O, 3 for CO_2, and 1 for the near–infra–red. The radiation scheme allows for gaseous absorption by O_3, CO_2, O_2, and H_2O. Molecular scattering and scattering/absorption by cloud droplets and aerosols are included. The model atmosphere consists of a discrete vertical set of horizontally homogeneous layers within which radiative heating rates are specified.

Each of these layers is considered to be a homogeneous combination of several radiatively active constituents. Five chemical species of aerosol are considered, including sea salt, soil dust, black and organic carbonaceous aerosols, sulphate, and volcanic sulphuric acid. Direct and semidirect effects of tropospheric aerosols on shortwave fluxes and heating rates are included. CAM 3.0 differentiates between the cloud drop effective radius for clouds diagnosed over maritime and continental regimes and over pristine surfaces (sea–ice, snow–covered land). The cloud parameterizations can treat random, maximum, or an arbitrary combination of overlap between clouds.

11.4.2.3 *ECHAM5 Scheme* This is the 5th generation of the ECHAM general circulation model produced by the Max Planck Institute for Meteorology in Hamburg, Germany. Its scheme for calculating radiative transfer of solar energy within the atmosphere uses the Eddington approximation for the integration over the zenith and azimuth angles and the delta–Eddington approximation for the reflectivity of a layer. The scheme includes Rayleigh scattering, absorption by water vapor and ozone, both varying in space and time, and CO_2, N_2O, CO, CH_4 and O_2 as uniformly mixed gases. Aerosols and cloud particles are effective through absorption and scattering. Interactions between scattering processes and absorption are considered for water vapor and for the uniformly mixed gases, but not for ozone, because gaseous absorption rather than scattering is assumed to dominate in the stratosphere.

The computation of transmissivities and reflectivities across a vertical column is split into two separate calculations for the clear–sky and cloudy–sky parts. The scheme has 4 spectral bands; 1 for the UV–visible (0.20–0.69 μm) and 3 for the near–infra–red (0.69–1.19, 1.19–2.38, 2.38–4.0 μm). The near–infra–red range is resolved by 3 bands to account for the wavelength dependence of the optical properties of cloud particles or aerosols. The interaction of scattering processes and gaseous absorption is included for the near–infra-red bands, but neglected in the visible band. Single scattering properties of clouds use Mie calculations for idealized size distributions for both cloud droplets and ice crystals. For each of the 4 spectral intervals polynomial fits are derived in order to express the respective single scattering properties as functions of the effective radii, which are parameterized in terms of the liquid and ice water content.

11.4.3 *Terrestrial radiation transfer in GCMs*

11.4.3.1 *HadAM3 Scheme* The longwave fluxes are calculated by a two–stream approximation, where the spectrum is divided into 8 bands with the following boundaries (in cm^{-1}): 0–400; 400–550; 550–800 (excluding 590–750); 590–750; 800–1200 (excluding 990–1120); 990–1120; 1200–1500; and 1500–3000. Gaseous absorption data are again derived from HITRAN and continuum absorption by water vapour is treated using a continuum model.

11.4.3.2 *NCAR CAM 3.0 Scheme* Longwave absorption by ozone and carbon dioxide is treated by the broadband absorptance technique. The broadband approach assumes that the spectral range of absorption by a gas is limited to a relatively small range in wave number, and hence can be evaluated at the band center. This broadband formalism is employed for CO_2, O_3, CH_4, N_2O, and minor absorption bands of CO_2, while for CFCs and stratospheric aerosols an exponential transmission approximation is employed. The longwave absorptance formulation includes Voigt line profile effects for CO_2 and O_3. For the mid–to–upper stratosphere spectral absorption lines are no longer Lorentzian in shape. Clouds are treated as grey bodies with emissivities that depend on cloud phase, condensed–water path, and the effective radius of ice particles.

11.4.3.3 *ECHAM5 Scheme* The scheme implemented in ECHAM5 is the so-called rapid radiative transfer model based on the correlated–k approximation. Absorption coefficients are derived from the LBLRTM line–by–line model and include the effect of the water-vapor continuum. The scheme computes 7 fluxes in the spectral range 10 cm^{-1} to 3000 cm^{-1} in 16 spectral bands and includes line absorption by H_2O, CO_2, O_3, CH_4, N_2O, CFC–11, and CFC–12. For cloud droplets, the mass absorption coefficient is a function of the respective effective radius with coefficients independent of wave number obtained from a polynomial fit to the results of Mie calculations. For ice clouds, an inverse dependence of the mass absorption coefficient on ice–crystal effective radius is assumed, and the coefficients vary with wave number. Maximum cloud overlap is assumed for contiguous cloud layers, random overlap elsewhere. Aerosols are considered in all spectral bands with emissivities derived from a climatology that distinguishes time–independent spatial distributions of the optical thickness of sea, land, urban, and desert aerosols, and well–mixed tropospheric and stratospheric background aerosols.

11.4.4 *Surface albedo and emissivity in GCMs*

11.4.4.1 *HadAM3 Scheme* The albedos of different types of surface are included explicitly, while their solar absorption and longwave emissivity assumes they are all blackbodies. The surface albedo of open ocean is a function of solar zenith angle. The albedo of sea ice varies between 0.60 and 0.85 as a linear function of the ice temperature above −5 °C, and is also modified by snow cover. Where there is partial coverage of a grid box by sea ice, the surface albedo is given by the fractionally weighted albedos of sea ice and open ocean. Albedos of snow–free and deeply snow–covered land are specified according to the different land-cover classes, where the deep–snow albedo decreases linearly with surface temperature above +2 °C. For intermediate snow depths, the land albedo approaches the deep-snow value exponentially, according to an e–folding depth of snow equivalent to 5 mm of water.

11.4.4.2 *NCAR CAM 3.0 Scheme* The surface albedo is specified in two wavebands (0.2–0.7 μm, and 0.7–5.0 μm) and distinguishes albedos for direct and diffuse incident radiation. Albedos for ocean surfaces, geographically varying land surfaces, and sea–ice surfaces are distinguished. Surface albedos depend on the solar zenith angle, the amount of leaf and stem material present, their optical properties, and the optical properties of snow and soil. For snow and ice the albedo depends upon the spectral band, snow thickness, ice thickness and surface temperature. The zenith–angle dependence of snow and ice is ignored. Ice thermal radiation emissivity is set to 0.95.

11.4.4.3 *ECHAM5 Scheme* For snow–free land surfaces, an annual mean background albedo derived from satellite data is allocated to a high–resolution global distribution of major ecosystem complexes. For most surfaces, the albedo of snow and ice is assumed to be a linear function of surface temperature, ranging between a minimum value at the melting point and a maximum value for 'cold' temperatures. The minimum and maximum values of snow/ice albedos are assigned according to the underlying surface type: land 0.30–0.80; canopy 0.20; land ice 0.60–0.80; sea ice 0.50–0.75; lake ice 0.50–0.75; snow on lake ice 0.60–0.80. For water surfaces, such as lakes and ocean, the albedo is set to 0.07. The longwave emissivity is set to 0.996 for all surfaces and spectral intervals.

11.5 GCM climate projections

The IPCC 2001 report distinguishes between *climate prediction* and *climate projection*. A climate projection describes the response of the climate system to emission or concentration scenarios of greenhouse gases and aerosols, or radiative-forcing scenarios, based upon simulations by climate models. Climate projections depend upon the forcing scenarios used, which are based on assumptions, concerning, e.g. future socio–economic and technological developments, that may or may not be realized, and are therefore subject to substantial uncertainty.

In Chapter 1 we discussed the IPCC Special Report on Emission Scenarios (SRES), scenarios that were used as a basis for climate projections in the IPCC 2001 report. Four scenario families, each with a similar demographic, societal, economic and technical–change storyline, comprise the baseline set. Probably the most important are A2 (medium–high emissions), which describes a world of self-reliance and preservation of local economic and technological identities, and B2, in which the emphasis is on local solutions to economic, social and environmental sustainability. Both feature a continuously increasing global population, B2 at a rate lower than A2.

11.5.1 *SRES emission scenarios*

Apart from CO_2, for which deforestation and land-use values are given, the SRES scenarios define only the changes in direct anthropogenic emissions of gases and do not specify the current magnitude of the natural emissions nor the concurrent changes in natural emissions due either to direct human activities such as land-use change or to the indirect impacts of climate change.

11.5.1.1 *Emissions* Table 11.9 gives the total anthropogenic emissions for the SRES A2 scenario. The CO_2 emission sources are fossil fuel, industrial, deforestation and land use. The projection by the end of the twenty–first century is for almost a trebling in CO_2, CH_4, CO and NO_x emissions, a doubling of N_2O emissions and an initial rise in SO_2 emissions followed by a decline to below present levels. The SRES emission scenarios show a lot of variation in projected emissions over the next century. The B2 scenario gives, as a general rule, a slower rise, reflecting the many uncertainties in industrial, technological and environmental developments to come.

Table 11.9 *Anthropogenic emissions for the SRES A2 scenario. CO_2 emission (PgC/year) sources are fossil fuel, industrial, deforestation and land use. Units of emission; CH_4 (Tg/year), N_2O (TgN/year), NO_x (TgN/year), CO (Tg/year), SO_2 (TgS/year), and black carbon BC (Tg/year). (Source: IPCC 2001)*

Year	CO_2	CH_4	N_2O	NO_x	CO	SO_2	BC
2000	7.97	323	7.0	32.0	877	69.0	12.4
2010	9.58	370	8.1	39.2	977	74.7	13.6
2020	12.25	424	9.6	50.3	1075	99.5	14.8
2030	14.72	486	10.7	60.7	1259	112.5	17.0
2040	16.07	542	11.3	65.9	1344	109.0	18.0
2050	17.43	598	12.0	71.1	1428	105.4	19.0
2060	19.16	654	12.9	75.5	1545	89.6	20.4
2070	20.89	711	13.9	79.8	1662	73.7	21.8
2080	23.22	770	14.8	87.5	1842	64.7	24.0
2090	26.15	829	15.7	98.3	2084	62.5	26.8
2100	29.09	889	16.5	109.2	2326	60.3	29.7

11.5.1.2 *Evolution of mixing ratios* The evolution of the mixing ratios of CO_2, CH_4 and N_2O over this century are given in Table 11.10, based on the SRES A2

Table 11.10 *Evolution greenhouse gas mixing ratios (ppmv) for the SRES A2 emission scenarios. The CO_2 mixing ratio is based on the Bern–CC (carbon cycle) model (high), while the CH_4 and N_2O mixing ratios are based on 3D chemistry–transport models. (Source: IPCC 2001)*

Year	CO_2	CH_4	N_2O
2000	367	1.760	0.316
2010	393	1.861	0.325
2020	431	1.997	0.335
2030	477	2.163	0.347
2040	533	2.357	0.360
2050	597	2.562	0.373
2060	670	2.779	0.387
2070	753	3.011	0.401
2080	848	3.252	0.416
2090	957	3.493	0.432
2100	1080	3.731	0.447

emission scenario.

While the trebling in the anthropogenic emission rate in CO_2 produced a trebling in the mixing ratio, based on a carbon–cycle model, the rise in the CH_4 anthropogenic emission rate to 2.75 times its present rate by the end of the century was translated by a 3D chemistry–transport model to a 2.1 times increase in the methane abundance, a translation ratio of 0.77. We note that the RC climate–photochemical model given in §11.3.5, gave similar results for methane; a doubling of an adopted total global emission rate (anthropogenic plus natural) equivalent to 6.69 times the present (A2 year 2000, Table 11.9) anthropogenic rate was translated to 4.74 times the present (Table 11.4) methane abundance, a translation ratio of 0.71. CH_4 and N_2O have large, but uncertain, sources of natural emissions, and the anthropogenic emissions of these gases are primarily associated with agricultural sources that are also difficult to quantify accurately.

11.5.1.3 *Radiative forcings and temperature change* The IPCC defines the radiative forcing of the surface–troposphere system due, for example to a change in greenhouse–gas concentrations, as the change in net downward (all–wave down minus all–wave up) irradiance at the tropopause, after allowing for stratospheric temperatures to readjust to radiative equilibrium, but with surface and tropospheric temperatures and state held fixed at the unperturbed values. The increase relative to the pre–industrial period in the mixing ratios of the greenhouse gases CO_2, CH_4 and N_2O, as well as the rise in black–carbon aerosols was translated via simple expressions to radiative forcing in the ICCP 2001 report.

The radiative forcing due to the increases in the well–mixed greenhouse gases from 1750 to 1998 was estimated to be 2.43 W m^{-2} (warming), comprising

CO_2 (1.46 W m^{-2}), CH_4 (0.48 W m^{-2}), N_2O (0.15 W m^{-2}) and halocarbons (halogen–containing compounds) (0.34 W m^{-2}), with an uncertainty of 10%. For aerosols, models were used to estimate the direct radiative forcing for five distinct species of anthropogenic origin. The global, annual mean radiative forcing is estimated as –0.4 W m^{-2} (–0.2 to –0.8 W m^{-2}, cooling) for sulphate aerosols; –0.2 W m^{-2} (–0.07 to –0.6 W m^{-2}) for biomass-burning aerosols; –0.10 W m^{-2} (–0.03 to –0.30 W m^{-2}) for fossil fuel organic–carbon aerosols; 0.2 W m^{-2} (0.1 to 0.4 W m^{-2}) for fossil–fuel black–carbon aerosols; and in the range –0.6 to 0.4 W m^{-2} for mineral–dust aerosols.

11.5.1.4 *Climate sensitivity parameter* This parameter is defined as the global mean surface temperature response, ΔT_g, to the radiative forcing, ΔF, (Dickinson 1982; WMO 1986; Cess *et al.* 1993) and is given by

$$\lambda = \Delta T_{\mathrm{g}}/\Delta F. \tag{11.22}$$

The concept of radiative forcing thus gives a quick estimate of the global mean annual surface temperature, T_{g}, response to a radiative forcing. In 1D RC models, where the concept was initiated, λ varied slowly with climatic parameters, with a value typically about 0.5 K W m^{-2} (Ramanathan *et al.* 1985), for a variety of radiative forcings. In eqn (11.5) we used the notion of radiation balance at the top of the atmosphere to derive a quick estimate of the Earth's effective temperature. We can also derive a similar expression for T_{g} via the radiative flux balance equation (erg cm^{-2} s^{-1})

$$\varepsilon\sigma T_{\mathrm{g}}^4 = \frac{(1-\alpha)}{4} S_{\odot}. \tag{11.23}$$

The planetary albedo is α, while ε represents the planetary emissivity that gives a measure of the effect of the surface–atmosphere (greenhouse gases, clouds, surface emissivity and temperature) on the outgoing radiation. Thus, we can translate the radiative forcing arising from changes in the outgoing longwave radiation, $\Delta F_{\mathrm{LW}}^{\uparrow}$, to temperature changes via

$$\lambda_{\mathrm{LW}} = \frac{\Delta T_{\mathrm{g}}}{\Delta F_{\mathrm{LW}}^{\uparrow}}, \tag{11.24}$$

$$= \frac{1}{4\varepsilon\sigma T_{\mathrm{g}}^3}, \tag{11.25}$$

and those due to changes in the solar constant, $\Delta S_{\odot}$, from

$$\lambda_{\odot} = \frac{\Delta T_{\mathrm{g}}}{\Delta S_{\odot}}, \tag{11.26}$$

$$= \frac{1-\alpha}{4}\lambda_{\mathrm{LW}}. \tag{11.27}$$

We can also show that the climate sensitivity parameter corresponding to changes in the reflected solar shortwave to space is $\lambda_{\mathrm{RSW}} = -\lambda_{\mathrm{LW}}/4$. For the present

Table 11.11 *Radiative forcings (W m^{-2}) for year 2000 relative to the pre-industrial period, for the SRES A2 scenario, for the greenhouse gases CO_2, CH_4 and N_2O, and total radiative forcing together with its corresponding estimated translation to global annual mean surface temperature change based on eqn (11.24). (Radiative forcings from IPCC 2001)*

Year	CO_2	CH_4	N_2O	Total	ΔT_g
2000	1.49	0.49	0.15	2.13	0.64
2010	1.85	0.53	0.18	2.56	0.77
2020	2.35	0.58	0.21	3.14	0.94
2030	2.89	0.63	0.25	3.77	1.13
2040	3.48	0.70	0.29	4.47	1.34
2050	4.09	0.76	0.33	5.18	1.55
2060	4.71	0.83	0.37	5.91	1.77
2070	5.33	0.89	0.41	6.63	1.99
2080	5.97	0.96	0.45	7.38	2.21
2090	6.61	1.02	0.49	8.12	2.44
2100	7.26	1.09	0.53	8.88	2.66

Earth we can set $\varepsilon \approx 0.61$, and $T_g = 288$ K, with $S_\odot = 1367$ W m^{-2} and using $\sigma = 5.669 \times 10^{-8}$ W m^{-2}K^{-4}, we obtain for the climate sensitivity parameters

$$\lambda_{LW} = 0.3 \quad \text{K/W m}^{-2} \tag{11.28}$$

$$\lambda_\odot = 0.05 \quad \text{K/W m}^{-2} \tag{11.29}$$

$$\lambda_{RSW} = 0.075 \quad \text{K/W m}^{-2}. \tag{11.30}$$

Thus, a radiative forcing $\Delta F_{LW}^{\uparrow} = 2.4$ Wm^{-2} translates to a warming of $\Delta T_g = 0.72$ K, a 1% change in the solar constant or $\Delta S_\odot = 13.67$ W m^{-2} translates to a corresponding change $\Delta T_g = 0.68$ K, and a 1% change in the outgoing shortwave radiation, typically 1 W m^{-2}, translates to a 0.075 K change.

In Table 11.11 are given the radiation forcings based on the reference year 1750 for the SRES A2 emission scenario. We see that the trebling in CO_2 mixing ratio over the twenty–first century is translated to almost 5 times in radiative forcing. The estimated change in global annual mean surface temperature to bring the planet back into radiative equilibrium, neglecting solar radiation and planetary albedo changes, is also given.

11.5.2 *Global change in temperature*

Most model experiments show broadly the same pattern for the change in annual mean surface air temperature. There is a maximum warming in the high latitudes of the Northern Hemisphere and a minimum in the Southern Ocean, due to ocean heat uptake (see the example in Fig. 11.6). Ocean heat uptake also contributes to a minimum of warming in the North Atlantic, while land

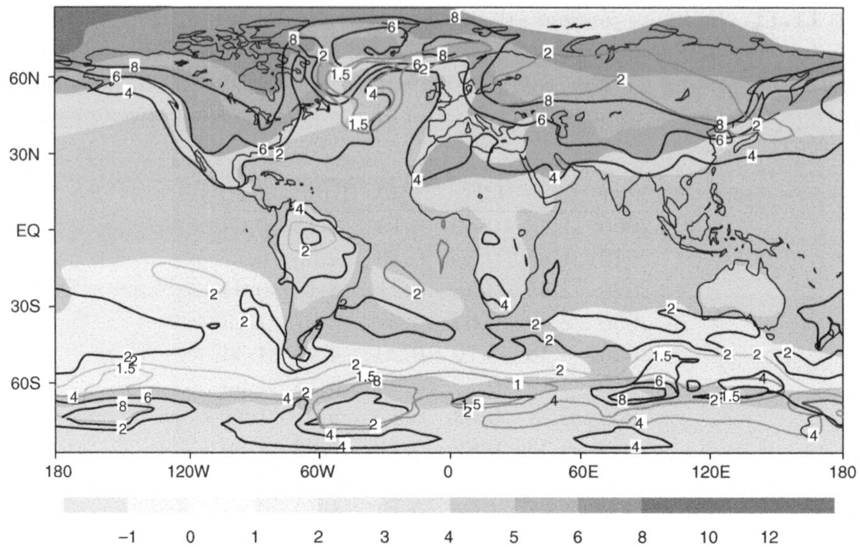

FIG. 11.6. The multimodel ensemble annual mean change of the temperature (shading), its range (thin isolines) (Unit: C) and the multimodel mean change divided by the multi–model standard deviation (solid isolines, absolute values) for the SRES A2 scenario. It shows the period 2071–2100 relative to the period 1961–1990. (Source: IPCC 2001)

warms more rapidly than ocean almost everywhere. The large warming in high latitudes of the Northern Hemisphere is connected with a reduction in the snow and sea-ice cover. There is also consistent midtropospheric tropical warming and stratospheric cooling. The range tends to increase with altitude, partly due to the variation in the level of the tropopause among the models.

For the SRES A2 scenario the mean change is 1.1 C with a range from 0.5 to 1.4 °C; while for B2, the mean is 1.2 °C with a range from 0.5 to 1.7 °C. By the end of the twenty–first century (2071 to 2100), the global average surface temperature change is 3.0 °C with a range 1.3 to 4.5 °C, while for B2 the mean SAT change is 2.2 °C with a range is 0.9 to 3.4 °C. The evolution of the temperature change for some of the models involved in these experiments appears in Fig. 11.7.

11.5.3 *Global change in precipitation*

Figure 11.8 shows the global distribution of the relative change in the mean precipitation for the SRES A2 emission scenario for the period 2071–2100 relative to the period 1961–1990. There is a general increase in the tropics, particularly the tropical oceans and parts of northern Africa and south Asia, and the mid- and high latitudes, while the rainfall generally decreases in the subtropical belts.

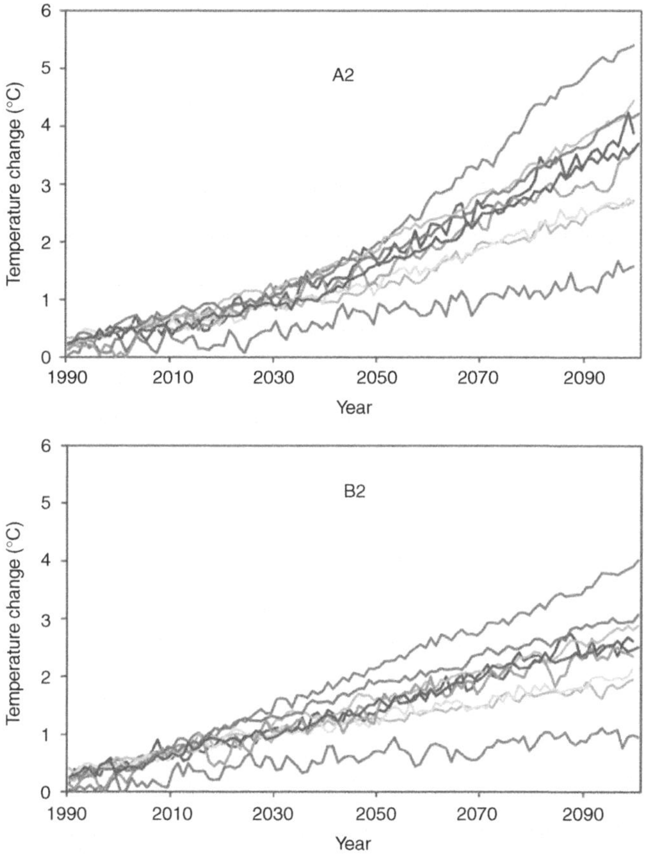

FIG. 11.7. The time evolution of the globally averaged temperature change relative to the baseline period 1961–1990 of the SRES simulations A2 (top) and B2 (bottom) (Unit: C), produced by various AOGCMs. (Source: IPCC 2001)

Over the central and eastern tropical Pacific, an El Niño-like surface temperature warming is associated with an eastward shift of precipitation. The A2 scenario experiment exhibits a relatively large percentage increase in precipitation over the Sahara and Arabia, although in absolute values the precipitation amount gained by these very dry regions is very small. Figure 11.9 shows the time evolution of the globally averaged precipitation change for the SRES A2 and B2 scenarios. For the 30-year average for the period 2071–2100 the increase that the A2 scenario gives is 3.9% with a range of 1.3 to 6.8%, while the B2 scenario gives an increase of 3.3% with a range of 1.2 to 6.1%. The lower precipitation increase values for the B2 scenario are consistent with less globally averaged warming for that scenario at the end of the twenty–first century compared to A2.

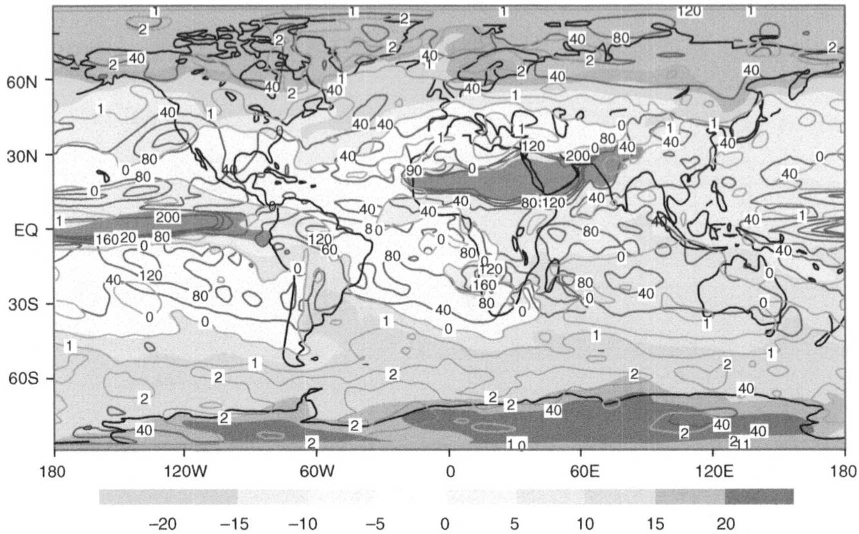

FIG. 11.8. The multimodel ensemble annual mean change of the precipitation (shading), its range (thin isolines) (Unit: %) and the multimodel mean change divided by the multimodel standard deviation (solid isolines, absolute values) for the SRES A2 scenario. Shown is the period 2071–2100 relative to the period 1961–1990. (Source: IPCC 2001)

11.5.4 *Global change in sea level*

The main processes that are expected to result in increased sea levels are thermal expansion, snow and ice sublimation, and melting and break–up of ice sheets, particularly that of Greenland. Snow and ice accumulation over Antarctica, due to lower temperatures there, are expected to contribute to a drop in sea level.

11.5.4.1 *Thermal expansion* In a warmer world, the volume of the ocean will increase and its density will decrease, at a rate that (for both atmosphere and ocean) depends strongly on the rate at which heat is stored in the deeper oceanic layers. The large heat–storage capacity of the oceans means that the oceans will not be in equilibrium on timescales of centuries, so that the global average sea level will initially be slow to rise but will continue to increase well after atmospheric greenhouse–gas concentrations have stabilized. The rise is not the same everywhere, since local sea levels are influenced by atmospheric pressure, air–sea exchange of mass, momentum and energy fluxes, and gains via land runoff and ice inputs. Salinity changes, in particular, affect the local sea-water density and thus local sea level. The geographical distribution of sea–level change due to density and circulation changes can be obtained from AOGCM results. The IPCC summary suggests that over the last hundred years the average rate of

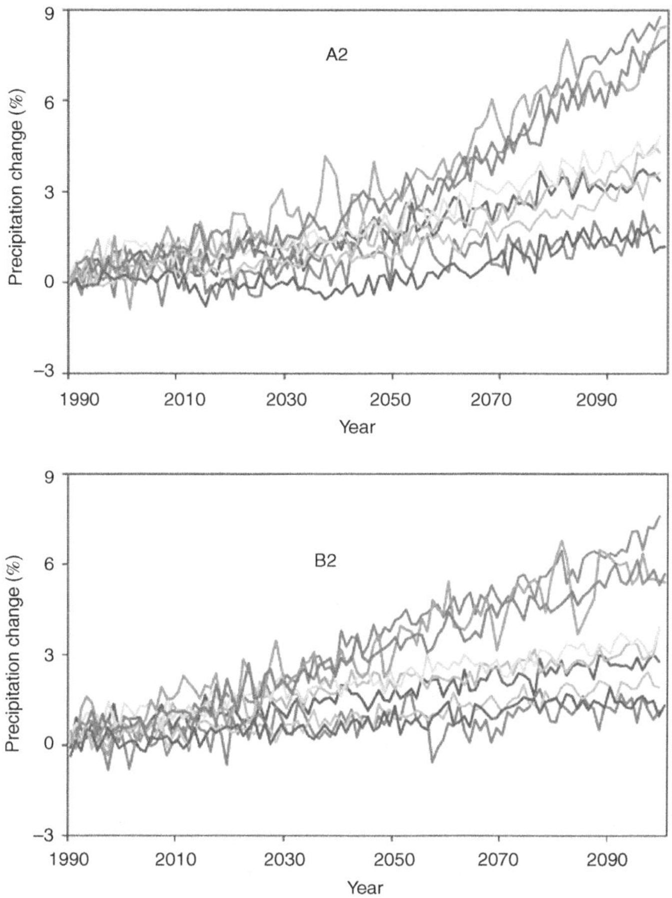

FIG. 11.9. The time evolution of the globally averaged precipitation change relative to the baseline period 1961–1990 of the SRES simulations A2 (top) and B2 (bottom) (Unit: %), produced by various AOGCMs. (Source: IPCC 2001)

sea–level rise due to thermal expansion was in the range 0.3–0.7 mm/year, rising to 0.6–1.1 mm/year in recent decades.

11.5.4.2 *Glaciers and ice sheets* The present Greenland and Antarctic ice sheets contain enough water to raise the sea level by almost 70 m, so that only a small fractional change in their volume would have a significant effect. The average annual snowfall onto the ice sheets is equivalent to 6.5 mm of sea level, this input being approximately balanced currently by the loss from melting and icebergs. Antarctic temperatures are very low, so melting is unimportant and water is lost mainly as floating ice shelves, which eventually break up to form

Table 11.12 *The average total sea–level change (mm) for the SRES A2 scenario, based on the results of various AOGCMs. Also shown are important components of sea–level change; thermal expansion (TE), glaciers and ice caps (GIC), Greenland (G), and Antarctica (A). Note that the sum of the various components of sea–level change do not add up to the average owing to the addition of other terms. (Source: IPCC 2001)*

Year	Total	TE	GIC	G	A
2000	17	10	4	0	–2
2010	38	23	10	1	–5
2020	61	39	16	2	–8
2030	88	57	23	4	–12
2040	120	81	31	5	–17
2050	157	109	41	7	–23
2060	201	142	52	10	–31
2070	250	180	65	13	–40
2080	304	224	79	16	–50
2090	362	272	93	20	–62
2100	424	325	108	25	–76

icebergs. In contrast, Greenland summer temperatures are high enough to cause widespread melting, which accounts for about half of the ice loss, the remainder being discharged as icebergs and small ice shelves.

For Greenland, estimates of the response to a 1 °C local warming over the ice sheet range from about 0.1 to 0.4 mm/year increase in global sea level (Table 11.12). This range mainly reflects differences in the predicted precipitation changes and the yearly distribution of temperature increase, predicted to be larger in winter than in summer. Some palaeoclimatic data from central Greenland ice cores indicate that variations in precipitation during the Holocene were related to changes in atmospheric circulation rather than directly to local temperature.

For Antarctica, the predicted response to a 1 °C warming is a sea–level decrease of about 0.4 mm/year. The role of melting is insignificant, even for summer temperature increases of a few degrees, so that only accumulation of snow/ice is important resulting in a reduction in sea level. In a warmer climate, changes in atmospheric circulation and increased moisture advection might become equally important, in particular close to the ice–sheet margin.

Antarctica is predicted to play a larger role than Greenland in a realistic scenario, so that the net effect of both polar ice sheets could be to contribute a net decrease to the sea level budget, which still increases overall (Table 11.12). The average total sea–level change evolution based on AOGCMs with climate sensitivities in the range 1.7 to 4.2 °C is shown in Table 11.13. For the complete range of AOGCMs and SRES scenarios, and including uncertainties in land–ice changes,

Table 11.13 *Final rise (m) in sea level due to thermal expansion in 2×CO_2 and 4×CO_2 experiments, produced by various climate models, after stabilization of CO_2 levels. (Source: IPCC 2001)*

Model	2×CO_2	4×CO_2
CLIMBER	0.78	1.44
ECHAM3/LSG	1.53	2.56
GFDL-R15-a	1.96	3.46
BERN2D GM	1.93	3.73
BERN2D HOR	1.28	4.30
UVic GM	0.53	1.24
UVic H	1.19	2.62
UVic HBL	0.65	1.78

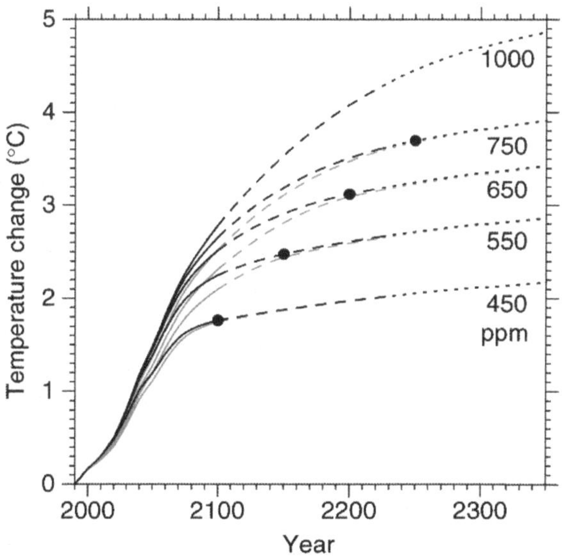

FIG. 11.10. Model results of projected global mean temperature changes when the concentration of CO_2 is stabilized. The broken lines after 2100 indicate increased uncertainty in the simple climate model results. The black dots indicate the time of CO_2 stabilization. (Source: IPCC 2001)

permafrost changes and sediment deposition, global average sea level is projected to rise by 0.09 to 0.88 m over the period 1990 to 2100, with a central value of 0.48 m. The central value gives an average rate of 2.2 to 4.4 times the rate that was recorded over the twentieth century.

Predictions beyond 2100 are very uncertain, not least because it is virtually impossible to guess what will be the levels of anthropogenic emissions. However, it is possible to estimate what will happen in the very long term if emissions are stabilized at the levels that might be reached by the end of the present century. As already noted, thermal expansion and ice sheet changes are likely to continue, since the processes involved have very long characteristic times of change. According to the modellers, the sea level would continue to rise for many centuries, perhaps reaching increases of 3 m or more over present levels (Table 11.13), even after greenhouse gas concentrations have stabilized (Fig. 11.10).

11.6 Bibliography

11.6.1 *Notes*

For an early review on radiative–convective (RC) modelling see Ramanathan and Coakley, and references therein.

For the convective adjustment approach to RC modelling see early work of Manabe and Strickler, Manabe and Wetherald. For parameterizations for solar and terrestrial radiation transfer see Ramanathan; and Vardavas and Carver.

GCM model descriptions can be found in Pope et al., Roeckner *et al.* and Collins *et al.*

11.6.2 *References and further reading*

Aspray, W. (1990). *John von Neumann and the origins of modern computing.* MIT Press, Cambridge, MA.

Cess, R. D. and coauthors (1993). Uncertainties in CO_2 radiative forcing in atmospheric general circulation models. *Science*, **262**, 1252–1255.

Charney, J. G., Fjörtoft, R. and von Neumann, J. (1950). Numerical integration of the barotropic vorticity equation. *Tellus*, **2**, 237–254.

Collins, W. D. and coauthors (2004). *Description of the NCAR Community Atmosphere Model (CAM 3.0).* NCAR/TN464, NCAR, Boulder, Colorado.

Dickinson, R. E. (1982). Modeling climate changes due to carbon dioxide increase. In: *Carbon dioxide review 1982.* Clark, W. C. (ed.). Oxford University Press, Oxford.

Goody, R. M. and Yung, Y. L. (1996). *Atmospheric radiation: Theoretical basis.* Oxford University Press, Oxford.

Houghton, J. T. (1997). *Global warming: The complete briefing.* 2nd edn., Cambridge University Press, Cambridge.

IPCC, 2001: *Climate change 2001: The scientific basis. Contribution of working group I to the Third Assessment Report of the Intergovernmental Panel on*

Climate Change. Houghton, J. T., Ding, Y., Griggs, D. J., Noguer, M., van der Linden, P. J., Dai, X., Maskell, K. and Johnson, C. A. (ed.), Cambridge University Press, Cambridge, United Kingdom and New York, NY, USA.

Iribarne, J. V. and Godson, W. L. (1981). *Atmospheric thermodynamics.* D. Reidel, Dordrecht.

Kiehl, J. T. and Ramanathan, V. (2006). *Frontiers in climate modeling.* ed., Cambridge University Press, Cambridge.

Lavvas, P. P., Coustenis, A. and Vardavas, I. M. (2007). Coupling photochemistry with haze formation in Titan's atmosphere. Part I: Model description. *Planet. Space Sci.*, in press.

López-Puertas, M. and Taylor, F. W. (2001). *Non-LTE radiation transfer in the atmosphere.* World Scientific, Singapore.

Manabe, S. and Bryan, K. (1969). Climate calculations with a combined ocean-atmosphere model. *J. Atmos. Sci.*, **26**, 786–789.

Manabe, S. and Stouffer, R. J. (1994). Multiple-century response of a coupled ocean-atmosphere model to an increase of atmospheric carbon dioxide. *J. Clim.*, **7**, 5–23.

Manabe, S., Smagorinsky, J. and Strickler, R. F. (1965). Simulated climatology of general circulation with a hydrologic cycle. *Mon. Weath. Rev.*, **93**, 769–798.

Manabe, S. and Wetherald, R. (1967). Thermal equilibrium of the atmosphere with a given distribution of relative humidity. *J. Atmos. Sci.*, **24**, 241–259.

Pope, V. D., Gallani, M. L., Rowntree, P. R. and Stratton, R. A. (2000). The impact of new physical parameterizations in the Hadley Centre climate model – HadAM3. *Climate Dynamics*, **16**, 123–146.

Ramanathan, V. (1976). Radiative transfer within the Earth's troposphere and stratosphere: A simplified radiative-convective model. *J. Atmos. Sci.*, **33**, 1330–1346.

Ramanathan, V. and Coakley, J. A. (1978). Climate modeling through radiative-convective models. *Rev. Geophys. Space. Phys.*, **16**, 465–490.

Ramanathan, V., Cicerone, R., Singh, H. and Kiehl, J. (1985). Trace gas trends and their potential role in climate change. *J. Geophys. Res.*, **90**, 5547–5566.

Richardson, L. F. (1922). *Weather prediction by numerical process.* Cambridge University Press, Cambridge.

Roeckner, E. and coauthors. (2003). *The atmospheric general circulation model ECHAM5.* Max Planck Institute for Meteorology, RN. 239, Hamburg.

Smagorinsky, J., Manabe, S. and Holloway, J. L. (1965). Numerical results from a nine-level general circulation model of the atmosphere. *Mon. Weath. Rev.*, **93**, 727–768.

Taylor, F. W. (1991). The greenhouse effect and climatic change. *Rep. Prog. Phys.*, **54,** 6, 881–918.

Taylor, F. W. (2002). The greenhouse effect revisited. *Reports on progress in physics*, **65**, 1–25.

Vardavas, I. M. and Carver, J. H. (1984a). Solar and terrestrial parameterizations for radiative-convective models. *Planet. Space Sci.*, **32**, 1307–1325.

Vardavas, I. M. and Carver, J.H. (1984b). Comments on the Newton-Raphson method for obtaining temperature profiles from radiative-convective models. *Planet. Space Sci.*, **32**, 803–807.

WMO (1986). *Atmospheric ozone: 1985, Global ozone research and monitoring project.* World Meteorological Organization, Report No. 16, Chapter 15, Geneva, Switzerland.

12

PLANETARY EVOLUTION AND COMPARATIVE CLIMATOLOGY

12.1 Introduction

Many of the radiative processes that govern the climate system of the Earth are also important on the other planets of the Solar System. The different boundary conditions that apply, especially distance from the Sun, give rise to different histories and surface and atmospheric properties, but parallels can often be drawn that test our understanding of the physics involved, which of course is everywhere the same. Comparisons between the planets are instructive, therefore, as are attempts to understand how the climate on each has changed over the lifetime of the Solar System. It is well known, for example, that the Earth has experienced numerous periods of dramatic climate change (the ice ages) over its history. Recent evidence from planetary probes confirms that Mars had a warm, wet epoch in its (probably distant) past, and that early Venus may have had oceans comparable in extent to those on the present–day Earth. It is vital that we understand the reasons for climate change on this sort of grand scale and the processes, balances and feedbacks involved in complete quantitative detail. By comparing the processes at work on Mars, Venus and Titan to those on our own planet we may gain a deeper understanding of present–day global change on the Earth, the origin and evolution over the long term of our world and its habitability, and the processes behind threats such as greenhouse warming.

A good place to start is to examine how well radiative–transfer models of the energy budget can explain the climatic conditions on the other planets, whose atmospheric composition and large–scale thermal structures are now quite well known. We begin with the regimes on the nearby, and similarly sized planets Mars and Venus, which are the best explored and bear the closest resemblance to the Earth. Saturn's giant moon Titan, although more distant and colder, has a number of Earth–like attributes, while even hot, virtually airless Mercury and the gas giant planets offer some instructive parallels, despite their more extreme contrasts with the terrestrial environment (Table 12.1).

All of the major bodies of the Solar System except Pluto have now been visited and we know in general terms what climatic conditions prevail there and, roughly at least, what physical processes sustain them. Comparative planetary climatology addresses the detailed theory of the physics that determines the environmental conditions, the stability in each case against climate change, and the

Table 12.1 *Data relevant to the climate on Venus, Earth, Mars and Titan. Composition is % by volume. For the Earth, water vapour is additional pressure of maximum of about 3% of the dry atmospheric pressure.*

	Venus	Earth	Mars	Titan
Distance from Sun (AU)	0.7	1	1.517	10
Radius (km)	6052	6376	3398	2575
Surface pressure (bar)	92	1	0.007	1.5
Surface temperature (K)	730	288	220	94
Day (Earth days)	243	1	1.029	16
Main atmospheric constituents	*CO_2 96.5%	N_2 79%	*CO_2 95%	N_2 95%
(*key greenhouse gases)	N_2 3.5%	O_2 21%	N_2 2.7%	*CH_4 5%
		Ar 1%	Ar 1.6%	
		*H_2O (3%)		
		*CO_2 0.035%		
Bond albedo	0.76	0.3	0.25	0.25
Gravity (m s^{-2})	8.60	9.82	3.74	1.34
Effective temperature (K)	230	255	210	84

practical development of new experiments to further investigate these. The comparative aspect asks not only what the climate systems on different planets have in common and what controls their stability, but what change has taken place and why, what is the chronology of past changes, and can we make plausible predictions of future change? From a consideration of the spatial and temporal bases for climate variability we can work out the appropriate measurement objectives for future space missions that seek to address these and other climate-related questions further.

At the present time, such objectives definitely include the diversity in climatic behaviour found between the Earth and even our closest neighbours. For example, why is Venus so hot and so dry? How are its thick, sulphurous cloud layers created and maintained? Is the surface of Venus shaped by plate tectonics like the Earth's? How was Mars able to have a thick atmosphere, and free water on its surface, in the past, when this would be impossible under the conditions we observe today? What mechanism maintains the Great Red Spot on Jupiter, and what is its relationship to the Earth–sized dark spot on Neptune? To what extent do these otherworldly storms resemble terrestrial hurricanes? Simple models and calculations, or attempts to draw analogies with the behaviour of the terrestrial ecosystem, give no definite answers to questions like these. Until we have developed a more complex and complete understanding of climate systems in general, we cannot have confidence in predictions of related behaviour on the Earth.

The First Law of Thermodynamics, determining how a planet achieves radiative balance, is the primary consideration in climate physics and this, of course, is not unique to the Earth. For instance, the greenhouse effect plays a major role in determining the surface environment on the other terrestrial planets with atmospheres, Mars and Venus, and the same greenhouse gases (principally carbon dioxide and water vapour) are responsible. The existence of these planets offers an important opportunity to see how the greenhouse effect works in situations other than that we observe on the Earth. The general message that Earth-like planets apparently can be subject to extremely large and variable greenhouse warmings is an important and often under–rated one.

12.1.1 *Origin of the solar system*

According to most current theories, the Sun, the planets, and the small bodies and orbiting debris in the Solar System all formed about 4.5 Byr (billion years) ago when a cloud of dust and gas known as the solar nebula underwent gravitational collapse. The potential energy released during this process resulted in heating, especially at the centre where the material was most compressed, and opaque enough to inhibit cooling by the escape of radiation.

Model calculations suggest that the initial collapse took less than 100000 years. At the end of this time, most of the mass was in the hot, vaporized 'protosun' at the centre, but a significant amount of cooler material remains in orbit around it, held by the centrifugal force of the rotating system. Unlike the protosun, this accretion disk was optically thin, and cooled relatively rapidly by radiation to space.

As a result of this cooling, metal, rock and, in the outer reaches of the Solar System, ice, all condensed out into dust–sized particles. These accreted by colliding and forming first loose aggregates, and then larger particles. The growth rate accelerated once these had enough mass to attract other particles by gravity, rather than by chance collisions, their final size depending on the distance from the star and the local density and composition of the protoplanetary nebula. In the inner Solar System, these 'planetesimals' probably had a size distribution similar to that found in the current asteroid belt, with a few as large as Earth's moon. In the outer Solar System, where condensation of ice provided additional material, there may have been planetesimals several times the mass of the Earth, which took from a few hundred thousand to about twenty million years to form.

Models of the early Solar System require a total mass during formation of the planets that is considerably larger than that which remains today. During the formation of planetesimals, the disk appears to have lost material, including most of its gas, perhaps during a 'T–Tauri' phase of the Sun, so called after a young star that can be seen exhibiting this behaviour today (Fig. 12.1). This phase of stellar evolution is characterized by an intense particle flux streaming outwards into space, a more vigorous version of the modern solar wind. Only the large, icy planetesimals in the outer Solar System that later formed the gas giant planets had enough mass to retain atmospheres at this point. These atmospheres, like that of the Sun, consisted of the most common elements, principally hydrogen and helium.

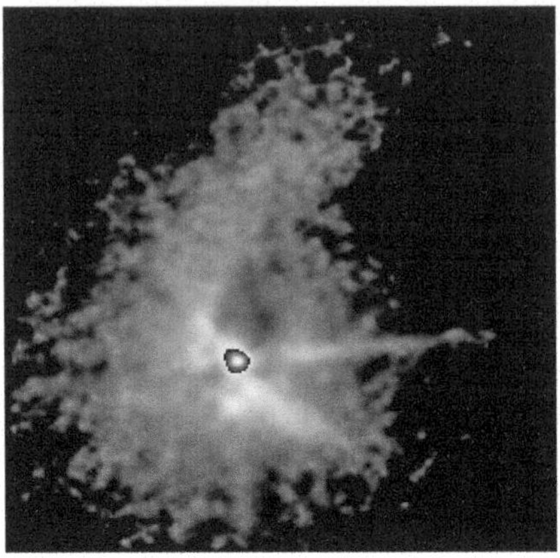

FIG. 12.1. The T–Tauri system, showing the accretion disk. Once thought to be a model for the early Solar System, T–Tauri is now known to be a double star. Nevertheless, the young Sun will have had a disk like this, containing material that, over several million years, either fell into the star, remained in orbit and aggregated into planetesimals, or was driven off into space by the solar wind.

Eventually, after perhaps another hundred million years, the planetesimals collided and combined until a small number of larger bodies – the planets we know today – were left in stable orbits. The process may not be entirely complete; the asteroid belt still contains planetesimals, and in the outermost reaches of the Solar System a large number of bodies of considerable size orbits in the Kuiper Belt. Some of these may eventually collide and form further planet-sized objects outside the orbit of Neptune.

In the effort to understand the details and veracity of the history outlined above, in addition to indepth exploration of our own solar system, we are now able to detect planetary systems around other stars. With many more samples to study, the theory of planetary system formation and evolution is advancing rapidly. Already, we run into difficulties when it comes to explaining why Jupiter–like planets are being found very close to their parent stars; it may be that the T–Tauri phase does not occur in all types of stars, for example.

12.1.2 *Evolution of planetary atmospheres*

The Earth probably retained very few of the gases that surrounded the solid bodies at the time the planetesimals condensed. Instead, the present atmosphere formed later as a result of outgassing from the crust, a process that would have been augmented by the infall of icy cometary material. Both processes are still happening, on a reduced scale, today.

The primitive atmosphere thus acquired would have had a composition quite different from that which exists now, probably consisting primarily of carbon dioxide, methane, ammonia, and water vapour with no, or very little, free nitrogen or oxygen. The large proportion of nitrogen we breathe today came about as a consequence of the photodissociation of ammonia into nitrogen and hydrogen. The latter gas is so light that it can escape from the Earth even at the present time, whereas most heavier gases are gravitationally trapped now that the Earth is cooler and the Sun less active.

12.1.3 *Escape processes*

Following the development of the kinetic theory of gases and Maxwell's work on the velocity distribution of the molecules in a gas of a given temperature and composition, Jeans formulated his theory of atmospheric escape rates in 1904. He reasoned that a molecule would escape if: a) the upward component of its velocity exceeded the escape velocity (11.2 km s^{-1} for the Earth), and b) it was above the level where the pressure is so low that collisions between molecules are negligible. The first condition applies when the thermal energy becomes greater than the gravitational potential energy, i.e.

$$kT \sim \frac{1}{2}mv^2 > \frac{mMG}{r}, \tag{12.1}$$

where v is the upward velocity of a molecule of mass m. The critical level for collisions, which we now call the exobase, falls between 500 and 600 km altitude on Earth. It occurs when the vertically integrated density accounts for one mean free path, i.e. at radius r_c from the centre of the planet where

$$\int_{r_c}^{\infty} N(r)\sigma \mathrm{d}r = \sigma N(r_c)H_c = 1, \tag{12.2}$$

where N is the number density, σ is the collisional cross-section, H is the scale height, and r_c defines the height of the exobase.

The Jeans' escape flux F_{Jeans} (mol cm^{-2} s^{-1}) is obtained by integrating upwards from this level and over the solid angle, assuming isothermal conditions and a Maxwellian distribution. The factor 1/2 allows for depletion of the fast tail of the distribution by the escape itself.

$$F_{\mathrm{Jeans}} = \frac{N(r_c)U}{2\sqrt{\pi}}(1+\lambda)\exp(-\lambda), \tag{12.3}$$

where $U = \sqrt{2kT/m}$ is the most probable (Maxwellian) velocity, $v_e = F_J/N$ is the mean expansion velocity, $H_c = kT/mg(r)$ is the scale height of the molecule or atom at the critical level, and λ is the escape parameter given by

$$\lambda = v_{\mathrm{esc}}^2/U^2 = r_c/H_c = GMm/kTr_c. \tag{12.4}$$

We usually also define a *characteristic escape time* given by $\tau_e = H_c/v_e$ (s).

Clearly this must be appreciably less than the age of the Solar System ($\sim 10^{17}$ s) for the loss of a given species from any particular planet to be important for the climate; on Earth, oxygen is safe, by this criterion at least, while hydrogen and helium are long gone (Table 12.2). On Jupiter, the combination of low temperature and high gravity means that this condition is not met, even for hydrogen (Table 12.3). Jeans' formula successfully explains why the larger bodies have denser atmospheres, and why the terrestrial planets are depleted in the lightest gases, despite production rates that can be high.

Table 12.2 *Exospheric escape time, τ_e (s), for reduction of exospheric density by Jeans escape for the Earth's atmosphere, assuming the exobase is at 500 km altitude, at temperature 1480 K and the gravitational acceleration is 8.43 ms $^{-2}$. Also given are the scale height, H (km), the most probable speed, U, and the mean expansion velocity, v_e, (km s^{-1}).*

Parameter	H	H_2	He	O
H	1460	730	365	91
U	4.96	3.51	2.48	1.24
v_e	7.32×10^{-2}	8.71×10^{-4}	9.94×10^{-8}	7.14×10^{-32}
τ_e	2.0×10^4	8.38×10^5	3.67×10^9	1.28×10^{33}

Table 12.3 *Thermal escape times , τ_e (s), according to Jeans theory for various gases on four planets and the Moon. $T(K)$ is the temperature and g is the gravity (m s^{-2}) at the critical level, r_c (km).*

Parameter	Moon	Mercury	Mars	Venus	Jupiter
T	300	600	365	700	155
r_c	1738	2439	3590	6255	69500
g	1.62	3.76	3.32	8.27	26.2
τ_e(H)	3.55×10^3	3.32×10^3	1.39×10^4	5.71×10^5	5.14×10^{617}
τ_e(He)	2.03×10^4	1.40×10^5	2.66×10^8	2.85×10^{16}	1.18×10^{2455}
τ_e(O)	2.25×10^9	7.37×10^{13}	1.04×10^{28}	7.87×10^{61}	1.03×10^{9820}
τ_e(Ar)	3.29×10^{20}	2.57×10^{32}	1.97×10^{68}	6.20×10^{153}	6.61×10^{24522}
τ_e(Kr)	3.53×10^{41}	9.09×10^{66}	4.45×10^{142}	4.67×10^{322}	3.72×10^{51445}

However, it is also the case that the Moon should have retained a significant atmosphere of the heavier gases if Jeans' formula is correct and if thermal escape is the only mechanism that applies. By the same argument, there should be much stronger fractionation of the noble gases than is actually observed. Clearly, processes other than Jeans escape are at work. In particular, it is now realized that a sufficiently rapid flow of a light molecule like hydrogen could carry off heavier gases as a result of collisional interactions, i.e. aerodynamic drag. The theory of this process, sometimes called 'blowoff', suggest a linear dependence on mass, making it potentially much more efficient than the exponential rate in the Jeans equation, provided of course that the carrier flow is sufficiently strong. This would not be the case in Earth's present atmosphere, but fits very well with some theoretical expectations for the T–Tauri phase of the Sun, when the output of solar ultraviolet radiation is thought to have been 100 times the present level. This would provide a copious supply of hydrogen, by the dissociation of water, and a high level of heating in the upper atmosphere, producing a large flux of hydrogen to space on Venus and Mars, as well as Earth. The fact that the life–free planets have no residual oxygen suggests that the flow of hydrogen may have been sufficient to remove the oxygen from the water, as well, or it may have reacted with minerals in the surface.

In addition to the escape processes that are possible for neutral molecules and atoms that have a Maxwellian distribution of velocities, a number of non-thermal escape mechanisms are also known to operate. Shortwave (λ <100 nm) solar radiation dissociates and ionizes the constituents of the tenuous upper layers of the atmosphere, and imparts to the fragments a non–Maxwellian or 'suprathermal' distribution of velocities. In addition to any enhanced thermal escape that then occurs, the fact that the particles are charged renders them subject to electromagnetic forces due to the solar wind and the magnetic fields surrounding the planet. Near the Earth's poles, for instance, where the magnetic field lines extend into space rather than reconnecting with the planet, a substantial outflow of charged particles, mainly protons, occurs (the 'polar wind'). These charged particles, and those in the solar wind itself, can also sweep neutral particles away.

A purely mechanical loss process that is thought to have been particularly important on Mars, and must also have occurred on Earth to some extent in the early life of the Solar System, is impact erosion. This is the atmospheric loss that occurs when a planet is bombarded with one or more large meteorites, particularly when the blow is oblique, directing the energy of the blast through the atmosphere and out into space. Clearly the actual mass of gas removed depends on the size and frequency of the colliding objects as well as the geometry of the collision and the mass of the planet, with more material being removed from a smaller planet, because of its smaller gravitational field. Thus impact erosion is relatively inefficient on Earth and Venus, while the impact of a sufficiently large and fast object (more than 3 km across and travelling at least 14 km/s) will create a plume on Mars that expands faster than the escape velocity and that could sweep away much of the atmosphere at a stroke.

12.2 The evolution of the Earth's climate

The age of the Earth from geological records is estimated at about 4.5 billion years (Byr) old. The Precambrian Era (4.5–0.54 Byr) is the longest geological record and comprises the early Precambrian, or Archean Eon (4.5–2.5 Byr), and the Proterozoic Eon (2.5–0.54 Byr). The Archean was a time extending back to the consolidation of the planet and covers the early chemical and mineralogical evolution of the Earth's crust and atmosphere. The initial atmosphere, a remnant of the Earth's formation (duration about 100 million years) by accretion from the Solar Nebula that rotated around the early sun was probably lost due to such processes as bombardment by other bodies, a stronger solar wind than today's, and diffusion to space.

12.2.1 *The Precambrian atmosphere*

The Earth's early atmosphere is thought to have outgassed from the Earth's interior and would have been reducing, composed of CH_4, H_2, with minor amounts of H_2O, CO, N_2, H_2, NH_3, Ar and He. Most of the H_2O condensed to form oceans that became a sink for soluble gases. This reducing atmosphere was thought to play an important role in the formation of amino acids, and hence to the origin of life on Earth, under the action of enhanced (see Chapter 5) solar ultraviolet radiation. The lifetime of such a reducing atmosphere is thought to be only about one million years as H_2 would have readily escaped to space. As the atmosphere was depleted of H_2, methane would have been oxidized to CO_2 by reactions with the products of H_2O photolysis (see Chapter 7), and NH_3 converted to N_2 by photolysis. Outgassing contributions of CO_2 and N_2 from volcanic and tectonic activity would have resulted in an atmosphere that basically consisted of CO_2 and N_2, with H_2O a trace gas in the atmosphere through condensation. This atmosphere lacked O_2 (only present as a trace gas), i.e. it was anoxic. The widespread occurrence of redbeds (oxidized subaerial deposists) at 1.8 Byr suggests significant oxygen production at that time, whilst observations from banded iron formations (BIFs) and paleosols indicate that atmospheric oxygen levels were low before 2 Byr, consistent with the small values of buried organic carbon. It is worth noting that today both Mars and Venus have CO_2–N_2 atmospheres, while Saturn's satellite Titan has an N_2–CH_4 atmosphere.

12.2.2 *The faint–young–Sun paradox*

Conventional theories of stellar evolution (see Chapter 5), imply that the Sun has brightened by about 30% since it first joined the main sequence some 4.6 Byr ago. If the Earth had the same atmosphere as today and similar continental–oceanic distribution, such a weak young Sun would have produced extensive and persistent glaciations throughout most of the Precambrian. Geological evidence suggests, on the contrary, that Precambrian glaciations were rare and confined to two periods, one in the early Proterozoic between 2.5 and 2 Byr and the other in the late Proterozoic between 1 and 0.54 Byr. Rather than a totally frozen Earth, the geological record indicates the presence of substantial expanses of liquid water on the Earth's surface as early as 3.8 Byr. For much of the Proterozoic, the Earth's surface may have been warmer rather than cooler, than its present mean value of 288 K. Although there are uncertainties in all palaeo–temperature estimates, chert and phosphate measurements suggested that, during the Archean, surface temperatures, at least at some locations, could have been as high as 350 K.

One way of resolving the faint–young–Sun paradox is to assume a compensating change in the Earth's atmosphere such as would be produced by a stronger CO_2 greenhouse effect than today. This stronger greenhouse would gradually diminish owing to the weathering of silicate rocks on the land and the deposition of carbonates as ocean sediments. On timescales of millions of years, the CO_2 cycle is completed by outgassing to the atmosphere as tectonic processes transport and transform the sediments at subduction zones (areas on Earth where two tectonic plates collide, the denser sliding underneath into the mantle). The compensation of the enhanced Precambrian greenhouse needs to be controlled by a mechanism that varies the atmospheric CO_2 as the Sun brightens, maintaining an approximate balance between the strength of the incoming solar radiation and the atmospheric CO_2–H_2O greenhouse.

12.2.3 *Greenhouse–weathering evolution model*

Walker *et al.* (1981) suggested that the required negative feedback controlling the Earth's surface temperature as the Sun brightened could have been provided by a temperature–dependent CO_2 weathering rate. Carver and Vardavas (1994) developed an evolutionary CO_2 weathering model incorporating the weathering rate mechanism and have shown that it can account, at least semiquantitatively, for one of the main features of Precambrian climate. This was the rarity in the geological record of extensive and persistent Precambrian glaciations, and their restriction to two time periods, one in the Early Proterozoic and the other in the late Proterozoic. According to their model, the Early Proterozoic glaciations were caused by increased surface-area weathering due to a major episode of continental land building that began about 3 Byr. The Late Proterozoic glaciations resulted from biologically enhanced weathering due to the proliferation of life forms that marked the transition from the Proterozoic to the Phanerozoic Eon (0.54 Byr–present).

12.2.3.1 *Weathering of* CO_2 The brightening of the Sun with time can be represented by (Gough 1981)

$$S(y) = 1/[1 + 0.4(1 - y/4.6)], \tag{12.5}$$

where y is the time in Byr since the Earth formed (today corresponds to $y =$ 4.6) and $S(y)$ is the solar flux in units of the present flux $S(4.6) = 1$. For an atmosphere with CO_2 content $A(y)$, the surface temperature $T(y)$ calculated using a one–dimensional radiative–convective climate model (with moist lapse rate, water–vapour feedback and global mean relative humidity set at 0.77, see Chapter 11) can be approximated by the simplified greenhouse equation

$$T(y) = 288 + [0.6/v^2][A^v(y) - 1] - 508[1 - r^{0.25} S^{0.25}(y)], \tag{12.6}$$

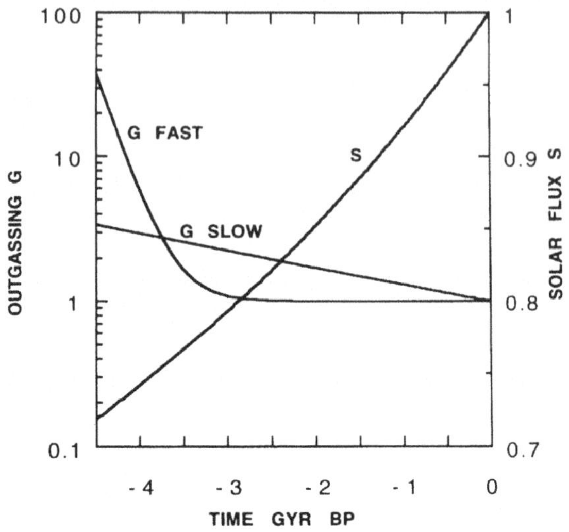

FIG. 12.2. Outgassing rate G for the fast and slow models with time in billion years before the present. Also shown is the incoming solar flux relative to today's value. (Carver and Vardavas 1994)

where $A(y)$ is measured in units of the present atmospheric CO_2 level (PAL), so that $A(4.6) = 1$, $r = (1 - \alpha(y))/(1 - 0.32)$, and $v = 0.3$ to fit the results of the radiative–convective model, and $\alpha(y)$ is the planetary albedo. The ice–albedo feedback is approximated by assuming a planetary albedo equal to 0.32 for $T \geq 288$ K, rising to 0.46 at $T = 260$ K, and to 0.6 for $T \leq 232$ K. This simplified model gives a global mean surface temperature rise of 1.5 K for a doubling of the present CO_2 level, whilst a 2% increase in the incoming solar radiation increases the global mean surface temperature by 2.5 K. The weathering rate and the organic–carbon burial rate determine the sequestering of atmospheric CO_2 by the land surface. The rate at which CO_2 is removed from the atmosphere increases with surface temperature owing to the increased rate of snowmelt runoff and the increased reaction of water in equilibrium with atmospheric CO_2 as it percolates through the soil and porous rocks. The weathering rate depends on some power ($0 \leq n \leq 1$) of the CO_2 atmospheric pressure $A(y)$, and upon the rate of dissolution of rocks by running water and on the runoff rate, both of which depend on the temperature $T(y)$ itself. The weathering rate also depends upon the land area $L(y)$, measured in units of the present land area $L(4.6) = 1$, and upon the weatherability $W(y)$ of the exposed land surface. Changes in weatherability occur as processes of biological evolution and physical erosion provide soil and vegetative cover to replace the original bare rock. Moist soil in continuous contact with rock will weather atmospheric CO_2 at a much faster rate than will bare rock alone. The presence of organisms in the soil assists in the retention of soil moisture while vegetation cover stabilizes soil formation against erosion. The proliferation of life forms thus leads to accelerated soil weathering.

The burial and decay of organic products enhances the weathering process through the release of CO_2 in the soil, while organic carbon burial is a sink for atmospheric CO_2 via photosynthesis. Episodes of continental rifting and orogeny also result in enhanced rates of weathering but also in soil erosion. Erosion can deliver nutrients for biological production but commonly exposes organic carbon resulting in its oxidation so that part of the organic carbon returns to the atmosphere.

We assume that these complex physical and biological processes can be characterized by a function $W(y)$ that measures the relative effectiveness in sequestering CO_2 from the atmosphere of ancient land surfaces with that of the present land surface. Thus, at present $W(4.6) = 1$ and throughout the Precambrian $W(y) < 1$. At equilibrium, the weathering rate $R(y)$ at which CO_2 is removed from the atmosphere must equal the rate $G(y)$ at which it is returned to the atmosphere by outgassing, so that

$$R(y) = A^n(y)L(y)W(y)\exp[(T(y) - 288)/T_c] = G(y), \tag{12.7}$$

where we take $n = 0.4$ and $T_c = 15$, based on estimates of these values. The coupled eqns (12.5, 12.6 and 12.7), involving the evolution of the solar flux, evolution of CO_2 outgassing–weathering and simplified greenhouse equation, define the feedback processes and determine the evolution of global mean surface temperature and atmospheric CO_2.

12.2.3.2 *Outgassing model* We consider both slow and fast outgassing models for the rate $G(y)$. In the slow model, $G(y)$ is assumed to follow the approximately exponential decline in the Earth's interior heat from radioactive decay. In the fast model it is assumed that after a period of rapid emission in the Early Archean, the rate $G(y)$ approached its present value at about 3.5 Byr and remained essentially constant thereafter. The higher outgassing during the Archean is assumed to have resulted from a greater juvenile emission that was more than sufficient to compensate for a smaller secondary emission due to reduced metamorphism. The two models for the outgassing rate along with the variation in the incoming solar flux are shown in Fig. 12.2.

12.2.3.3 *Land-formation model* The representations of the development of the land surface area are based on isotopic and geochemical analyses that support formation as opposed to no-growth models that suggest that the crust was formed early in the Earth's history (prior to 4 Byr). The growth models propose extensive episodic land–surface growth in the Early Proterozoic with only minor additions since that time. These conclusions are based in part on the analysis of zircon populations in the Late Archean (3.0–2.4 Byr) sediments. Geological evidence indicates rapid continental growth starting at about 3 Byr with only

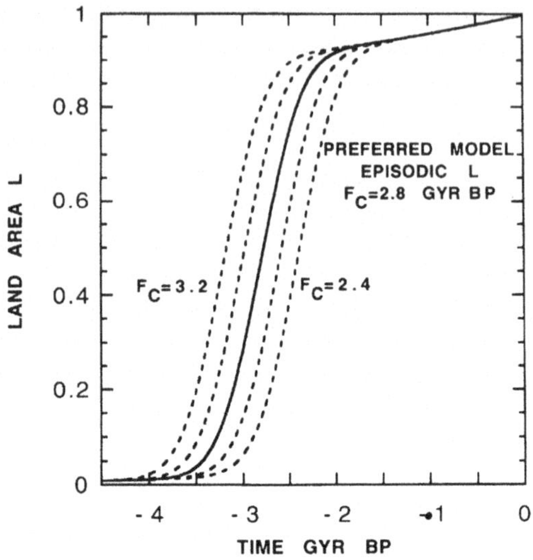

FIG. 12.3. Rapid–land formation models with F_c representing the time of maximum continental land formation, as a function of time in billion years before the present. The preferred model for surface–temperature computations was taken to correspond to $F_c = 2.8$. (Carver and Vardavas 1994)

small areas of continental crust before that time and minor later additions. In Fig. 12.3 are shown rapid land formation models with F_c representing the time of maximum continental land formation.

12.2.3.4 *Weatherability model* The assumed variation of the weatherability due to biological evolutionary factors $W(y)$ is based on the work of Schartzman and Volk (1991) who suggested that $W(y)$ was of the order of 0.001–0.01 for the abiotic Earth, rising to about 0.2 with the microbial colonization of the Archean Earth that marked the transition from abiotic to biotic land surfaces some 3.9 Byr. The value of W is assumed to have remained fairly constant throughout the Early Proterozoic rising during the late Proterozoic to approach unity at the end of the Phanerozoic. The Late Proterozoic (1.0–0.54 Byr) was a geologically and biologically important period of change. Geological evidence suggests that during the early part of this period there was increased tectonic activity with episodes of global rifting and orogeny, conditions favourable to subsequent weathering, erosion, and through runoff to sedimentation. These processes result in increased nutrient release that can promote enhanced biological activity and organic–carbon burial both on land and in the oceans. During most of the Riphean Epoch (0.9–0.6 Byr) $^{13}C/^{12}C$ enrichment in sedimentary carbonates indicates high rates of organic carbon burial (reduced availability of ^{12}C for the formation of sedimentary rocks) suggesting enhanced marine biological activity.

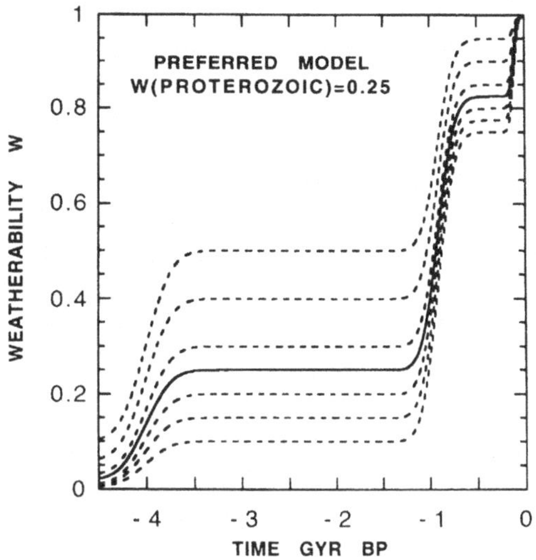

FIG. 12.4. Weatherability $W(y)$ models, where W is the weatherability arising from biological evolution through geological time in billion years before the present. (Carver and Vardavas 1994)

We note that plants preferentially take up ^{12}C during photosynthesis. It is not unreasonable to assume that if the nutrient release from the land led to increased biological activity in the oceans there was also enhanced biological activity on the land itself. It may be noted that specific glacial episodes (e.g. Varangian about 0.6 Byr, Sturtian about 0.78 Byr) during this period are correlated with dips in the ^{13}C enrichment and hence with a reduced rate of organic–carbon burial. This can be explained by a scenario in which the biologically enhanced weathering was the precursor to a reduction in atmospheric CO_2 that led to a weaker greenhouse that triggered the specific glacial episodes. During the glacial episodes, runoff would have been reduced resulting in a drop in organic–carbon burial and hence in ^{13}C enrichment.

We interpret the overall enhanced carbon burial rate during the Late Proterozoic as indicating increased biological activity leading to more effective soil weathering of CO_2. The changes that begun about 1 Byr involved the gradual extinction of the Proterozoic biosphere and its replacement by evolving multicellular organisms in the approach to the Phanerozoic. Accordingly, we allow W to increase from about 1 Byr to approach $W = 1$ at the end of the Phanerozoic with steps of 0.4 Byr, corresponding to the development of higher land plants and a final increase to present values following the evolution of flowering plants, the angiosperms, about 0.1 Byr. In Fig. 12.4 we show various models for the weatherability variation with time. The preferred model is that with $W(2) = 0.25$, where $W(2)$ corresponds to Proterozoic weatherability.

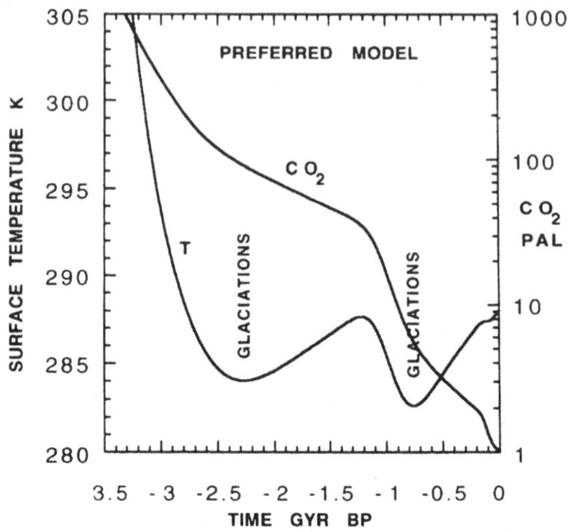

FIG. 12.5. Evolution of surface temperature and atmospheric CO_2 amount (PAL = present atmospheric level) with time in billion years before the present. (Carver and Vardavas 1994)

12.2.4 *Surface temperature and CO_2 evolution*

The evolution of atmospheric CO_2 and the consequent climate change have been calculated for the preferred model parameters. We have chosen the fast outgassing rate. The maximum in the rate of continental land formation was set at 2.8 Byr and the weatherability was taken to correspond to the curve with a value of 0.25 during the Proterozoic. The calculated evolution of the atmospheric CO_2 level and the evolution of the global mean surface temperature are shown in Fig. 12.5. After initially hot conditions, the surface temperature falls as the land area increases and the model suggests that glaciations become possible for the first time only during the period 2.5–2.0 Byr. We note that the model gives a CO_2 level of about 100 PAL at about 2.5 Byr. After about 1 Byr there is a rapid drop in atmospheric CO_2 that results in a second surface-temperature minimum near 0.75 Byr, suggesting that the glaciations found between 1–0.54 Byr were caused by the fall in the atmospheric CO_2 content produced by the increased weatherability of the land surface. There is a strong correlation between the minima of the model temperature curve and the main periods of Precambrian glaciations indicating that the model can account for the broad features of the Precambrian climate if we associate the early Proterozoic glaciations with a major episode of continental land building and the late Proterozoic glaciations with the enhanced weatherability of the land surface due to the proliferation of new life forms that marked the transition from the Proterozoic to the Phanerozoic Eon.

12.3 Comparative climatology of the terrestrial planets

12.3.1 *Mercury*

The surface temperature on Mercury is dominated by the interaction of the solar flux directly with the surface, with the thin atmosphere playing a negligible role. The axis of rotation has a very small obliquity, which normally would mean the absence of seasons; however, the climate retains an interesting degree of complexity due to the large eccentricity of the orbit, and the unusual synchronization of the rotation of the planet with its orbit, such that two days last three years (Table 12.4). If we simplify the real situation and assume that the surface of

Table 12.4 *Physical data for Mercury, compared to the same quantities for the Earth.*

	Mercury	Earth
Orbital and rotational data		
Mean distance from Sun (10^8 km)	0.579	1.496
Eccentricity	0.2056	0.0167
Obliquity (deg)	0	23.45
Siderial period (days)	87.97	365.26
Rotational period (hs)	1407.5	23.93
Solar day (days)	115.88	1
Solar constant (kW m^{-2})	6.3 to 14.5	1.366
Solid body data		
Mass (10^{24} kg)	0.3302	5.976
Radius (km)	2439	6373
Surface gravity (m s^{-2})	3.63	9.82
Atmospheric data		
Principal composition	He, Na	N_2, O_2
Mean surface temperature (K)	$\sim$ 400	288
Mean surface pressure (atm)	$< 10^{-12}$	1
Mass (kg)	$\sim$ 1300	5.1×10^{18}

Mercury is a smooth sphere, and has the same albedo (reflectivity integrated over wavelength) everywhere, then a straightforward energy–balance calculation gives the results shown in Fig. 12.6 for the variation of temperature with time at different locations around the equator. The odd shape of the curves is a result of the eccentric orbit and the coupling between the orbit and spin of the planet. At higher latitudes, similar curves apply but with a smaller range of temperatures. The maximum does not fall below 100 °C until about 5° from the pole, however, while the nighttime minimum is well below the freezing point of CO_2 everywhere. In reality, variations in composition and topography introduce additional variations in the temperature distribution, one intriguing consequence of which is that Mercury apparently has thick deposits of ice in craters near the poles that are permanently shaded from the Sun.

Another interesting feature of Mercury that relates to its origin and evolution, and therefore also to those of the Earth, is its high mean density. The planet must consist predominantly of metals, probably mostly iron since this has the

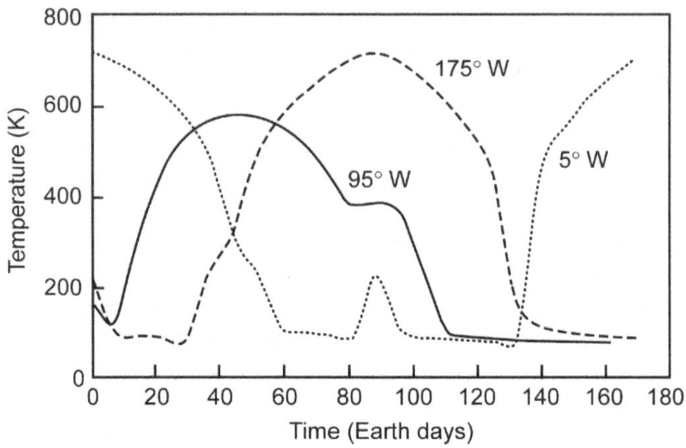

FIG. 12.6. Surface temperatures at the equator on Mercury, calculated from a radiative energy balance model with a simple 'billiard ball' planet with no topography and uniform albedo and emissivity. Results for three different longitudes (degrees West relative to local noon at perihelion) are shown. (Source: ESA)

highest cosmic abundance. This massive core may still be partially liquid, like the Earth's, despite calculations indicating that Mercury is small enough to have cooled throughout by now, since it exhibits a magnetic field that is too strong to exist without an active dynamo. This suggests that there must be an active heat source in the core, presumably a relatively high concentration of radioactive elements. Inferences like these provide valuable information about the distribution of elements in the circumsolar accretion disk when the planetary system formed.

Mercury, in terms of its geology and place in the inner Solar System is undoubtedly a terrestrial planet, but its atmosphere is so thin it more resembles the terrestrial exosphere, a tenuous region that has only a long-term relationship to the climate at the Earth's surface, mainly through atmospheric escape processes. There are also interesting questions concerning the volatile inventory on Mercury, since polar deposits, apparently of water ice at least several tens of metres thick, were discovered by Earth-based observers using radar mapping at large radio telescopes. There exists no obvious source for water in these amounts, and the investigation of the problem is likely to yield answers that will lead to a better understanding of the origin of water on all the planets.

12.3.2 *Venus*

On Venus, an extreme case of the greenhouse effect raises the surface temperature to around 730 K, which is higher than the melting points of lead, zinc and tin, in spite of the fact that the net solar input is significantly less than for the Earth. Apparently, the very high albedo of Venus (76% compared to around 30% for

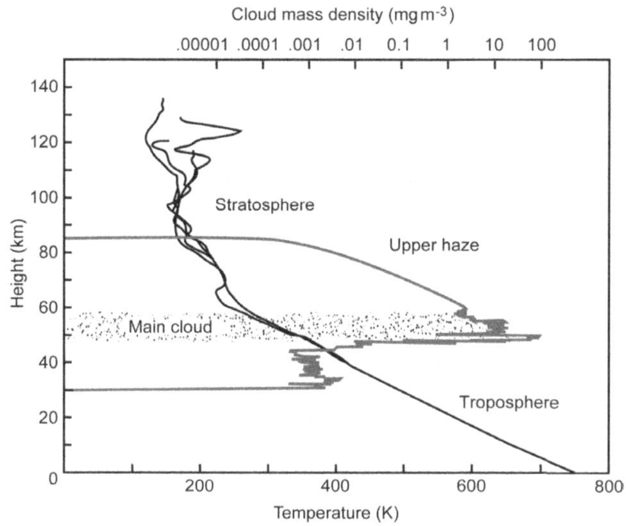

FIG. 12.7. Atmospheric temperature and cloud density profiles on Venus, as measured by the Pioneer Venus probes in December 1979. The main cloud layers on Venus have a high reflectivity in the near-infra-red and visible parts of the spectrum, and a high absorptivity in the thermal infrared. The atmosphere is 100 times more massive than Earth's and is 95% carbon dioxide. For the amount of energy escaping as thermal radiation from below the clouds to match that absorbed from the Sun, the temperature at the surface has to become very high.

the Earth) more than offsets its greater proximity to the Sun (Table 12.1 and Fig. 12.7). The sulphuric acid of which the Venusian clouds are composed forms droplets that scatter very conservatively, diffusing a fraction of the incoming sunlight down to the surface. At the same time, they are opaque in the infra–red, so clouds made of concentrated sulphuric acid droplets have a higher optical depth in the thermal infra–red than they do in the near infra–red and visible part of the spectrum. The cloud blanket therefore makes a contribution to the backwarming effect, which outweighs the albedo effect overall when combined with the contribution of around a million times as much CO_2 as the Earth, plus substantial traces of water vapour and other greenhouse gases.

The solar constant at Venus is about twice that at the Earth, at 2626 W m^{-2}. Assuming an albedo of $A = 0.76$ and dividing by the ratio of the area of the spherical planet to its cross-sectional area, the mean solar power absorbed by the atmosphere is therefore 157 W m^{-2}. Equating this to the outgoing blackbody flux σT_{E}^4, where σ is the Stefan–Boltzmann constant, gives $T_{\mathrm{E}} = 230$ K for the effective radiating temperature of the planet. This is much lower than the surface temperature of Venus, which is 730 K, and is in fact approximately the temperature at the cloud tops (Fig. 12.8). This is to be expected since the clouds

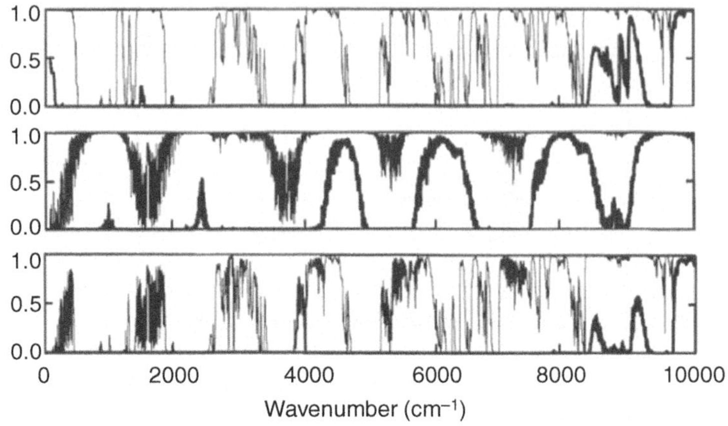

FIG. 12.8. The atmospheric transmission of a vertical column on Venus, above the 50 mb level (fine line) and above the surface (bold line). The top frame is for CO_2 only; the middle frame for H_2O only, and the bottom frame is for all gases, including CO, OCS, HCl, and HF, as well as CO_2 and H_2O. The cloud opacity is not included in any of these calculations. (After Crisp and Titov 1997)

are optically thick in the thermal infra–red, and their effective upper boundary is at the relatively low pressure of about 50 mb, i.e. above about 99.95% of the total mass of the atmosphere.

While the cloud opacity dominates the net optical depth of the atmosphere above about 50 km, and scattering by cloud particles is the main factor producing the high albedo of Venus, in the lower atmosphere the gaseous constituents, especially carbon dioxide and water vapour, take over. Figure 12.8 shows the atmospheric transmission for a vertical path from the surface to space, and from the 50 mb level to space, computed separately for two of the principal greenhouse gases, and then for all of the infra–red active gases known to have significant abundances. Even before the contribution of absorption in the cloud and haze layers is considered, it can be seen that Venus' atmosphere is opaque at every wavelength except for some narrow transparency windows in the near-infra-red. As with the Earth, this sort of calculation can be incorporated into a full radiative transfer model, which allows the energy balance at every level in the atmosphere to be computed. An added difficulty for Venus is that the cloud properties are not known in the deep atmosphere (and indeed they seem to be very variable) which makes it difficult, in particular, to calculate the penetration of sunlight through the clouds, and the energy deposited at each level, including the surface. Some current models avoid this problem by constraining their codes to give the intensity at solar wavelengths that has been measured near the surface by Soviet and American probes. Another problem is the fact that even quite well–studied gases like CO_2 and H_2O have spectral properties that are poorly determined at the very high temperatures and pressures that prevail on Venus. Finally, the abundances and variability of all of the greenhouse gases except CO_2 remain poorly known in the lower atmosphere.

Despite all of these difficulties, a number of studies have shown that models can be constructed that, with some reasonable assumptions, do account for the observed thermal structure of Venus' atmosphere by a combination of thermal balance at the surface, with space, and in the stratosphere, and convective equilibrium in the troposphere. The high surface temperature results from the very deep and thermally opaque atmosphere containing large quantities of efficient greenhouse absorbers, in particular CO_2, H_2O, and the H_2SO_4 clouds.

It still remains to explain why the Venusian atmosphere is so massive. It has been estimated that the amount of CO_2 that once was in the Earth's atmosphere and is now locked in coral reefs and other carbonate rocks, after having first dissolved in the ocean, is of the same order as that which is still in the atmosphere on Venus. The implication is that the Earth, by being cool enough to condense liquid water on its surface, escaped having a much thicker CO_2 atmosphere. Also, Venus may be much more volcanically active than Earth, and even today volcanoes may be pumping large amounts of carbonic and sulphurous gases into the atmosphere, continually reinforcing the greenhouse effect. This could, in time, subside.

As long ago as 1952, when it was known that the atmosphere was predominantly CO_2, although not that it was so hot or so dense, Harold Urey suggested that the reaction of atmospheric carbon dioxide with common surface minerals might have a role in determining the surface pressure. On theoretical and abundance grounds, the most likely process to dominate is the equilibrium balance between calcium carbonate, silica and calcium silicate

$$CaCO_3 + SiO_2 \leftrightarrow CaSiO_3 + CO_2, \tag{12.8}$$

and indeed we now know that the temperature and pressure on the surface of Venus at the present time fall on the equilibrium phase curve for this reaction. Not only that, but if we also plot the surface temperature as a function of surface pressure in the simple radiative–convective equilibrium model described in outline above, it can be seen that the two intersect at a point that is strikingly close to the surface conditions measured by the Venera landers (Fig. 12.9).

This agreement strongly suggests that the high–pressure, high–temperature state of the climate on Venus can be attributed to the balance for a pure CO_2 atmosphere in equilibrium with the common minerals likely to be abundant on Venus. Although possibly no more than a coincidence, the existence of a straightforward account for the extreme state of Venus' climate, which is otherwise very difficult to explain, makes it hard to set it aside, at least without a further examination using more complex models and more data from new missions to the planet, both unavailable at present.

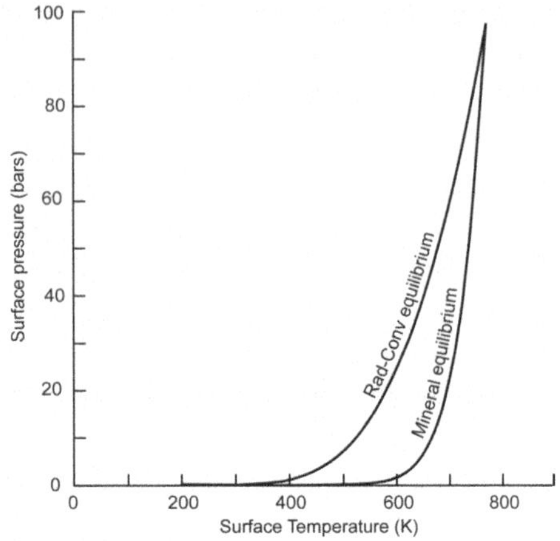

FIG. 12.9. Phase curves for the surface temperature and pressure on Venus, corresponding to radiative–convective equilibrium in the atmosphere (Taylor 2006), and to chemical equilibrium between CO_2 in the atmosphere and in surface minerals. (Calculated using the data provided by Adamcik and Draper 1963)

However, it has often been noted that the temperature dependence of the mineral equilibrium is such that the present climate would be unstable if it was produced in this way. The strong dependence of atmospheric CO_2 amount on surface temperature represented by the phase curve shown in Fig. 12.9 means that any small perturbation in the surface temperature, either positive or negative, leads to positive feedback and dramatic changes in atmospheric CO_2, resulting in runaway heating or cooling, respectively. Also, even if the regolith on Venus has the necessary supply of silicate minerals, it is not clear how they can be in sufficiently intimate contact with the atmosphere. Finally, the surface of Venus is not at a single uniform temperature; topography alone leads to temperature contrasts of more than 100 K, corresponding to differences in CO_2 pressure for Urey equilibrium that would be far out of hydrostatic balance. Other surface–atmosphere reactions are undoubtedly involved, and may be important. Some aspect of these and other complexities not yet understood may act to stabilise the climate, at least against moderate perturbations, somewhat as the complexity of Earth's climate achieves an overall quasistable regime despite the disequilibrium state of most of its components. The strong suggestion that heterogeneous reactions, involving the atmosphere and the regolith, control the climate gives a high priority to plans for long-lived missions to the surface of Venus, possibly including mobility on the surface and the return of samples to the Earth. The climate itself makes that extremely difficult, of course, and it is not clear how many years or decades must pass before the space agencies will be up to the challenge. It may be premature to think too deeply about global change on Venus while there is much still to be understood about the present climate, but recent data do suggest that major change has occurred. The ratio of heavy to normal hydrogen

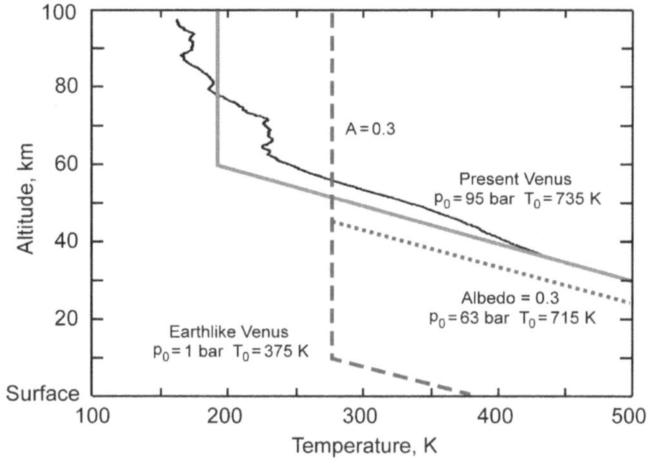

FIG. 12.10. Simple climate models for Venus, corresponding to; a) the present day, b) a decrease in albedo to 0.3, and c) atmospheric loss of CO_2 such that the surface pressure falls to 1 bar. A measured profile, obtained by the radio-occultation experiment on the Magellan orbiter on 5 October 1991 at latitude 67N, is shown for comparison. (Taylor 2006)

isotopes on Venus is more than 100 times greater than that on Earth, a fact that is generally interpreted in terms of the loss of a global ocean from Venus over the first few billion years of its existence. In this paradigm, water vapour near the top of the atmosphere is dissociated by solar ultraviolet radiation and the hydrogen is lost to space, with the lighter isotope lost at a higher rate due to gravitational fractionation of the water vapour, enriching the residue we now observe. Some have speculated that the long, deep, sinuous channels cut into the surface of Venus by some unknown agent and seen in radar images obtained by the Magellan spacecraft must have been produced by flowing water, suggesting that Venus was cooler in the not-too-distant past. Finally, the amount of sulphur dioxide detected spectroscopically in Venus' atmosphere has shown very large variations on the timescale of a decade, suggesting that the level of volcanic activity on Venus varies considerably. The effect of this on the global climate is not known, but model calculations suggest it could be considerable. Like the Earth, we would expect Venus to respond to changes in external forcing, such as any long-term variation in total solar irradiance. However, again like the Earth, the Sun in the recent epoch has been stable enough that this factor is probably negligible compared to others. For a change of 0.1% in solar output, similar to those measured currently by spacecraft orbiting the Earth, a Venus climate model predicts a change in mean surface temperature of only 0.05 K. The spread in surface temperatures expected due to topographical variations is around three orders of magnitude greater than this. Volcanism seems much more likely to be an important agent for climate change in the relatively short term, particularly since Venus appears to be much more volcanically active than Earth. This evidence includes the observed disequilibrium in the atmosphere of the sulphurous gases that play a large role in the production and maintenance of the cloud layer, and consequently the high planetary albedo. If, at some point in the future, volcanism ceased on Venus and the cloud regime relaxed to one much more like the Earth, the albedo may fall from the present 0.76 to an Earth–like 0.3. Neglecting for the moment all other consequences, such as atmospheric composition changes (except those in CO_2 due to adjustment to the Urey equilibrium at the surface), the model stratospheric temperature may increase to around 273 K, while the surface temperature and pressure fall to about 715 K and 63 bar, respectively (Fig. 12.10).

A further consequence of a complete cessation in volcanic activity is that the greenhouse effect would become less efficient through the removal from the atmosphere of some of the more chemically active minor constituents, such as SO_2, H_2S and COS, as they move closer to their equilibrium values with the surface. The loss of atmospheric water due to dissociation and exospheric escape of hydrogen may no longer be compensated by emission from the interior, as presumably it is at present. Any loss of atmospheric infra-red opacity, due to weaker absorption in the bands of these species, moves the tropopause downwards to a higher pressure level, resulting in falling surface temperatures and pressures as radiative-convective equilibrium is maintained. This large cooling perturbation leads to the steady take-up of atmospheric CO_2 by the surface, although it remains to be shown that this is physically possible in the absence of liquid water to facilitate the exchange, particularly since the carbon dioxide content of the current atmosphere is sufficient to produce a layer of carbonate around 1 km thick if entirely converted to surface rock.

Venus without active volcanism might thus become more Earth–like. Venus has about the same amount of N_2 in its present–day atmosphere as the Earth; the selective removal of CO_2 could at some point leave a composition of 99% N_2, 0.5% Ar, 0.5% H_2O, and 0.035% CO_2, with a surface pressure in the region of 1 bar. In this case, the mean surface temperature would be a relatively balmy 70 C. This is the Venus that was imagined by our forebears, by extrapolation from the Earth, before the importance of the greenhouse effect and the true state of Venus' climate had been realized.

12.3.3 *Mars*

The present–day atmosphere of Mars is $\sim$10000 times thinner than that of Venus, with a mean surface pressure of about 6.5 mbar, and, like Venus, is composed also mainly of CO_2. The atmosphere of Mars is also warmed by the interaction of both solar and thermal radiation with airborne particles, in this case windblown dust from the surface. This is present in variable quantities, depending on dust-storm activity and other meteorological factors, often in significant quantities at heights up to 50 km above the surface. The dust opacity is in fact the main contributor to the Martian greenhouse effect, exceeding even that of carbon dioxide under most conditions. The role of water vapour in the radiative energy balance of the Martian atmosphere is very small, since the amounts present at the the low prevailing temperatures on Mars are generally insignificant in terms of their contribution to the net opacity of the atmosphere. Under normal conditions, the greenhouse effect on Mars due to all constituents produces a much smaller enhancement of surface temperature than on Earth or Venus. Using $A = 0.25$ for the albedo of Mars, and 593 W m^{-2} for the mean solar flux reaching Mars' orbit, a global energy-balance calculation finds that the effective radiating temperature of the planet is T = 210 K. Compared to an observed mean surface temperature of 218 K, this indicates a small greenhouse effect of about 8 K. Under very dusty conditions, this can increase five- or even tenfold.

Of all the planets, including Earth, the clearest evidence for large-scale climate change in the past is found on present–day Mars. Today's frozen rocks and desert bear unmistakable features of rivers, lakes and seas. It is now generally accepted, on the basis of extensive orbital imaging of fluvial features (like the example in Fig. 12.11) and mineralogical evidence (most recently from the Mars Exploration Rovers, Spirit and Opportunity, operating as robot geologists on the surface), that Mars must have had a warm, wet climate in the past with liquid water on the surface. Precisely how warm, how wet, what produced these conditions and when they prevailed, all remain subjects of intense debate. The leading theory is that Mars may in the past have had a much thicker atmosphere, probably still consisting primarily of CO_2, but, being warmer, holding much more water vapour, and possibly increased levels of other constituents including methane. Together with an enhanced cloud cover, these might have produced a greenhouse effect large enough to raise the temperature and pressure to values that could support liquid water on the surface.

The large change since then may have originated in the high eccentricity of Mars' orbit and the large fluctuations in sunfall that result from resonances between this and other orbital parameters. Alternatively, the heating could have been primarily due to volcanic activity, which has since subsided. In this case, water vapour, CO_2 and other gases and particles from the interior of Mars could have warmed the planet by greenhouse action and produced lakes and rivers for as long as the volcanoes were sufficiently active. Yet another possibility is that Mars lost enough of its early atmosphere in a collision with a large asteroid to cause temperatures to fall below the freezing point of water. The resulting loss of water vapour and clouds from the remaining atmosphere would further reduce the surface temperature and eventually lead to the situation we find today, where the water that remained after the collision mostly lies frozen beneath the surface and in the polar caps. We still do not know whether the present Martian climate is stable or simply a stage in a gradual decline in surface pressure and temperature that has been going on since volcanic activity ceased.

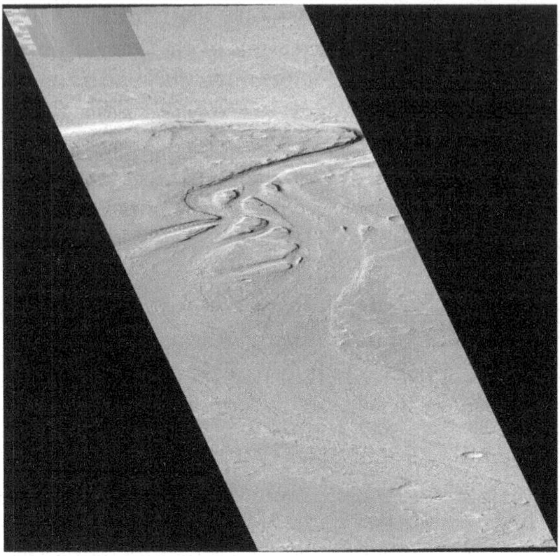

FIG. 12.11. Features are seen all over Mars that, like these river–valley networks, were formed by flowing water, which also apparently collected into large bodies or seas, whose coastlines can still be discerned. Their existence implies a warmer, wetter climate on Mars in the past, although the details of what this was like, and how and when Mars changed to its present cold, dry state, remain to be understood.

The development of global mean models for Mars is more capricious than for Venus because the temperature profile is very variable, with season, time of day, and location on the planet. It also depends strongly on the amount of airborne dust, which again is highly variable and dependent on the wind field, including unpredictable factors such as the onset of dust storms and the frequency of occurrence of dust 'devils'. The latter occur in desert regions of the Earth and are now known to be extremely common on Mars, to the extent that they have been belatedly recognized as a key factor in maintaining a high level of airborne dust even when the atmosphere is otherwise relatively quiescent.

The results of an attempt to represent the temperature profile on Mars using a simple radiative–convective model are shown in Fig. 12.12. A solar constant of 593 W m^{-2} and global mean albedo 0.25, give 210 K for the effective radiating temperature of Mars, a predicted stratospheric temperature of 177 K, and a surface temperature, calculated from a single–slab greenhouse model, of 250 K. Since the adiabatic lapse rate g/c_p is 4.5 K km^{-1} on Mars, the tropopause should occur at approximately 16.2 km above the surface in the absence of dust.

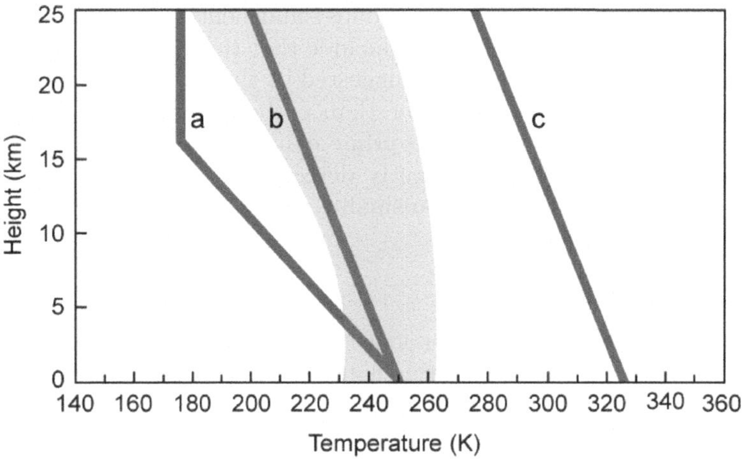

FIG. 12.12. Temperature profiles for the atmosphere of Mars, calculated assuming; a) an atmosphere without suspended dust; b) including the radiative effects of a model dust profile, and c) possible past conditions where the surface pressure was higher, and the temperature was warm enough for liquid water to be present. The shaded area encloses the range of measured temperature profiles obtained by Mariner 9 *circa* 1971.

This is not what is normally observed, however, since airborne dust is ubiquitous. Its effect, for a typical level of dust loading, is to reduce the lapse rate to about 1 K km^{-1}, which raises the tropopause dramatically to 36.6 km. With this refinement, the model curve b) falls within the shaded area representing observed temperatures on Mars. Retaining the same model dust distribution, it is now straightforward to run models in which the surface pressure is increased, to simulate the conditions thought to have prevailed on early Mars. Curve c in Fig. 12.12 shows the result of assuming the same surface pressure on Mars as on present-day Earth, while retaining a composition of pure CO_2 on the smaller planet.

The result suggests that global mean temperatures, well above the freezing point of liquid water, were possible if this amount of CO_2 were present then. It might be argued that the possibility of rain would reduce the dust loading of the air and tend to restore the CO_2 adiabatic lapse rate; in this case the model predicts even higher surface temperatures, and considerably less than 1 bar of surface pressure would be sufficient to permit the existence of liquid water.

Of course, simple model predictions have to be treated with caution since other factors, not represented in the model, can be important and, may radically alter the conclusions drawn. First, the Sun is thought to have been less bright by 25 to 30% during its early life, possibly including the period when Mars seems to have been warmer. Second, a thicker early atmosphere probably was much cloudier (with CO_2 as well as water clouds) and hence Mars may have had a much higher albedo. Some modellers conclude that the greenhouse effect cannot have produced the warm, early Mars suggested by the geological evidence, unless the atmosphere contained gases other than carbon dioxide and water vapour in substantial quantities. One of the prime candidates to produce the missing opacity is methane, which was recently detected in the Martian atmosphere, although in tiny quantities much too small to affect the climate.

12.4 The giant planets

The outer regions of the atmospheres of the gas giant planets Jupiter, Saturn, Uranus, and Neptune, have almost Earth–like temperature and pressure regimes, where the vertical profile is controlled by radiative–convective equilibrium much as on the terrestrial planets (Fig. 12.13). The key processes are those that are familiar to us on our planet, for instance, the absorption and transmission of sunlight in gaseous absorbers and clouds, and the transfer of latent heat as condensates. Along with quite different atmospheric compositions (mainly hydrogen and helium, with substantial traces of ammonia, methane, and water) the important differences from the Earth–like inner planets is that they lack solid surfaces (except possibly in their deep interiors) and they have internal sources of energy that are of the same order as that arriving externally from the Sun. The origin of the latter is uncertain, but it is probably produced by the conversion of potential energy to heat as the planet slowly contracts under its own gravitational field and heavier elements move towards the core. The required rate of contraction is a few cm per year, which of course is quite undetectable.

Taking Jupiter as the largest, closest and therefore best–studied example, energy–balance calculations based solely on its distance from the Sun (five times that of Earth) and its albedo (0.35) predict that its effective emitting temperature should be about 110 K. If we add the internal power of about 30 W m^{-2}, estimated from spacecraft measurements of the total infra–red flux leaving the planet, then this increases to 130 K. The predicted stratospheric temperature based on this is 109 K, which is close to the value observed by spacecraft radio occultation measurements and during the descent of the Galileo entry probe. The heat flux from the interior, along with the high opacity of the atmosphere in the thermal infra–red at pressures higher than about 0.25 bar, guarantees that the deep atmosphere is convective, with a lapse rate $g/c_p = 2$ K km^{-1}, and this is also close to that observed (Fig. 12.13).

The smaller giant planets have smaller internal energy sources than Jupiter (that on Uranus has so far proved undetectable), and different concentrations of methane and other gases that are radiatively active in the infra–red. When these differences are taken into account, they are found to behave in broadly the same way as Jupiter, with a quasi–isothermal stratosphere overlying a virtually bottomless, convective troposphere. All have complex cloud structures,

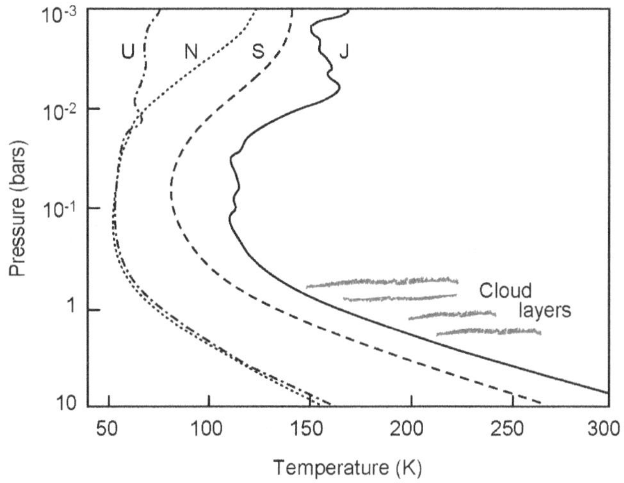

FIG. 12.13. Temperature profiles for the four giant outer planets, measured by the NASA Voyager spacecraft using infra–red sounding and the radio-occultation method. (Taylor 2005)

as would be expected from the combination of deep convection and low upper–tropospheric temperatures, with water, ammonia, and (except on Jupiter that is too warm) methane, among the species that condense. The existence of multiple cloud and aerosol layers, the long path lengths and high pressures, compared to Earth, which mean that processes such as pressure–induced dipole and quadrupole absorption bands of hydrogen dominate the opacity below the tropopause, and limited knowledge of the detailed composition, all make the construction of detailed radiative–dynamical equilibrium models for the giant planets difficult. Given also that there are few reliable measurements with which to compare, climate studies involving radiative models are in their infancy for the outer planets.

12.5 Titan's atmosphere and haze

Saturn's biggest moon, Titan, is the only natural satellite in the Solar System that is surrounded by a dense atmosphere. Titan has been the focus of scientific investigation from its first observation in the seventeenth century, by the Dutch mathematician, physicist and astronomer Christian Huygens, to the present space missions and the dedicated ground-based observations. Titan is larger than the planet Mercury, and not much smaller than Mars, but has a thicker atmosphere even than that of the Earth with a surface pressure of around 1.5 bar, and with the same primary constituent, nitrogen. Titan is ten times more distant from the Sun than Earth and so very cold. The principal minor constituents are a few per cent of methane and other hydrocarbons, some of which

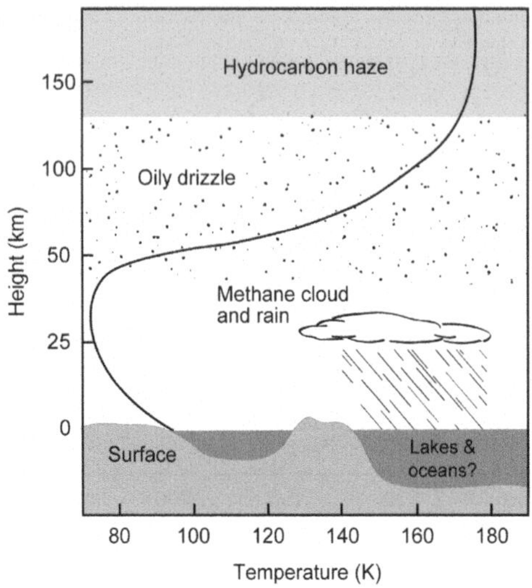

FIG. 12.14. The main features of the climate system on Titan include an atmospheric temperature profile with a distinct troposphere, stratosphere and thermosphere and at least two distinct types of cloud and haze. Instruments on the Cassini spacecraft recently discovered fluvial features and lakes on the surface, probably primarily created and filled by liquid methane rainfall.

form oily haze layers composed of light hydrocarbons. A steady drizzle from these, plus seasonal rain from the lower–level methane clouds, contributes to the production of lakes, rivers and shorelines on the surface (Fig. 12.14). The cloud cover, with a global, high-altitude haze of organic material produced by photochemistry and patchy, lower–level methane clouds, is reminiscent in structure and total N_2 content (if not overall composition), of that of Venus. Methane is the principal greenhouse gas and, allowing also for contributions by the pressure–induced bands of nitrogen and hydrogen, and by the clouds, its presence raises the surface temperature by about 12 K above the radiative equilibrium temperature of 82 K. The Huygens probe recorded a mean lapse rate of about 1.2 K km^{-1} during its descent (the dry lapse rate g/c_p is about 1.3 K km^{-1}) and a temperature of $93.65 + 0.25$ K at a pressure of $1467 + 15$ hPa at its landing site.

At Saturn's mean distance from the Sun, the solar constant is only 15 W m^{-2}, or about 1% of that at Earth. Titan has an albedo of 0.25, which works out, using a simple energy balance model to an effective emitting temperature of 84 K. A one-layer greenhouse model estimates 100 K for the surface temperature and the optically thin approximation gives 71 K for the stratospheric temperature. The dry adiabatic lapse rate $g/c_p = 1.3$ K km^{-1} on Titan, so the tropopause (based on the definition of the troposphere having a negative temperature gradient, and not on convective equilibrium) should fall at approximately 22.3 km above the surface. These global mean values compare reasonably well to the profile in

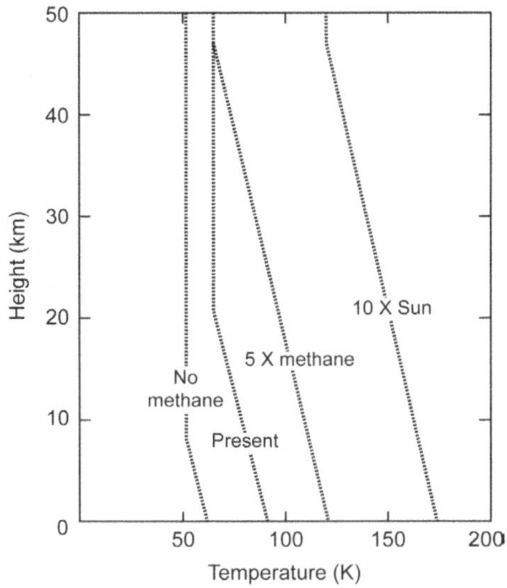

FIG. 12.15. Temperature profiles corresponding to various climate–change scenarios for Titan; a) the present day; b) after the methane reservoir is completely depleted by photolysis; c) with abundant methane in a 'runaway greenhouse' scenario; and d) 6 Byr hence, when the Sun is a red giant. (From Taylor 2006, based on calculations by Lorenz *et al.* 1997a,b and 1999.)

Fig. 12.14, which is based on the radio–occultation profile measured near the equator on Titan by Voyager 2 in 1980 and the Huygens descent measurements. The profiles in Fig. 12.15 show radiative–convective models calculated for various parameters that represent various climate–change scenarios for Titan. These assume mainly that the amount of methane presently in Titan's atmosphere is unstable and may quite substantially increase, or decrease, or even vary cyclically. Since methane is continuously removed from Titan's atmosphere by photolysis, there must be a reservoir on the surface or in the interior with access to the surface. Should the reservoir become depleted, methane would be removed by photolysis in a few million years and the surface temperature would drop below the condensation point for nitrogen, so Titan's atmosphere would collapse. The albedo at present is rather low at $\sim$0.25, but this could increase to 0.8 as aerosol and Rayleigh scattering become less important and fresh ice forms on the surface, further accelerating the cooling. The residual opacity of nitrogen itself would allow a surface pressure of perhaps one tenth the present value, and a temperature of 65 K, as shown in the left-hand profile in Fig. 12.15. If the absorptivity of the surface subsequently increases with time due to the photochemical conversion of material in the ice or accumulation of dark, organic material from space, as appears to have occurred on the surfaces of some of the other Saturnian moons, the surface temperature could increase again, perhaps to the point where nitrogen and methane could re-evaporate and build up the atmosphere once more.

Conversely, if the methane supply is abundant, a perturbation such as a small increase in insolation over the present value could produce a dramatic warming of Titan's climate. For example, if the methane reservoir referred to above is in the form of liquid on the surface, small increases in heating would raise the atmospheric mixing ratio of methane rapidly and produce an enhanced greenhouse effect that would further increase the temperature of the surface, leading to more evaporation. In models, surface temperatures of 130 K and pressures of 7 bar are reached if the total amount of methane is about 5 times that currently found in the atmosphere (Fig. 12.15). If the solar irradiance increases, as it will eventually, on the way to the red giant phase of the Sun in about 6 Byr, the irradiance might be about 10 times the present level, when the model predicts a surface temperature of around 177 K (Fig. 12.15). In this temperature range, other volatiles such as ammonia, which is likely to be abundant on Titan but currently frozen, could start to contribute to the greenhouse opacity leading to further warming. Conceivably, near Earth–like temperatures could prevail on Titan in the very distant future.

12.5.1 *Physical properties*

Titan is at an average distance of 20 Saturnian radii from Saturn. The radius of its solid surface is R_T=2575 km, while its atmosphere extends to hundreds of kilometres above the surface, making its apparent size significantly bigger. The first species detected on Titan in large amounts was methane (Kuiper 1944) while from the Voyager missions it became clear that molecular nitrogen is the major atmospheric gas. Detailed analysis of the IR spectra of the Voyager mission (Coustenis *et al.* 1989) showed that a significant number of complex organic hydrocarbons and nitriles are produced in Titan's atmosphere as a result of the dissociation of CH_4 and N_2 from solar UV photons and cosmic–ray energetic particles (Table 12.5). The photodissociation of the product species and their chemical interactions is believed to lead to the production of complex polymer structures that constitute the haze (discussed later).

The spatial resolution of Titan's disk by the Voyager observations showed that there was significant latitudinal variation in species concentrations (Coustenis and Bézard 1995) and that the South Pole had a significantly higher albedo at visible wavelengths than the North Pole. This North–South asymmetry has reversed since then as Titan moved from northern spring at Voyager's time to today's northern winter, making it clear that Titan's atmosphere is perturbed by seasonal effects. In addition, a secondary detached haze layer was prominent from the Voyager observations, a characteristic that can only be explained by dynamics.

The products of methane photolysis in the atmosphere and the consequent production of its haze did not allow the observation of the surface in almost all the spectrum from UV to IR by the instruments on board the Voyager spacecraft.

Table 12.5 *Physical properties of Titan and Earth.*

	Titan	Earth
Mass (10^{23} kg)	1.346	59.8
Equatorial radius (km)	2575	6371
Rotational period (days)	15.945	1
Period around Sun (years)	29.46	1
Mean distance from Sun (AU)	9.546	1
Orbital eccentricity	0.0292	0.017
Obliquity	27°	23.5°
Average albedo	0.25	0.30
Solar constant (W m^{-2})	15	1366
Surface pressure (bar)	1.467	1.01325
Surface temperature (K)	94	288
Escape velocity (km s^{-1})	2.65	11.2

Only in specific and narrow regions of the spectrum where methane absorption is very weak ('methane windows') do photons reach the surface of Titan and return back to detectors outside Titan. Titan's hidden surface has resulted in great interest in its dense atmosphere and fuelled speculation on the possible presence of seas or lakes of liquid methane and other hydrocarbons and surface processes such as cryovolcanism.

The evolution of observation techniques and the discovery of adaptive optics technology have allowed observations from the Earth's surface resulting in the first maps of Titan's surface, produced at wavelengths in the methane windows. Even though the resolution of these maps was constrained by the large distance of Titan from the Earth, they clearly showed that the satellite's surface was inhomogeneous with varying albedo.

The success of the Cassini/Huygens mission has provided answers to many of the outstanding questions regarding Titan's atmosphere. Initial results have shown that Titan's atmosphere is similar to that of the Earth. The temperature and pressure conditions near Titan's surface, 93.65 K, 1.467 bar from the HASI instrument at the Huygens landing region, are close to the triple point of methane. This means that methane on Titan can condense leading to cloud formation and rain, as does water vapour on Earth.

The question still remains regarding the possible source of the methane in the atmosphere. As photodissociation in Titan's atmosphere is the major loss mechanism for methane, the total abundance of methane would be irreversibly destroyed within about 10 Myr if it is not recyled or supplied from a surface reservoir. Seas or lakes of hydrocarbons from which methane can evaporate have yet to be conclusively identified.

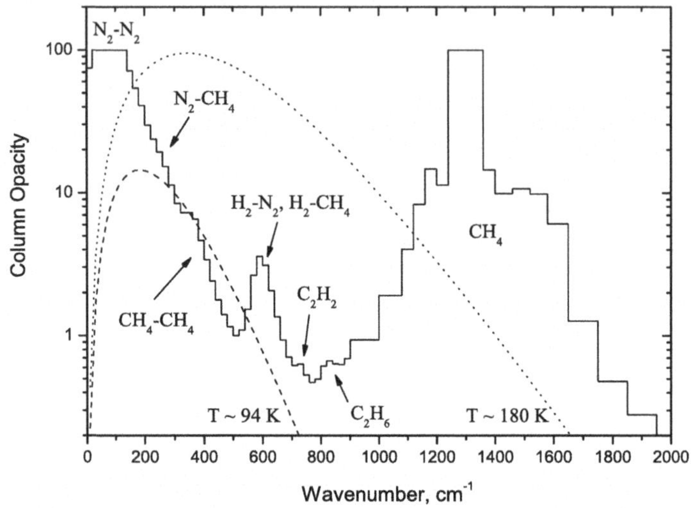

FIG. 12.16. Model column opacities in the thermal infra-red. Up to $\sim$ 700 cm^{-1} the main contribution comes from the collision–induced absorption between the $N_2/CH_4/H_2$ pairs with the spectral dominance of each pair shown. At higher wave numbers the opacity is controlled by methane, acetylene and ethane. The dashed and dotted lines are Planck functions (in arbitrary units) near 94 and 180 K, respectively, representing the spectral range of Titan's thermal radiation. (Lavvas *et al.* 2007a)

12.5.2 *Collisionally induced absorption*

In Titan's atmosphere, the high–pressure conditions give rise to collision induced absorption (CIA) involving collisions between N_2, CH_4 and H_2 molecules. Even though N_2 and H_2 are homopolar molecules and under normal conditions they don't interact strongly with the radiation field, the collisions between these molecules and methane, disturb temporally their dipolar torsion and render these pairs strong absorbers of the thermal radiation. Figure 12.16 presents the CIA column opacity in Titan's atmosphere with wave number for the thermal infra–red together with the main collision pairs that are dominant. The low opacity between 400 and 600 cm^{-1} is the only region where thermal radiation is allowed to escape from the surface of Titan, referred to as the 'thermal window'. CIA acts in a similar way as the CO_2 greenhouse effect, allowing solar radiation to penetrate the atmosphere but trapping the thermal radiation emitted from the surface.

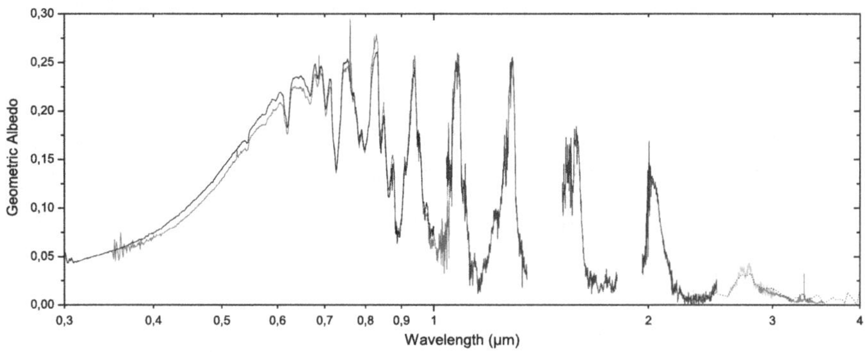

FIG. 12.17. Titan's geometric albedo from various measurements: Neff *et al.* (1984) 0.35–1.05 μm, Karkoscka (1994) 0.3–1.0 μm, Coustenis *et al.* (1995) 0.8–2.5 μm, Geballe *et al.* (2003) 2.9–3.5 μm, Coustenis *et al.* (2003), 2.6–3 μm.

12.5.3 *Geometric albedo*

One of the most important measurements used in the interpretation of the physical processes in Titan's atmosphere is the spectrally resolved geometric albedo (Fig. 12.17). This is the fraction of the incident solar radiation, scattered by the atmosphere in the direction of an observer on Earth. Since at different wavelengths the interaction of the radiation field with the atmosphere corresponds to different altitude regions of the atmosphere characterized by different constituents, the geometric albedo is a strong diagnostic tool for the investigation of atmospheric optical properties. The geometric albedo has thus contributed significantly in the constrain of the haze optical properties.

Photons below 0.3 μm control the photochemistry in Titan's atmosphere. Most of these photons are absorbed at high altitudes by the hydrocarbons and nitriles, while few photons are reflected back to space through Rayleigh scattering by the main gases, nitrogen and methane. As we move towards the visible part of the spectrum, photons reach deeper regions of the atmosphere and are mainly affected by the haze that absorbs and/or scatters depending on wavelength. Although the gas opacity is very small between 0.3 and 0.6 μm, the haze particles are highly absorbing and determine the slope of the geometric albedo with wavelength.

At longer wavelengths, the haze absorption efficiency becomes significantly smaller and scattering by the particles dominates, while at the same time as we move to larger wavelengths the total haze extinction becomes smaller due to the decreasing values of the size parameter. Above 0.6 μm, methane exhibits many and

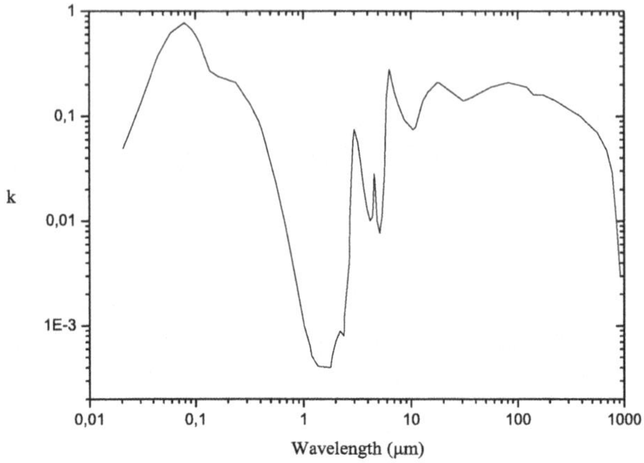

FIG. 12.18. Spectral variation for the measured imaginary part of the tholin refractive index that controls absorptivity of the material. The sharp decrease of absoprtivity in the visible region is similar to the one inferred for Titan's haze. (Data from Khare *et al.* 1984)

strong absorption bands that produce the well–defined deeps in the geometric albedo. The domains of high albedo in between the methane bands, correspond to the methane 'windows'. Photons inside these windows can reach the surface, hence the geometric albedo in these regions is affected significantly by surface reflectivity. In this way, spectral information about the optical properties of the haze, the size distribution of the particles, their production rate and the spectral surface albedo can be retrieved from the geometric albedo.

12.5.4 *Laboratory and in–situ measurements*

Many laboratory measurements have contributed to the interpretation of haze properties and its formation pathways. The usual approach is to irradiate a mixture of nitrogen, methane and possibly other hydrocarbons (in amounts equivalent to those present in Titan's atmosphere) with energy sources analogous to those found in Titan (UV radiation, magnetospheric electrons and protons, cosmic rays). The energy sources usually chosen are electric discharge, cold plasma discharge and UV radiation. The results of such experiments were the production of complex organic solids called tholins. The most well–established laboratory simulation of tholins was that of Khare *et al.* (1984) using a mixture of 90% N_2 – 10% CH_4 and DC electrical discharge as an energy source. They provided for the first time measurements of the complex refractive index that exhibited the correct wavelength dependence of the imaginary part, necessary to fit the geometric albedo. The haze is the main absorber in the UV region and responsible for the sharp decrease in the geometric albedo below 0.6 μm. The slope of the geometric albedo is directly correlated with the slope of the imaginary part of the refractive index in this region (Fig. 12.18).

Many methods have also been used for the chemical analysis and elemental composition of the produced haze analogues. Their structure was found to be a complex mixture of hydrocarbon and nitriles, including aromatic compounds and polycyclic aromatic hydrocarbons (PAH), but the results between different experiments were not always similar: although the C/H ratio measured was always constrained over a small range of values (0.6 – 1.0) among the experiments, the C/N ratio, which indicates the extent of nitrogen incorporation in the analogues, exhibited large variability (0.75 – 11.0).

Pressure is a critical parameter in the chemical and hence optical properties of the derived analogues, with nitrogen being more efficiently incorporated into the structure of the particles at low deposition pressures (increasing C/N ratio with increasing pressure). In addition, the presence of aromatic compounds and nitrogen–containing polycyclic aromatic compounds (N–PACs) was found to increase with decreasing pressure, resulting in more absorbing tholins in the UV/visible region due to the delocalized π electrons of these structures. These conclusions explained some of the differences between experiments and proved that the haze particles found in Titan's atmosphere can have different properties depending on the pathways of formation that correspond to more than one pressure region. The ACP instrument on board the Huygens probe, provided the first results regarding the chemical composition of the haze particles in Titan's atmosphere. The pyrolysis of the refractory core of the particles, indicated the presence of HCN and NH_3 in their structural form.

12.5.5 *Haze formation*

Based on these laboratory simulations for the chemical production of the haze particles, possible pathways have been suggested (see Lavvas *et al.* 2007a,b, and references therein). The particles are believed to form from the the chemical growth of certain polymer structures. A possible mechanism for the production of these polymers could be the photolysis of acetylene leading to polyyne ($C_{2n+2}H_2$) formation. This process starts with the photolysis of C_2H_2, which produces the ethynyl radical (C_2H) via

$$C_2H_2 + h\nu \rightarrow C_2H + H \ .$$

The latter reacts readily with another acetylene molecule, giving diacetylene (C_4H_2). The process continues giving higher-order polyacetylenes (C_6H_2, C_8H_2, etc.)

$$C_{2n}H_2 + C_2H \rightarrow C_{2n+2}H_2 + H \ .$$

In addition to the above, the photolysis of the polyynes formed leads to the production of polyyne radicals (C_4H, C_6H, C_8H) that also enhance the polymerization. The reaction for the first step of the polymerization leading to C_4H_2 has been measured down to very low temperatures and found to be very efficient for Titan's conditions. Yet, the production of polyynes is constrained by the acetylenic recycling of the C_2H and other polyyne radicals through reactions with molecular hydrogen and the saturated hydrocarbons found in the atmosphere, from which methane is the more important

$$C_2H + H_2 \rightarrow C_2H_2 + H$$
$$C_2H + CH_4 \rightarrow C_2H_2 + CH_3 \ .$$

This recycling is very efficient and the contribution of increasing–order polyynes has a decreasing effect on the total haze production.

Similar to polyynes, cyanopolyynes were also suggested as possible haze precursors. These are formed in the same way as polyynes but with the ethynyl radical replaced by CN that reacts with hydrocarbons and nitriles giving cyanoacetylene (HC_3N), acrylonitrile (C_2H_3CN) and dicyanoacetylene (C_4N_2). The polymerization proceeds with further reaction of the CN radical with the produced nitriles. The above pathways lead to the production of polymers, which due to their structural form, are characterized by large C/N and C/H ratios. When these are compared with the laboratory measured ratios suggests that if indeed the haze in Titan's atmosphere is similar to the analogues produced in the laboratory, the above polymers must have a small contribution to the total production.

In addition, heteropolymer formation mechanisms were suggested in which the polymerization proceeds through the addition of both CN radicals and RCN chains to nitriles. Also polymers of HCN may lead to C/N ratios closer to laboratory values. In these, the polymerization starts with the production of an $(HCN)_2$ dimer and subsequent addition of HCN molecules leads to poly–HCN structures. Even though HCN structures were defined in the haze analogues produced, the optical properties of pure poly–HCN structures do not exhibit the correct wavelength dependence inferred from Titan's geometric albedo and hence they cannot have a major contribution to bulk haze production. Another possible polymerization pathway could be from the irradiation of HC_3N/C_2H_2 and HC_3N/C_2H_6 mixtures. HC_3N was found to be much more reactive in polymerization than HCN and that the copolymers formed could exhibit the necessary wavelength dependence, something that the pure polymers of either case did not do.

Pathways that include aromatic hydrocarbons have also been suggested. Since benzene, C_6H_6, the simplest aromatic hydrocarbon has been detected on Titan, polycyclic aromatic hydrocarbons (PAH) could contribute to haze formation. The production of PAH is considered to proceed through the reaction of benzene molecules with acetylene. PAHs are known to exist in the interstellar medium

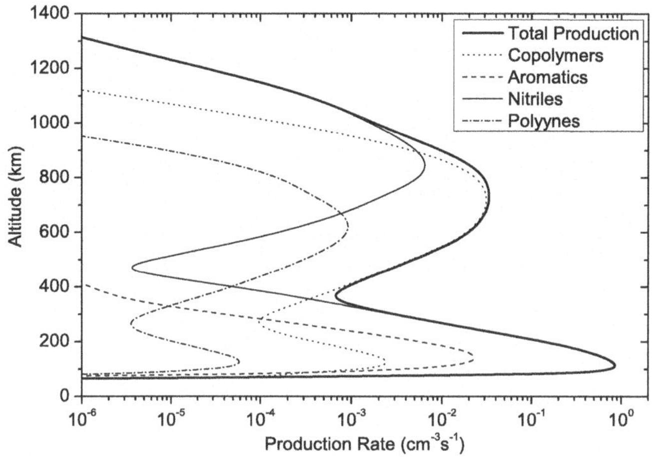

FIG. 12.19. Vertical production profile of haze particles, based on the model of Lavvas *et al.* 2007b.

in significant amounts. Recently, Hudgins *et al.* (2005) have found that other types of polycyclic aromatic compounds, in which one or more nitrogen atoms are substituted in the carbon skeleton, are also abundant in the universe and are produced around dying stars. These are called polycyclic aromatic nitrogen heterocycles (PANH) and are considered to be more important than simple PAH due to their connection with the chemistry of life. Chlorophyll, for example, is such a compound.

Another compound that could be important for haze formation is $CH_2{=}NH$, known as methyleneimine. The latter has been detected by the INMS instrument on board Cassini in the upper atmosphere and it has also been found in the low pressure haze analogues. $CH_2{=}NH$ is known as a suitable compound for polymerization and might be a precursor for haze production. Although reaction rates for this species have not been reported in the literature so far, its increased importance will definitely hasten its laboratory investigation.

The growth of the above polymers proceeds through chemical addition until a size beyond which nucleation is initiated. At this limit the coagulation between the monomers produced, is more efficient than the chemical growth and the evolution in size of the particles is determined by the laws of microphysics.

In order to simulate the coupling between photochemistry and haze, it is necessary to have a model that will be able to simulate the three main coupled processes in Titan's atmosphere: radiation transfer – photochemistry – micro-

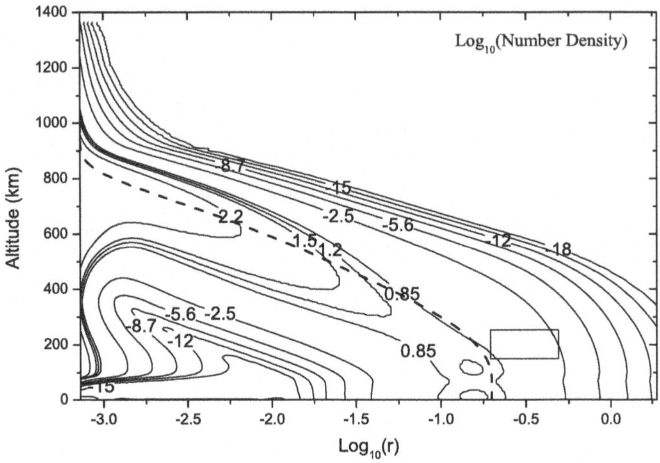

FIG. 12.20. Contour plot of the size distribution for the haze particles from Lavvas *et al.* 2007b. Size in μm, number density in cm^{-3}. Dashed curve corresponds to the optically effective radius of the particles. Numbers on contours correspond to the log10 number density. Box is Voyager I estimate.

physics. The interaction of solar radiation with the atmosphere (photochemistry) and haze determines the chemical species that are formed and the temperature structure. The chemical rates that depend on the temperature, control the pathways that lead to the formation of the haze precursors. The chemical growth of the latter, proceeds until a size beyond which the microphysical growth of the particles is more efficient. From this point on, the coagulation rate between the particles (which depends on the atmospheric temperature) defines their size and hence the optical properties of the haze layers. Finally, the interaction of the haze layers with the radiation field controls the solar radiation flux and the temperature structure. Under this self-consistent picture it is possible to validate the possible pathways through comparison with the observed temperature structure, the geometric albedo, the chemical structure and other observed parameters of the atmosphere. Such a model was developed by Lavvas *et al.* (2007a,b). The following paragraph presents some of the results of this simulation which are compared with the Cassini/Huygens observations.

Based on the chemical precursors of the suggested pathways the latter can be subdivided into four main families:

Pure nitrile polymers

$H_2CN + HCN \rightarrow$ polymer
$CN + C_4N_2 \rightarrow$ polymer
$CN + C_2H_3CN \rightarrow$ polymer
$CN + C_2N_2 \rightarrow$ polymer
$H_2CN + CH_2NH \rightarrow$ polymer

Polyynes

$C_4H + C_6H_2 \rightarrow$ polymer
$C_6H + C_4H_2 \rightarrow$ polymer
$C_6H + C_6H_2 \rightarrow$ polymer

Copolymers

$HC_3N + C_4H \rightarrow$ polymer
$C_4H_2 + C_3N \rightarrow$ polymer
$HC_3N + C_4H_3 \rightarrow$ polymer

Polyaromatics PAHs and N-PACs

$C_6H_5 + C_2H_2 \rightarrow$ polymer
$C_6H_5 + HC_3N \rightarrow$ polymer
$C_6H_5 + C_6H_6 \rightarrow$ polymer.

The photochemical description of the atmospheric composition along with the above pathways, provide a vertical production profile for the haze monomers. This is presented in Fig. 12.19 where the contribution of each family is shown along with the cumulative production at each altitude. Below 300 km contributions from aromatics, pure nitriles and copolymers are comparable, while at higher altitudes copolymers dominate and nitriles become important above 800 km. The microphysical evolution of the produced monomers leads to the production of bigger particles, the final size of which is constrained only by the electrostatic repulsion among them. Figure 12.20 presents the calculated size distribution of the particles based on the latter production profile. The average size of the particles is constant below 200 km and between 0.1 and 0.2 μm.

Using measurements for the refractive index, the opacity of the produced particles can be calculated with the use of Mie theory (see §6.3). The resulting column haze opacity along with the vertical extinction profile at 0.5 μm, are presented in Fig. 12.21 and Fig. 12.22, respectively, and compared with the measured values retrieved by the DISR instrument on board the Huygens probe (Tomasko *et al.* 2005). The measurements correspond to Titan's atmosphere through which the probe descended, while the model results correspond to the disk average opacity necessary to match the geometric albedo. UV photons are absorbed at

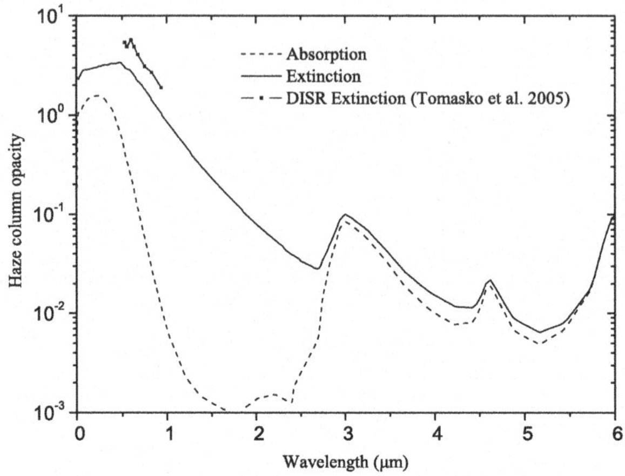

FIG. 12.21. The column absorption opacity for the produced haze particles based on the model of Lavvas *et al.* (2007b) along with the measured extinction by the DISR instrument on board the Huygens probe (Tomasko *et al.* 2005).

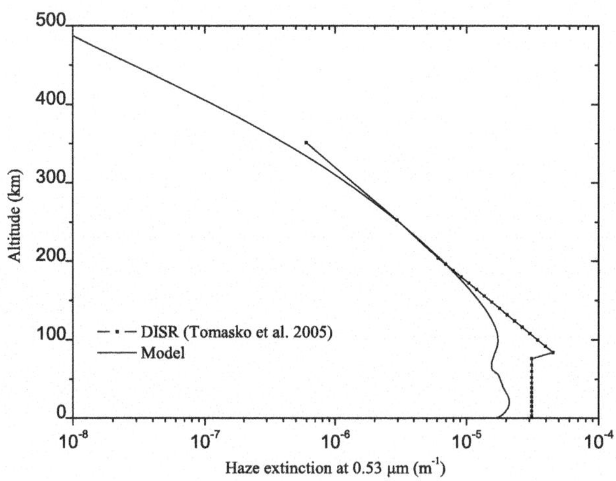

FIG. 12.22. Extinction profile of the haze particles at 0.53 μm compared with the DISR retrieved profile. (Lavvas *et al.* 2007b)

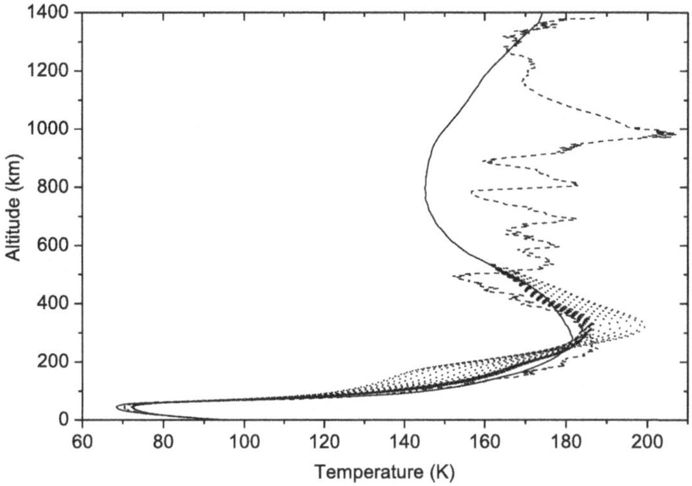

FIG. 12.23. Calculated temperature profile (solid line) from Lavvas *et al.* 2007b and retrieved temperature profiles of HASI (dashed line, Huygens probe) and CIRS (dotted lines, Cassini) instruments.

high altitudes while up to 0.6 μm haze opacity determines the altitude at which photons are absorbed. At longer wavelengths photons can reach the surface only inside the methane windows where photon reflection depends significantly on the surface albedo.

12.5.6 *Thermal structure*

The resulting temperature profile, based on the above description of haze production and evolution is presented in Fig. 12.23. For comparison the retrieved vertical temperature profiles from HASI (F. Ferri, private communication) and CIRS (A. Coustenis, private communication) are presented. The bundle of CIRS profiles give the variation of the temperature structure at different latitudes from pole to pole. The variation of the temperature is rather small, only a few degrees for a large part of the atmosphere until the winter pole where the stratopause temperature increases and the stratosphere cools. This is related to the seasonal variation of the atmospheric structure, which enhances the production of haze particles in the winter pole. The calculated vertical profile, which corresponds to midlatitude conditions, provides a good fit to the observed profiles, close to the

Table 12.6 *Species composition from Voyager and Cassini.*

Species		Voyager IRIS (7°S)[a]	Cassini CIRS (15°S,160 km)[b]	Cassini INMS (∼1200 km)[c]
Acetylene	C_2H_2	$2.7 \pm 0.2 \times 10^{-6}$	3.0×10^{-6}	2.8×10^{-4}
Ethylene	C_2H_4	$1.5 \pm 0.3 \times 10^{-7}$	1.65×10^{-7}	6×10^{-3}
Ethane	C_2H_6	$1.65 \pm 0.1 \times 10^{-5}$	1.5×10^{-5}	1.2×10^{-4}
Methylacetylene	C_3H_4	$8.0 \pm 0.6 \times 10^{-9}$	9.0×10^{-9}	4.0×10^{-6}
Propane	C_3H_8	$9.0 \pm 1.6 \times 10^{-7}$	4.55×10^{-7}	2.3×10^{-6}
Diacetylene	C_4H_2	$1.7 \pm 0.2 \times 10^{-9}$	1.8×10^{-9}	6.0×10^{-5}
Hydrogen cyanide	HCN	$2.2 \pm 0.2 \times 10^{-7}$	1.76×10^{-7}	2.0×10^{-4}
Cyanogen	C_2N_2	$<1 \times 10^{-9}$		
Cyanoacetylene	HC_3N	$<1.5 \times 10^{-9}$		2×10^{-5}
Ammonia	NH_3			7×10^{-6}
Methyleneimine	CH_2NH			$<10^{-5}$
Carbon dioxide	CO_2	$1.5 \pm 0.15 \times 10^{-8}$		

a. Coustenis and Bézard 1995, b. Flasar *et al.* 2005, c. Waite *et al.* 2005, Vuitton *et al.* 2006

HASI profile in the lower atmosphere and approaches the CIRS profile above the stratosphere.

In the stratosphere, the strong temperature increase is due to absorption of solar radiation by methane and most importantly, by the haze, similarly to the Earth's ozone layer. The haze has an 'antigreenhouse' effect as it is optically thin for thermal radiation but reflects solar radiation. In this way the heating from the absorption of the solar radiation is balanced partially by the escape to space of the thermal radiation. In the mesosphere, the strong negative temperature slope is due to thermal emission, mainly by methane and other hydrocarbons present in smaller amounts (C_2H_2, C_2H_6 and HCN) but enough to affect the thermal structure to a smaller degree. In the thermosphere, the photodissociation by solar UV radiation of the main species produces the temperature increase. The observed temperature oscillations are still under investigation.

12.5.7 *Atmospheric chemistry*

Table 12.6 presents a comparison of the chemical composition of Titan's atmosphere, based on measurements from the Voyager missions and the recent results of the Cassini/Huygens mission. The latter has verified the results of the former but also has provided new nitrile species that make Titan's atmospheric chemical laboratory even more interesting. The simulated chemical composition of the atmosphere from Lavvas *et al.* (2007a,b), for some of the main hydrocarbons (Fig. 12.24) and nitriles (Fig. 12.25) found in Titan's atmosphere, are presented together with the latest results from the CIRS and INMS instruments.

The most abundant hydrocarbon found in Titan's atmosphere after methane, is ethane (C_2H_6), which is mainly produced in the collisional addition of two methyl radicals

$$2\,CH_3 + M \rightarrow C_2H_6 + M\,.$$

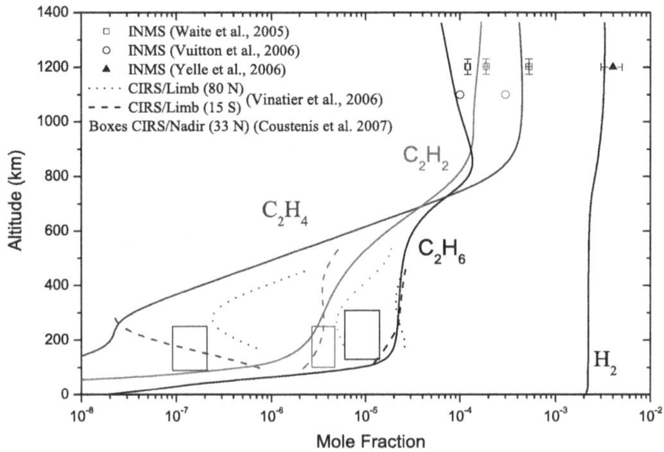

FIG. 12.24. Calculated vertical profiles of the main hydrocarbon species found in Titan's atmosphere. (Lavvas *et al.* 2007b)

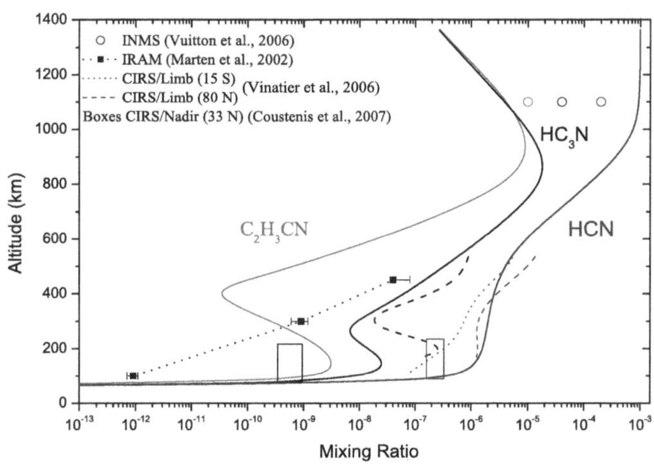

FIG. 12.25. Calculated vertical profiles of the main nitrile species found in Titan's atmosphere. (Lavvas *et al.* 2007b)

The chemical destruction of methane molecules by the methylene radicals enhances the production of methyl radicals and hence the total ethane production

$$^1CH_2 + CH_4 \rightarrow 2\ CH_3\ .$$

Because ethane's absorption cross–section falls in the same wavelength region as that of methane, its photolysis is constrained by the high amounts of the latter. Hence its destruction is mainly due to chemical reactions with radicals, of which the most important are with excited methylene

$$^1CH_2 + C_2H_6 \rightarrow CH_3 + C_2H_5,$$

and ethynyl (C_2H) that is produced by the photolysis of acetylene (C_2H_2)

$$\begin{gathered} C_2H_2 + h\nu \rightarrow C_2H + H \\ C_2H + C_2H_6 \rightarrow C_2H_2 + C_2H_5\ . \end{gathered}$$

Acetylene and ethylene (C_2H_4) are the next most abundant hydrocarbons after ethane and their formation is initiated in the upper atmosphere directly from the products of methane photolysis

$$\begin{gathered} 2\ {}^{1,3}CH_2 \rightarrow C_2H_2 + 2\ H \\ {}^{1,3}CH_2 + CH_3 \rightarrow C_2H_4 + H \\ CH + CH_4 \rightarrow C_2H_4 + H\ . \end{gathered}$$

Destruction of ethylene through photolysis acts as the major source of acetylene in the upper atmosphere

$$C_2H_4 + h\nu \rightarrow C_2H_2 + 2H/H_2,$$

while in the lower atmosphere acetylene recycles back to ethylene through the vinyl radical (C_2H_3) according to the following scheme

$$\begin{gathered} C_2H_2 + H + M \rightarrow C_2H_3 + M \\ \underline{C_2H_3 + H + M \rightarrow C_2H_4 + M.} \\ C_2H_2 + 2\ H \rightarrow C_2H_4 \end{gathered}$$

The main loss of acetylene in the upper atmosphere is from reaction with methylene that leads to the formation of propargyl radicals (C_3H_3, which is the main precursor of benzene formation)

$$^1CH_2 + C_2H_2 \rightarrow C_3H_3 + H,$$

while in the lower atmosphere the main loss is due to photolysis. Acetylene's absorption extends well beyond methane's dissociation limit, up to ~230 nm and due to its high concentration it is the main absorber in this part of the spectrum (along with the haze). The photolysis rate is significant down to the stratosphere. Hence, there is large production of the radicals C_2 and C_2H, by acetylene's photolysis in this region, which, as in the case of ethane, leads to the catalytic destruction of saturated hydrocarbons through hydrogen abstraction. This mechanism affects methane

$$\begin{array}{c} C_2H_2 + h\nu \rightarrow C_2 + H_2 \\ C_2 + CH_4 \rightarrow C_2H + CH_3 \\ C_2H + CH_4 \rightarrow C_2H_2 + CH_3 \\ \hline CH_4 \rightarrow CH_3 + H, \end{array}$$

or

$$\begin{array}{c} C_2H_2 + h\nu \rightarrow C_2H + H \\ C_2H + CH_4 \rightarrow C_2H_2 + CH_3 \\ \hline CH_4 \rightarrow CH_3 + H\ . \end{array}$$

The overall destruction of methane is dominated by its catalytic destruction by the above two schemes along with loss due to excited methylene radicals. The three molecules C_2H_6, C_2H_4, and C_2H_2 together with molecular hydrogen correspond to the most abundant hydrocarbons observed in Titan's atmosphere since the Voyager missions. From these, more complex species are produced, such as propane (C_3H_8), diacetylene (C_4H_2) and benzene (C_6H_6), which finally lead to the formation of the haze precursors.

The basis of the nitrile chemistry is the hydrogen cyanide molecule (HCN) that is formed directly by the reaction of atomic nitrogen with the methyl radical produced by methane photolysis

$$N + CH_3 \rightarrow HCN + H_2\ .$$

This pathway is not dominant as the H_2CN radical, also produced in the above reaction, reacts readily with atomic hydrogen to give HCN with an overall larger rate

$$\begin{array}{c} N + CH_3 \rightarrow H_2CN + H \\ H_2CN + H \rightarrow HCN + H_2 \\ \hline N + CH_3 \rightarrow HCN + H_2\ . \end{array}$$

Since the majority of nitrogen atoms is produced in the upper atmosphere, the production of HCN has a maximum in this region. The photolysis of HCN leads to the production of cyano radicals (CN)

$$HCN + h\nu \rightarrow CN + H\ ,$$

which have a double role in the atmospheric chemistry. Like C_2H radicals, they enhance the catalytic destruction of molecular hydrogen and saturated hydrocarbons through the hydrogen-abstraction mechanism

$$\mathrm{CN + H_2 \rightarrow HCN + H}$$
$$\mathrm{CN + CH_4 \rightarrow HCN + CH_3}$$
$$\mathrm{CN + C_2H_6 \rightarrow HCN + C_2H_5},$$

leaving a free radical and a HCN molecule, but also lead to the production of other nitrile species in reaction with unsaturated hydrocarbons through the abstraction/addition mechanism. In the latter, a hydrogen atom is abstracted from the unsaturated hydrocarbon and the CN radical is added in its place. Typical examples are the formation of cyanoacetylene (HC_3N) and acrylonitrile (C_2H_3CN)

$$\mathrm{CN + C_2H_2 \rightarrow HC_3N + H}$$
$$\mathrm{CN + C_2H_4 \rightarrow C_2H_3CN + H}\ .$$

Although acrylonitrile has not been observed in Titan's atmosphere before the Cassini/Huygens mission, it was believed to be formed and was taken into consideration in photochemical models that included nitrile chemistry. Recently, the analysis of the ionospheric composition by the Ion Neutral Mass Spectrometer (INMS) on the Cassini spacecraft has verified the presence of this species in the upper atmosphere together with other nitrile species.

Excited nitrogen atoms produced in the upper atmosphere from the photoionization of N_2, lead to the production of other nitriles and amines. Reaction with methane has two possible products, imidogen (NH) and methylenimine (CH_2NH) with the second having the largest yield

$$\begin{aligned}\mathrm{N(^2D) + CH_4} &\rightarrow \mathrm{NH + CH_3} \quad (20\%)\\ &\rightarrow \mathrm{CH_2NH + H} \quad (80\%)\ .\end{aligned}$$

Methylenimine has been detected by the INMS instrument and is an intermediate in HCN production and is a possible candidate for haze formation.

Ammonia (NH_3) is another species recently detected in Titan's atmosphere but also in the solid core of the haze particles. Its presence was anticipated due to the large abundance of nitrogen in the atmosphere and its formation was believed to be controlled by ionospheric chemistry. Yet, the measured abundance is significantly higher than suggested by models. Lavvas *et al.* (2007b) suggested another possible contribution to its formation from the neutral chemistry originating from the photodissociation of ethyleneimine (C_2H_5N), which is a possible product from the reaction of $N(^2D)$ with ethane

$$\mathrm{N(^2D) + C_2H_6 \rightarrow C_2H_5N + H}\ .$$

Finally, reaction of excited nitrogen atoms with ethylene and acetylene leads to the production of acetonitrile (CH_3CN) and the CHCN radical, respectively

$$\mathrm{N(^2D) + C_2H_4 \rightarrow CH_3CN + H}$$

$$N(^2D) + C_2H_2 \rightarrow CHCN + H,$$

while production of the latter is dominated in the same region by the replacement of H by CH in HCN

$$CH + HCN \rightarrow CHCN + H\ .$$

Yung (1987) suggested that the CHCN radical could be the basis for the production of cyanogen (C_2N_2) and dicyanoacetylene (C_4N_2)

$$N + CHCN \rightarrow C_2N_2 + H$$
$$CHCN + CHCN \rightarrow C_4N_2 + H_2,$$

from which the first has been detected in Titan's atmosphere but the second, although not detected, is included in photochemical models since it is believed to be a possible precursor of haze formation.

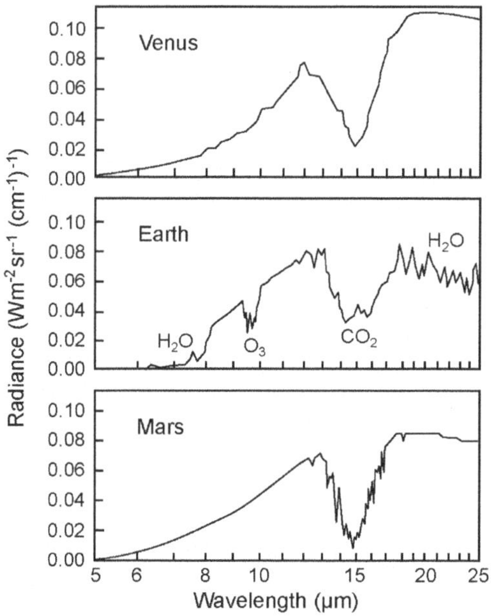

FIG. 12.26. Low–resolution infra-red spectra of Venus, Earth and Mars, based on actual observations from spectrometers on NASA spacecraft. In the next few decades, we should have spectra of this quality from planets orbiting other stars, and from them we should be able to say whether ozone (hence oxygen) and water are present, and possibly to estimate the surface pressures and temperatures.

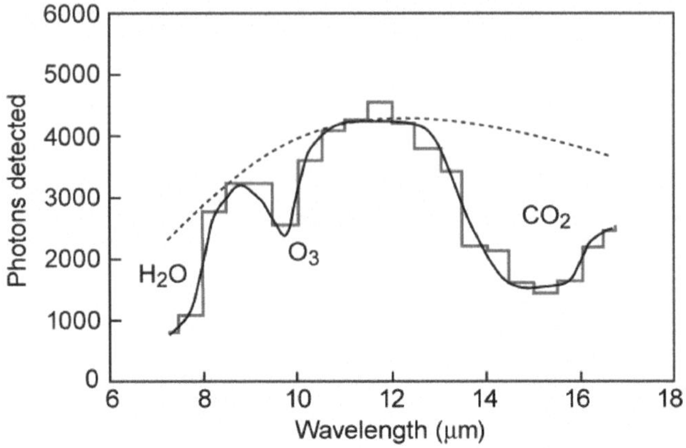

FIG. 12.27. The calculated infra–red spectrum of an Earth–like planet around a Sun–like star, as viewed by a large, but feasible, multimirror space telescope from a distance of 10 parsecs. The dashed line is the blackbody spectrum of a planet with no atmosphere at 300 K, the solid line is the signal from the planet, and the stepped line is the simulated spectrum, with noise sources included, binned at the resolution of the spectrum. (After ESA)

12.6 Extrasolar planets

Because our own Solar System exists, and given the size of the Universe with an estimated 50 billion galaxies, some containing thousands of billions of stars, there is a high probablility that planetary systems exist in large numbers. Some of these have already been detected, more than 100 at the time of writing, although with a bias towards those that contain very large planets – more like Jupiter than Earth – that are easier to detect. It is only a matter of time until we are able to detect planets similar to the Earth, orbiting stars that resemble the Sun in size and luminosity. Very large telescopes, floating in space, have already been designed that have the capability to obtain infra–red spectra of these planets, allowing techniques like those described in Chapter 9 to be employed to infer the existence of atmospheres, and their compositions and temperatures. Assuming reasonable technical progress, and the expenditure of sufficient funds, simple models of the climate on a number of these new Earths may be available to us in a few decades from now. It is likely that these will be sufficiently detailed to reveal the frequency with which Earth–like life accompanies an Earth-like environment.

Figure 12.26 shows what Venus, Earth and Mars would look like when viewed from outside the Solar System with a spectrometer attached to a suitably large telescope. The dominant feature in all three is the carbon dioxide band at 15 μm, while the 6.3 and 9.6 μm water and ozone bands only show up strongly on the Earth. Figure 12.27 is a simulation of how the terrestrial spectrum would appear from a distance of 10 pc through the DARWIN telescope currently planned by the European Space Agency. An 800–h (33 days) exposure by Darwin is required to obtain a spectral resolution of 0.5 μm per pixel. This is very low resolution, but is just enough to permit the detection of water and ozone with reasonable certainty. This in turn might be a strong indication of the presence of active, oxygen–producing life on the planet.

12.7 Bibliography

12.7.1 *Notes*

For the origin and evolution of planetary atmospheres see text by Lewis and Prinn, and for the the Earth's atmospheric evolution see texts by Holland, and by Walker.

For the early work on the origin of life on Earth see Miller, and Miller and Urey.

For anoxic conditions on the primitive Earth see early work of Cloud; Holland; and review of Kasting and references therein. For the role of organic burial on atmospheric oxygen levels see Holland *et al.*; Des Marais *et al.*

See the early work on chert and phosphate measurements by Knauth and Epstein; Karhu and Epstein that suggested 'warm' surface temperatures on the primitive Earth.

For isotopic and geochemical analyses that support growth models of continental crust formation see references in Carver and Vardavas.

For geological/biological conditions in the Late Proterozoic see Knoll; and Des Marais. For the evolution of the continental crust see Taylor and McLennan.

For references related to Titan's atmosphere, see for example; Yung *et al.*, Coustenis *et al.*, McKay *et al.*, Rannou *et al.*, Yelle *et al.*, Imanaka *et al.*, Fulcignoni *et al.*, Israel *et al.*, Wilson and Atreya. See also the text by Coustenis and Taylor.

12.7.2 *References and further reading*

Adamcik, J. A. and Draper, A. L. (1963). The temperature dependence of the Urey Equilibrium and the problem of the carbon dioxide content of the atmosphere of Venus. *Planet. Space Sci.*, **11**, 1303–1307.

Bullock, M. A., and D. G. Grinspoon (1996). The stability of climate on Venus. *J. Geophys. Res.* **101**, 7521–7529.

Bullock, M. A. and D. H. Grinspoon (2001). The recent evolution of climate on Venus. *Icarus*, **150**, 19–37.

Carlson, R. W, Kamp, L., Baines, K., Pollack, J. B., Grinspoon, D., Encrenaz, Th., Drossart, P. and Taylor, F. W. (1993). Variations in Venus cloud-particle properties: A new view of Venus's cloud morphology as observed by the Galileo near infrared mapping spectrometer. *Planet. Space Sci.*, **41**, 477–486.

Carver, J. H. and Vardavas, I. M. (1994). Precambrian glaciations and the evolution of the atmosphere. *Ann. Geophys.*, **12**, 674–682.

Carver, J. H. and Vardavas, I. M. (1995). Atmospheric carbon dioxide and the long term control of the Earth's climate. *Ann. Geophys.*, **13**, 782–790.

Cloud Jr, P. (1972). A working model of the primitive Earth. *Am. J. Sci.*, **272**, 537–548.

Coustenis, A., Bézard, B. and Gautier, D. (1989). The gas composition of Titan's equatorial region. *Icarus*, **80**, 54–76.

Coustenis, A., Lellouch, E., Maillard, J. P., McKay, C. P. (1995). Titan's surface: composition and variability from the near-infrared albedo. *Icarus*, **118**, 87–104.

Coustenis, A. and B. Bézard, 1995. Titan's atmosphere from Voyager infrared observations. IV. Latitudinal variations of temperature and composition. *Icarus*, **115**, 126–140.

Coustenis, A. and Taylor, F. W. (1999). *Titan, the Earthlike moon.* World Scientific Publishing, Singapore.

Coustenis, A., Salama, A., Schulz, B., Ott, S., Lellouch, E., Encrenaz, Th., Gautier, D., Feuchtgruber, H. (2003). Titan's atmosphere from ISO mid-infrared spectroscopy. *Icarus*, **161**, 383–403.

Crisp, D. and Titov, D. V. (1997). The thermal balance of Venus atmosphere. In *Venus II.* S. W. Bougher, D. M. Hunten and R.J. Philips, ed. The University of Arizona Press, Tucson.

Des Marais, D. J. (1985). Carbon exchange between the mantle and the crust and its effect upon the atmosphere: Today compared to Archean time. In *The carbon cycle and atmospheric* CO_2*: Natural variations Archean to present.* ed. E. T. Sundquist and W. S. Broecker. Geophys. Monograph **32**, 602–611, American Geophysical Union, Washington, DC.

Des Marais, D. J., Strauss, H., Summons, R. E. and Hayes, J. M. (1992). Carbon isotope evidence for the stepwise oxidation of the Proterozoic environment. *Nature*, **359**, 605–609.

Donahue, T. M., Grinspoon, D. H., Hartle, R. E. and Hodges, R. R. (1997). Ion/neutral escape of hydrogen and deuterium: Evolution of water. In *Venus II.* S. W. Bougher, D. M. Hunten and R.J. Philips, eds. The University of Arizona Press, Tucson.

Esposito, L. W. (1984). Sulfur dioxide: Episodic injection shows evidence for active Venus volcanism. *Science*, **223**, 1072–1074.

Flasar, F. M. and coauthors (2005). Titan's atmospheric temperatures, winds, and composition. *Science*, **308**, 975-978.

Fulcignoni, M. and coauthors (2005). In situ measurements of the physical characteristics of Titan's environment. *Nature*, doi:10.1038/nature04314.

Geballe, T. R., Kim, S. J., Noll, K. S., Griffith, C. A. (2003). High-resolution 3 micron spectroscopy of molecules in the mesosphere and troposphere of Titan. *Astrophys. J.*, **583**, L39–L42.

Gough, O. (1981). Solar interior structure and luminosity variations. *Sol. Phys.*, **74**, 21–34.

Grinspoon, D. H. (1993). Implications of the high D/H ratio for the sources of water in Venus' atmosphere. *Nature*, **363**, 428–431.

Hashimoto, G. L., and Abe, Y. (2005). Climate control on Venus: Comparison of the carbonate and pyrite models. *Planet. Space Sci.*, **53**, 839–848.

Hashimoto, G. L., Abe, Y. and Sasakis, S. (1997). CO_2 amount on Venus constrained by a criterion of topographic-greenhouse instability. *Geophys. Res. Lett.*, **24**, 289–292.

Holland, H. D. (1984). *The chemical evolution of the atmosphere and oceans.* Princeton University Press, Princeton.

Holland, H. D., Feakes, C. R. and Zbiden, E. A. (1989). The Flin Flon paleosol and the composition of the atmosphere 1.8 BYBP. *Am. J. Sci.*, **289**, 362–389.

Hudgins, D. M., Bauschlicher, C. W. and Allamandola, L. J. (2005). Variation in the peak position of the 6.2 μm interstellar emission future: a tracer of N in the interstellar polycyclic aromatic hydrocarbon population. *Astrophys. J.*, **632**, 316–332.

Imanaka, H., Khare, B. N., Elsila, J. E., Bakes, E. L. O., McKay, C. P., Cruikshank, D. P., Sugita, S., Matsui, T. and Zare, R. N. (2004). Laboratory experiments of Titan tholin formed in cold plasma at various pressures: implications for nitrogen-containing polycyclic aromatic compounds in Titan haze. *Icarus*, **168**, 344–366.

Israël G. and coauthors (2005). Complex organic matter in Titan's atmospheric aerosols from in-situ pyrolysis and analysis. *Nature*, doi:10.1038/nature04349.

Karhu, J. and Epstein, S. (1986). The implications of oxygen isotope ratios in coexisting cherts and phosphates. *Geochim. Cosmochim. Acta.*, **50**, 1745–1756.

Karkoschka, E. (1994). Spectrophotometry of the jovian planets and Titan at 300 to 1000 nm wavelength: The methane spectrum. *Icarus*, **111**, 174–192.

Kasting, J. F. (1993). Earth's early atmosphere. *Science*, **259**, 920–926.

Khare, B. N., Sagan, C., Arakawa, E. T., Suits, F., Calcott, T. A., and Williams, M. W. (1984). Optical constants of organic tholins produced in a simulated Titanian atmosphere: from soft X-ray to microwave frequencies. *Icarus*, **60**, 127–137.

Knauth, L. P. and Epstein, S. (1976). Hydrogen and oxygen isotope ratios in nodular and bedded cherts. *Geochim. Cosmochim. Acta.*, **40**, 1095–1108.

Knoll, A. K. (1991). End of the Proterozoic Eon. *Sci. Am.*, **265**, 42–49.

Knollenberg, R. G. and Hunten, D. M. (1980). The microphysics of the clouds of Venus: Results of the Pioneer Venus particle size spectrometer experiments. *J. Geophys. Res.*, **85**, 8039–8058.

Lavvas, P. P., Coustenis, A. and Vardavas, I. M. (2007a). Coupling photochemistry with haze formation in Titan's atmosphere. Part I: Model description. *Planet. Space Sci.*, in press.

Lavvas, P. P., Coustenis, A. and Vardavas, I. M. (2007b). Coupling photochemistry with haze formation in Titan's atmosphere. Part II: Results and validation with Cassini/Huygens data. *Planet. Space Sci.*, in press.

Lewis, J. S. and Prinn, R. G. (1984). *Planets and their atmospheres. Origin and evolution.* Academic Press, New York.

Lorenz, R. D., McKay, C. P., Lunine, J. I. (1997a). Photochemically-driven collapse of Titan's atmosphere. *Science*, **275**, 642–644.

Lorenz, R. D., Lunine, J. I., McKay, C. P. (1997b). Titan under a red giant Sun: A new kind of 'habitable' world. *Geophys. Res. Lett.*, **24**, 2905–2908.

Lorenz, R. D., McKay, C. P., Lunine, J. I. (1999). Analytic investigation of climate stability on Titan: sensitivity to volatile inventory. *Planet. Space Sci.*, **47**, 1503–1515.

Marten, A., Hidayat, T., Biraud, Y. and Moreno, R. (2002). New millimeter heterodyne observations of Titan: Vertical distributions of HCN, HC_3N, CH_3CN, and the isotopic ratio $^{15}N/^{14}N$ in its atmosphere. *Icarus*, **158**, 532–544.

McKay, C. P., Pollack, J. B., and Courtin, R. (1991). The greenhouse and antigreenhouse effects on Titan. *Science*, **253**, 1118–1121.

McKay, C. P., Pollack, J. B., Lunine, J. I., and Courtin, R. (1993). Coupled atmosphere-ocean models of Titan's past. *Icarus*, **102**, 88–98.

Mennesson, B. and Mariotti, J. M. (1997). Array configurations for a space infrared nulling interferometer dedicated to the search for Earthlike extrasolar planets. *Icarus*, **128**, 202–212.

Miller, L. (1955). Production of some organic compounds under possible primitive Earth conditions. *J. Am. Chem. Soc.*, **77**, 2351–2361.

Miller, L. and Urey, H. C. (1959). Organic compound synthesis on the primitive Earth. *Science*, **130**, 245–251.

Neff, J. S., Humm, D., Bergstralh, J. T., Cochran, A. L., Cochran, W. D., Barker, E. S., Tull, R. G. (1984). Absolute spectrophotometry of Titan, Uranus, and Neptune 3500-10,500 *Å*. *Icarus*, **60**, 221–235.

Rannou, P., Cabane, M., Chassefiere, E., Botet, R., McKay, C. P., Courtin, R. (1995). Titan's geometric albedo: Role of the fractal structure of the aerosols. *Icarus*, **118**, 355–372.

Schwartzman, D. W. and Volk, T. (1991). Biotic enhancement of weathering and surface temperatures on Earth since the origin of life. *Palaeogeogr. Palaeoclimat. Palaeocol.*, **90**, 357–374.

Taylor, F. W. (2006). Comparative planetary climatology. *Surveys Geophys.*, **27**, 149–167.

Taylor, S. R. and McLennan, S. M. (1985). *The continental crust: its composition and evolution.* Blackwell, Oxford.

Tomasko, M.G. and coauthors (2005). Rain, winds, and haze during the Huygens probe's descent to Titan's surface. *Nature*, doi:10.1038/nature04126.

Urey, H. C. (1952). *The planets*, Yale University Press, New Haven.

Vinatier, S. and coauthors (2006). Vertical abundance profiles of hydrocarbons in Titan's atmosphere at 15°S and 80 °N retrieved from Cassini/CIRS spectra. in press.

Vuitton, V., Yelle, R. and Anicich, V. G. (2006). The nitrogen chemistry of Titan's upper atmosphere revealed. *Ap.J.*, **647**, L175–L178.

Waite Jnr, J. H. and coauthors (2005). Ion neutral mass spectrometry results from the first flyby of Titan. *Science*, **308**, 982–986.

Walker, J. C., Hays, P. B. and Kasting, J. F. (1981). A negative feedback mechanism for the long-term stabilization of the Earth's surface temperature. *J. Geophys. Res.*, **86**, 9776–9782.

Walker, J. C. (1977). *Evolution of the atmosphere.* Macmillan, New York.

Wilson, E. H. and Atreya, S. K. (2004). Current state of modeling the photochemistry of Titan's mutually dependent atmosphere and ionosphere. *J. Geophys. Res.*, **109**, doi:10.129/2003JE002181.

Yelle R. V., Strobel, D. F., Lellouch, E., Gautier, D. (1997). Engineering models for Titan's atmosphere, ESA SP–1177, 243–256, European Space Agency, Noordwijk.

Yung, Y. L., Allen, M. and Pinto, J. P. (1984). Photochemistry of the atmosphere of Titan: Comparison between model and observations. *Astrophys. J. Supp.* , **55**, 465–506.

Yung, Y. L. (1987). An update of nitrile photochemistry on Titan. *Icarus*, **72**, 468–472.

APPENDIX A

PHYSICAL CONSTANTS

Table A.1

Constant	Symbol	Value	Units
Speed of light	c	2.99793×10^{8}	m s^{-1}
Gravitational constant	G	6.670×10^{-11}	N m ^{2} kg^{-2}
Planck constant	h	6.6262×10^{-34}	J s
Boltzmann constant	k	1.3806×10^{-23}	J K^{-1}
Stefan–Boltzmann constant	σ	5.66956×10^{-8}	J m $^{-2}$ K^{-4} s^{-1}
Energy of 1 eV	E_{o}	1.6022×10^{-19}	J
Wavelength of 1 eV	λ_{o}	1239.85	nm
Atomic mass unit (^{12}C scale)	amu	1.66053×10^{-27}	kg
Astronomical unit	AU	1.49598×10^{11}	m
Parsec	pc	3.0857×10^{16}	m
Light year	ly	9.46053×10^{15}	m
Solar radius (mean)	$R_{\odot}$	6.9599×10^{8}	m
Solar mass	$M_{\odot}$	1.989×10^{30}	kg
Solar effective temperature	$T_{\odot}$	5777	K
Earth radius (mean)	$R_{\oplus}$	6.371×10^{6}	m
Earth mass	$M_{\oplus}$	5.9736×10^{24}	kg
Earth land area	A_l	1.49×10^{14}	m^{2}
Earth ocean area	A_o	3.62×10^{14}	m^{2}
Earth rotational period	day	23.934	h
Earth orbital period	year	365.26	day
Standard surface gravity	g	9.80665	m s^{-2}
Avogadro number	N_{A}	6.0224×10^{23}	mole^{-1}
Universal gas constant	R	8.3143	J mole^{-1} K^{-1}
Loschmidt number	N_{L}	2.687×10^{25}	m^{-3}
Standard pressure	P_{o}	1013.25	mb
Standard temperature	T_{o}	273.15	K
Triple point of water	T_t	273.16	K
Freezing point of water	T_f	273.15	K
Atmospheric molecular weight	M_{d}	28.964	g mole^{-1}
Water vapour molecular weight	M_{v}	18.015	g mole^{-1}
Specific heat capacity of dry air	c_{pd}	1005	J kg^{-1} K^{-1}
Specific heat capacity of H_2O	c_{pv}	1850	J kg^{-1} K^{-1}
IT calorie	cal	4.1868	J
Latent heat of fusion at 0 °C	L_{f}	79.7	cal g^{-1}
Latent heat of vaporization at 0 °C	L_{v}	597.3	cal g^{-1}
Latent heat of sublimation at 0 °C	L_{s}	677.0	cal g^{-1}

APPENDIX B

TABLES OF REACTIONS

Table B.1 *Photoionization rates and products of atmospheric species, based on the quiet-Sun irradiance in the ultraviolet and visible spectral regions at the top of the atmosphere, for global mean conditions.*

Photoionization	j (s^{-1})
$O_2 \rightarrow O_2^+ + e$	9.4×10^{-8}
$O_2 \rightarrow O + O^+ + e$	5.5×10^{-7}
$N_2 \rightarrow N_2^+ + e$	2.0×10^{-7}
$N_2 \rightarrow N(^2D) + N^+ + e$	2.7×10^{-7}
$NO \rightarrow NO^+ + e$	1.7×10^{-6}
$O \rightarrow O^+ + e$	3.2×10^{-7}
$N \rightarrow N^+ + e$	2.3×10^{-7}
$N(^2D) \rightarrow N^+ + e$	2.3×10^{-7}
$He \rightarrow He^+ + e$	1.0×10^{-7}

Table B.2 *Total photodissociation and photoionization rates of atmospheric species, based on the quiet-Sun irradiance in the ultraviolet and visible spectral regions at the top of the atmosphere, for global mean conditions.*

Molecule	Region (nm)	j (s^{-1})
photodissociation		
O_2	100–242	3.7×10^{-6}
O_3	100–850	8.6×10^{-3}
N_2	80–103	5.9×10^{-7}
NO	181 & 191	4.1×10^{-6}
NO_2	200–425	1.7×10^{-2}
NO_3	500–670	1.8×10^{-1}
N_2O	175–240	1.1×10^{-6}
N_2O_5	200–380	6.9×10^{-4}
CO_2	105–200	7.7×10^{-7}
HO_2	190–250	6.1×10^{-4}
H_2O	69–190	1.0×10^{-5}
H_2O_2	190–350	1.2×10^{-4}
HONO	300–400	2.7×10^{-3}
HNO_3	190–350	1.4×10^{-4}
HNO_4	190–325	4.4×10^{-4}
NH_3	100–230	1.9×10^{-4}
CH_4	100–160	6.4×10^{-6}
CH_3OOH	220–360	5.6×10^{-5}
CH_2O	300–360	3.0×10^{-4}
Cl_2	260–470	5.4×10^{-3}
ClO	220–320	7.4×10^{-3}
ClO_2	220–280	1.2×10^{-2}
$ClONO_2$	196–432	7.3×10^{-4}
$ClNO_2$	190–370	1.2×10^{-2}
HCl	140–220	4.9×10^{-6}
HOCl	200–420	1.1×10^{-3}
CH_3Cl	186–216	5.5×10^{-7}
CCl_4	174–275	4.3×10^{-5}
photoionization		
O_2	8–102.8	6.4×10^{-7}
N_2	8–79.6	4.7×10^{-7}
NO	121.6	7.2×10^{-7}
NO	8–102.6	1.0×10^{-6}
O	8–91.0	3.2×10^{-7}
N	8–85.2	2.3×10^{-7}
He	8–50.4	1.0×10^{-7}

Table B.3 *Photodissociation rates and products of atmospheric species, based on the quiet-Sun irradiance in the ultraviolet and visible spectral regions at the top of the atmosphere, for global mean conditions.*

Photolysis	j (s^{-1})
$O_2 \rightarrow 2\ O$	3.7×10^{-6}
$O_3 \rightarrow O + O_2$	1.5×10^{-3}
$O_3 \rightarrow O(^1D) + O_2$	7.1×10^{-3}
$NO \rightarrow N + O$	4.1×10^{-6}
$NO_2 \rightarrow NO + O$	1.7×10^{-2}
$NO_3 \rightarrow NO + O_2$	2.0×10^{-2}
$NO_3 \rightarrow NO_2 + O$	1.6×10^{-1}
$N_2O \rightarrow N_2 + O(^1D)$	1.1×10^{-6}
$N_2 \rightarrow N(^2D) + N$	5.9×10^{-7}
$N_2O_5 \rightarrow NO_2 + NO_3$	4.8×10^{-4}
$N_2O_5 \rightarrow 2\ NO_2 + O$	2.1×10^{-4}
$CO_2 \rightarrow CO + O$	2.7×10^{-8}
$CO_2 \rightarrow CO + O(^1D)$	7.5×10^{-7}
$HO_2 \rightarrow OH + O$	6.1×10^{-4}
$H_2O \rightarrow O(^1D) + H_2$	5.1×10^{-7}
$H_2O \rightarrow 2\ H + O$	5.1×10^{-7}
$H_2O \rightarrow H + OH$	3.6×10^{-6}
$H_2O_2 \rightarrow 2\ OH$	1.2×10^{-4}
$HONO \rightarrow NO + OH$	2.7×10^{-3}
$HNO_3 \rightarrow NO_2 + OH$	1.4×10^{-4}
$HNO_4 \rightarrow NO_2 + HO_2$	4.4×10^{-4}
$NH_3 \rightarrow NH_2 + H$	1.9×10^{-4}
$CH_4 \rightarrow CH_3 + H$	3.8×10^{-6}
$CH_4 \rightarrow CH_2 + H_2$	1.3×10^{-6}
$CH_4 \rightarrow CH_2 + 2\ H$	1.3×10^{-6}
$CH_3OOH \rightarrow CH_3 + OH$	5.6×10^{-5}
$CH_2O \rightarrow HCO + H$	1.5×10^{-4}
$CH_2O \rightarrow CO + H_2$	1.5×10^{-4}
$Cl_2 \rightarrow 2\ Cl$	5.4×10^{-3}
$ClO \rightarrow Cl + O$	6.5×10^{-3}
$ClO \rightarrow Cl + O(^1D)$	9.1×10^{-4}
$ClO_2 \rightarrow ClO + O$	7.6×10^{-3}
$ClO_2 \rightarrow ClO + O(^1D)$	4.1×10^{-3}
$ClONO_2 \rightarrow ClO + NO_2$	7.3×10^{-4}
$ClNO_2 \rightarrow Cl + NO_2$	1.2×10^{-2}
$HCl \rightarrow H + Cl$	4.9×10^{-6}
$HOCl \rightarrow Cl + OH$	1.1×10^{-3}
$CH_3Cl \rightarrow CH_3 + Cl$	5.5×10^{-7}
$CCl_4 \rightarrow CCl_3 + Cl$	4.3×10^{-5}

Table B.4 *Oxygen and nitrogen reactions. The bimolecular forward rate constant, k_f, is given, or the termolecular low pressure rate constant, k_0, the high--pressure rate constant, k_∞ and the parameter f. (Sources: JPL 2006, IUPAC 2006, NIST 2006)*

Chemical reaction	k_f or k_0, k_∞, f
$O + O + M \rightarrow O_2 + M$	$5.21\times10^{-35}\exp(900/T)$
$O + O_2 + M \rightarrow O_3 + M$	$6.0\times10^{-34}(300/T)^{2.4}$
$O(^1D) + O_2 \rightarrow O + O_2$	$3.3\times10^{-11}\exp(55/T)$
$O(^1D) + N_2 \rightarrow O + N_2$	$1.8\times10^{-11}\exp(110/T)$
$O_3 + O \rightarrow 2\ O_2$	$8.0\times10^{-12}\exp(-2060/T)$
$O_3 + O(^1D) \rightarrow 2\ O_2$	1.2×10^{-10}
$O_3 + O(^1D) \rightarrow O_2 + 2\ O$	1.2×10^{-10}
$O_3 + N \rightarrow NO + O_2$	2.0×10^{-16}
$O_3 + NO \rightarrow NO_2 + O_2$	$3.0\times10^{-12}\exp(-1500/T)$
$O_3 + NO_2 \rightarrow NO_3 + O_2$	$1.2\times10^{-13}\exp(-2450/T$
$O_3 + OH \rightarrow HO_2 + O_2$	$1.7\times10^{-12}\exp(-940/T)$
$O_3 + HO_2 \rightarrow OH + 2\ O_2$	$1.0\times10^{-14}\exp(-490/T)$
$O_3 + H \rightarrow OH + O_2$	$1.4\times10^{-10}\exp(-470/T)$
$N + O_2 \rightarrow NO + O$	$1.5\times10^{-11}\exp(-3600/T)$
$N + O + M \rightarrow NO + M$	$5.46\times10^{-33}\exp(156/T)$
$NO + O + M \rightarrow NO_2 + M$	$1.0\times10^{-31}(300/T)^{1.6}$
	$3.0\times10^{-11}(300/T)^{-0.3}$, 0.85
$NO + N \rightarrow N_2 + O$	$2.1\times10^{-11}\exp(100/T)$
$NO_2 + O \rightarrow NO + O_2$	$5.6\times10^{-12}\exp(180/T)$
$NO_2 + O + M \rightarrow NO_3 + M$	$1.3\times10^{-31}(300/T)^{1.5}$
	$2.3\times10^{-11}(300/T)^{-0.24}$, 0.60
$NO_2 + N \rightarrow N_2O + O$	$5.8\times10^{-12}\exp(221/T)$
$NO_2 + H \rightarrow NO + OH$	$4.0\times10^{-10}\exp(-340/T)$
$N_2O + H \rightarrow N_2 + OH$	$9.22\times10^{-14}\exp(-2988/T)$
$NO_3 + O \rightarrow O_2 + NO_2$	1.0×10^{-11}
$NO_3 + OH \rightarrow HO_2 + NO_2$	2.2×10^{-11}
$NO_3 + NO_2 + M \rightarrow N_2O_5 + M$	$3.6\times10^{-30}(300/T)^{4.1}$
	$1.9\times10^{-12}(300/T)^{-0.2}$, 0.35
$NO_3 + NO \rightarrow 2\ NO_2$	$1.5\times10^{-11}\exp(170/T)$
$NO_3 + NO_3 \rightarrow 2\ NO_2 + O_2$	$8.5\times10^{-13}\exp(-2450/T)$
$N_2 + O(^1D) + M \rightarrow N_2O + M$	$3.5\times10^{-37}(300/T)^{0.6}$
$N_2O + O(^1D) \rightarrow N_2 + O_2$	$4.7\times10^{-11}\exp(20/T)$
$N_2O + O(^1D) \rightarrow 2\ NO$	$6.7\times10^{-11}\exp(20/T)$
$N_2O_5 + M \rightarrow NO_2 + NO_3 + M$	$1.33\times10^{-3}(T/298)^{3.5}\exp(-11000/T)$
	$9.7\times10^{14}(T/298)^{-0.1}\exp(-11080/T)$, 0.35
$N_2O_5 + H_2O \rightarrow 2\ HNO_3$	2.5×10^{-22}
$N(^2D) + O_2 \rightarrow NO + O$	$9.7\times10^{-12}\exp(-186/T)$
$N(^2D) + O \rightarrow N + O$	1.0×10^{-12}
$N(^2D) + CO_2 \rightarrow NO + CO$	3.6×10^{-13}
$N(^2D) + CO \rightarrow N + CO$	6.0×10^{-12}
$N(^2D) + H_2 \rightarrow NH + H$	$4.2\times10^{-11}\exp(-881/T)$
$N(^2D) + N_2 \rightarrow N + N_2$	1.7×10^{-14}
$N(^2D) + NO \rightarrow N_2 + O$	6.0×10^{-11}
$N(^2D) + N_2O \rightarrow N_2 + NO$	$1.5\times10^{-11}\exp(-567/T)$
$N(^2D) \rightarrow N$	1.07×10^{-5}

Table B.5 *Ion-neutral reactions. The bimolecular forward rate constant, k_f, is given. (Source: UDfA 2005)*

Ion-neutral reaction	k_f
$N(^2D) + O_2^+ \rightarrow N^+ + O_2$	4.0×10^{-10}
$N(^2D) + e \rightarrow N + e$	$6.0\times10^{-10}(300/T_e)^{-0.5}$
$NO^+ + e \rightarrow N(^2D) + O$	$3.57\times10^{-7}(300/T_e)^{0.85}$
$NO^+ + e \rightarrow N + O$	$0.63\times10^{-7}(300/T_e)^{0.85}$
$N_2^+ + O \rightarrow NO^+ + N(^2D)$	$1.4\times10^{-10}(300/T)^{-0.44}$
$N_2^+ + O \rightarrow O^+ + N_2$	1.3×10^{-10}
$N_2^+ + O_2 \rightarrow O2^+ + N_2$	5.0×10^{-11}
$N_2^+ + N \rightarrow N^+ + N_2$	1.0×10^{-11}
$N_2^+ + NO \rightarrow NO^+ + N_2$	4.4×10^{-10}
$N_2^+ + e \rightarrow N + N(^2D)$	$1.17\times10^{-7}(300/T_e)^{0.3}$
$N_2^+ + CO_2 \rightarrow CO_2^+ + N_2$	7.7×10^{-11}
$N_2^+ + CO \rightarrow CO^+ + N_2$	7.4×10^{-11}
$N^+ + O_2 \rightarrow O + NO^+$	2.63×10^{-10}
$N^+ + O_2 \rightarrow O_2^+ + N$	3.11×10^{-10}
$N^+ + NO \rightarrow NO^+ + N$	4.51×10^{-10}
$N^+ + NO \rightarrow N_2^+ + O$	7.9×10^{-11}
$N^+ + N \rightarrow N_2^+$	$3.71\times10^{-18}(300/T)^{-0.24}$
$N^+ + CO_2 \rightarrow CO_2^+ + N$	7.5×10^{-10}
$N^+ + CO_2 \rightarrow CO^+ + NO$	2.5×10^{-10}
$N^+ + CO \rightarrow CO^+ + N$	8.25×10^{-10}
$N^+ + CO \rightarrow NO^+ + C$	1.45×10^{-10}
$N^+ + e \rightarrow N$	$4.0\times10^{-12}(300/T_e)^{0.58}$
$O_2^+ + NO \rightarrow NO^+ + O_2$	4.6×10^{-10}
$O_2^+ + N \rightarrow NO^+ + O$	1.8×10^{-10}
$O_2^+ + C \rightarrow C^+ + O_2$	5.2×10^{-11}
$O_2^+ + C \rightarrow CO^+ + O$	5.2×10^{-11}
$O_2^+ + e \rightarrow O + O$	$1.9\times10^{-7}(300/T_e)^{0.7}$
$O^+ + O_2 \rightarrow O_2^+ + O$	1.9×10^{-11}
$O^+ + N_2 \rightarrow NO_+ + N$	1.2×10^{-12}
$O^+ + NO \rightarrow NO^+ + O$	1.3×10^{-12}
$O^+ + CO_2 \rightarrow O_2^+ + CO$	9.4×10^{-10}
$O^+ + CO \rightarrow CO^+ + O$	$4.9\times10^{-12}(300/T)^{-0.5}$
$O^+ + C \rightarrow CO^+$	$4.69\times10^{-11}(300/T)^{3.08}$
$O^+ + e \rightarrow O$	$3.24\times10^{-12}(300/T_e)^{0.66}$
$CO_2^+ + O \rightarrow O_2^+ + CO$	1.64×10^{-10}
$CO_2^+ + O \rightarrow O^+ + CO_2$	9.62×10^{-11}
$CO_2^+ + O_2 \rightarrow O_2^+ + CO_2$	5.3×10^{-11}
$CO_2^+ + NO \rightarrow CO_2 + NO^+$	1.2×10^{-10}
$CO_2^+ + e \rightarrow CO + O$	$3.80\times10^{-7}(300/T_e)^{0.5}$
$CO^+ + N \rightarrow NO^+ + C$	2.0×10^{-11}
$CO^+ + NO \rightarrow NO^+ + CO$	3.3×10^{-10}
$CO^+ + O_2 \rightarrow O_2^+ + CO$	1.2×10^{-10}
$CO^+ + O \rightarrow O^+ + CO$	1.4×10^{-10}
$CO^+ + CO_2 \rightarrow CO_2^+ + CO$	1.0×10^{-9}
$CO^+ + C \rightarrow CO^+ + CO$	1.1×10^{-10}
$CO^+ + e \rightarrow C + O$	$2.0\times10^{-7}(300/T_e)^{0.48}$
$C^+ + CO_2 \rightarrow CO^+ + CO$	1.1×10^{-9}
$C^+ + e \rightarrow C$	$4.67\times10^{-12}(300/T_e)^{0.6}$
$He^+ + CO \rightarrow C^+ + O + He$	1.6×10^{-9}
$He^+ + CO_2 \rightarrow CO^+ + O + He$	8.7×10^{-10}
$He^+ + CO_2 \rightarrow O^+ + CO + He$	1.0×10^{-10}
$He^+ + CO_2 \rightarrow CO_2^+ + He$	1.21×10^{-10}
$He^+ + CO_2 \rightarrow O_2^+ + C + He$	1.1×10^{-11}
$He^+ + CO_2 \rightarrow C^+ + O_2 + He$	4.0×10^{-11}
$He^+ + C \rightarrow C^+ + He$	6.3×10^{-15}
$He^+ + e \rightarrow He$	$2.36\times10^{-12}(300/T_e)^{0.64}$

Table B.6 *Hydrogen reactions. The bimolecular forward rate constant, k_f, is given, or the termolecular low pressure rate constant, k_0, the high-pressure rate constant, k_∞ and the parameter f. (Sources: JPL 2006, IUPAC 2006, NIST 2006)*

Chemical reactions	k_f or k_0, k_∞, f
$H + O_2 + M \rightarrow HO_2 + M$	$5.4\times10^{-32}(300/T)^{1.8}$
	$9.5\times10^{-11}(300/T)^{-0.44}$, 0.5
$H + H + M \rightarrow H_2 + M$	$8.85\times10^{-33}(T/298)^{-0.6}$
$H + OH + M \rightarrow H_2O + M$	$6.87\times10^{-31}(T/298)^{-2}$
$H + HO_2 \rightarrow 2\ OH$	7.2×10^{-11}
$H + HO_2 \rightarrow H_2 + O_2$	6.9×10^{-12}
$H + HO_2 \rightarrow H_2O + O$	1.6×10^{-12}
$OH + O \rightarrow O_2 + H$	$2.2\times10^{-11}\exp(120/T)$
$OH + NO + M \rightarrow HONO + M$	$7.4\times10^{-31}(300/T)^{2.4}$
	$3.3\times10^{-11}(300/T)^{0.3}$, 0.81
$OH + NO_2 + M \rightarrow HNO_3 + M$	$3.0\times10^{-30}(300/T)^{3.0}$,
	4.1×10^{-11}, 0.4
$2\ OH \rightarrow H_2O + O$	$6.2\times10^{-14}(T/298)^{2.6}\exp(946/T)$
$2\ OH + M \rightarrow H_2O_2 + M$	$6.9\times10^{-31}(300/T)$
	2.6×10^{-11}
$HO_2 + O \rightarrow OH + O_2$	$3.0\times10^{-11}\exp(200/T)$
$HO_2 + OH \rightarrow H_2O + O_2$	$4.8\times10^{-11}\exp(250/T)$
$HO_2 + NO \rightarrow NO_2 + OH$	$3.5\times10^{-12}\exp(250/T)$
$HO_2 + NO_2 + M \rightarrow HNO_4 + M$	$2.0\times10^{-31}(300/T)^{3.4}$
	$2.9\times10^{-12}(300/T)^{1.1}$, 0.6
$HO_2 + CO \rightarrow CO_2 + OH$	$5.96\times10^{-11}(T/298)^{0.5}\exp(-11498/T)$
$2\ HO_2 \rightarrow H_2O_2 + O_2$	$3.5\times10^{-13}\exp(430/T)$
$2\ HO_2 + M \rightarrow H_2O_2 + O_2 + M$	$1.9\times10^{-33}\exp(980/T)$
$H_2 + O \rightarrow OH + H$	$1.54\times10^{-9}\exp(-9742/T)$
$H_2 + OH \rightarrow H_2O + H$	$5.5\times10^{-12}\exp(-2000/T)$
$H_2 + O(^1D) \rightarrow OH + H$	1.1×10^{-10}
$H_2 + HO_2 \rightarrow H_2O_2 + H$	$5.0\times10^{-11}\exp(-13098/T)$
$H_2O + O(^1D) \rightarrow 2\ OH$	$1.63\times10^{-10}\exp(60/T)$
$H_2O_2 + O \rightarrow OH + HO_2$	$1.4\times10^{-12}\exp(-2000/T)$
$H_2O_2 + OH \rightarrow H_2O + HO_2$	$2.9\times10^{-12}\exp(-160/T)$
$HONO + OH \rightarrow H_2O + NO_2$	$1.8\times10^{-11}\exp(-390/T)$
$HNO_3 + OH \rightarrow H_2O + NO_3$	1.5×10^{-13}
$HNO_4 + M \rightarrow HO_2 + NO_2 + M$	$4.1\times10^{-5}\exp(-10600/T)$
	$4.8\times10^{15}\exp(-111170/T)$, 0.6
$HNO_4 + O \rightarrow OH + NO_2 + O_2$	$7.8\times10^{-11}\exp(-3400/T)$
$HNO_4 + OH \rightarrow H_2O + NO_2 + O_2$	$1.9\times10^{-12}\exp(272/T)$
$NH_3 + OH \rightarrow NH_2 + H_2O$	$1.7\times10^{-12}\exp(-710/T)$
$NH_3 + O(^1D) \rightarrow NH_2 + OH$	2.5×10^{-10}
$NH_2 + H + M \rightarrow NH_3 + M$	3.0×10^{-30}
$NH_2 + O \rightarrow HNO + H$	$9.15\times10^{-11}(T/298)^{-0.49}\exp(-166/T)$
$NH_2 + NO \rightarrow H_2O + N_2$	$4.0\times10^{-12}\exp(450/T)$
$NH_2 + O_2 \rightarrow H_2O + NO$	6.0×10^{-21}
$NH_2 + NO_2 \rightarrow H_2O + N_2O$	$2.1\times10^{-12}\exp(650/T)$
$NH_2 + H \rightarrow H_2 + NH$	$1.05\times10^{-10}\exp(-4448/T)$
$NH + O \rightarrow N + OH$	1.0×10^{-11}
$NH + NO \rightarrow N_2 + O + H$	4.9×10^{-11}
$NH + NO_2 \rightarrow N_2O + OH$	$3.5\times10^{-13}\exp(1140/T)$
$HNO + H \rightarrow NO + H_2$	$4.48\times10^{-11}(T/298)^{0.72}\exp(-327/T)$
$HNO + OH \rightarrow NO + H_2O$	5.0×10^{-11}
$HNO + O \rightarrow OH + NO$	3.8×10^{-11}

Table B.7 *Chlorine reactions. The bimolecular forward rate constant, k_f, is given, or the termolecular low pressure rate constant, k_0, the high-pressure rate constant, k_∞ and the parameter f. (Sources: JPL 2006, IUPAC 2006, NIST 2006)*

Chemical reactions	k_f or k_0, k_∞, f
$Cl + O_3 \rightarrow ClO + O_2$	$2.3\times10^{-11}\exp(-200/T)$
$Cl + O_2 + M \rightarrow ClO_2 + M$	$2.2\times10^{-33}(300/T)^{3.1}$
	1.8×10^{-10}, 0.6
$Cl + NO_2 + M \rightarrow ClNO_2 + M$	$1.8\times10^{-31}(300/T)^{2.0}$
	$1.0\times10^{-10}(300/\mathrm{T})$, 0.6
$Cl + CH_4 \rightarrow HCl + CH_3$	$7.3\times10^{-12}\exp(-1280/T)$
$Cl + HO_2 \rightarrow OH + ClO$	$4.1\times10^{-11}\exp(-450/T)$
$Cl + HO_2 \rightarrow HCl + O_2$	$1.8\times10^{-11}\exp(171/T)$
$Cl + H_2O_2 \rightarrow HCl + HO_2$	$1.1\times10^{-11}\exp(-980/T)$
$Cl + H_2 \rightarrow HCl + H$	$3.05\times10^{-11}\exp(-2270/T)$
$Cl + HNO_4 \rightarrow HCl + NO_2 + O_2$	1.0×10^{-13}
$Cl + NO_3 \rightarrow ClO + NO_2$	2.4×10^{-11}
$Cl + CH_2O \rightarrow HCl + HCO$	$8.1\times10^{-11}\exp(-30/T)$
$Cl + HOCl \rightarrow Cl_2 + OH$	$2.5\times10^{-12}\exp(-130/T)$
$Cl + CH_3Cl \rightarrow CH_2Cl + HCl$	$2.17\times10^{-11}\exp(-1130/T)$
$Cl + ClONO_2 \rightarrow Cl_2 + NO_3$	$6.5\times10^{-12}\exp(136/T)$
$Cl + CH_3OH \rightarrow CH_3O + HCl$	5.5×10^{-11}
$Cl_2 + OH \rightarrow Cl + HOCl$	$1.4\times10^{-12}\exp(-900/T)$
$Cl_2 + O(^1D) \rightarrow ClO + Cl$	1.99×10^{-10}
$ClO + NO_2 + M \rightarrow ClONO_2 + M$	$1.8\times10^{-31}(300/T)^{3.4}$
	$1.5\times10^{-11}(300/T)^{1.9}$, 0.6
$ClO + O \rightarrow Cl + O_2$	$2.8\times10^{-11}\exp(85/T)$
$ClO + O_3 \rightarrow ClO_2 + O_2$	$1.0\times10^{-12}\exp(-5000/T)$
$ClO + NO \rightarrow NO_2 + Cl$	$6.4\times10^{-12}\exp(290/T)$
$ClO + NO_3 \rightarrow NO_2 + ClO_2$	4.7×10^{-13}
$ClO + HO_2 \rightarrow HOCl + O_2$	$2.7\times10^{-12}\exp(220/T)$
$ClO + OH \rightarrow HO_2 + Cl$	$7.4\times10^{-12}\exp(270/T)$
$ClO + OH \rightarrow HCl + O_2$	$6.0\times10^{-13}\exp(230/T)$
$ClO + H_2 \rightarrow HOCl + H$	1.0×10^{-19}
$ClO + ClO \rightarrow ClO_2 + Cl$	$3.0\times10^{-11}\exp(-2450/T)$
$ClO + ClO \rightarrow O_2 + Cl_2$	$1.0\times10^{-12}\exp(-1590/T)$
$ClO_2 + M \rightarrow Cl + O_2$	$2.8\times10^{-10}\exp(-1820/T)$
$ClO_2 + O \rightarrow ClO + O_2$	$2.4\times10^{-12}\exp(-960/T)$
$ClO_2 + NO \rightarrow ClO + NO_2$	$2.5\times10^{-12}\exp(-600/T)$
$ClO_2 + OH \rightarrow HOCl + O_2$	$4.5\times10^{-13}\exp(800/T)$
$ClO_2 + Cl \rightarrow 2\ ClO$	1.2×10^{-11}
$ClO_2 + Cl \rightarrow Cl_2 + O_2$	2.3×10^{-10}
$ClONO_2 + O \rightarrow ClO + NO_3$	$2.9\times10^{-12}\exp(-800/T)$
$ClONO_2 + OH \rightarrow HOCl + NO_3$	$1.2\times10^{-12}\exp(-330/T)$
$ClNO_2 + OH \rightarrow HOCl + NO_2$	$2.4\times10^{-12}\exp(-1250/T)$
$HCl + O(^1D) \rightarrow Cl + OH$	1.5×10^{-10}
$HCl + O \rightarrow OH + Cl$	$1.0\times10^{-11}\exp(-3300/T$
$HCl + OH \rightarrow H_2O + Cl$	$2.6\times10^{-12}\exp(-350/T)$
$HOCl + O \rightarrow ClO + OH$	1.7×10^{-13}
$HOCl + OH \rightarrow ClO + H_2O$	$3.0\times10^{-12}\exp(-500/T$
$CH_3Cl + OH \rightarrow H_2O + CH_2Cl$	$2.4\times10^{-12}\exp(-1250/T)$
$CH_2Cl + O \rightarrow CH_2O + Cl$	$1.49\times10^{-10}\exp(-176/T)$
$CCl_4 + O(^1D) \rightarrow 4\ Cl + CO$	3.3×10^{-10}

Table B.8 *Carbon reactions. The bimolecular forward rate constant, k_f, is given, or the termolecular low pressure rate constant, k_0, the high-pressure rate constant, k_∞ and the parameter f. P_0 = 1 bar. (Sources: JPL 2006, IUPAC 2006, NIST 2006)*

Chemical reactions	k_f or $_0$, k_∞, f
$HCO + O_2 \rightarrow CO + HO_2$	5.2×10^{-12}
$CO + OH \rightarrow H + CO_2$	$1.3\times10^{-13}(1 + 0.6\ P/P_0)(300/T)$
$CO + O + M \rightarrow CO_2 + M$	$1.7\times10^{-33}\exp(-1510/T)$
$CO + H + M \rightarrow HCO + M$	$5.29\times10^{-34}\exp(-372/T)$
$CO_2 + O(^1D) \rightarrow CO_2 + O$	$7.4\times10^{-11}\exp(120/T)$
$CO_2 + O(^1D) \rightarrow CO + O_2$	2.01×10^{-10}
$CO_2 + O \rightarrow CO + O_2$	$2.81\times10^{-11}\exp(-26499/T)$
$CO_2 + H \rightarrow CO + OH$	$2.51\times10^{-10}\exp(-13300/T)$
$CO_2 + N \rightarrow CO + NO$	$3.2\times10^{-13}\exp(-1711/T)$
$CO_2 + C \rightarrow 2\ CO$	1.0×10^{-15}
$C + O_2 \rightarrow CO + O$	1.0×10^{-9}
$CO_2 + CH_2 \rightarrow CH_2O + CO$	1.4×10^{-14}
$CH_4 + O(^1D) \rightarrow CH_3 + OH$	1.125×10^{-10}
$CH_4 + O(^1D) \rightarrow CH_3O + H$	3.0×10^{-11}
$CH_4 + O(^1D) \rightarrow CH_2O + H_2$	7.5×10^{-12}
$CH_4 + O \rightarrow CH_3 + OH$	$5.63\times10^{-10}\exp(-6230/T)$
$CH_4 + OH \rightarrow CH_3 + H_2O$	$2.45\times10^{-12}\exp(-1775/T)$
$CH_3 + O \rightarrow CH_2O + H$	1.1×10^{-10}
$CH_3 + O_3 \rightarrow CH_3O + O_2$	$5.4\times10^{-12}\exp(-220/T)$
$CH_2 + H_2 \rightarrow CH_3 + H$	6.9×10^{-15}
$CH_3 + O_2 + M \rightarrow CH_3O_2 + M$	$1.0\times10^{-30}(300/T)^{3.3}$
	$1.8\times10^{-12}(300/T)^{-1.1}$, 0.27
$CH_3O + O_2 \rightarrow CH_2O + HO_2$	$3.9\times10^{-14}\exp(-900/T)$
$CH_3O + NO \rightarrow CH_2O + HNO$	8.0×10^{-12}
$CH_3O + NO_2 \rightarrow CH_2O + HONO$	$1.1\times10^{-11}\exp(-1200/T)$
$CH_3O_2 + O \rightarrow CH_3O + O_2$	4.3×10^{-11}
$CH_3O_2 + NO \rightarrow NO_2 + CH_3O$	$2.8\times10^{-12}\exp(287/T)$
$CH_3O_2 + HO_2 \rightarrow CH_3OOH + O_2$	$3.8\times10^{-13}\exp(780/T)$
$CH_3O_2 + CH_3O_2 \rightarrow 2\ CH_3O + O_2$	$5.94\times10^{-14}\exp(390/T)$
$CH_3O_2 + CH_3O_2 \rightarrow CH_2O + CH_3OH + O_2$	$3.56\times10^{-14}\exp(390/T)$
$CH_3OH + OH \rightarrow CH_3O + H_2O$	$2.9\times10^{-12}\exp(-345/T)$
$CH_3OOH + OH \rightarrow CH_3O_2 + H_2O$	$3.8\times10^{-12}\exp(200/T)$
$CH_2O + O \rightarrow HCO + OH$	$3.4\times10^{-11}\exp(-1600/T)$
$CH_2O + OH \rightarrow H_2O + HCO$	$5.5\times10^{-12}\exp(125/T)$

INDEX

The manufacturer's authorised representative in the EU for product safety is Oxford University Press España S.A. of El Parque Empresarial San Fernando de Henares, Avenida de Castilla, 2 - 28830 Madrid (www.oup.es/en or product.safety@oup.com). OUP España S.A. also acts as importer into Spain of products made by the manufacturer.
Printed and bound by CPI Group (UK) Ltd, Croydon, CR0 4YY
14/09/2026
02216307-0017